Springer Finance

Springer Finance

Springer Finance is a programme of books aimed at students, academics and practitioners working on increasingly technical approaches to the analysis of financial markets. It aims to cover a variety of topics, not only mathematical finance but foreign exchanges, term structure, risk management, portfolio theory, equity derivatives, and financial economics.

Ammann M., Credit Risk Valuation: Methods, Models, and Application (2001)
Back K., A Course in Derivative Securities: Introduction to Theory and Computation (2005)
Barucci E., Financial Markets Theory. Equilibrium, Efficiency and Information (2003)
Bielecki T.R. and Rutkowski M., Credit Risk: Modeling, Valuation and Hedging (2002)
Bingham N.H. and Kiesel R., Risk-Neutral Valuation: Pricing and Hedging of Financial Derivatives (1998, 2nd ed. 2004)
Brigo D. and Mercurio F., Interest Rate Models: Theory and Practice (2001, 2nd ed. 2006)
Buff R., Uncertain Volatility Models – Theory and Application (2002)
Carmona R.A. and Tehranchi M.R., Interest Rate Models: An Infinite Dimensional Stochastic Analysis Perspective (2006)
Dana R.-A. and Jeanblanc M., Financial Markets in Continuous Time (2003)
Deboeck G. and Kohonen T. (Editors), Visual Explorations in Finance with Self-Organizing Maps (1998)
Delbaen F. and Schachermayer W., The Mathematics of Arbitrage (2005)
Elliott R.J. and Kopp P.E., Mathematics of Financial Markets (1999, 2nd ed. 2005)
Fengler M.R., Semiparametric Modeling of Implied Volatility (2005)
Geman H., Madan D., Pliska S.R. and Vorst T. (Editors), Mathematical Finance – Bachelier Congress 2000 (2001)
Gundlach M., Lehrbass F. (Editors), CreditRisk$^+$ in the Banking Industry (2004)
Jondeau E., Financial Modeling Under Non-Gaussian Distributions (2007)
Kellerhals B.P., Asset Pricing (2004)
Külpmann M., Irrational Exuberance Reconsidered (2004)
Kwok Y.-K., Mathematical Models of Financial Derivatives (1998)
Malliavin P. and Thalmaier A., Stochastic Calculus of Variations in Mathematical Finance (2005)
Meucci A., Risk and Asset Allocation (2005)
Pelsser A., Efficient Methods for Valuing Interest Rate Derivatives (2000)
Prigent J.-L., Weak Convergence of Financial Markets (2003)
Schmid B., Credit Risk Pricing Models (2004)
Shreve S.E., Stochastic Calculus for Finance I (2004)
Shreve S.E., Stochastic Calculus for Finance II (2004)
Yor M., Exponential Functionals of Brownian Motion and Related Processes (2001)
Zagst R., Interest-Rate Management (2002)
Zhu Y.-L., Wu X., Chern I.-L., Derivative Securities and Difference Methods (2004)
Ziegler A., Incomplete Information and Heterogeneous Beliefs in Continuous-time Finance (2003)
Ziegler A., A Game Theory Analysis of Options (2004)

Gianluca Fusai · Andrea Roncoroni

Implementing Models in Quantitative Finance: Methods and Cases

 Springer

Gianluca Fusai

Dipartimento di Scienze Economiche
e Metodi Quantitativi
Facoltà di Economia
Università del Piemonte
Orientale "A. Avogadro"
Via Perrone, 18
28100 Novara
Italy
E-mail: gianluca.fusai@eco.unipmn.it

Andrea Roncoroni

Finance Department
ESSEC Graduate Business School
Avenue Bernard Hirsch BP 50105
Cergy Pontoise Cedex
France
E-mails: roncoroni@essec.fr
        roncoroni@gmail.com

Mathematics Subject Classification (2000): 35-01, 65-01, 65C05, 65C10, 65C20, 65C30, 91B28

JEL Classification: G11, G13, C15, C22, C63

Library of Congress Control Number: 2007931341

ISBN 978-3-540-22348-1 Springer Berlin Heidelberg New York

Springer is a part of Springer Science+Business Media

springer.com

© Springer-Verlag Berlin Heidelberg 2008

Cover design: WMX Design GmbH, Heidelberg
Typesetting by the authors and VTEX using a Springer LATEX macro package

Printed on acid-free paper   41/3100 VTEX - 5 4 3 2 1 0

To our families

To Nicola

Contents

Part I Methods

Preface

Introduction

This book presents and develops major numerical methods currently used for solving problems arising in quantitative finance. Our presentation splits into two parts.

Part I is methodological, and offers a comprehensive toolkit on numerical methods and algorithms. This includes Monte Carlo simulation, numerical schemes for partial differential equations, stochastic optimization in discrete time, copula functions, transform-based methods and quadrature techniques.

Part II is practical, and features a number of self-contained cases. Each case introduces a concrete problem and offers a detailed, step-by-step solution. Computer code that implements the cases and the resulting output is also included.

The cases encompass a wide variety of quantitative issues arising in markets for equity, interest rates, credit risk, energy and exotic derivatives. The corresponding problems cover model simulation, derivative valuation, dynamic hedging, portfolio selection, risk management, statistical estimation and model calibration.

We provide algorithms implemented using either Matlab® or Visual Basic for Applications® (VBA). Several codes are made available through a link accessible from the Editor's web site.

Origin

Necessity is the mother of invention and, as such, the present work originates in class notes and problems developed for the courses "Numerical Methods in Finance" and "Exotic Derivatives" offered by the authors at Bocconi University within the Master in Quantitative Finance and Insurance program (from 2000–2001 to 2003–2004) and the Master of Quantitative Finance and Risk Management program (2004–2005 to present).

The "Numerical Methods in Finance" course schedule allots 14 hours to the presentation of Monte Carlo methods and dynamic programming and an additional 14 hours to partial differential equations and applications. These time constraints seem to be a rather common feature for most academic and professional programs in quantitative finance.

The "Exotic Derivatives" course schedule allots 14 hours to the introduction of pricing and hedging techniques using case-studies taken from energy and commodity finance.

Audience

Presentations are developed at an intermediate-advanced level. We wish to address those who have a relatively sound background in the theoretical aspects of finance, and who wish to implement models into viable working tools.

Users typically include:

A. Junior analysts joining quantitative positions in the financial or insurance industry;
B. Master of Science (MS) students;
C. Ph.D. candidates;
D. Professionals enrolled in programs for continuing education in finance.

Our experience has shown that, instead of more "novel-like" monographs, this audience usually succeeds with short, precise, self-contained presentations. People also ask for focused training lectures on practical issues in model implementation. In response, we have invested a considerable amount of time in writing a book that offers a "hands-on" educational approach.

Prerequisites

We assume the user is acquainted with basic derivative pricing theory (e.g., pay-off structuring, risk-neutral valuation, Black–Scholes model) and basic portfolio theory (e.g., mean-variance asset allocation), standard stochastic calculus (e.g., Itô formula and martingales) and introductory econometrics (e.g., linear regression).

Style

We strive to be as concise as possible throughout the text. This helps us minimize ambiguities in the methodological part, a pitfall that sometimes arises in nontechnical presentations of technical subjects. Moreover, it reflects the way we covered the presented material in our courses. An exception is made for chapters on copulas and Laplace transforms, which have been included due to their fast-growing relevance to the practice of quantitative finance.

We present cases following a constructive path. We first introduce a problem in an informal way, and then formalize it into a precise problem statement. Depending on the particular problem, we either set up a model or present a specific methodology in a self-contained manner. We proceed by detailing an implementation procedure, usually in the form of an algorithm, which is then coded into a programming language. Finally, we discuss empirical results stemming from the execution of the corresponding code.

Our presentation is modular. Thus, chapters in Part I offer systematic and self-contained presentations coupled with an extensive bibliography of published articles, monographs and working papers.

For ease of comparison, the notation adopted in each case has been kept as close as possible to the one employed in the original article(s). Note that this choice requires the reader to have a certain level of flexibility in handling notation across cases.

What's missing here?

By its very nature, a treatment on numerical methods in finance tends to be encyclopedic. In order to prevent textual overflow, we do not include certain topics. The most apparent missing topic is perhaps "discrete time financial econometrics". We insert a few cases on basic and advanced econometrics, but ultimately direct the reader to other more comprehensive treatments of these issues.

Content

Part I: Methods

Static Monte Carlo; Dynamic Monte Carlo; Dynamic Programming for Stochastic Optimization; Finite Difference Methods; Numerical Solution of Linear Systems; Quadrature Methods; The Laplace Transform; Structuring Dependence Using Copula Functions.

Part II: Cases

Portfolio Selection: 'Optimizing an Error'; Alpha, Beta and Beyond; Automatic Trading: Winning or Losing in a kBit; Estimating the Risk Neutral Density; An 'American' Monte Carlo; Fixing Volatile Volatility; An Average Problem; Quasi-Monte Carlo; Lookback Options: A Discrete Problem; Electrifying the Price of Power; A Sparkling Option; Swinging on a Tree; Floating-Rate Mortgages; Basket Default Swaps; Scenario Simulation using Principal Components; Parametric Estimation of Jump-Diffusions; Nonparametric Estimation of Jump-Diffusions; A Smiling GARCH.

The cases included are not necessarily a mechanical application of the methods developed in Part I. Conversely, some topics in Part I may not have a direct application in cases. We have, nevertheless, decided to include them both for the sake of completeness and given their importance in quantitative finance. We selected cases based on our research interests and (or) their importance in the practice of quantitative finance. More importantly, all methods lead to nontrivial implementation algorithms, reflecting our ambition to deliver an effective training toolkit.

Use

Given the modular structure of the book, readers can use its content in several ways. We offer a few sample sets of coursework for different types of users:

A. Six Hour MS Courses

A1. Quadrature methods for finance

Chapter "Quadrature Methods" (Newton–Cotes and Gaussian quadrature); inversion of the characteristic function and the Fast Fourier Transform (FFT); pricing using Lévy processes.

A2. Transform methods

Laplace and Fourier transforms; examples on pricing using Lévy processes and the CIR model; cases "Fixing Volatile Volatility" and "An Average Problem".

A3. Copula functions

Chapter "Structuring Dependence Using Copula Functions". Case "Basket Default Swaps".

A4. Portfolio theory

Cases "Portfolio Selection: Optimizing an Error", "Alpha, Beta and Beyond" and "Automatic Trading: Winning or Losing in a kBit".

A5. Applied financial econometrics

Cases "Scenario Simulation Using Principal Components", "Parametric Estimation of Jump-Diffusions", "Nonparametric Estimation of Jump-Diffusions" and "A Smiling GARCH".

B. Ten to Twelve Hour MS Courses

B.1. Monte Carlo methods

Chapters "Static Monte Carlo" and "Dynamic Monte Carlo". Cases "An 'American' Monte Carlo", "Lookback Options: A Discrete Problem", "Quasi-Monte Carlo", "A Sparkling Option" and "Basket Default Swaps".

B.2. Partial differential equations

Chapters "Finite Difference Methods" and "Numerical Solution of Linear Systems"; Cases "An Average Problem" and "Lookback Options: A Discrete Problem".

B.3. Advanced numerical methods for exotic derivatives

Chapters "Finite Difference Methods" and "Quadrature Methods"; Cases "An Average Problem", "Quasi-Monte Carlo: An Asian Bet", "Lookback Options: A Discrete Problem", and "A Sparkling Option".

B.4. Problem solving in quantitative finance

Presentation of various problems across different areas such as derivative pricing, portfolio selection, and financial econometrics; key cases are "Portfolio Selection: Optimizing an Error"; "Alpha, Beta and Beyond"; "Estimating the Risk Neutral Density"; "A Sparkling Option"; "Scenario Simulation Using Principal Components"; "Parametric Estimation of Jump-Diffusions"; "Nonparametric Estimation of Jump-Diffusions"; "A Smiling GARCH".

Abstracts

Portfolio Selection: Optimizing an Error

We assess the impact of sampling errors on mean-variance portfolios. Two alternative solutions (shrinkage and resampling) to the resulting issue are proposed. An out-of-sample comparison of the two methods is also presented.

Alpha, Beta and Beyond

We compare statistical procedures for estimating the beta coefficient in the market model. Statistical procedures (OLS regression, shrinkage, robust regression, exponential smoothing, Kalman filter) for measuring the Value at Risk of a portfolio are studied and compared.

Automatic Trading: Winning or Losing in a kBit

We present a technical analysis strategy based on the cross-over of moving averages. A statistical assessment of the strategy performance is developed using a nonparametric procedure (bootstrap method). Contrasting results are also presented.

Estimating the Risk-Neutral Density

We describe a lognormal-mixture based method to infer the risk-neutral probability density from option quotations in a given market. The model is tested by examining a trading strategy grounded on mispriced options.

An 'American' Monte Carlo

American option pricing requires the identification of an optimal exercise policy. This issue is usually cast as a backward stochastic optimization problem. Here we implement a forward method based on Monte Carlo simulation. This technique is particularly suited for pricing American-style options written on complex underlying processes.

Fixing Volatile Volatility

We propose a calibration of the celebrated Heston stochastic volatility model to a set of market prices of options. The method is based on the Fast Fourier algorithm. Extension to jump-diffusions and analysis of the parametric estimation stability are also presented.

An Average Problem

We describe, implement and compare several alternative algorithms for pricing Asian-style options, namely derivatives written on an average value in the Geometric Brownian framework.

Quasi-Monte Carlo: An Asian Bet

Quasi-Monte Carlo simulation is based on the fact that "wisely" selected deterministic sequences of numbers performs better in simulation studies than sequences produced by standard uniform generators. The method is presented and applied to the pricing of exotic derivatives.

Lookback Options: A Discrete Problem

We compare three algorithms (PDE, Monte Carlo and Transform Inversion) for pricing discretely monitored lookback options written on the minimum and the maximum attained by the underlying asset.

Electrifying the Price of Power

We illustrate a multi-agent competitive-equilibrium model for pricing forward contracts in deregulated electricity markets. Simulations are provided for sample price paths.

A Sparkling Option

A real option problem concerns the valuation of physical assets using a formal representation in terms of option pricing. We price co-generation power plants as an option written on the spark spread, namely the difference between electricity and gas prices.

Swinging on a Tree

A swing option allows the buyer to interrupt delivery of a given flow commodity, such as gas or electricity. Interruption can occur several times on a given time period. We cast this as a multiple-exercise American-style option and evaluate it using Dynamic Programming.

Floating Mortgages

An outstanding debt can be refinanced a fixed number of times over a larger set of dates. We compute the value of this option by solving for the corresponding multidimensional optimal stopping rule in a discrete time stochastic framework.

Basket Default Swaps

We price swaps written on a basket of liabilities whose default probability is modeled using copula functions. Alternative pricing methods are illustrated and compared.

Scenario Simulation Using Principal Components

We perform an approximate simulation of market scenarios defined by high-dimensional quantities using a reduction method based on the statistical notion of Principal Components.

Parametric Estimation of Jump-Diffusions

A simulation-based method for estimating parameters of continuous and discontinuous diffusion processes is proposed. This is particularly useful for asset valuation under high-dimensional underlying quantities.

Nonparametric Estimation of Jump-Diffusions

We estimate a jump-diffusion process using a kernel-based nonparametric method. Efficiency tests are performed for the purpose to assess the quality of the results.

A Smiling GARCH

We calibrate a GARCH model to the volatility surface by combining Monte Carlo simulation with a local optimization scheme.

Acknowledgements

It is a great pleasure for us to thank all those who helped us in improving both content
and format of this book during the last few years. In particular, we wish to express
our gratitude to:

- Our direct *collaborators*, who contributed at a various degree of involvement
  to the achievement of most problem-solving cases through the development of
  viable working tools:
    Mariano Biondelli (Mediobanca SpA, mariano.biondelli@mediobanca.it)
    Matteo Bissiri (Cassa Depositi e Prestiti, matteo.bissiri@fastwebnet.it)
    Giovanna Boi (Consob, giovanna.boi@inwind.it)
    Andrea Bosio (Zero11 SRL, a.bosio@zero11.it)
    Paolo Carta (Royal Bank of Scotland plc, Paolo.CARTA@rbos.com)
    Gianna Figà-Talamanca (Università di Perugia, giannaft@unipg.it)
    Paolo Ghini (Green Energies, paolo.ghini@greenenergies.eu)
    Riccardo Grassi (MPS Alternative Investments SGR SpA, grassi@
    mpsalternative.it)
    Michele Lanza (Banca IMI, michele.lanza@bancaimi.it)
    Giacomo Le Pera (CREDARIS CPM, giacomo.lepera@credaris.com)
    Samuele Marafin (samuele.marafin@fastwebnet.it)
    Francesco Martinelli (Banca Lombarda, francesco.martinelli@bancalombarda.
    it)
    Davide Meneguzzo (Deutsche Bank, davide.meneguzzo@db.com)
    Enrico Michelotti (Dresdner Kleinwort, enrico.michelotti@dkib.com)
    Alessandro Moro (Morgan Stanley, alessandro.moro@morganstanley.com)
    Alessandra Palmieri (Moody's Italia SRL, alessandra.palmieri@moodys.com)
    Federico Roveda (Calyon, super fede <super_fede@email.it>)
    Piergiacomo Sabino (Dufenergy SA, piergiacomo.sabino@gmail.com)
    Marco Tarenghi (Banca Leonardo, marco.tarenghi@bancaleonardo.com)
    Igor Toder (Dexia, igor.toder@clf-dexia.com)
    Valerio Zuccolo (Banca IMI, valerio.zuccolo@polimi.it)
- Our *colleagues* Emanuele Amerio (INSEAD), Laura Ballotta (Cass Business
  School), Mascia Bedendo (Bocconi University), Enrico Biffis (Cass Business
  School), Rossano Danieli (Endesa SpA), Margherita Grasso (Enel SpA), Lorenzo
  Liesch (UBM), Daniele Marazzina (Università degli Studi del Piemonte
  Orentale), Marina Marena (Università degli Studi di Torino), Attilio Meucci
  (Lehman Brothers), Pietro Millossovich (Università degli Studi di Trieste), Maria
  Cristina Recchioni (Università Politecnica delle Marche), Simona Sanfelici (Uni-
  versità degli Studi di Parma), Antonino Zanette (Università degli Studi di Udine),
  for carefully revising parts of preliminary drafts of this book and making skilful
  comments that significantly improved the final outcome.
- Our *colleagues* Emilio Barucci (Politecnico di Milano), Hélyette Geman (ESSEC
  and Birckbek College), Stewart Hodges (King's College), Giovanni Longo (Università degli Studi del Piemonte Orientale), Elisa Luciano (Università degli Studi

di Torino), Aldo Tagliani (Università degli Studi di Trento), Antonio Vulcano (Deutsche Bank), for supporting our work and making important suggestions on our project during these years.

- *Text reviewers*, including Aine Bolder, Mahwish Nasir, David Papazian, Robert Rath, Brian Glenn Rossitier, Valentin Tataru and Jennifer Williams. A particular thanks must be addressed to Eugenia Shlimovich and Jonathan Lipsmeyer, who sacrificed hours of more interesting reading in the English classics to revise the whole manuscript and figure out ways to adapt our Anglo-Italian style into a more readable presentation.
- The three *content reviewers* acting on behalf of our Editor, for precious comments that substantially improved the final result of our work.
- The *editor*, in particular Dr. Catriona Byrne and Dr. Susanne Denskus for the time spent all over the editing and production processes. Their moral support during the various steps of the writing of this book has been of great value to us.
- All *institutions*, and their *representatives*, who supported this initiative with insightful suggestions and strong encouragement. In particular,

    Erio Castagnoli, Donato Michele Cifarelli and Lorenzo Peccati, Institute of Quantitative Methods, Bocconi University, Milan;
    Francesco Corielli, Francesca Beccacece, Davide Maspero and Fulvio Ortu, MaFinRisk (previously, MQFI), Bocconi University, Milan;
    Stewart Hodges and Nick Webber, Financial Options Research Centre (FORC), Warwick Business School, University of Warwick;
    Sandro Salsa, Department of Mathematics, Politecnico di Milano, Milan.
- A special thanks goes to CERESSEC and its Director, Radu Vranceanu, for providing us with funding to financially support part of this work.
- Part of the book has been written while Andrea Roncoroni was Research Visiting at IEMIF-Bocconi; a particular appreciation goes to its Director, Paolo Mottura, and to the Director of the Finance Department, Francesco Saita.
- Our *assistant* Sophie Lémann at ESSEC Business School for precious help at formatting preliminary versions of the draft and compiling useful information.
- Federica Trioschi at Bocconi University for arranging our classes at MaFinRisk.
- Our *students* Rachid Id Brik and Antoine Jacquier for helpful comments and experiment design on some parts of the main text.

    Clearly, all errors, omissions and "bugs" are our own responsibility.

Disclaimer

We accept no liability for any outcome of the use of codes, pseudo-codes, algorithms and programs included in the text nor for those reported in a companion web site.

Part I

Methods

1

Static Monte Carlo

This chapter introduces fundamental methods and algorithms for simulating samples of random variables and vectors, and provides illustrative examples of these techniques in quantitative finance. Section 1.1 introduces the simulation problem and the basic Monte Carlo valuation. Section 1.2 describes several algorithms for implementing a simulation scheme. Section 1.3 treats some methods for reducing the variance in Monte Carlo valuations.

1.1 Motivation and Issues

Monte Carlo is a beautiful town on the Mediterranean coast near the border between France and Italy. It is known for hosting an important casino. Since gambling has been long considered as the prototype of a repeatable statistical experiment, the term "Monte Carlo" has been borrowed by scientists in order to denote computational techniques designed for the purpose of simulating statistical experiments. A simulation algorithm is a sequence of deterministic operations delivering possible outcomes of a statistical experiment. The input usually consists of a probability distribution describing the statistical properties of the experiment and the output is a simulated sample from this distribution. Simulation is performed in a way that reflects probabilities associated with all possible outcomes. As such, it is a valuable device whenever a given experiment cannot be repeated, or it only can be repeated at a high cost. In this case, first a model of the conditions defining the original experiment is established. Then, a simulation is performed on this model and taken as an approximate sampling of the true experiment. This method is referred to as a Monte Carlo simulation. For instance, one may generate scenarios about the future evolution of a financial market variable by simulating samples of a market model defining certain distributional assumptions. Monte Carlo methods are very easy to implement on any computer system. They can be employed for financial security valuation, model calibration, risk management, scenario analysis and statistical estimation, among others. Monte Carlo delivers numerical results in most cases where all other numerical methods fail to. However, compared to alternative methods, computational speed is often slower.

Example (Arbitrage pricing by partial differential equations) Arbitrage theory is a relative pricing device. It provides equilibrium values for financial contingent claims written on prices $S_1, \ldots, S_n$ of tradeable securities. Equilibrium is ensured by the law of one price. Broadly speaking, two financial securities sharing a future pay-off stream must have the same current market value. Otherwise, by buying the cheapest and selling the dearest one would incur a positive profit today and no net cash-flow in the future: that is an arbitrage. The current arbitrage-free value of a claim is the minimum amount of wealth x we should invest today in a portfolio whose future cash-flow stream matches the one stemming from holding the claim, that is, its pay-off. The number x can be computed by the first fundamental theorem of asset pricing. If t_0 denotes current time and $B(t)$ represents the time t value of 1 Euro invested in the risk-free asset, i.e., the money market account, over $[t_0, t]$, the pricing theorem states the existence of a probability measure $\mathbb{P}^*$, which is equivalent[1] to the historical probability $\mathbb{P}$, under which price dynamics are given, such that relative prices S_i/B are all martingales under $\mathbb{P}^*$. This measure is commonly referred to as a risk neutral probability. The martingale property leads to an explicit expression for any security price:[2]

$$V(t_0) = \mathbb{E}^*_{t_0}\left(e^{-\int_{t_0}^T r(s)\,ds} V(T)\right). \tag{1.1}$$

If the random variable $V(t)$ is a function $F(t, \mathbf{x}) \in C^{1,2}(\mathbb{R}^+ \times \mathbb{R}^k)$ of a k-dimensional state variable $\mathbf{X} = (X_1, \ldots, X_k)$ satisfying the stochastic differential equation (s.d.e.)

$$d\mathbf{X}(t) = \boldsymbol{\mu}(t, \mathbf{X}(t))\,dt + \Sigma(t, \mathbf{X}(t)) \cdot d\mathbf{W}(t), \tag{1.2}$$

and the risk-free asset is driven by $dB(t) = B(t)r(t, \mathbf{X}(t))\,dt$, the martingale property of relative prices $\frac{V(t)}{B(t)} = \frac{F(t, \mathbf{X}(t))}{B(t)}$ implies their $\mathbb{P}^*$-drift must vanish for all $t \in [0, T]$ and for $\mathbb{P}^*$-almost surely all ω in Ω. This drift can be computed by the Itô formula. If D denotes the support of the diffusion $\mathbf{X}$, we obtain a partial differential equation (p.d.e.)

$$0 = \left[\partial_t + \boldsymbol{\mu}(t, \mathbf{x}) \cdot \nabla_x + \frac{1}{2}\operatorname{Tr}\left(\Sigma(t, \mathbf{x})\Sigma(t, \mathbf{x})^\top \operatorname{He}[\cdot]\right) - r(t, \mathbf{x})\cdot\right]F(t, \mathbf{x}), \tag{1.3}$$

for all $\mathbf{x} \in D$. This equation, together with the boundary condition $F(T, \mathbf{x}) = V(T) = h(\mathbf{x})$, delivers a pricing function $F(t, \mathbf{x})$ and a price process $V(t) = F(t, \mathbf{X}(t))$. Numerical methods for p.d.e.'s allow us to compute approximate solutions to this equation in most cases. There are at least two important instances where these methods are difficult, if not impossible, to apply:

(1) Non-Markovian processes.

---

[1] Broadly speaking, $\mathbb{P}^*$ is equivalent to $\mathbb{P}$, and we write $\mathbb{P}^* \sim \mathbb{P}$, if there is a unique (up to measure equivalence) function f such that the probability $\mathbb{P}^*$ of any event A can be computed as:

$$\mathbb{P}^*(A) = \int_A f(\omega)\mathbb{P}(d\omega).$$

[2] $\mathbb{E}^*_{t_0}$ is a short form for the conditional expectation under $\mathbb{P}^*$, that is $\mathbb{E}^{\mathbb{P}^*}(\cdot|\mathcal{F}_{t_0})$.

- *Case I.* The state variable $\mathbf{X}$ is not Markovian, i.e., its statistical properties as evaluated today depend on the entire past history of the variable. This happens whenever $\boldsymbol{\mu}, \boldsymbol{\Sigma}$ are path-dependent, e.g., $\boldsymbol{\mu}(t, \omega) = f(t, \{\mathbf{X}(s), 0 \leq s \leq t\})$.
- *Case II.* The pay-off $V(T)$ is path-dependent: then F is a functional and Itô formula cannot be applied.

(2) High dimension. The state variable dimension k is high (e.g. basket options). Numerical methods for p.d.e.'s may not provide reliable approximating solutions.

In each of these situations, Monte Carlo delivers a reliable approximated value for the price V in formula (1.1).

1.1.1 Issue 1: Monte Carlo Estimation

We wish to estimate the expected value $\theta = \mathbb{E}(X)$ of a random variable (r.v.) X with distribution $\mathbb{P}_X$.[3] A sample mean of this variable is any random average

$$\widehat{\theta}_n(\mathbf{X}) := \frac{1}{n} \sum_{i=1}^{n} X^{(i)},$$

where $\mathbf{X} = (X^{(1)}, \dots, X^{(n)})$ is a random vector with independent and identically distributed (i.i.d.) components with common distribution $\mathbb{P}_X$. If $\mathbf{x} = (x_1, \dots, x_n)$ is a sample of this vector,[4] then the number $\widehat{\theta}_n(\mathbf{x})$ can be taken as an approximation to the target quantity θ for at least two reasons. First, simple computations show that this quantity has mean θ and variance $\mathrm{Var}(X)/n$. This suggests that for n sufficiently large, the estimation $\widehat{\theta}_n(\mathbf{x})$ converges to the target quantity. Indeed the strong law of large numbers states that this is the case. Second, the central limit theorem states that the normalized centered sample means converge in distribution to a standard normal variable, i.e.,

$$z_n := \frac{\widehat{\theta}_n(\mathbf{X}) - \theta}{\widehat{\sigma}_n/\sqrt{n}} \xrightarrow{\mathrm{d}} \mathcal{N}(0, 1) \quad \text{as } n \to \infty. \tag{1.4}$$

This expression means that the cumulative distribution function (c.d.f.) of the r.v. z_n converges pointwise to the c.d.f. of a Gaussian variable with zero mean and unit variance. The normalization can be indifferently performed by using either the exact mean square error $\sigma = \sqrt{\mathrm{Var}(X)}$, which is usually unknown, or its unbiased estimator

$$\widehat{\sigma}_n(\mathbf{X}) := \sqrt{\frac{1}{n-1} \sum_{i=1}^{n} (X_i - \widehat{\theta}_n(\mathbf{X}))^2}$$

as is shown in formula (1.4). This statement says a lot about the way the sample mean converges to the target number. In particular, the estimation error $\widehat{\theta}_n(\mathbf{X}) -$

---

[3] We suppose there is an underlying probability space $(\Omega, \mathcal{F}, \mathbb{P})$. The distribution of X is defined by $\mathbb{P}_X(X \leq x) := \mathbb{P}(\{\omega \in \Omega : X(\omega) \leq x\})$.

[4] In mathematical terms $\mathbf{x} = \mathbf{X}(\omega)$ for some $\omega \in \Omega$.

θ is approximately distributed as a normal $\mathcal{N}(0, \widehat{\sigma}_n^2/n)$, which allows us to build confidence intervals for the estimated value.

Example (Empirical verification of the central limit theorem) Let $X^{(i)} \overset{\text{i.i.d.}}{\sim} \mathcal{U}[0, 1]$ with $i = 1, \ldots, n$. Figure 1.1 shows the empirical distribution of z_n for $n = 2, 10, 15$ as computed by simulation. To do this, we first generate 1,000 samples for each $X^{(i)}$, that is $1,000 \times n$ random numbers. We then compute a first sample of z_n by summing up the first n numbers, a second sample of z_n by summing up the next n numbers, and so on, until we come up to 1,000 samples of z_n. After partitioning the interval $[-4, 4]$

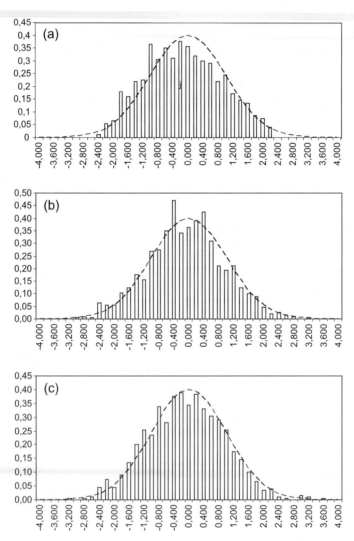

Fig. 1.1. Convergence of sample histograms to a Normal distribution.

Table 1.1. List of symbols

Name	Label	Type	Definition
State variable	X	r.v.	$X : \Omega \to \mathbb{R}$ measurable
Sample of X	x	det.	$x = X(\omega)$ for some $\omega \in \Omega$
State vector	$\mathbf{X}$	r.v.	$\mathbf{X} = (X^1, \ldots, X^n)$, X^i i.i.d.
Sample state	$\mathbf{x}$	det.	$\mathbf{x} = \mathbf{X}(\omega)$ for some $\omega \in \Omega$
Sample mean estimator	$\widehat{\theta}_n$	det.	$\mathbf{y} \to \widehat{\theta}_n(\mathbf{y}) = \frac{1}{n}\sum_{i=1}^n y_i$
Sample mean	$\widehat{\theta}_n(\mathbf{X})$	r.v.	$\omega \in \Omega \to \widehat{\theta}_n(\mathbf{X}(\omega))$
Sample mean estimation	$\widehat{\theta}_n(\mathbf{x})$	det.	$\widehat{\theta}_n(\mathbf{x})$, x_i = indep. samples

into bins of length 0.2, we finally compute the histogram of relative frequencies by counting the proportion of those samples falling into each bin. Figures 1.1(a), 1.1(b), and 1.1(c) display the resulting histograms for z_2, z_{10} and z_{15}, respectively, vs. the theoretical density function (dotted line). We see that convergence to a Gaussian law occurs moderately rapidly. This property is exploited in Sect. 1.2.5 for the purpose of building a quick generator of normal random samples.

Notice that the sample mean is the random variable obtained as the composite function of the sample mean estimator[5] $\widehat{\theta}_n : \mathbf{y} = (y_1, \ldots, y_n) \in \mathbb{R}^n \to n^{-1}\sum_{i=1}^n y_i$ and the random vector $\mathbf{X} = (X^1, \ldots, X^n)$ with $X^i \overset{\text{i.i.d.}}{\sim} X$, whereas a sample mean estimation is the value taken by the sample mean at one particular sample. We are led to the following:

Algorithm (Monte Carlo method)

1. Fix n "large".
2. Generate $\mathbf{x} = (x_1, \ldots, x_n)$ where the x_i's are samples of independent copies of X.
3. Return the sample mean estimation $\widehat{\theta}_n(\mathbf{x})$:

$$\mathbb{E}(X) \approx \widehat{\theta}_n(\mathbf{x}).$$

This is the simplest Monte Carlo estimation. Table 1.1 summarizes the terminology introduced hereby.

1.1.2 Issue 2: Efficiency and Sample Size

The simplest Monte Carlo estimator is unbiased for all $n \geq 1$, i.e., $\mathbb{E}(\widehat{\theta}_n(\mathbf{X})) = \theta$. We suppose from now on that the variance $\text{Var}(X)$ is finite. The convergence argument illustrated in the previous paragraph makes no explicit prescription about the size of the sample. Increasing this size improves the performance of a given estimator, but also increases the computational cost of the resulting procedure.

---

[5] We recall that an estimator of a quantity $\theta \in \Theta$ is a deterministic function of a sample space into Θ. A sample space of a random element $X \in \Xi$ is any product space Ξ^n.

We examine the problem of choosing between two estimators A and B given the computational budget T indicating the time units we decide to allocate for the purpose of performing the estimation procedure. Let τ_A and τ_B denote the units of time required for accomplishing the two estimations above, respectively. We indicate the corresponding mean square errors by σ_A and σ_B. Of course, if an estimation takes less time to be performed and shows a lower variability than the other, then it is preferable to this one. The problem arises whenever one of the two estimators requires lower time per replication, but displays a higher mean square error than the other. Without loss of generality, we may assume that $\tau_A < \tau_B$ and $\sigma_B < \sigma_A$. Which estimator should we select then? Given the computational budget T, we can perform as many as

$$n(T, \tau_i) = [T/\tau_i]$$

replications of the estimation procedure $i = A, B$. Here $[x]$ denotes the integer part of x. The error stemming from the estimation is obtained by substituting this number into formula (1.4):

$$\sqrt{T}\left(\widehat{\theta}^{(i)}_{n(T, \tau_i)} - \theta\right) \approx \mathcal{N}\left(0, \sigma_i^2 \tau_i\right), \quad i = A, B. \tag{1.5}$$

This expression provides the error in terms of the computational budget T and the time per replication τ_i. The best estimator between A and B is thus the one leading to the smallest estimation error represented by the product $\sigma_i^2 \tau_i$. We are led to the following:

Rule (Estimation selection by efficiency)

1. Fix a computational budget T.
2. Choose the estimator $i = A, B$ minimizing:

$$\text{Efficiency}(i) := \sigma_i^2 \tau_i.$$

This measure of efficiency is intuitive and does not depend on the way a replication is constructed. Indeed, if we change the definition of replication and say that one replication in the new sense is given by the average of two replications in the old sense, then the cost per replication doubles and the variance per replication halves, leaving the efficiency measure unchanged, as was expected. After all we have simply renamed the steps of a same algorithm.

Sometimes the computational time τ is random. This is the case when the chain of steps leading to one replication depends on intermediate values. For instance, in the evaluation of a barrier option the path simulation is interrupted whenever the barrier is reached. If τ is random, then formula (1.5) still holds with τ replaced by $\mathbb{E}(\tau)$ or any unbiased estimation of it.

1.1.3 Issue 3: How to Simulate Samples

No truly random number can be generated by a computer code as long as it can only perform sequences of deterministic operations. Moreover, the notion of randomness is somehow fuzzy and has been debated for long by epistemologists. However, there are deterministic sequences of numbers which "look like" random samples from independent copies of the uniform distribution on the unit interval. There are also well-established tests for the statistical quality of these uniform generators. Each of these numbers is a uniform "pseudo-random sample". From uniform pseudo-random samples we can obtain pseudo-random samples drawn from any other distributions by applying suitable deterministic transformations. More precisely, if $X \sim F_X$, then there exists a number n and function $G_X : [0, 1]^n \to \mathbb{R}$ such that for any sequence of mutually independent uniform copies $U^{(1)}, \ldots, U^{(n)}$, the compound r.v. $G_X(U^{(1)}, \ldots, U^{(n)})$ has distribution F_X. It turns out that the function G_X can be determined by the knowledge of the distribution $\mathbb{P}_X$, which is usually assigned through:

- A cumulative distribution function (c.d.f.) $F_X(x)$;
- A density function (d.f.) $f_X(x) = \frac{d}{dx} F_X(x)$ (if F_X is absolutely continuous);
- A discrete distribution function (d.d.f.) $p_X(x) = F_X(x) - F_X(x^-)$ (if F_X is discrete);
- A hazard rate function (h.r.f.) $h_X(x) = f_X(x)/(1 - F_X(x))$.

Methods for determining the transformation G_X (and thus delivering random samples from $\mathbb{P}_X$) are available for each of these assignments. Section 1.2 below is entirely devoted to this issue.

Numbers generated by any of these methods are called pseudo-random samples. Monte Carlo simulation delivers pseudo-random samples of a statistical experiment given its distributional properties. In the rest of the book, the terms "simulated sample" and "sample" are used as synonyms of the more proper term "pseudo-random sample".

1.1.4 Issue 4: How to Evaluate Financial Derivatives

Derivative valuation involves the computation of expected values of complex functionals of random paths. The Monte Carlo method can be applied to compute approximated values for these quantities. For instance, we consider a European-style derivative written on a state variable whose time t value is denoted by $X(t)$. At a given time T in the future, the security pays out an amount corresponding to a functional F of the state variable path $\{X(s), t \leq s \leq T\}$ between current time t and the exercise time T. For notational convenience, this path is denoted by $X_{t,T}$.

History between t and T	$\to$	Pay-off at time T
$X_{t,T} := \{X(s), t \leq s \leq T\}$		$F(X_{t,T})$

The arbitrage-free time t price of this contingent claim is given by the conditional expectation of the present value of its future cash-flow under the risk-neutral probability $\mathbb{P}^*$, that is[6]

$$V(t) = \mathbb{E}_t^{\mathbb{P}^*} \left(e^{-\int_t^T r(u, X(u))\, du} F(X_{t,T}) \right).$$

Notice that the state variable may enter into the determination of the short rate of interest r.

---

[6] The risk-neutral probability $\mathbb{P}^*$ makes all discounted security prices martingales. In other words, $X := V(t)/\exp(\int_0^t r(s)\, ds)$ is a $\mathbb{P}^*$-martingale for any security price process V.

Example (Options) In a European-style call option position, the holder has the right to buy one unit of the underlying state variable at time T for a strike price K. Here $F(X_{t,T}) = \max(0, X(T) - K)$, where $K > 0$ is the strike price and $T > 0$ is the exercise date. In an Asian option position, the holder receives the arithmetic average of all values assumed by the underlying state variable over an interval $[t, T]$. Here $F(X_{t,T}) = \int_t^T X(s)\,ds/(T - t)$, where $T > 0$ is the exercise date. In an up-and-out call option position, the holder has the right to exercise a call option $C(T, K)$ provided that the underlying state variable has always stayed below a threshold Γ over the option lifetime $[t, T]$. Here $F(X_{t,T}) = (X(T) - K)_+ \mathbf{1}_{\mathbf{E}}(X_{t,T})$, $T > t$, is the exercise date, and the set $\mathbf{E} = \{g \in \mathbb{R}^{[t,T]} : g(s) < \Gamma, \forall s \in [t, T]\}$ identifies all paths never crossing the threshold Γ on the interval $[t, T]$.[7]

If we can somehow generate i.i.d. samples $x_{t,T}^{(1)}, \ldots, x_{t,T}^{(n)}$ of the random price path X_{tT}, the simple Monte Carlo estimation gives us

$$V(t) \simeq \frac{1}{n} \sum_{i=1}^n e^{-\int_t^T r(u, x_{t,u}^{(i)})\,du} F\left(x_{t,T}^{(i)}\right).$$

This method can be implemented as follows:

Algorithm (Path-dependent Monte Carlo method)

1. Fix n "large".
2. Generate n independent paths $x_{t,T}^{(1)}, \ldots, x_{t,T}^{(n)}$ of process X on $[t, T]$.
3. Compute the discount factor and the pay-off over each path $x_{t,T}^{(i)}$.
4. Store the present value of the pay-off over each path, that is $V^{(i)} = \exp(-\int_t^T r(u, x_{t,u}^{(i)})\,du) \times F(x_{t,T}^{(i)})$.
5. Return the sum of all $V^{(1)}, \ldots, V^{(n)}$ divided by n.

In most cases paths need not, or simply cannot, be simulated in continuous time. Therefore we may carry out a dimension reduction of the problem by identifying a path $(g(t), 0 \le t \le T)$ through a finite number of its value increments on consecutive intervals of length, say, Δt:

$$\Delta g_1, \ldots, \Delta g_N \to \tilde{g}_{\Delta g_1, \ldots, \Delta g_N}(t) := \Delta g_1 + \cdots + \Delta g_{[t/\Delta t]},$$

for all $t \le \Delta t \times N =: T$. We say that paths are discretely monitored. In these cases, the expected value of a functional of a continuous time path $g \in \mathbb{R}^{[0,T]}$ with respect to the probability measure $\mathbb{P}^X$ induced by a stochastic process X over the path space $\mathbb{R}^{[0,T]}$ can be approximately evaluated as an integral over the finite-dimensional space where a finite sample of increments in X is simulated:

$$\mathbb{E}^{\mathbb{P}^X}(F) = \int_{\mathbb{R}^{[0,T]}} F(g)\mathbb{P}^X(dg)$$

$$= \int_{\mathbb{R}^N} F\left(\left(\tilde{g}_{x_1, \ldots, x_N}(t), 0 \le t \le T\right)\right) f_{\Delta X_1, \ldots, \Delta X_N}(x_1, \ldots, x_N)\,dx_1 \cdots dx_N,$$

---

[7] The symbol $\mathbb{R}^{[t,T]}$ denotes the class of all paths between t and T.

where $f_{\Delta X_1, \dots, \Delta X_N}$ denotes the distribution density of the vector of process increments $X((k+1)\Delta t) - X(k\Delta t)$, $k = 0, \dots, N-1$. Consequently, Monte Carlo can
be seen as a method for numerically computing multidimensional integrals. This idea
is particularly useful each time the dimension of the problem is so large that neither
analytical evaluation nor standard numerical discretization procedures deliver reliable results.

Example (Basket option) Consider an option on a basket consisting of k many assets. The underlying process is k dimensional and a vector $\mathbf{x}_i := (x_{i,1}, \dots, x_{i,k})$ is
sampled at each time t_i. A Monte Carlo estimation reads as

$$\mathbb{E}(F) \simeq \frac{1}{n} \sum_{m=1}^{n} F^* \big(x_{1,1}^{(m)}, \dots, x_{1,k}^{(m)}, x_{2,1}^{(m)}, \dots, x_{N-1,k}^{(m)}, x_{N,1}^{(m)}, \dots, x_{N,k}^{(m)} \big),$$

where $x_{i,j}^{(m)}$ denotes the mth sample of the jth component of the state variable at
time t_i. This corresponds to an integral in $N \times k$ dimensions. For a two year option, hedged once every week on the calendar year (52 weeks), on three indices, the
dimension is $2 \times 52 \times 3 = 312$!!!.

1.1.5 The Monte Carlo Simulation Algorithm

Figure 1.2 provides a general Monte Carlo algorithm for financial applications. The
static branch deals with methods for generating pseudo-random samples from distributions assigned by any of the functions listed in Sect. 1.1.3. This topic is developed
in the rest of the present chapter. Chapter 2 tackles the issue of sampling random
paths of a given stochastic process. We suppose a stochastic model is given in a
continuous time diffusion framework. The corresponding dynamics are then approximated using processes that depend on a finite number of random variables. We use
any of the methods illustrated in Chapter 1 to generate samples from these distributions and then deliver pseudo-random sample paths. Monte Carlo estimation can be
improved by adopting variance reduction techniques that will be introduced at the
end of Chapter 2.

1.2 Simulation of Random Variables

We present some algorithms for generating independent and uniformly distributed
pseudo-random numbers on the unit interval $[0, 1]$. We then describe three general methods for transforming uniform numbers into random samples with assigned
probability distributions. Inverse transformation methods take a c.d.f. as input, the
acceptance–rejection method assumes the d.f. is known, and the hazard rate method
moves from a hazard rate function.

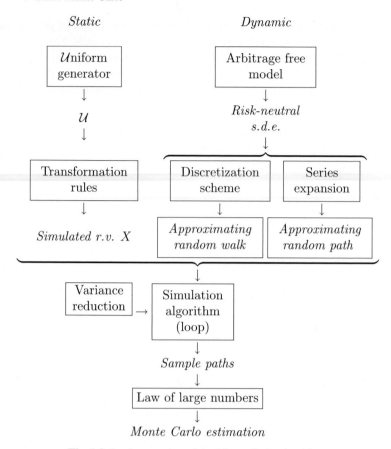

Fig. 1.2. Implementation of the Monte Carlo algorithm.

1.2.1 Uniform Numbers Generation

A congruential generator is a recursive formula returning a sequence of pseudo-random numbers. It starts with a "seed" value x_0. Recall that for any positive integer m, the modulus-m of a real number x is defined as the residual of the fraction x/m and is denoted by $x \pmod m$. We consider linear congruential generators.

Algorithm (Linear congruential generator)

1. Fix positive integers m (modulus), a (multiplier), and c (increment).
2. Set up a seed $x_0 \in \{0, \ldots, m-1\}$.
3. Run the recursive rule $x_{i+1} = (ax_i + c)(\mathrm{mod}\, m)$.
4. Return $u_{i+1} = x_{i+1}/m \in [0, 1]$.

The modulus-m of $ax_i + c$ is given by $x_{i+1} = ax_i + c - k_i m$, where k_i is the number of times the integer m can be "plugged" into $ax_i + c$, i.e., $k_i = \left[\frac{ax_i + c}{m}\right]$. (Here, square brackets denote the integer part of the argument.) As an example, if

Table 1.2. Linear congruential generators with maximal period

Name	m	a	c
1	$2^{37} - 1$	16,807	0
2	2,147,483,399	40,692	0
3	2,147,483,563	40,014	0
4	$2^{31} - 1$	39,373	0
5	$2^{31} - 1$	742,938,285	0
6	$2^{31} - 1$	950,706,276	0
7	$2^{31} - 1$	1,226,874,159	0
8	2^{35}	$2^7 + 1$	1
9	10^8	31,415,821	1

$x_0 = 2, a = 1, c = 3, m = 2$, we have $ax_0 + c = 5$ and $k_1 = [\frac{5}{2}] = 2$, so
that $x_1 = 5 - 2 \times 2 = 1$. This is exactly the residual of the ratio $(ax_i + c)/m$.
If constants m, a, and c are correctly chosen, the elements in the sequence $(x_i)_{i\in\mathbb{N}}$
are good approximation of i.i.d. uniformly distributed samples on $[0, 1]$. Notice that
given $u_i \sim \mathcal{U}[0, 1]$, the r.v. $u_i \times M$ is uniform on $[0, M]$.

The sequence $(x_i)_{i\in\mathbb{N}}$ provided by any congruential generator is periodic. The
maximal period is m, since the sequence must self-intersect, i.e., meet twice the
same value in no more than m steps. Indeed, any sample value x_i lies in the set
$\{0, \dots, m - 1\}$ by construction. Consequently, any sequence $(x_0, \dots, x_m)$ cannot
display distinct elements and is thus self-intersecting. This result suggests that one
should (1) select a relatively large m and (2) find conditions on the input coefficients
m, a, and c ensuring that the period p is maximal, i.e., $p = m$. Table 1.2 reports a
list of input values for which the maximality condition holds true. If the following
conditions are met, the corresponding generator can be proven to be maximal: m is
any power 10^k, $a(\mathrm{mod}\,200) = 21$ or $\max\{m/100, \sqrt{m}\} < a \le m/100$, c is an odd
number not a multiple of 5. It is recommended that the seed be reset to an arbitrarily
chosen value at each run of the code.

Below we report the relative performance of all generators indicated in Table 1.2.
Ranking is done according to the sample L^2 deviation from the exact uniform density. (Error is computed as the difference in area underlying the two graphs.) Execution time for all generators is about the same across all generators.

Figure 1.3 compares the exact density to histograms for the best and the worst
generators reported in Table 1.3 according to the error criterion. These are generator 2 and generator 7, respectively.

Figure 1.4 shows the L_2 error for each bin.

Example (Stratified sampling) To improve the uniformity of generated numbers we
may employ stratified sampling. This technique introduces a bias in the allocation of
sampled points by forcing samples to stick into subintervals refining $[0, 1]$. Let n be
the number of samples we want to generate. We may divide $[0, 1]$ into M stratifying
bins $[\frac{i}{M}, \frac{i+1}{M}]$, $i = 0, \dots, M - 1$, and force the first sample to stick into $[0, \frac{1}{M}]$,
the second sample into $[\frac{1}{M}, \frac{2}{M}]$, and so on until the Mth sample has been generated

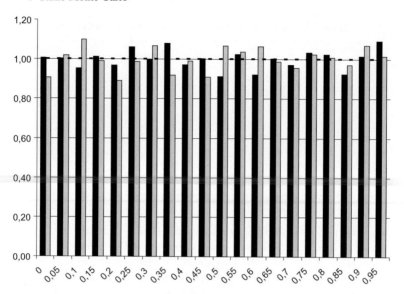

Fig. 1.3. Theoretical vs. sample uniform densities.

Table 1.3.

Generator	1	2	3	4	5	6	7	8
L_2^{error} $(\times 10^{-2})$	2.67	2.22	2.52	2.44	2.75	2.30	5.28	2.48

in the last subinterval $[\frac{M-1}{M}, 1]$. Then, the following number is generated   within $[0, \frac{1}{M}]$ and so on. The general rule states that $u_k \in [\frac{i}{M}(\mathrm{mod}\, M), \frac{i+1}{M}(\mathrm{mod}\, M)]$ for $k = 1, \ldots, n$. If $n = k \times M$, this method ensures that k samples fall into each interval $[\frac{i}{M}, \frac{i+1}{M}]$. Table 1.4 displays a pseudo-code for sampling pseudo-random numbers from stratified bins.

1.2.2 Transformation Methods

A. Inverse Transformation

This is the simplest method for simulating a r.v. with assigned c.d.f. F.

Idea Given a uniform r.v. U on $[0, 1]$, we look for a transformation f of U such that $f(U)$ has c.d.f. given by F, that is:

$$\mathbb{P}\big(f(U) \leq x\big) = F(x). \tag{1.6}$$

If f is bijective and monotonically increasing, the inverse function f^{-1} is well defined and we may write:

$$\mathbb{P}\big(f(U) \leq x\big) = \mathbb{P}\big(U \leq f^{-1}(x)\big)$$
$$= f^{-1}(x) \tag{1.7}$$

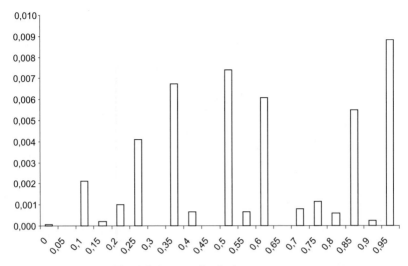

Fig. 1.4. L_2-error for the best generator.

Table 1.4. Pseudo-code for stratified sampling and histogram of sample density

```
h[1] = h[2] = ... = h[m] = 0;
M = number of stratifying bins; /* bin length=1/M */
n = number of samples /* of the form k*M, k = integer */;
m = number of histogram bins /* of form p*M, bin length=1/m */
for (j = 1, j <= k, j++){
        for (i = 1, i <= M, i++){
                v = Uniform[0,1];/* sampling a uniform in [0,1]*/
                u = (v+i-1)/M;/* zoom [0,1] into [(i-1)/M,i/M]*/
                b = integerPart[(u*m)+1];
                h[b]+=1;
        }
};
Plot[(h[i]/n)/(1/m)] over i = 1,...,m;
```

because the probability that a uniformly distributed variable on the unit interval is less than or equal to a given number is the number itself. Comparing expressions (1.6) and (1.7) suggests that any function f whose inverse f^{-1} matches F is a candidate transformation. We consider three cases.

Case 1 F is continuous and strictly increasing (Figure 1.5). Then F is bijective and $f = F^{-1}$ satisfies the required properties.

Case 2 F is continuous (Figure 1.6). Then, it need not be injective and F^{-1} may not even be defined. Let $F^{-1}(y) = \min\{x\colon F(x) = y\}$ be the generalized inverse function of F, which always exists because F is right-continuous. Note that this definition recovers the traditional definition of inverse in the case of strictly monotone functions. Since $F \circ F^{-1} = \mathrm{Id}$, the function $f = F^{-1}$ satisfies $\mathbb{P}(f(U) \le x) = F(x)$.

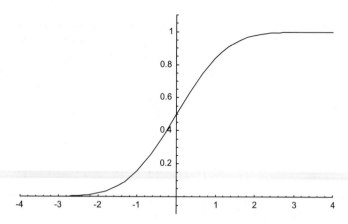

Fig. 1.5. Strictly monotone cumulative distribution function.

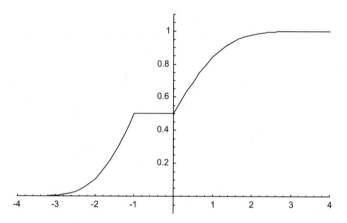

Fig. 1.6. Continuous cumulative distribution function displaying weak monotony.

Case 3 F is discontinuous, e.g., $F(x) = \sum_{i:x_i \leq x} p_i$ (Figure 1.7). Here p_i is interpreted as the probability of x_i. This function is neither injective nor surjective, so the generalized inverse $F^{-1}(y)$ may not even be defined. This is actually the case for points $y \in [0, 1] \setminus \mathrm{Im}(F)$. Let $F^{-1}(y) = \min\{x : F(x) \geq y\}$ be a further generalization of the notion of inverse function. Again, the right continuity of F ensures the well-definiteness of F^{-1} and this notion matches the two definitions above in their corresponding cases. In general $F \circ F^{-1} \neq F^{-1} \circ F \neq$ Identity. For any $u \in [0, 1]$, the set $\{x : u \leq F(x) < F \circ F^{-1}(u)\}$ is always empty. Consequently, it has zero probability. We may then write:

$$
\begin{aligned}
\mathbb{P}\big(F^{-1}(U) \leq x\big) &= \mathbb{P}\big(F \circ F^{-1}(U) \leq F(x)\big) \\
&= \mathbb{P}\big(F \circ F^{-1}(U) \leq F(x)\big) + \mathbb{P}\big(U \leq F(x) < F \circ F^{-1}(U)\big) \\
&= \mathbb{P}\big(U \leq F(x)\big) = F(x).
\end{aligned}
$$

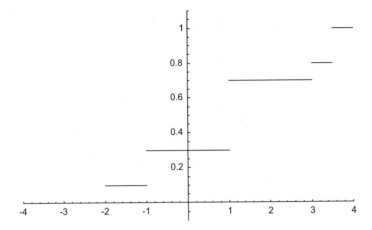

Fig. 1.7. Piecewise constant cumulative distribution function.

These considerations justify the following procedure for generating a random sample from a distribution F.

Algorithm (Inverse transformation)

1. Simulate $U \sim \mathcal{U}[0, 1]$.
2. Return $F^{-1}(U)$.

Example (Exponential sampling) The c.d.f. of an exponentially distributed r.v. X with unitary mean is $F(x) = 1 - e^{-x}$. Thus $X = F^{-1}(U) = -\ln(1 - U) \sim \mathcal{E}(1)$. Since $1 - U$ is also a uniform on $[0, 1]$, it suffices to take:

$$X = -\ln U.$$

Note that cX is an exponential with mean c. An exponential with parameter λ is one with mean λ^{-1}, that is $X = -\lambda^{-1} \ln U$, where U is uniform on $[0, 1]$.

B. Multidimensional Inverse Transformation

This method extends univariate inverse transformation to the case of random vectors.

Idea Let $\mathbf{X} = (X_1, \ldots, X_k)$ be a random vector. If $X_1, \ldots, X_k$ are mutually independent and their c.d.f.'s $F_1, \ldots, F_k$ are known, we may generate a sample of $\mathbf{X}$ by applying the inverse transformation method to each component of $\mathbf{X}$:

$$X_i = F_i^{-1}(U_i).$$

If vector components are not mutually independent, we must follow a slightly different route. First, we fix a permutation $i(1), \ldots, i(k)$ of the first k positive integers. For any arbitrary k-uple $u_1, \ldots, u_k$ in $[0, 1]^k$, we solve the following system for $x_{i(1)}, \ldots, x_{i(k)}$:

$$\begin{cases} F_{i(1)}(x_{i(1)}) = u_1 \\ F_{i(2)}(x_{i(2)}|x_{i(1)}) = u_2 \\ \vdots \\ F_{i(k)}(x_{i(k)}|x_{i(1)}, \ldots, x_{i(k-1)}) = u_k, \end{cases} \qquad (1.8)$$

and arrive at $\mathbf{x} = \mathbf{G}(u_1, \ldots, u_k)$. Then, we sample k independent uniform numbers $U_1, \ldots, U_k$ and return $\mathbf{G}(U_1, \ldots, U_k)$.

Note that there are $k!$ admissible systems, each one corresponding to a selected permutation. To see this, suppose that distributions admit densities. For each permutation $i(1), \ldots, i(k)$ of the first k indexes, we may write:

$$f_{\mathbf{X}}(x_1, \ldots, x_k)$$
$$= f_{i(1)}(x_{i(1)}) \cdot \prod_{j=1}^{k-1} f_{i(j+1)}(x_{j+1}|X_{i(1)} = x_{i(1)}, \ldots, X_{i(j)} = x_{i(j)}). \qquad (1.9)$$

Each $f_{i(n)}$ is a density with corresponding c.d.f. $F_{i(n)}$ determining a system of the above form. Of course, there are $k!$ many representations of the form (1.9). Some of these systems may be easier to solve than others. No methods are available to select *a priori* any of these systems.

Algorithm (Multidimensional inverse transformation)

1. Solve any of the admissible systems of form (1.8) and get to $\mathbf{G}(u_1, u_2, \ldots, u_k)$.
2. Simulate $U_1, U_2, \ldots, U_k \overset{\text{i.i.d.}}{\sim} \mathcal{U}[0, 1]$.
3. Return $\mathbf{G}(U_1, \ldots, U_k)$.

Example   Let (X_1, X_2) be a two-dimensional random vector with density $f_{X_1,X_2}(x_1, x_2) = 6x_1 \mathbf{1}_{\{x_1+x_2 \leq 1, x_1 \geq 0, x_2 \geq 0\}}(x_1, x_2)$. We show that representing f in terms of conditional densities in those ways corresponding to $i(1) = 1$, $i(2) = 2$ and $i(1) = 2$, $i(2) = 1$, respectively, does matter for solving the resulting system (1.8). If $f = f_1(x_1)f_2(x_2|x_1)$, the marginal c.d.f. is:

$$F_1(x_1) = \int_0^{x_1} f_1(y_1)\,\mathrm{d}y_1$$
$$= \int_0^{x_1} \left(\int_0^{1-y_1} f_{X_1,X_2}(y_1, y_2)\,\mathrm{d}y_2 \right) \mathrm{d}y_1$$
$$= 3x_1^2 - 2x_1^3,$$

where $0 \leq x_1 \leq 1$ and condition $[0, 1 - y_1]$ $y_2 \in$ defines the support of the distribution on the y_2 axis. The conditional c.d.f. is:

$$F_2(x_2|x_1) = \int_0^{x_2} \left(\frac{f_{X_1,X_2}(x_1, y_2)}{\int_0^{1-x_1} f_{X_1,X_2}(x_1, y_2)\,\mathrm{d}y_2} \right) \mathrm{d}y_2 = x_2(1 - x_1)^{-1},$$

where $0 \leq x_2 \leq 1 - x_1$. The system becomes:

Table 1.5. Performance of alternative multivariate inversion schemes

Method	Execution time (s)
Numerical inversion (system (1.10))	11.861
Numerical inversion (system (1.11))	13.349
Direct substitution	0.101

$$\begin{cases} F_1(X_1) = 3X_1^2 - 2X_1^3 = U_1, \\ F_2(X_2|X_1) = X_2(1 - X_1)^{-1} = U_2. \end{cases} \tag{1.10}$$

The alternative representation $f = f_2(x_2)f_1(x_1|x_2)$ leads to:

$$\begin{cases} F_2(X_2) = 1 - (1 - X_2)^3 = U_1, \\ F_1(X_1|X_2) = X_1^2(1 - X_2)^{-2} = U_2. \end{cases} \tag{1.11}$$

We note that (1.10) is difficult to solve whereas (1.11) is statistically equivalent to $(1 - X_2)^3 = U_1$ and $X_1^2(1 - X_2)^{-2} = U_2$. This system has a solution:

$$X_2 = 1 - U_1^{1/3}, \qquad X_1 = U_1^{1/3}U_2^{1/2}. \tag{1.12}$$

There is no known method to determine a priori which among the $k!$ representations of the joint d.f. leads to the easiest to solve within the set of systems (1.8). Table 1.5 compares computational times for generating samples by numerically solving system (1.10), system (1.11), or using the explicit solution (1.12).

C. Multivariate Direct Transformation

We present this method in the bivariate case.

Idea We wish to simulate a random vector (Y_1, Y_2) which is a known one-to-one transformation $\mathbf{g}$ of another random vector (X_1, X_2) with a known density $f_{X_1,X_2}(x_1, x_2)$. Let $\boldsymbol{\psi}$ be the inverse of $\mathbf{g}$, namely:

$$\boldsymbol{\psi} : (y_1, y_2) \rightarrow \big(\psi_1(y_1, y_2), \psi_2(y_1, y_2)\big) : \mathbf{g}\big(\psi_1(y_1, y_2), \psi_2(y_1, y_2)\big) = (y_1, y_2)$$

and J its Jacobian:

$$J = \det \begin{pmatrix} \dfrac{\partial \psi_1}{\partial y_1} & \dfrac{\partial \psi_1}{\partial y_2} \\ \dfrac{\partial \psi_2}{\partial y_1} & \dfrac{\partial \psi_2}{\partial y_2} \end{pmatrix}.$$

An application of the change of variable formula for double integrals gives:

$$f_{Y_1,Y_2}(y_1, y_2) = J \times f_{X_1,X_2}(x_1, x_2)|_{x_1=\psi_1(y_1,y_2), x_2=\psi_2(y_1,y_2)}. \tag{1.13}$$

Example (Normal distribution) Suppose that (Y_1, Y_2) is a standard normally distributed random vector. We look for r.v.'s X_1, X_2 and functions g_1, g_2 such that each

Y_i is equal to $g_i(X_1, X_2)$ and both X_1 and X_2 are easy to sample. The strategy is to look for ψ bringing Y_1, Y_2 somewhere, X_1 and X_2 say, and then define $\mathbf{g} = \psi^{-1}$. On a hunch, we check for ψ transforming Cartesian into polar coordinates, that is:

$$x_1 := \psi_1(y_1, y_2) := y_1^2 + y_2^2,$$
$$x_2 := \psi_2(y_1, y_2) := \tan^{-1}(y_2/y_1).$$

We apply formula (1.13) to the function $\psi : (y_1, y_2) \rightarrow (x_1, x_2)$ defined above. As the Jacobian of ψ is $J = 1/2$, we have:

$$f_{X_1,X_2}(x_1, x_2) = \frac{1}{2\pi}e^{-x_1/2} \times \frac{1}{2} = \frac{1}{2}e^{-x_1/2} \times \frac{1}{2\pi}$$

for $x_1 \in (0, \infty)$ and $x_2 \in (0, 2\pi)$. By integrating over appropriate domains, we can compute the marginals of X_1 and X_2. We easily read that X_1 is exponential with rate $1/2$ and X_2 is uniformly distributed on the interval $[0, 2\pi]$. As a result, if X_1 and X_2 are respectively samples from an exponential distribution with rate $1/2$ (i.e., with mean equal to 2) and from a uniform variate on the interval $[0, 2\pi]$, then the r.v.'s Y_1, Y_2 defined by:

$$Y_1 = g_1(X_1, X_2) = \psi_1^{-1}(X_1, X_2) = \sqrt{X_1} \cos X_2,$$
$$Y_2 = g_2(X_1, X_2) = \psi_2^{-1}(X_1, X_2) = \sqrt{X_1} \sin X_2,$$

are independent standard normals. Furthermore, if $U_1, U_2 \overset{\text{i.i.d.}}{\sim} \mathcal{U}[0, 1]$, then $-2 \ln U_1$ is exponential with rate $1/2$ and $2\pi U_2$ is uniform on $[0, 2\pi]$. We have proven the following:

Algorithm (Box–Müller)

1. Generate U_1, U_2 from $\mathcal{U}[0, 1]$.
2. Return $\sqrt{-2 \ln U_1} \cos(2\pi U_2)$ and $\sqrt{-2 \ln U_1} \sin(2\pi U_2)$.

1.2.3 Acceptance–Rejection Methods

This method is due to John Von Neumann.

Idea Let $\mathbb{P}_X$ be a target distribution on $\mathcal{D} := \text{Im}(X)$. A random sample Y is first drawn from a distribution $\mathbb{P}_Y$ with support $\mathcal{D}$ and then undergoes a random test. If the test is successful, Y is returned; otherwise, it is rejected. The test is designed so that accepted samples drawn from $\mathbb{P}_Y$ have distribution equal to $\mathbb{P}_X$. This scheme is advantageous whenever sampling from $\mathbb{P}_Y$ is easier than sampling from $\mathbb{P}_X$. However, the method may be largely time consuming as long as several test may be required before a number is accepted. More precisely, a test consists of drawing a sample from the r.v. $\mathbf{1}_{\{U \leq g(Y)\}}$, where $U \sim \mathcal{U}[0, 1]$ and g is a suitable function onto $[0, 1]$. If the outcome is 1, this is interpreted as a "success"; otherwise, it is meant as a "failure". Consequently, an acceptance–rejection scheme for a target distribution $\mathbb{P}_X$ is unequivocally defined by the selection of two quantities, namely: (1) a sampling distribution $\mathbb{P}_Y$ and (2) a test function g. The problem can be stated as follows:

Problem Given a distribution $\mathbb{P}_X$, find a distribution $\mathbb{P}_Y$ and a test function g : Support(Y) $\to$ $[0, 1]$, such that the distribution of Y, conditional to a successful outcome for the test $\mathbf{1}_{\{U \le g(Y)\}}$ (i.e., an independent draw $U \sim \mathcal{U}[0, 1]$ does not exceed $g(Y)$) matches $\mathbb{P}_X$. In terms of densities, this condition reads as

$$f_X(x) = f_{Y|U \le g(Y)}(x), \tag{1.14}$$

for all x in the support $\mathcal{D}$ of X.

Solution One condition for two unknowns (f_Y and g) generally leads to several solutions. We may shed light on this issue by expanding the right-hand side in (1.14) as

$$\begin{aligned} f_{Y|U \le g(Y)}(x) &= \frac{\mathbb{P}(U \le g(Y)|Y = x) f_Y(x)}{\mathbb{P}(U \le g(Y))} \\ &= \frac{\mathbb{P}(U \le g(x)) f_Y(x)}{\int \mathbb{P}(U \le g(Y)|Y = y) f_Y(y) \, dy} \\ &= \frac{g(x) f_Y(x)}{\int \mathbb{P}(U \le g(y)) f_Y(y) \, dy} = \frac{g(x) f_Y(x)}{\int g(y) f_Y(y) \, dy} \end{aligned}$$

where the Bayes' formula, the rule of total probabilities and a property of uniform cumulative distribution functions are used in this order.[8] By combining this expression with (1.14), we see that if f_X is required to be decomposed into the product of a test function $0 \le g \le 1$, a density function f_Y, and a (normalizing) constant $C := (\int g(y) f_Y(y) \, dy)^{-1}$, i.e.,

$$f_X(x) = C \times g(x) \times f_Y(x) \quad \text{on } \mathcal{D}. \tag{1.15}$$

Then condition (1.14) is fulfilled and the following procedure delivers a sample from f_X. To implement this program, we may select a d.f. f_Y satisfying

$$f_X \le C f_Y \tag{1.16}$$

on $\mathcal{D}$, for some constant C. Thus, decomposition (1.15) holds true for g defined by:

$$g := f_X/(C f_Y) \quad \text{on } \mathcal{D}. \tag{1.17}$$

Algorithm (Acceptance–rejection)

1. Select a constant C, a d.f. f_Y and a function g as in (1.15).
2. Simulate $U \sim \mathcal{U}[0, 1]$.
3. Simulate $Y \sim f_Y$.
4. If $U > g(Y)$, go to Step 2.
5. Return Y.

---

[8] Bayes' formula: $\mathbb{P}(A|B) = \mathbb{P}(A \cap B)/\mathbb{P}(B) = \mathbb{P}(B|A)\mathbb{P}(A)/\mathbb{P}(B)$.
Rule of total probabilities: $\mathbb{P}(X < Y) = \int \mathbb{P}(X < Y|Y = y) f_Y(y) \, dy$.
Uniform property: $\mathbb{P}(U \le x) = x$ if $U \sim \mathcal{U}[0, 1]$.

Example (Testing sampled uniforms) Drawing samples from $\mathbb{P}_Y$ should be as easy as possible to perform. It is thus natural to try and simulate uniformly distributed r.v.'s. Since the target distribution $\mathbb{P}_X$ and the simulated one must share a common support, an approximation of infinite domains is required. For the sake of clarity, we assume an interval $[a, b]$ is a common support for both $\mathbb{P}_X$ and $\mathbb{P}_Y$. That is Y is uniformly distributed $\mathcal{U}[a, b]$ and X assumes values in $[a, b]$. This corresponds to setting $f_Y(x) := (b - a)^{-1} \mathbf{1}_{[a,b]}(x)$. If $M \geq \sup_{x \in D} f_X(x)$, then inequality (1.16) holds true for $C = M(b - a)$. Expression (1.17) provides us with $g := f_X / (M(b - a) f_Y)$ and decomposition (1.15) reads as

$$f_X(x) = M(b - a) \times \frac{f_X(x)}{M} \times \frac{1}{b - a} \mathbf{1}_{[a,b]}(x).$$

The parameter M provides a control on the constant C which turns out to be related to the efficiency of the resulting algorithm. Indeed, efficiency can be measured by the inverse of the number of trials N, before a successful pair (U, Y) is drawn: the smaller the number of trials required to deliver a sample, the more efficient the algorithm. Since trial pairs (U, Y) are statistically independent of each other, the probability of success is given by:

$$\mathbb{P}(\mathbf{1}_{\{U \leq g(Y)\}} = 1) = \mathbb{P}(U \leq g(Y))$$
$$= \int \mathbb{P}(U \leq g(Y) | Y = y) f_Y(y) \, dy$$
$$= \int \mathbb{P}(U \leq g(y)) f_Y(y) \, dy$$
$$= \int g(y) f_Y(y) \, dy = C^{-1}.$$

It can be shown that the number of trials before a success occurs is a geometrically distributed r.v. N, i.e., $f_N(n) = p(1 - p)^n$, with expectation C. This latter attains the minimum value 1 provided that $M = (b - a)^{-1}$.

Example (Absolute normal distribution) The exponential density is $g_{\mathcal{E}(1)}(x) := e^{-x} \mathbf{1}_{\mathbb{R}_+}(x)$. The absolute value of a standard normal $\mathcal{N}(0, 1)$ has density:

$$f_{|\mathcal{N}(0,1)|}(x) := \frac{d}{dx} \mathbb{P}(|Z| \leq x) = \frac{d}{dx} \left[\int_{-x}^{x} f_{\mathcal{N}(0,1)}(x) \, dx \right] = \frac{2}{\sqrt{2\pi}} e^{-x^2/2}.$$

An upper bound for the ratio between $f_{|\mathcal{N}(0,1)|}$ and $g_{\mathcal{E}(1)}$ is given by

$$\frac{f_{|\mathcal{N}(0,1)|}(x)}{f_{Exp}(x)} = \frac{2}{\sqrt{2\pi}} e^{-x^2/2+x} = \sqrt{\frac{2e}{\pi}} e^{-(x-1)^2/2} \leq \sqrt{\frac{2e}{\pi}} =: C.$$

The method provides a sample from the absolute value of a normal by sampling from the exponential distribution.

1.2.4 Hazard Rate Function Method

Let τ be a positive r.v.; the ratio between its d.f. f_τ and 1 minus its c.d.f. F_τ defines the hazard rate function (h.r.f.) of τ:

$$\lambda_\tau(t) := \frac{f_\tau(t)}{1 - F_\tau(t)}.$$

This name is motivated by the following:

Example (Bond default) If τ represents the time a bond defaults, the corresponding h.r.f. computed at time t is the probability that default occurs in a "small" interval $(t, t + dt)$ conditional to the event that the bond has not defaulted before time t:

$$\begin{aligned} \mathbb{P}\big(\tau \in (t, t+dt)|\tau > t\big) &= \frac{\mathbb{P}(\tau \in (t, t+dt), \tau > t)}{\mathbb{P}(\tau > t)} \\ &= \frac{\mathbb{P}(\tau \in (t, t+dt))}{\mathbb{P}(\tau > t)} \\ &= \frac{f_\tau(t)\, dt}{1 - F_\tau(t)}. \end{aligned}$$

We now show how to sample τ given its h.r.f. $\lambda_\tau(\cdot) > 0$. For this, we need to make the following:

Assumptions (1) $\lambda_\tau(\cdot)$ is bounded from above by a constant λ^* (e.g., $\lambda^* = \max_{t \geq 0} \lambda_\tau(t) < +\infty$); (2) $\int_0^\infty \lambda_\tau(s)\, ds = \infty$.

Idea We draw sample jump times $\tau_1, \tau_2, \ldots$, of a homogeneous Poisson process until we meet the first time τ_{i^*} for which a random test provides a positive outcome. Our task is to design a test in such a way that the *selected* jump time τ_{i^*} has h.r.f. $\lambda_{\tau_{i^*}}(t)$ equal to $\lambda_\tau(t)$ for all $t \geq 0$, and can therefore be taken as a sample of τ. Let $\tau_1, \tau_2, \ldots$ denote jump times of a Poisson process with intensity λ^* and starting at time 0. We recall that each time τ_i can be obtained as a sum $\sum_{k=1}^{i} T_k$ of independent samples $T_1, \ldots, T_i$ with exponential distribution $\mathcal{E}(\lambda^*)$. A time t random test is an experiment taking value "s" (success) with probability $\alpha(t)$, and value "f" (failure) otherwise. A test function is a set of random tests, one for each time $t \geq 0$. As $\mathbb{P}(U \leq \alpha(t)) = \alpha(t)$ for uniform r.v.'s $U \sim \mathcal{U}[0, 1]$, a test function can be represented as a random process $(Y_\alpha(t), t \geq 0)$ with $Y_\alpha(t) = \mathbf{1}_{U \leq \alpha(t)}$, and is therefore unequivocally determined by a function $(\alpha(t), t \geq 0)$. We thus look for a time dependent function $(\alpha^*(t), t \geq 0)$ defining a test function $(Y_{\alpha^*}(t), t \geq 0)$, such that the conditional r.v. τ_{i^*} representing the first Poisson jump time for which the test succeeds (i.e., $i^* := \min\{i \geq 1 : Y_{\alpha^*}(\tau_i) = 1\}$) equals τ in the distributional sense, i.e., the h.r.f.'s of τ_{i^*} and τ agree for all times $t \geq 0$.

Problem Given a hazard rate function $(\lambda_\tau(t), t \geq 0)$, find a test function $(\alpha^*(t), t \geq 0)$, such that

$$\lambda_{\tau_{i^*}}(t) = \lambda_\tau(t),$$

for all $t \geq 0$; here $i^* := \min\{i \geq 1 : Y_{\alpha^*}(\tau_i) = 1\}$, $Y_{\alpha^*}(\tau_i) := \mathbf{1}_{U_i \leq \alpha^*(\tau_i)}$, $U_i \overset{\text{i.i.d.}}{\sim} \mathcal{U}[0, 1]$, $\tau_i = \sum_{k=1}^{i} T_k$, and $T_i \overset{\text{i.i.d.}}{\sim} \mathcal{E}(\lambda)$.

Solution We first clarify the following terms of the problem:

(1) The event $\{\tau \in (t, t + dt)\}$ corresponds to the existence of a Poisson jump time τ_i lying in $(t, t + dt)$, for which the test succeeds, i.e., $Y_\alpha(\tau_i) = 1$;
(2) The event $\{\tau > t\}$ means that "the test fails for all Poisson jump times $\tau_i \leq t$";
(3) Poisson processes have independent and stationary increments;
(4) Any test $Y_\alpha(t)$ is statistically independent of the Poisson process and of all tests $Y_\alpha(s)$, with $s < t$.

Under these assumptions, we may compute the h.r.f. of τ_{i*} as

$$\lambda_{\tau_{i*}}(t)\, dt \quad := \quad \mathbb{P}\big(\tau_{i*} \in (t, t + dt)|\tau_{i*} > t\big)$$

$$\overset{(1)\wedge(2)}{=} \mathbb{P}\big(\exists i : \tau_i \in (t, t + dt), Y_{\alpha*}(\tau_i) = 1 | Y_{\alpha*}(\tau_k) = 0, k \leq i - 1\big)$$

$$\overset{(3)\wedge(4)}{=} \mathbb{P}\big(\exists i : \tau_i \in (t, t + dt), Y_{\alpha*}(\tau_i) = 1\big)$$

$$= \quad \mathbb{P}\big(\tau_i \in (t, t + dt)\big) \times \mathbb{P}\big(Y_{\alpha*}(\tau_i) = 1 | \tau_i \in (t, t + dt)\big)$$

$$\overset{dt \text{ small}}{=} \big(\lambda\, dt + o(dt)\big) \times \alpha^*(t),$$

where $o(dt)$ denotes a function converging to 0 quicker than t. Consequently,

$$\lambda_{\tau_{i*}}(t) = \lambda_\tau(t) \quad \Longleftrightarrow \quad \alpha^*(t) := \frac{\lambda_\tau(t)}{\lambda},$$

up to a term of order dt. This method can be implemented using the following:

Algorithm (Hazard rate function)

1. Let $i = \tau = 0$ and $\lambda^* : \lambda(t) \leq \lambda^*$ for all $t \geq 0$.
2. Set $i = i + 1$.
3. Sample $T_i \sim \mathcal{E}(\lambda)$, and define $\tau = \tau + T_i$.
4. Sample $U_i \sim \mathcal{U}[0, 1]$.
5. If $\lambda \times U_i > \lambda(\tau)$, then go to Step 2.
6. Return τ.

The heuristic argument detailed above is turned into a rigorous proof in the next chapter, where inhomogeneous Poisson processes are simulated using a similar technique.

1.2.5 Special Methods

A. Univariate Normal Distribution

We describe a "quick-and-dirty" algorithm to generate independent samples from a standard normal distribution $\mathcal{N}(0, 1)$.

Idea The central limit theorem states that the ratio $(\sum_{i=1}^n X_i - n\mu)/(\sqrt{n}\sigma)$ asymptotically converges to $\mathcal{N}(0, 1)$. Set $X_i = U_i \sim \mathcal{U}[0, 1]$. Then $\mu = 1/2$, $\sigma = 1/\sqrt{12}$ and, for n sufficiently large, the ratio $(\sum_{i=1}^n U_i - n/2)/\sqrt{n/12}$ is approximately normal. The assignment $n = 12$ provides a sufficiently accurate approximation for most simulation purposes.

Algorithm (Normal distribution by summing up uniforms)

1. Generate $U_1, \ldots, U_{12} \overset{\text{i.i.d.}}{\sim} \mathcal{U}[0, 1]$.
2. Return $\sum_{i=1}^{12} U_i - 6$.

Let us conclude with a few remarks. For a log-normal r.v., we may return e^X, with X normal. We cannot use the inverse transformation method to generate a normal sample because the c.d.f. admits no analytical expression. Analytical approximations to this c.d.f. actually exist, but inverting them involves a root searching algorithm that may make the resulting sampling method lower than alternative algorithms.

We have simulated Gaussian variables $\mathcal{N}(\mu = 2, \sigma^2 = 3)$ by four methods: (1) numerical inversion of the c.d.f., (2) acceptance–rejection of an absolute normal with probability 0.5, (3) Box–Müller algorithm, (4) summing up 12 uniforms. For sample sizes $n = 1{,}000; 10{,}000; 100{,}000; 1{,}000{,}000;$ and $10{,}000{,}000$, we have computed execution time (in seconds) and the corresponding deviations of the sampled histogram from the theoretical normal density. Two metrics have been implemented for the purpose of evaluating this figure. The first one is the L_2 metric: it computes the difference in area between the two graphs. The second one is the Sup metrics: it determines the maximum distance between ordinates of the two graphs.[9] Table 1.6 reports results for all cases. These results depend on the refinement of the sample range. Table 1.6 reports values corresponding to splitting the interval $[-6, 10]$ into $N = 100$ bins. We have performed the same analysis under thinner refinements, that is $N = 200$ and $N = 300$. In general, error numbers increase by refining the sample range. Table 1.7 displays errors for the cases of 10,000 and 100,000 samples generated by numerical inversion of the c.d.f. However, the trend measured along increasing sample sizes is the same: errors tend to decrease as the sample size becomes larger and larger. Figures 1.8 and 1.9 display the L_2-error across different values in the abscissa under the hypothesis that the interval $[-6, 10]$ has been partitioned into 300 evenly spaced bins. The graphs support labeling the uniform-based generator as a "quick-and-dirty" method.

B. Multivariate Normal Distribution

The density of a random vector $\mathbf{X}$ with multivariate normal distribution $\mathcal{N}(\mu, \Sigma)$ is

---

[9] The L_2-distance between functions f and g on the interval $[0, T]$ is a measure of the area underlying their difference:

$$d_{\mathcal{L}_2}(f, g) := \sqrt{\int_0^T |f(t) - g(t)|^2}.$$

The Sup-distance between functions f and g on the interval $[0, T]$ is the least upper bound of the absolute value of their difference:

$$d_{\sup}(f, g) := \sup_{t \in [0, T]} |f(t) - g(t)|.$$

Table 1.6. Performance of alternative sampling schemes for normal variables

Method	Criterion	$n = 10^3$	$n = 10^5$	$n = 10^7$
Numer.–Invers.	L_2	0.20831	0.02073	0.00179
	Sup	0.09516	0.00615	0.00077
	Time	0	0	5
Accept.–Reject.	L_2	0.16413	0.01845	0.00208
	Sup	0.05041	0.00624	0.00069
	Time	0	0	10
Box–Müller	L_2	0.14496	0.01699	0.00210
	Sup	0.05763	0.00612	0.00105
	Time	0	0	5
Uniform Sum	L_2	0.16833	0.02027	0.01140
	Sup	0.07493	0.00697	0.00315
	Time	0	1	11

Table 1.7. Performance of the numerical inversion method for normal variables

Sample size	Criterion	Ref. = 100	Ref. = 200	Ref. = 300
10,000	L_2	0.056044	0.122860	0.174221
	Sup	0.023736	0.038525	0.038613
100,000	L_2	0.020734	0.041032	0.064318
	Sup	0.006150	0.016850	0.021737

$$f_{\mathbf{X}}(\mathbf{x}) = \frac{1}{\sqrt{(2\pi)^k \det \boldsymbol{\Sigma}}} e^{-1/2(\mathbf{x}-\mu)^\top \boldsymbol{\Sigma}^{-1}(\mathbf{x}-\mu)}.$$

Here $\mu \in \mathbb{R}^k$ is the mean vector and $\boldsymbol{\Sigma} \in \mathcal{M}(k)$ is the covariance matrix, which is positive semi-definite and symmetric. The class of normal distributions is closed under linear transformations, meaning that for any matrix $\mathbf{C} \in \mathcal{M}(k)$, the product vector $\mathbf{Cx}$ is $\mathcal{N}(\mathbf{C}\mu, \mathbf{C}\boldsymbol{\Sigma}\mathbf{C}^\top)$. This leads to the following:

Algorithm (Multivariate normal distribution)

1. Decompose $\boldsymbol{\Sigma}$ as $\mathbf{CC}^\top$ for a suitable matrix $\mathbf{C} \in \mathcal{M}(k)$.
2. Simulate $\mathbf{Z} = (Z_1, \ldots, Z_n) \sim \mathcal{N}(\mathbf{0}, \mathbf{I})$ by generating samples $Z_i \overset{\text{i.i.d.}}{\sim} \mathcal{N}(0, 1)$.
3. Return $\mathbf{CZ} + \mu$ as a sample of $\mathcal{N}(\mu, \boldsymbol{\Sigma})$.

This method admits as many variants as the number of possible factorizations of the covariance matrix $\boldsymbol{\Sigma}$. We examine two of them.

Cholesky factorization

We look for a decomposition $\boldsymbol{\Sigma} = \mathbf{CC}^\top$ where $\mathbf{C}$ is a lower triangular matrix. For any positive definite symmetric matrix this decomposition always exists and can be obtained by a recursive procedure. Given semi-rows $c_{i1}, \ldots, c_{ij-1}$ and $c_{j1}, \ldots, c_{jj-1}$, the cell (i, j) is defined by

Error Inverse Method vs. Exact Density

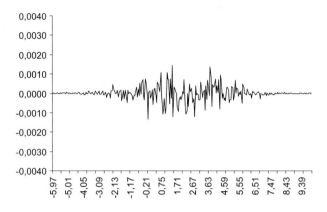

Error Acceptance/Rejection Method vs. Exact Density

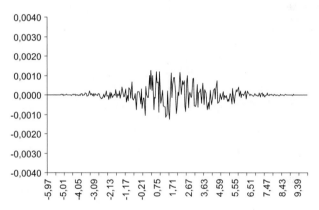

Fig. 1.8. Sample errors for approximated inverse function and acceptance–rejection methods vs. the exact normal density.

$$c_{ij} := \frac{\sigma_{ij} - \sum_{k=1}^{j-1} c_{ik} c_{jk}}{\sqrt{(\sigma_{jj} - \sum_{k=1}^{j-1} c_{jk}^2)}},$$

where the sum $\sum_{i=1}^{0} \ldots$ is set to zero.

Principal components decomposition

Any covariance matrix admits a spectral decomposition $\boldsymbol{\Sigma} = \mathbf{U} \boldsymbol{\Lambda} \mathbf{U}^{\top}$, where $\mathbf{U}$ is an orthogonal matrix[10] and $\boldsymbol{\Lambda}$ is a diagonal matrix $\mathrm{diag}(\lambda_1, \ldots, \lambda_k)$. The ith element λ_i in the diagonal of $\boldsymbol{\Lambda}$ is an eigenvalue of $\boldsymbol{\Sigma}$ and the ith column $\mathbf{u}^i$ in

---

[10] A matrix $\mathbf{U} = (\mathbf{u}^1 | \cdots | \mathbf{u}^k)$ is orthogonal if distinct columns have zero inner product: $\langle \mathbf{u}^i, \mathbf{u}^j \rangle := \sum_{m=1}^{k} u_m^i u_m^j = 0$. This implies $\mathbf{U}^{-1} = \mathbf{U}^{\top}$.

Error Box-Muller Method vs. Exact Density

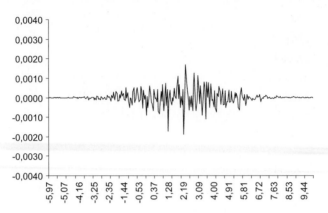

Error Uniform Method vs. Exact Density

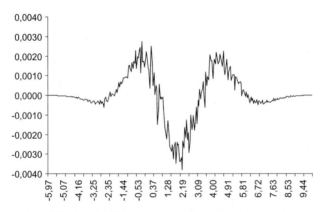

Fig. 1.9. Sample errors for Box–Müller and uniform-based methods vs. the exact normal density.

matrix $\mathbf{U}$ is its corresponding orthonormal eigenvector.[11] Since $\boldsymbol{\Sigma}$ is symmetric, then all eigenvalues are real. This leads to a decomposition $\boldsymbol{\Sigma} = \mathbf{CC}^\top$ where $\mathbf{C} := \mathbf{U}\sqrt{\boldsymbol{\Lambda}} = \mathbf{U} \cdot \mathrm{diag}(\sqrt{\lambda_1}, \ldots, \sqrt{\lambda_k})$. If $\mathbf{X} \sim \mathcal{N}(\boldsymbol{\mu}, \boldsymbol{\Sigma})$, then

$$\mathbf{X} \sim \boldsymbol{\mu} + \mathbf{Cz} = \boldsymbol{\mu} + z_1\sqrt{\lambda_1}\mathbf{u}^1 + \cdots + z_k\sqrt{\lambda_k}\mathbf{u}^k, \quad \text{where } z_i \overset{\text{i.i.d.}}{\sim} \mathcal{N}(0, 1).$$

---

[11] A number $\lambda \in \mathbb{C}$ is an eigenvalue of a matrix $\boldsymbol{\Sigma}$ and $\mathbf{u} \in \mathbb{R}^k$ is an eigenvector corresponding to λ if they fulfill the following equation:

$$\boldsymbol{\Sigma}\mathbf{u} = \lambda\mathbf{u}.$$

Eigenvectors corresponding to distinct eigenvalues are orthogonal. If their norm $\|\mathbf{u}\| := \sum_{m=1}^{k} u_m^i$ is equal to 1, they constitute an orthonormal basis in $\mathbb{R}^k$.

This expression is particularly useful whenever k is very large and a reduction in dimension is required. In this case, we first arrange the eigenvalues in decreasing order, i.e., $\lambda_1 \geq \cdots \geq \lambda_k$. Then, we fix a threshold $\mathcal{T}$ representing the percentage of variance embodied by the original vector $\mathbf{X}$ to be explained by the reduced vector. The former being the sum of all variances of the components of $\mathbf{X}$, that is $\text{Trace}(\boldsymbol{\Sigma}) = \sum_{i=1}^{k} \Sigma_{ii} = \sum_i \lambda_i$,[12] we select the smallest index k^* such that the ratio $\sum_{i=1}^{k^*} \lambda_i / \sum_{i=1}^{k} \lambda_i \geq \mathcal{T}$. Finally the vector $\mu + z_1\sqrt{\lambda_1}\mathbf{u}^1 + \cdots + z_{k^*}\sqrt{\lambda_{k^*}}\mathbf{u}^{k^*}$ is the reduced vector with the required approximating property:

$$\mathbf{X} \sim \mu + z_1\sqrt{\lambda_1}\mathbf{u}^1 + \cdots + z_{k^*}\sqrt{\lambda_{k^*}}\mathbf{u}^{k^*},$$

where variables z_i are as above.

Stable Distributions

These distributions arise in the simulation of Levy processes. A r.v. X has an α-stable distribution if any linear combination of two independent copies of X is distributed as X up to a linear transformation. More formally, $\forall a_1, a_2 > 0$ and $X_1, X_2 \overset{\text{i.i.d.}}{\sim} X$, $\exists c > 0, b \in \mathbb{R}: a_1 X_1 + a_2 X_2 \overset{d}{=} cX + b$. In particular, the sum of independent copies of X display the same property. It can be shown this distribution has characteristic function given by

$$\varphi(x) = \exp\{\sqrt{-1}\mu x - \sigma^\alpha |x|^\alpha [1 + \sqrt{-1}\beta \times \text{Sign}(x)h(x, \alpha)]\},$$

$$h(x, \alpha) = \begin{cases} -\tan\left(\alpha\frac{\pi}{2}\right), & \alpha \neq 1, \\ \frac{2}{\pi}\ln x, & \alpha = 1. \end{cases}$$

Here $\mu \in \mathbb{R}$, $\sigma \geq 0$, $\beta \in [-1, 1]$, $\alpha \in (0, 2]$ have the interpretation of shift, scale, skewness, and stabilization parameters, respectively. We write $S_\alpha(\sigma, \beta, \mu)$ to denote this distribution. It is not difficult to see that $S_2(1, 0, 0)$ is a standard normal and that $S_1(1, 0, 0)$ is a Cauchy distribution. The attribute "α-stable" stems from the following property involving the characteristic function of X: $\forall a > 0$, $\exists \alpha \in (0, 2]$ and $c = c(a) \in \mathbb{R}$ such that $\varphi(x)^\alpha = \varphi(\sqrt[\alpha]{a}x)\exp\{\sqrt{-1}cx\}$.

To generate a sample of $X \sim S_\alpha(\sigma, \beta, \mu)$, it is sufficient to simulate $Y \sim S_\alpha(1, \beta, 0)$, apply the stable property and return

$$\sigma X + \mu \quad \text{if } \alpha \neq 1,$$

$$\sigma X + \mu + \frac{2}{\pi}\beta\sigma\ln\sigma \quad \text{if } \alpha = 1.$$

We may apply the following:

Algorithm (α-stable distribution)

1. Generate $U \sim \mathcal{U}[0, 1]$.

---

[12] The trace of a matrix is invariant under orthogonal transformations: $\text{Trace}(\boldsymbol{\Sigma}) = \text{Trace}(\mathbf{U}^\top \boldsymbol{\Sigma})$ for any orthogonal matrix $\boldsymbol{\Sigma}$.

2. Generate $\varepsilon \sim \mathcal{E}xp(1)$ independently of U.
3. Set $V = \pi U - \pi/2$ and $c = -(2\alpha)^{-1}\pi\beta(1 - |1 - \alpha|)$.
4. Return

$$
\begin{cases}
\sin[\alpha(U - c)]\big(\varepsilon^{-1}\cos[U - \alpha(U - c)]\big)^{(1-\alpha)/\alpha}\cos(U)^{-\alpha}, & \text{if } \alpha \neq 1, \\
\frac{2}{\pi}\big[\big(\frac{\pi}{2} + \beta U\big)\tan(U) - \beta \ln\big(\frac{\frac{\pi}{2}\varepsilon\cos(U)}{\frac{\pi}{2} + \beta U}\big)\big], & \text{if } \alpha = 1.
\end{cases}
$$

Gamma and Chi-Square Distributions

The gamma distribution with shape parameter $\alpha > 0$ and scale parameter $\beta > 0$ has density:

$$
f_{\Gamma(\alpha,\beta)}(x) = \frac{1}{\Gamma(\alpha)\beta^{\alpha}}x^{\alpha-1}e^{-x/\beta} \quad (x > 0),
$$

where Γ is the gamma function defined by:

$$
\Gamma(\alpha) := \int_0^{\infty} x^{\alpha-1}e^{-x}\,dx.
$$

The moment generating function of a gamma distributed r.v. X can be computed as follows:

$$
m_X(t) = \int_0^{\infty} e^{tx}\frac{1}{\Gamma(\alpha)\beta^{\alpha}}x^{\alpha-1}e^{-x/\beta}\,dx = \left(\frac{\beta^{-1}}{\beta^{-1} - t}\right)^{\alpha}.
$$

By differentiating one and two times this expression with respect to t and computing the obtained results at $t = 0$, we come up to $\mathbb{E}(X) = \alpha\beta$, $\mathbb{E}(X^2) = \alpha(\alpha + 1)\beta^2$ and $\mathrm{Var}(X) = \alpha\beta^2$.

It can be proved that $\Gamma(\alpha, \beta)$ is homogeneous with respect to β, i.e., $\beta\Gamma(\alpha, 1) \overset{d}{=} \Gamma(\alpha, \beta)$. Consequently, if we wish to simulate a random sample from $X \sim \Gamma(\alpha, \beta)$, it suffices to generate a sample $Y \sim \Gamma(\alpha, 1)$ and return βX. The following algorithm allows us to simulate variables $\Gamma(\alpha, 1)$ with $\alpha \geq 1$ a case covering most applications in finance.[13]

Algorithm (Gamma distribution $\Gamma(\alpha, 1)$, with $\alpha \geq 1$)

1. Set $b = \alpha - 1$ and $c = 3\alpha - 3/4$.
2. Generate $U, V \overset{\text{i.i.d.}}{\sim} \mathcal{U}[0, 1]$.
3. Set $Z = U(1 - U)$, $Y = \sqrt{\frac{c}{Z}}(U - \frac{1}{2})$, and $X = b + Y$.
4. If $X < 0$, then go to Step 2.
5. If $\ln(64Z^3V^3) > 2(b\ln(\frac{X}{b}) - Y)$, then go to Step 2.
6. Return X.

Example (Central chi-square distribution) A chi-square distribution $\chi^2(\nu)$ with $\nu \in \mathbb{N}_+$ degrees of freedom is defined as the distribution of the sum of ν squared i.i.d. Gaussian variables $Z_1, \ldots, Z_{\nu}$:

---

[13] The case $0 < \alpha < 1$ is treated in, e.g., Cont and Tankov (2004) and Glasserman (2004).

$$X_{\chi^2(\nu)} = Z_1^2 + \cdots + Z_\nu^2.$$

It can be proved that $\chi^2(\nu)$ is also a gamma with shape parameter $\alpha = \nu/2$ and scale parameter $\beta = 2$; we may then generate a sample $Y \sim \Gamma(\nu/2, 1)$ and finally return $X = 2Y$.

Example (Noncentral chi-square distribution) A noncentral chi-square distribution $\chi^2(\nu, \lambda)$ with ν degrees of freedom and noncentrality parameter $\lambda = \sum_{i=1}^{\nu} \alpha_i^2$ is defined as the distribution of the sum of a number ν of α_i-shifted squared i.i.d. Gaussian variables $Z_1 + \alpha_1, \ldots, Z_\nu + \alpha_\nu$:

$$X_{\chi^2(\nu, \lambda)} = (Z_1 + \alpha_1)^2 + \cdots + (Z_\nu + \alpha_\nu)^2.$$

From this expression we read:

$$\begin{aligned} \chi^2(\nu, \lambda) &= \chi^2(1, \lambda) + \chi^2(\nu - 1) \\ &= \left(\mathcal{N}(0, 1) + \sqrt{\lambda}\right)^2 + \chi^2(\nu - 1), \end{aligned}$$

namely the sum of the square of a Gaussian variable with mean $\sqrt{\lambda}$ and a standard chi-square variable with $\nu - 1$ degrees of freedom. This method works for any integer $\nu > 1$. It turns out that a noncentral chi-square can be defined for any real number $\nu > 0$ by showing that the expression of the distribution function for integer ν makes sense for arbitrary ν. In this case, Cont and Tankov (2004) provide us with a way to simulate a random sample through an alternative algorithm.

1.3 Variance Reduction

We present three techniques for reducing the variance in the basic Monte Carlo algorithm. Other methods gathered around the label "Quasi-Monte Carlo" are developed as a case-study in Part II of this treatise.

1.3.1 Antithetic Variables

Let the target θ be the expected value of a *monotone* function g of a random variable X. We know that the sample mean $\widehat{\theta}_n(\mathbf{X}) := \frac{1}{n} \sum_{i=1}^{n} g(X^{(i)})$ is an unbiased estimator of θ whose precision can be measured by

$$\mathrm{Var}(\widehat{\theta}_n) = \frac{\mathrm{Var}(g(X))}{n}, \tag{1.18}$$

or by its unbiased estimator $\widehat{\sigma}_n^2(g(\mathbf{X})) := \frac{1}{n-1} \sum_{i=1}^{n} (g(X_i) - \widehat{\theta}_n(\mathbf{X}))^2$.

Problem Find an unbiased estimator of θ with greater precision, that is smaller variance than (1.18).

Idea Let $n = 2$. If X^1 and X^2 are samples from common c.d.f. F, the variance computes as

$$\mathrm{Var}\left(\frac{g(X^1) + g(X^2)}{2}\right) = \frac{1}{2}\big[\mathrm{Var}\big(g(X^1)\big) + \mathrm{Cov}\big(g(X^1), g(X^2)\big)\big],$$

since $\mathrm{Var}(g(X^1)) = \mathrm{Var}(g(X^2))$. If these samples are statistically independent, this quantity is $\mathrm{Var}(\widehat{\theta}_2)$. If we set g such that $g(X^1)$ and $g(X^2)$ are negatively correlated, then:

$$\mathrm{Var}\left(\frac{g(X^1) + g(X^2)}{2}\right) \le \mathrm{Var}(\widehat{\theta}_2).$$

According to the inverse function method, we may generate X^1 as $F^{-1}(U^1)$ and X^2 as $F^{-1}(U^2)$, where U^1 and U^2 are uniform r.v.'s. If these two uniforms are negatively correlated, we obtain negatively correlated samples X^1 and X^2. Given independent uniform samples U^1 and U^2, we may introduce negative correlation by substituting U^1 with $\iota(U^1)$. Here ι can be any decreasing function from $[0, 1]$ to itself such that $\iota(U_1) \sim \mathcal{U}[0, 1]$. Because the covariance of an increasing function (here F^{-1}) of a random variable and a decreasing function (here $F^{-1} \circ \iota$) of the same variable is always less than or equal to zero, we have:

$$\mathrm{Cov}\big(g(X^1), g(X^2)\big) = \mathrm{Cov}\big(\underbrace{g \circ F^{-1}(U_1)}_{\text{increasing}}, \underbrace{g \circ F^{-1} \circ \iota(U_1)}_{\text{decreasing}}\big) \le 0.$$

A simple choice for ι is $\iota(u) = 1 - u$. For an n-sized sample of a k-dimensional random vector $\mathbf{X}$ with c.d.f. $F = (F_1, \ldots, F_k)$ and statistically independent components, this method generalizes to the following algorithm.

Algorithm (Antithetic variables)

1. Simulate $\mathbf{U}_1, \ldots, \mathbf{U}_n \overset{\text{i.i.d.}}{\sim} \mathcal{U}[0, 1]^k$ ($n \times k$ uniforms);
2. Estimate θ by:

$$\widehat{\theta}_{\mathrm{AV}} := \frac{1}{2n} \sum_{i=1}^{n} \big[g\big(F_1^{-1}(U_{i1}), \ldots, F_k^{-1}(U_{ik})\big)$$
$$+ g\big(F_1^{-1}(1 - U_{i1}), \ldots, F_k^{-1}(1 - U_{ik})\big)\big].$$

Example In the Black–Scholes model the monotone function g transforming input $Z \sim \mathcal{N}(0, 1)$ into a call option discounted pay-off is given by

$$Z \to S(T) = S_0 e^{(r - \sigma^2/2)T + \sigma\sqrt{T}Z}$$
$$\to C(T, S(T); T, K) = e^{-rT}\big(S(T) - K\big)_+ =: g(Z).$$

Taking independent sample $Z_1, \ldots, Z_n$, the standard Monte Carlo estimator is $\widehat{\theta}_n :=$ $\frac{1}{n}\sum_{i=1}^n g(Z_i)$. We obtain the estimator

$$\widehat{\theta}_{AV} := \frac{1}{n} \sum_{i=1}^{n} \frac{g(Z_i) + g(-Z_i)}{2},$$

To implement this estimator, we may compute $\frac{1}{n} \sum_{i=1}^{n} \frac{H(U_i) + H(1 - U_i)}{2}$, where $H :=$ $g \circ F_{\mathcal{N}(0,1)}^{-1}(u)$ and U_i independent uniform variables.

Example In a short interest rate model, we may consider antithetic paths generated by:

$$r_{i+1}^{\pm} = r_i^{\pm} + \mu(t_i, r_i^{\pm}) \times \Delta t + \sigma(t_i, r_i^{\pm}) \times \sqrt{\Delta t} \times (\pm 1) \times \mathcal{N}(0, 1).$$

Here, for each set of N i.i.d. standard normal r.v.'s, we simulate two antithetic paths. As an alternative we may employ:

$$r_{i+1}^{\pm} = r_i^{\pm} + \mu(t_i, r_i^{\pm}) \times \Delta t + \sigma(t_i, r_i^{\pm}) \times \sqrt{\Delta t} \times \mathcal{N}(0, 1, \pm \mathcal{U}[0, 1]),$$

where the last factor denotes normals generated by antithetic uniforms, U and $1 - U$.

We stress the importance of the monotony hypothesis for g. For instance, a derivative having pay-off of form $|S(T) - K|$ does not satisfy this requirement. Therefore we do not expect a variance reduction by antithetic variables in this case.

1.3.2 Control Variables

If we have some information about functionals of the underlying process, we may foresee the possibility to exploit it in order to build up a new unbiased estimator which is more effective, either in terms of variance reduction or efficiency, than a naive Monte Carlo valuation.

Remember that the aim is to estimate:

$$\theta = \mathbb{E}(g(X)),$$

where g denotes a pay-off functional computed over a sample of the underlying stochastic process X. An antithetic variable estimator is:

$$\widehat{\theta}_{AV} = \frac{g(X) + g(Y)}{2},$$

where Y is a random path identically distributed as X displaying negative covariance with respect to Y.

Suppose we know the expected value of another functional f of X, such that f and g are one "close" to the other. For instance, we may take g to be the pay-off profile of an Asian option with arithmetic average, for which no closed form solution exists in the Black–Scholes framework, and f to be the pay-off cash flow of an Asian option with geometric average. This latter has a closed-form solution.

We may consider the following unbiased and consistent (can you prove it?) estimator:

$$\widehat{\theta}_{CV}^{\alpha} = g(X) + \alpha\big(f(X) - \mathbb{E}\big(f(X)\big)\big),$$

where the expected value is a known term.

The variance of $\widehat{\theta}_{CV}^{\alpha}$ is:

$$\mathrm{Var}\big(\widehat{\theta}_{CV}^{\alpha}\big) = \mathrm{Var}\big(g(X)\big) + \alpha^2\,\mathrm{Var}\big(f(X)\big) + 2\alpha\,\mathrm{Cov}\big(g(X), f(X)\big),$$

which is minimal for:

$$\alpha^* = -\frac{\mathrm{Cov}(g(X), f(X))}{\mathrm{Var}(f(X))}.$$

The optimal estimator:

$$\widehat{\theta}_{CV}^{\alpha} = g(X) - \frac{\mathrm{Cov}(g(X), f(X))}{\mathrm{Var}(f(X))} \times \big(f(X) - \mathbb{E}\big(f(X)\big)\big)$$

is merely of a theoretical importance. Indeed, it would seem very strange to know $\mathrm{Cov}(g(X), f(X))$ without knowing $\mathbb{E}(g(X))$! Anyway, we may follow two routes. First, we can estimate such a covariance by regression over past observed data. This amounts to computing the slope of the least-square regression line through the set of points $(g(X_i), f(X_i))$, $i = 1, \ldots, n$, i.e.,

$$\alpha_n^* = \frac{\sum_{i=1}^{n}(g(X_i) - n^{-1}\sum g(X_i))(f(X_i) - n^{-1}\sum f(X_i))}{\sum_{i=1}^{n}[g(X_i) - n^{-1}\sum g(X_i)]^2},$$

and adopting the estimator $\widehat{\theta}_{CV}^{\alpha_n^*}$. Alternatively, we may as well have a broad idea about the value taken by α^* according to our prior experience on the relation between $f(X)$ and $g(X)$. That is the case of two derivatives with similar payoffs.

The optimal estimator has variance:

$$
\begin{aligned}
\mathrm{Var}\big(\widehat{\theta}_{CV}^{\alpha^*}\big) &= \mathrm{Var}\big(g(X)\big) - \frac{\mathrm{Cov}(g(X), f(X))^2}{\mathrm{Var}(f(X))} \\
&= \mathrm{Var}\big(g(X)\big) - \frac{\mathrm{Corr}(g(X), f(X))^2(\sqrt{\mathrm{Var}(g(X))})^2(\sqrt{\mathrm{Var}(f(X))})^2}{\mathrm{Var}(f(X))} \\
&= \mathrm{Var}\big(g(X)\big)\big(1 - \mathrm{Corr}\big(f(X), g(X)\big)^2\big),
\end{aligned}
$$

showing that under the optimal parameter variance reduction is assured provided that f and g are one "close" to the other.

Example Let g be a functional transforming a sample price path into a time T payoff. We may take f to be the final value of a discretely rebalanced Δ-hedging strategy within the Black–Scholes model:

$$S_{0,T} \xrightarrow{f} V(S_{0,T}) = C_{BS}(0) + \sum_{i=0}^{N-1}\left[\partial_S C_i \times \Delta S_i + \frac{C_i - \partial_S C_i S_i}{B_i}\Delta B_i\right]P_{i+1}.$$

Here $\partial_S C_i = \partial_S C(t_i, S_{0,T}(t_i))$ is the option price delta, $\Delta S_i = S_{0,T}(t_{i+1}) - S_{0,T}(t_i)$ is the absolute variation of the stock price S, $C_i = C(t_i, S_{0,T}(t_i))$ is the option price, $S_i = S_{0,T}(t_i)$ is the stock price, $B_i = B(t_i)$ is the risk free asset price, $\Delta B_i = B(t_{i+1}) - B(t_i)$ is the risk free asset price variation, and $P_{i+1} = P_{t_{i+1}}(0)$ is the discount factor from time t_{i+1} to the current date. We can easily verify that $\mathbb{E}(f(X_{0,T})) = C_{BS}(0)$, that is the Black–Scholes price. Since the value process above is a proxy of the true replicating strategy for the option defined by g, we may argue that $\alpha = -1$ is likely to work well. For the case of a call option on $S(T)$, the estimator would be:

$$\widehat{\theta}_{CV}^{\alpha=-1} = \left(S_{0,T}(T) - K\right)_+ - 1 \times \left(V - C_{BS}(0)\right).$$

1.3.3 Importance Sampling

The general goal of any Monte Carlo method is to provide an estimation of the expected value of a function h of a random variable X: $\theta = E^f(h(X))$. We assume that the distribution of X is absolutely continuous with respect to the Lebesgue measure on the range of X and denote the density of the distribution function P_X of X by $f = f_X$. The Monte Carlo estimate is $\widehat{\theta}_n = n^{-1} \sum_{i=1}^n h(X^{(n)})$, where $X^{(1)}, \ldots, X^{(n)}$ is a sequence of independent and identically distributed samples from $\mathbb{P}_X$.

This estimator may be particularly inefficient whenever the region where h assumes relatively important values has a low probability. This is the case of a deep out-of-the-money European digital option written on S with maturity T and pay-off $N \times \mathbf{1}_{\{S(T)>K\}}$. For instance, the underlying price distribution can be concentrated on a region below the triggering threshold K and the probability for the option to generate a positive pay-off turns out to be relatively low. However, if the amount N is sufficiently big, then the option may have a nonnegligible value. It turns out that a sample generator may require several trials before meeting a number exceeding K. The Monte Carlo estimate is extremely sensitive to these circumstances and its convergence may be unreasonably slow.

Importance sampling consists of (1) replacing f with a density g whose samples are more likely to fall into the desired region, and (2) performing a Monte Carlo estimates by sampling from the new distribution and weighing the resulting samples appropriately. Intuitively, samples occurring more often under g than under f should be cut by a fraction of the value. This is the case if each sample $h(X^{(i)})$ is weighed by the ratio $f(X^{(i)})/g(X^{(i)})$. A simple computation shows that this conjecture is correct:

$$\theta := \mathbb{E}^f(h(X))$$
$$= \int h(x) f(x) \, dx$$
$$= \int h(x) \frac{f(x)}{g(x)} g(x) \, dx$$
$$= \mathbb{E}^g\left(h(X) \frac{f(X)}{g(X)}\right).$$

For this to hold true, we must require that $g > 0$. For more general distributions, the two probabilities must assign zero mass to the same events. We say that they are mutually equivalent probability measures. If we denote by $\mathbb{G}_X$ the probability distribution having density g, the new Monte Carlo estimate is

$$\widehat{\theta}_n^g = \frac{1}{n}\sum_{i=1}^{n} h(X^{(n)})\frac{f(X^{(n)})}{g(X^{(n)})},$$

where $X^{(1)}, \ldots, X^{(n)}$ is a sequence of independent and identically distributed samples from $\mathbb{G}_X$. In other words, after choosing a suitable $g > 0$, samples are drawn from it and plugged into the classical Monte Carlo estimator where the target function is now $h \times f/g$.

Properties It is easy to check that $\widehat{\theta}_n^g$ is an unbiased estimator of θ. However, its variance differs from the one exhibited by $\widehat{\theta}_n$. To show this, we assume without loss of generality that $\theta = 0$. (Otherwise, we may subtract θ from $\widehat{\theta}_n^g$ and proceed as follows.) In this case, variance coincides with the moment of order two. For the new estimator, this figure is easily computed as

$$\mathbb{E}^g\big[(\widehat{\theta}_n^g)^2\big] = \mathbb{E}^g\left[\left(\frac{1}{n}\sum_{i=1}^{n}h\frac{f}{g}\right)^2\right]$$

$$= \frac{1}{n^2}\sum_{i=1}^{n}\mathbb{E}^g\left[\left(h\frac{f}{g}\right)^2\right]$$

$$= \frac{1}{n}\mathbb{E}^g\left[\left(h\frac{f}{g}\right)^2\right], \qquad (1.19)$$

which we compare to the second moment of the traditional estimator $\widehat{\theta}_n$:

$$\mathbb{E}^g\big[\widehat{\theta}_n^2\big] = \frac{1}{n}\mathbb{E}^f\big[h^2\big]. \qquad (1.20)$$

We should select g, i.e., the new sampling measure, in a way such that (1.19) be smaller than (1.20):

$$\mathbb{E}^g\left[\left(h\frac{f}{g}\right)^2\right] < \mathbb{E}^f\big[h^2\big]. \qquad (1.21)$$

Clearly, the ideal situation of zero variance obtains for a density $g = \frac{hf}{\int hf\,dx}$, meaning that the best estimator of θ is θ itself! This tautology is useful however in providing us with a guidance towards a wise selection for g.

Prescription The importance sampling density g should be selected to be as close as possible to a proportion of the product between the contract pay-off h and the probability density function f of the underlying asset at maturity:

$$g \propto h \times f.$$

Example Consider a call option written on a standard normally distributed index W with strike price k. Then $\theta = \mathbb{E}[(W - k)_+]$ and the standard Monte Carlo estimate is $\widehat{\theta}_n^1 = n^{-1} \sum (W^{(i)} - k)_+$, where $W^{(i)} \overset{\text{i.i.d.}}{\sim} \mathcal{N}(0, 1)$. If the option is deeply out-of-the-money, then it may be useful to sample from a normal distribution targeting the exercise region of the underlying asset space more thoroughly. An upward shift for the price mean can do the job. Specifically, we change the density from $\mathcal{N}(0, 1)$ to $\mathcal{N}(\mu, 1)$ and compute

$$\widehat{\theta}_n^2 = n^{-1} \sum_{i=1}^{n} (W^{(i)} - k)_+ e^{-\mu W^{(i)*} + \mu^2/2},$$

where $W^{(i)*} \overset{\text{i.i.d.}}{\sim} \mathcal{N}(\mu, 1)$. An estimate of the sample error can be obtained by computing a Monte Carlo estimate of the variance reported on the left-hand side in formula (1.21), i.e.,

$$\text{Var}(\widehat{\theta}_n) \simeq \frac{1}{n} \sum_{i=1}^{n} [(W^{(i)} - k)_+ e^{-\mu W^{(i)*} + \mu^2/2}]^2.$$

The path-dependent case Consider a European-style derivative paying-off $h(x_1, \ldots, x_N)$ if the underlying index is worth $x_1, \ldots, x_N$ at intermediate monitoring times $t_1, \ldots, t_N$. Assuming that, for the sake of simplicity, the time value of money is zero, then the derivative value can be written as:

$$\theta = \mathbb{E}(h) = \int h(x_1, \ldots, x_N) f_{X_1, \ldots, X_N}(x_1, \ldots, x_N) \, dx_1 \cdots dx_N$$

$$= \int h \times f_{X_N | X_1, \ldots, X_{N-1}} \times f_{X_1, \ldots, X_{N-1}} \, dx_1 \cdots dx_N$$

$$= \int h \times f_{X_N | X_1, \ldots, X_{N-1}} \times f_{X_{N-1} | X_1, \ldots, X_{N-2}} \times f_{X_1, \ldots, X_{N-2}} \, dx_1 \cdots dx_N$$

$$= \int h(x_1, \ldots, x_N) \prod_{i=1}^{N} f_{X_i | X_1, \ldots, X_{i-1}}(x_1, \ldots, x_N) \, dx_1 \cdots dx_N \quad \text{(recursion)}$$

$$= \int h(x_1, \ldots, x_N) \prod_{i=1}^{N} f_{X_i | X_{i-1}}(x_i, x_{i-1}) \, dx_1 \cdots dx_N \quad \text{(Markov property),}$$

where all integrands are evaluated at $(x_1, \ldots, x_N)$ and $f_{X_i | X_1, \ldots, X_{i-1}} |_{i=1} = f_{X_1}$.

If we denote the transition density $f_{X_i | X_{i-1}}(x_i, x_{i-1})$ by f_i, then the importance sampling estimator corresponding to a new probability measure with transition densities g_i reads as:

$$\widehat{\theta}_n = \mathbb{E}^{\mathbf{g}} \left[h(X_1, \ldots, X_N) \prod_{i=1}^{N} \frac{f_i(X_i, X_{i-1})}{g_i(X_i, X_{i-1})} \right],$$

with $\mathbf{g} := (g_1, \ldots, g_N)$.

Example Consider a standard Brownian motion W moving on a Δ-spaced time refinement $0 = t_0 \leq t_1 \leq \cdots \leq t_N$. The transition density is

$$\mathbb{P}\big(W(t_i) \in \mathrm{d}x_i | W(t_{i-1}) = x_{i-1}\big) = f_i(x_i, x_{i-1})$$
$$= f_{\mathcal{N}(x_{i-1}, \Delta)}(x_i)\,\mathrm{d}x_i.$$

Let $g_i(x_i, x_{i-1}) = f_{\mathcal{N}(x_{i-1}+\mu_i\sqrt{\Delta}, \Delta)}(x_i)\,\mathrm{d}x_i$ be the transition density of a motion drifting with a rate μ_i on each interval $[t_i, t_{i+1}]$. The weighing ratio is

$$\prod_{i=1}^{N} \frac{f_i(x_i, x_{i-1})}{g_i(x_i, x_{i-1})} = \prod_{i=1}^{N} \frac{\frac{1}{\sqrt{2\pi\Delta}} \exp\{-[(x_i - x_{i-1})^2/(2\Delta)]\}}{\frac{1}{\sqrt{2\pi\Delta}} \exp\{-[(x_i - x_{i-1} - \mu_i\sqrt{\Delta})^2/(2\Delta)]\}}$$

$$= \prod_{i=1}^{N} \exp\left\{ \frac{(\sqrt{\Delta}\mu_i)^2 - 2\sqrt{\Delta}\mu_i(x_i - x_{i-1})}{2\Delta} \right\}$$

$$= \exp\left\{ -\sum_{i=1}^{N} \mu_i\left(\frac{x_i - x_{i-1}}{\sqrt{\Delta}}\right) + \frac{1}{2}\sum_{i=1}^{N} \mu_i^2 \right\}.$$

Consider a general path-dependent option on W. The pay-off can be written as $h(W(t_1), \ldots, W(t_N))$. An importance sampling estimator reads as

$$\widehat{\theta}_n^{\mu} = \frac{1}{n} \sum_{j=1}^{n} h\big(W^{(j)}(t_1), \ldots, W^{(j)}(t_N)\big)$$

$$\times \exp\left\{ -\sum_{i=1}^{N} \mu_i\left(\frac{W^{(j)}(t_i) - W^{(j)}(t_{i-1})}{\sqrt{\Delta}}\right) + \frac{1}{2}\sum_{i=1}^{N} \mu_i^2 \right\}, \qquad (1.22)$$

where $(W^{(j)}(t_1), \ldots, W^{(j)}(t_N))$ denotes the jth simulated path of the drifted random walk

$$W(t_i) = W(t_{i-1}) + \mu\sqrt{\Delta} + \mathcal{N}(0, \Delta).$$

Approximating the Zero Variance Estimator The importance sampling estimator (1.22) can be succinctly written as:

$$\widehat{\theta}^{\mu} = \mathbb{E}^{\mu}\left(h(\mathbf{x}) \exp\left(-\boldsymbol{\mu}^{\top}\mathbf{x} + \frac{1}{2}\boldsymbol{\mu}^{\top}\boldsymbol{\mu} \right) \right),$$

for an N-dimensional Gaussian random vector $\mathbf{x}$. If we write $h(\mathbf{x}) = \exp(H(\mathbf{x}))$, then the exponent in the previous expectation is $H(\mathbf{x}) - \boldsymbol{\mu}^{\top}\mathbf{x} + \frac{1}{2}\boldsymbol{\mu}^{\top}\boldsymbol{\mu}$. Under the μ-drifted measure, $\mathbf{x}$ has mean $\boldsymbol{\mu}$. We may therefore replace it by a standard normal vector $\mathbf{x}_0$ plus a constant mean vector $\boldsymbol{\mu}$, and get to

$$\widehat{\theta}^{\mu} = \mathbb{E}\left[\exp\left(H(\mathbf{x}_0 + \boldsymbol{\mu}) - \boldsymbol{\mu}^{\top}(\mathbf{x}_0 + \boldsymbol{\mu}) + \frac{1}{2}\boldsymbol{\mu}^{\top}\boldsymbol{\mu} \right) \right]$$

$$\simeq \mathbb{E}\left[\exp\left(H(\boldsymbol{\mu}) - \nabla H(\boldsymbol{\mu})^{\top}\mathbf{x}_0 - \boldsymbol{\mu}^{\top}\mathbf{x}_0 - \frac{1}{2}\boldsymbol{\mu}^{\top}\boldsymbol{\mu} \right) \right],$$

where the value of H at point $\mathbf{x}_0 + \boldsymbol{\mu}$ has been approximated by its linear expansion around $\boldsymbol{\mu}$. This expression becomes deterministic provided that

$$\nabla H(\boldsymbol{\mu}) = \boldsymbol{\mu}. \tag{1.23}$$

This observation leads to the following algorithm.

Algorithm (Optimal importance sampling)

1. Solve the fixed-point condition (1.23) and get to $\boldsymbol{\mu}^*$.
2. Compute $\widehat{\theta}_n^{\mu^*}$.

The quality of this estimator depends on the proximity of H at point $\boldsymbol{\mu}$ to its linear differential.

1.4 Comments

The literature on Monte Carlo methods in finance is exceptionally vast. General treatments can be found in Rubinstein (1981), Press et al. (1992), Fishman (1996), Ross (1997), and Grigoriu (2003), among others. Boyle (1977) was the first ever paper on Monte Carlo methods in finance. Approximations of probability distributions and their funcitonals can be found in Johnson and Kotz (1995). Clewlow and Strickland (1998) and Glasserman (2004) explore a wide variety of simulation methods in finance and provide a complete selection of additional references. Synthetic descriptions of simulation methods are contained in Lamberton and Lapeyre (1996), Boyle, Broadie and Glasserman (1997), Broadie and Glasserman (1997), Clewlow and Strickland (1998), James and Webber (2000), Duffie (2001), and Cont and Tankov (2004). Seminal papers on simulation include Box and Muller (1958), Marsaglia (1972), Marsaglia and Bray (1964) and Siegmund (1976), among others. Further technical issues together with a wide variety of simulation methods are explored in Devroye (1986), L'Ecuyer (1988), Glynn and Iglehart (1989), Glynn and Whitt (1992), L'Ecuyer (1994), Gentle (1998), L'Ecuyer, Simard and Wegenkittl (2002). Special topics are investigated in Clewlow and Carverhill (1994), Dupire (1998), Rogers and Talay (1997), and Bouchaud, Potters and Sestovic (2000), among others. The density arising in importance sampling has an interesting dynamic counterpart in finance. It represents the likelihood process associated to a change of measure as noticed by Jarrow (1986) and Jamshidian (1987, 1989). Applications of this process can be found in Chen and Scott (1993), Geman, El Karoui and Rochet (1995), Jamshidian (1990, 1991c, 1993, 1995, 1996, 1997, 1999), Musiela and Rutkowski (1997), among others.

2

Dynamic Monte Carlo

This chapter presents both existing and new algorithms for simulating paths of a random process. Section 2.1 introduces the main issue of sampling from probability measures on a path space. Section 2.2 focuses on continuous path diffusion processes: four methods are presented and illustrated using a comprehensive example on derivative pricing. Section 2.3 details methods for simulating pure and mixed-jump diffusions. Section 2.4 sketches procedures that have been designed for special classes of continuous time processes.

2.1 Main Issues

Let us address the problem of simulating samples of continuous time random processes. There are two main differences from the simpler case of sampling random variables (r.v.'s):

(1) No analytical expression exists for the distribution of the "object" we wish to simulate.
(2) Sample paths are a continuum of values, whereas any simulation algorithm can only provides us with a finite amount of pseudo random numbers.

The mathematics underlying the first of these items goes well beyond the scope of the present treatment. However, it is possible to provide the reader with an intuitive explanation of this fact. Recall that a stochastic process aims at describing the random evolution of a state variable X over time. More precisely, for each date $t \geq 0$, a random variable $X(t)$ is defined on a common probability space $(\Omega, \mathcal{F}, \mathbb{P})$ and a sample path is the ordered set $X_{0T} = (X(t))_{0 \leq t \leq T}$ of outcomes for these variables corresponding to a particular result ω in the sample space Ω. Consequently, the distribution of a continuous time process is a probability measure on the set of sample paths.[1] There are several ways we can build a continuous time process and each

---

[1] More formally, let $\mathbb{R}^{[0,T]}$ be the Cartesian product of infinitely many copies of $\mathbb{R}$, one for each point in $[0, T]$: that is the space of all functions $f : [0, T] \to \mathbb{R}$. A stochastic

method implicitly offers such a probability measure. Because the space of sample paths cannot be ordered, this measure does not allow for a cumulative distribution function. This absence, along with the fact that the Lebesgue measure[2] defined on this space does not exist, implies that density functions are not available for stochastic processes.

For these reasons, we must search for approximations of continuous time process involving a finite number of random variables that can be simulated. This introduces a further source of error beyond the one stemming from the impossibility of generating truly random samples from any distribution: this is the error induced by replacing a continuous time process with one among all possible approximations.

There are essentially two methodologies for approximating a continuous time process that can serve the scope of simulation. They share the property of producing a discretized version of the process in the sense that randomness enters through a finite number of random variables.

The first methodology consists of replacing a whole continuous time path $X_{t,T}$ by the vector $X^{(0)}, \ldots, X^{(N)}$ representing values of $X_{t,T}$ at fixed points in time $0 = t_0 < t_1 < \cdots < t_N = T$. Simulation is performed on these variables and the entire path, if needed, is obtained by interpolation methods.

We illustrate this idea by simulating paths of a standard Brownian motion W on the positive real axis. Our goal is to compute the expected value of a functional F of the process path W_{0T}, that is $\mathbb{E}(F(W_{0T}))$. We may approximate W by a suitable process taking constant values over finitely many consecutive intervals.

The first part of this program is to split $[0, T]$ into N evenly spaced time lags with a common length Δ and consider the law of the finite-dimensional vector $(W_0, W_{\Delta t}, \ldots, W_{N\Delta t})$ of Brownian increments. This vector is distributed as $(0, \sqrt{\Delta t} g_1, \ldots, \sqrt{\Delta t}(g_1 + \cdots + g_N))$, where the g_i's are independent standard normal variables. We may thus approximate $W(t)$ by

$$g(t) = W_{[t/\Delta t]\Delta t} \sim \sqrt{\Delta t}(g_1 + \cdots + g_{[t/\Delta t]}), \qquad (2.1)$$

process is a mapping from Ω to $\mathbb{R}^{[0,T]}$: an elementary event ω is mapped into the set of all values $\{f(t), t \in [0, T]\}$, which is the function $f \in \mathbb{R}^{[0,T]}$ itself. Given a probability space $(\Omega, \mathcal{F}, \mathbb{P})$, the law $\mathbb{P}_X$ of a process X is a probability measure induced by X on its image space $\mathbb{R}^{[0,T]}$ through the assignment

$$\mathbb{P}_X(A) = \mathbb{P}(\{\omega \in \Omega \colon X(\omega) \in A\}),$$

for all Borel sets $A \in \mathcal{B}(\mathbb{R}^{[0,T]})$.

[2] Broadly speaking, the Lebesgue measure on a given space is defined as a sigma-additive measure which is invariant up to rotations and translations of subsets in the space. To illustrate this property, we may think of the area of a given square in the plane. This number does not change if the square is translated and rotated. The Lebesgue measure on the real line (resp. plane; resp. 3D space) measures the length (resp. area; resp. volume) of a set. A definition given in Chapter 1 states that the density function of a probability measure is the Radon–Nikodym derivative of its cumulative distribution function with respect to the Lebesgue measure.

where $[x]$ denotes the integer part of x. The expected value above mentioned can be computed as:

$$\mathbb{E}(F[g]) \approx \mathbb{E}\big(F\big[\big(\sqrt{\Delta t}(g_1 + \cdots + g_{[t/\Delta t]})\big)_{0 \leq t \leq T}\big]\big), \qquad (2.2)$$

which is a standard $[t/\Delta t]$-dimensional integral. A Monte Carlo estimation can be performed using the following algorithm.

Algorithm (Monte Carlo valuation)

1. Fix n equal to a "large" value; set $i = 1$ and $N = [t/\Delta t]$;
2. Simulate $g_1^{(i)}, \ldots, g_N^{(i)} \overset{\text{i.i.d.}}{\sim} \mathcal{N}(0, 1)$;
3. Compute $F^{(i)} = F[(W^{(i)}(t))_{0 \leq t \leq T}]$ by evaluating the payoff functional F on the approximate Brownian path $W^{(i)}(\cdot) = \sqrt{\Delta t}(g_1^{(i)} + \cdots + g_{[\cdot/\Delta t]}^{(i)})$;
4. Let $i = i + 1$. If $i < n$, then go to Step 2;
5. Return the average sampled payoff $n^{-1} \sum_{i=1}^{n} F^{(i)}$.

Example (Call option) Consider a call option $C(T, K)$ on a market index whose dynamics are described using a Brownian motion W. Here T is the option maturity and K denotes the strike price. Let $\Delta t = T/N$ and, for $i = 1, \ldots, n$ and $k = 1, \ldots, N$, let $g_k^{(i)} \overset{\text{i.i.d.}}{\sim} \mathcal{N}(0, 1)$. Setting $x^{(i)} = \sqrt{\Delta t}(g_1 + \cdots + g_N)$, we obtain a sequence of sampled payoffs $(x^{(1)} - K)_+, \ldots, (x^{(n)} - K)_+$. Assuming a constant short rate of interest r, the Monte Carlo estimate of the option value reads as $n^{-1} \sum_{m=1}^{n} e^{-tT}(x^{(m)} - K)_+$. (Of course, this example is trivial and no one would use a Monte Carlo to compute the analytically solvable integral $\int (x - K)_+ f_{\mathcal{N}(0,T)}(x) \, dx!$.)

Another possible implementation of the same program is to consider the joint law of a finite number of Brownian points $W(t_1), \ldots, W(t_n)$. We know that any vector $\mathbf{W} = (W(t_1), \ldots, W(t_n))$ follows a normal distribution $\mathcal{N}(\mathbf{0}, \boldsymbol{\Sigma})$ with $\Sigma_{ik} = \min\{t_i, t_k\}$ $(i, k = 1, \ldots, n)$. By applying the Principal Components decomposition detailed in chapter "Static Monte Carlo", a sample $\mathbf{W}^i$ can be obtained as $\sum_{j=1}^{n} z_j \sqrt{\lambda_j} \mathbf{u}^j$, where λ_j is the jth greatest eigenvalue of $\boldsymbol{\Sigma}$ and $\mathbf{u}^j$ denotes the corresponding normalized eigenvector. If W_k^i denotes the kth coordinate of this sample, an approximate Brownian path can be obtained as $W^{(i)}(t) = W_{[t/\Delta t]}^i$.

The second methodology for simulating a continuous time process starts from the representation of a path in terms of a series expansion with respect to selected basis functions. More precisely, the process is written as the sum of a series $X(t) = \sum_{k=1}^{\infty} \alpha_k \varphi_k(t)$ with respect to a system of deterministic functions $\varphi_k(t)$ for appropriate random coefficients α_k. A simulation is then performed according to the following algorithm.

Algorithm (Series expansion method)

1. Fix a sample size K;
2. Generate samples $\alpha_1, \ldots, \alpha_K$;
3. Return $X(t) \approx \sum_{k=1}^{K} \alpha_k \varphi_k(t)$.

In order to illustrate this method, let us consider the problem of sampling a random path $(W(t))_{0 \leq t \leq 1}$ of a Brownian motion. For the sake of simplicity, we assume the simulation occurs on the unit interval $[0, 1]$. We examine three possible methods for implementing this program by means of a series expansion.

The first method is based on the continuous time extension of the eigenvalue decomposition reported above for the finite-dimensional case. We diagonalize the covariance operator $C(t, u) = \min\{t, u\}$ by solving the problem

$$\int \min\{t, u\} \varphi(t) \, dt = \lambda \varphi(t)$$

with respect to the scalar λ and the function φ. It can be shown that this equation produces the following system of solutions:

$$\lambda_i = \left(\frac{2}{(2\sqrt{-1} + 1)\pi} \right)^2,$$

$$\varphi_i(t) = \sqrt{2} \sin\left(\frac{(2\sqrt{-1} + 1)\pi t}{2} \right) \quad (i \geq 1).$$

This system leads to the well-known Karhounen–Loeve expansion of a standard Brownian motion

$$W(t) = \sum_{i=1}^{\infty} z_i \sqrt{\lambda_i} \varphi_i(t),$$

where $z_i \overset{\text{i.i.d.}}{\sim} \mathcal{N}(0, 1)$.

The second method expands a path in terms of a basis in the space $\mathcal{L}_2([0, 1])$ of square-integrable functions on the closed interval $[0, 1]$. Consider the system of Haar functions defined as follows: for each integer $n \geq 1$, let $k \in I(n)$ be the set of odd integers between 0 and 2^n, and for $t \in [0, 1]$, set

$$H_{k,n}(t) = \begin{cases} 2^{(n-1)/2} & \text{for } t \in \left[\frac{k-1}{2^n}, \frac{k}{2^n} \right), \\ -2^{(n-1)/2} & \text{for } t \in \left[\frac{k}{2^n}, \frac{k+1}{2^n} \right), \\ 0 & \text{elsewhere.} \end{cases}$$

The system of Schauder functions is defined by

$$S_{k,n}(t) = \int_0^t H_{k,n}(u) \, du.$$

Then, the following representation holds true:

$$W(t) = \sum_{n=0}^{\infty} \sum_{k \in I(n)} z_{k,n} S_{k,n}(t),$$

where $z_{k,n} \overset{\text{i.i.d.}}{\sim} \mathcal{N}(0, 1)$.

The third method is based on the first ever construction of a Brownian motion. Paley and Wiener used a trigonometric basis and obtained the following expression:

$$W(t) = z_0 t + \sum_{n=1}^{\infty} \sum_{k=2^{n-1}}^{2^n-1} \sqrt{2}\frac{\sin(k\pi t)}{k\pi} z_k,$$

where $z_k \overset{\text{i.i.d.}}{\sim} \mathcal{N}(0, 1)$.

Compared to path generation by the accumulation of consecutive increments, this series expansion has two disadvantages: first, it involves $\mathcal{O}(n^2)$ operations; second it produces a whole path in one shot. This may not be advisable whenever an entire path over the option horizon need not be computed, as is the case with a barrier option for which path simulation is to be interrupted each time the barrier is hit. Methods based on time discretization can be applied to all diffusion processes and generate simulated paths even for very complicated processes. These considerations explain the higher popularity of these methods in finance literature compared to the ones based on series expansions. These latter are particularly suitable for stationary processes, such as Levy processes, for which the distribution of coefficients α_k is known. (See Sect. 2.5 for these extensions.) One further reason for their attractiveness lies in their independence from interpolation procedures for the purpose of obtaining paths in continuous time.

2.2 Continuous Diffusions

We consider four methods for sampling solutions of stochastic differential equations (s.d.e.):

$$dX(t) = \mu(t)\,dt + \sigma(t)\,dW(t).$$

These methods correspond to simulating samples:

(1) from the exact transition density $p(t, dx; s, y) = \mathbb{P}(X(s) \in dx | X(t) = x)$;
(2) of the solution of the exact dynamics followed by the process;
(3) of the solution of approximate dynamics to the original s.d.e.;
(4) of the coefficients in a truncated series expansion of the process.

The first three of these methods simulate first values for a sample path at a discrete set of times $t_1, \ldots, t_N$ and then make an interpolation to produce a continuous time trajectory. Throughout, we make the following assumptions:

- the starting time is $t_0 = 0$;
- the starting state is a known value x_0;
- the sampling interval $[0, T]$ splits into N equally Δt-sized intervals $[t_{i+1}, t_i]$.

2.2.1 Method I: Exact Transition

This method can be performed whenever the transition distribution of the process is known for any pair of consecutive times.

Algorithm (Transition distribution)

1. Set $X_0 = x_0$;
2. For $i = 1, \ldots, N$, sample $X_i \sim p(t_i, \cdot; t_{i-1}, X_{i-1})$;
3. Return $(X_{[t/\Delta t]}, 0 \le t \le T)$ as a sample of the process X on $[0, T]$.

Example (Vasicek model) In the Vasicek term structure model the one-dimensional state variable is represented by the short rate of interest r, whose dynamics are given by $dr(t) = \alpha(\beta - r(t)) \, dt + \gamma \, dW(t)$. Solving for r, we see that $r(t)$ has a normal transition density with conditional mean $\mu(t; s, y) = \beta + e^{-\alpha(t-s)}(x - \beta)$ and conditional variance $\sigma(t; s, y) = \frac{\sigma^2}{2\alpha}(1 - e^{-2\alpha(t-s)})$.

2.2.2 Method II: Exact Solution

Recall that the strong solution X of an s.d.e. is explicit if it can be written as an analytic functional F of time t and the driving random noise W until that time, i.e., $X(t) = G(t, W_{0t})$. This method can be performed whenever dynamics are given by an s.d.e. whose strong solution is explicit. The method consists of discretizing the underlying noise over a finite set of sampling times. An instance of this approximation has been developed for the case of a Brownian noise in formula (2.1).

Algorithm (Exact solution)

1. Set $X_0 = x_0$;
2. For $i = 1, \ldots, N$, sample random noise $W(t_i)$ and set

$$X_i = G\big(t_i, \{W(t_1), \ldots, W(t_i)\}\big);$$

3. Return $(X_{[t/\Delta t]}, 0 \le t \le T)$ as a sample of the process X on $[0, T]$.

Example (Geometric Brownian motion) The stock price dynamics in the Black–Scholes model are given by $dS(t) = S(t)(r \, dt + \sigma \, dW(t))$, with $S(0) = x_0$. The strong solution of this equation is

$$S(t) = x_0 \exp\left(\left(r - \frac{\sigma^2}{2}\right)t + \sigma W(t)\right).$$

Discretizing the underlying Brownian noise leads to a discrete time process

$$S_{i+1} = S_i \exp\left(\left(r - \frac{\sigma^2}{2}\right)\Delta t + \sigma \sqrt{\Delta t} \times g_{i+1}\right).$$

Here the g_i's are independent samples from a standard normal distribution and the corresponding sample path may be obtained using the rule $(S_0, S_1, \ldots, S_N) \to S^N(t) = S([t/\Delta t])$.

2.2.3 Method III: Approximate Dynamics

Both the exact transition and exact solution methods introduced above simulate an *approximate* solution of given dynamics. If neither transition probabilities, nor the explicit solution are available, we may look for simulating the *exact* solution of a discrete time process *approximating* the system dynamics:

$$dX(t) = \mu\big(t, X(t)\big)\,dt + \sigma\big(t, X(t)\big)\,dW(t). \qquad (2.3)$$

The method amounts to solving a stochastic *difference* equation obtained by discretizing the above equation. There are several ways for doing this and we shall focus on the two most popular among practitioners, namely the Euler and the Milstein schemes.

A. Euler Scheme

S.d.e. (2.3) is discretized into the finite difference system:

$$X_{i+1} = X_i + \mu(t_i, X_i)\Delta t + \sigma(t_i, X_i)\sqrt{\Delta t} \times g_i, \qquad (2.4)$$

where $g_i \overset{\text{i.i.d.}}{\sim} \mathcal{N}(0, 1)$.

Algorithm

1. Set $X_0 = x_0$;
2. For $i = 0, \ldots, N-1$, sample $g_i \overset{\text{i.i.d.}}{\sim} \mathcal{N}(0, 1)$ and set X_{i+1} as in (2.4);
3. Return $(X_{[t/\Delta t]}, 0 \leq t \leq T)$ as a sample of the process X on $[0, T]$.

The Euler scheme gives an approximate solution which is pathwise convergent to the exact solution of the original equation according to a mixed L^2-sup norm; that is, $\forall T > 0, \exists C = C(T)$:

$$\mathbb{E}\left(\sup_{t\in[0,T]} \left|X^N(t) - X(t)\right|^2\right) \leq C \times \Delta t.$$

Actually, there is no need to adopt a time grid with evenly-spaced subintervals. We may as well consider partitions with a finer grid on certain regions of the time axis. This would improve the precision of our approximate process on some key portions of the time spectrum, for instance around coupon payment dates of a bond. We just have to replace Δt with $t_{i+i} - t_i$ at the ith step.

Example (A convergent scheme in law) If we take $g_i = 2b_i - 1$, where $b_i \overset{\text{i.i.d.}}{\sim}$ Ber$(1/2)$ are independent Bernoulli variables, the resulting scheme provides the user with an approximate process whose trajectories do not necessarily converge to those of the exact solution. Yet, the law induced by the resulting discrete process on the path space converges to the of the true process. For practical purposes, this means that the expected value of any regular functional F of the approximate process X^N converges as $N \to \infty$ (i.e., $\Delta t \to 0$) to the true expectation $\mathbb{E}(F(X_{0T}))$. For derivative evaluation purposes, one may thus adopt a weakly convergent scheme such as the one involving Bernoulli samples.

Example (Fong–Vasicek model) We discretize a two-factor interest rate model where r is the short rate of interest and v represents its instantaneous volatility:

$$dr(t) = \alpha\big(\mu - r(t)\big)\,dt + \sqrt{v(t)}\,dW^1,$$

$$dv(t) = \beta\big(\bar{\mu} - v(t)\big)\,dt + \sigma\sqrt{v(t)}\,dW^2,$$

$$r_{i+1} = r_i + \alpha(\mu - r_i)\Delta t + \sqrt{v_i\,\Delta t} \times n_i,$$

$$v_{i+1} = v_i + \beta(\bar{\mu} - v_i)\Delta t + \sigma\sqrt{v_i\,\Delta t} \times m_i,$$

where (n_i, m_i) denotes a sample from $\mathcal{N}_2(0, \Sigma)$ and $\Sigma\,dt = \mathrm{Cov}(dW^1, dW^2)$.

B. Milstein Scheme

The idea is to add a second order term in the series expansion of the true solution. The convergence order of the resulting scheme is the same as the one stemming from a Euler discretization. However, this comes at the expense of increased computational complexity.

$$X_{i+1} = X_i + \mu(t_i, X_i)(t_{i+1} - t_i) + \sigma(t_i, X_i)\sqrt{t_{i+1} - t_i} \times g_i$$
$$+ \frac{1}{2}\sigma(t_i, X_i)\sigma(t_i, X_i)^\top \times (t_{i+1} - t_i) \times \big[g_i^2 - 1\big].$$

Notice that the last term has zero expected value.

2.2.4 Example: Option Valuation under Alternative Simulation Schemes

Let us compute the arbitrage-free value of four European options. Each option is written on the security price process S and expires at $T = 0.4$ years (i.e., 100 days for a day-count convention assuming 250 days per year). The strike price is $K = 1.9$ Euros, the current spot price is $S(t) = 2.1$ Euros, and the risk-free rate is 2% *per annum*. We consider a European call, a digital, a barrier and an Asian option on the geometric average of past prices. For the barrier option, the up-and-out threshold is set to 2.5 Euros. The underlying process S is a geometric Brownian motion with volatility $\sigma = 0.2$. The option value is first computed using the exact formula and then compared to the value obtained by simulating the underlying process through the following five alternative methods:

1. Recursive sampling from the analytical transition density over intervals of length Δ until either the expiration time is reached or the up-and-out threshold is attained. (Method I.)
2. Discrete time simulation of the exact solution of underlying process. (Method II.)
3. Simulation of the random walk stemming from discretizing the s.d.e. followed by the process S according to the Euler scheme. (Method III.A.)
4. As in Step 3, with a Milstein discretization scheme. (Method III.B.)
5. Simulation of the geometric Brownian motion under a Fourier representation of the driving standard Brownian motion.

Each Monte Carlo evaluation involves 100,000 sample paths. Each path is obtained by sampling the process 500 times between the outset and the option maturity, unless

Table 2.1. Comparison of option prices obtained using alternative simulation schemes. First round parentheses contain percentage error with respect to the exact Black–Scholes price. Second round parentheses indicate standard errors

	Call	Digital	Barrier	Asian
Exact formula	0.242559	0.779335	0.141267	0.210084
Analyt. transition	0.242946	0.779632	0.141416	0.210336
	(0.39%)	(0.15%)	(0.32%)	(0.27%)
	(−0.16%)	(0.04%)	(0.11%)	(0.12%)
Discret. solution	0.242763	0.779677	0.143879	0.210208
	(0.24%)	(0.16%)	(0.43%)	(0.17%)
	(0.08%)	(0.04%)	(1.85%)	(0.06%)
Discr. SDE (Euler)	0.242572	0.779492	0.144098	0.210099
	(0.22%)	(0.12%)	(0.31%)	(0.16%)
	(0.01%)	(0.02%)	(2.00%)	(0.01%)
Discr. SDE (Milstein)	0.242636	0.779669	0.144328	0.210148
	(0.20%)	(0.04%)	(0.27%)	(0.12%)
	(0.03%)	(0.04%)	(2.17%)	(0.03%)
Fourier expansion	0.242669	0.779287	0.145251	0.210133
	(0.24%)	(0.08%)	(0.39%)	(0.15%)
	(0.05%)	(−0.01%)	(2.82%)	(0.02%)

the barrier in attained before. For case 5, the Fourier series has been computed up to the 200th term. Each option price is calculated 10 times for the purpose of computing the numerical standard deviation of the reported option value. Table 2.1 reports simulated option prices for all options and computational methods. The numbers within round brackets are the standard deviation and the pricing error with respect to the exact option value as expressed in percentage over this latter.

2.3 Jump Processes

A *pure* jump process is one whose trajectories vary according to discontinuities only. In the previous section, we presented several methods for simulating samples of continuous diffusions. In this section, we introduce simulation algorithms for *compound jump processes*. The following section combines continuous and compound jump diffusion processes and develops algorithms to simulate the resulting *mixed-jump diffusions*. The last section is devoted to methods expressly conceived for Gaussian processes.

2.3.1 Compound Jump Processes

We consider a stochastic process of the following form:

$$J(t) = \sum_{j=1}^{N(t)} Y_j. \tag{2.5}$$

This expression involves two terms. The first term N governs the jump *occurrence*: for a given elementary event ω (usually identified with a sample trajectory), $N(t, \omega)$ counts the number of jumps between the initial time 0 and current time t (both included). We call N a counting process and denote the corresponding jump times by $\tau_1(\omega), \ldots, \tau_{N(t,\omega)}(\omega)$. The second term $(Y_i)_{i \geq 1}$ determines the jump *magnitudes*: Y_i represents the ith jump size. The joint effect of the two terms is described by expression (2.5) defining a compound jump process J.

Sampling paths of a compound jump process J can be split into simulating jump times corresponding to the counting process N and then generating sample jump sizes $Y_i, i \geq 1$. The latter are usually assigned through their distributions. Consequently, methods described in chapter "Static Monte Carlo" can be employed for the purpose of sampling jump amplitudes. Simulating jump times is less trivial. Actually, the relationship between counting processes and the associated jump times needs to be made more explicit.

Example (Counting process) A counting process $N = (N(t), t \geq 0)$ is any nondecreasing process taking values in the set $\mathbb{N}$ of natural numbers $0, 1, \ldots$. This process can be used to model the number of jumps occurring on a time period $[0, T]$, i.e., $N(t) = $ "# jumps until time t". Counting processes can be sampled as follows.

Algorithm (Simulation of a counting process)

1. Generate r.v.'s $T_1, \ldots$ with distributions supporting the positive real axis; (T_i represents the ith *interarrival time*, that is the time between the ith jump and the following one.)
2. Jump times $\tau_1, \tau_2, \ldots$ are obtained by summing up interarrival times:

$$\tau_k = \sum_{i=1}^{k} T_i;$$

3. For each time t, $N(t)$ records the number of jumps that have occurred since the beginning, that is

$$N(t) = \sum_{n=1}^{\infty} \mathbf{1}_{\{\tau_n \leq t\}} = \sum_{n=1}^{\infty} n \mathbf{1}_{\{\tau_n \leq t < \tau_{n+1}\}}, \tag{2.6}$$

where the indicator function $\mathbf{1}_A(\omega)$ is equal to 1 if $\omega \in A$, and 0 otherwise. The first expression sums up as many 1's as the number of jump times occurring until time t. It is clear that a counting process is a compound jump process with unit jumps, namely $Y_i = 1$ for all i. We note that the second expression in formula (2.6) selects n whenever the nth jump occurs no later than t and the $(n + 1)$th jump occurs after time t. The second expression in (2.6) can be used for simulating a Poisson random variables.

Example (Poisson process) Let the interarrival times be modeled by independent and identically distributed exponential r.v.'s. The corresponding counting process is called "homogeneous Poisson process". Simulation reads as follows:

$$T_i \overset{\text{i.i.d.}}{\sim} \text{Exp}(\lambda) \rightarrow \tau_n = \sum_{k=1}^{n} T_k \rightarrow N(t) = \sum_{n=1}^{\infty} \mathbf{1}_{\{\tau_n \leq t\}} \in \mathbb{Z}. \qquad (2.7)$$

Recall the exponential density $f_{\text{Exp}(\lambda)}(x) = \lambda e^{-\lambda x}$, for $X \geq 0$. It can be shown that the r.v. $N(t)$ is distributed according to a Poisson law $\text{Po}(\lambda t)$ with parameter λt, i.e., $f_{N(t)}(n) = e^{-\lambda t} (\lambda t)^n / n!$, for all $n \in \mathbb{Z}_+$. Coefficient λ represents the expected number of jumps per time unit, i.e., $\mathbb{E}(N(t)) = \lambda t$. Notice that this quantity is the same for all times t and that $\text{Var}(N(t)) = \lambda t$ as well.

2.3.2 Modelling via Jump Intensity

A compound jump process $\sum_{j=1}^{N(t)} Y_j$ is determined by specifying jump occurrence and size. Occurrence can be modeled in two distinct yet equivalent ways.

The first method was described in the previous section: the ith jump time can be obtained by summing up the first i interarrival times. Since these variables are sampled from exogenously assigned distributions, this method is advantageous for simulation purposes as long as their distribution is known by assumption. However, we may wish to model jump occurrence as a function of time and, possibly, the underlying state variable. In this case it is not clear how interarrival time distributions ought to be selected.

The second method serves the purpose of explicitly linking jump occurrence to both time and state variables. We begin by assigning a *jump intensity*, or frequency $(\lambda(t), t \geq 0)$, which may be constant, time dependent or random as well. Heuristically, this process defines the number of jumps per time unit in a "small" neighborhood of each point t in the time interval $[0, T]$, i.e., $\lambda(t) \, dt = d_s \mathbb{E}_t(N(s))|_{s=t}$. The integrated intensity $\int_0^t \lambda(s) \, ds$ is referred to as the *compensator* of the jump process N. This term stems from the property that $N(t) - \int_0^t \lambda(s) \, ds$ is a martingale. For instance, the intensity of a homogeneous Poisson process with parameter λ is λ itself. It can be shown that the intensity process unequivocally determines a corresponding counting process N. The main advantage of this method is that the jump occurrence can be easily modulated over the time horizon and can be made random, e.g., state dependent. This goal can be achieved by assigning a specific functional form to the intensity process. We consider four possible specifications for the jump intensity of a given process X:

(1) λ constant;
(2) $\lambda(t)$ time dependent;
(3) $\lambda(t, Y(t))$ random and dependent on a process Y that is statistically independent of X;
(4) $\lambda(t, X(t))$ random and dependent on the underlying process X.

The first two cases are developed in this section for pure jump processes. Models driven by random intensity will be treated in the more general context of mixed-jump diffusions.

Example (Electricity price modelling, Roncoroni *(2002))* Poisson processes have constant intensity: jump occurrence is uniformly spread over the time axis. Electricity markets often display spikes during certain periods of the calendar year. A spike is a sequence of upward jumps followed by a sequence of downward jumps. If spikes tend to occur during the warm season, we may use a periodic intensity function

$$\lambda(t) = \theta \times \left[\frac{2}{1 + |\sin(\pi(t - \tau)/k)|} - 1 \right]^d. \tag{2.8}$$

Figure 2.1 displays the graph of intensity function (2.8) across varying levels of the squeezing coefficient d and for a fixed $\theta = 1$. This function concentrates jumps around a precise portion of the time axis and has been proven to be quite effective for describing the empirical jump occurrence in most U.S. electricity markets. If jump occurrence is linked to other quantities than time, we may adopt a random intensity function depending on the values assumed by these quantities. For instance, spikes in electricity markets are linked to the prevailing temperature in the region where power is delivered. This suggests to first model the temperature as an independent variable $F(t)$ and then consider an intensity function $\lambda(t, F(t))$. This turns out to be an effective way of simulating upward jumps featuring the ascendant movements during a price spike. However modeling the descendent side of a spike requires a more "clever" intensity. Whenever prices reach high values compared to those prevailing under normal market conditions, they tend to revert to their mean value. This effect is usually modeled by introducing a smooth mean reversion effect in the drift of the underlying diffusion process. Occasionally, this reversion consists of downward jumps. To force the occurrence of reverting jumps, we can consider an intensity that also depends on the standing market price $E(t)$. For instance, by selecting

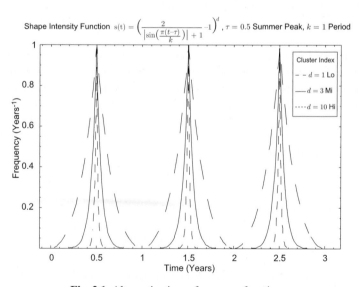

Fig. 2.1. Alternative jump frequency functions.

$$\lambda(t, E(t)) = \lambda(t) \times [1 + \max(0, E(t^-) - b(t))],$$

with $\lambda(t)$ as in expression (2.8), we may amplify the time dependency of jump occurrence in a way that is proportional to the "distance" between the current price
$E(t^-)$ and the average price $b(t)$ computed at the same time under normal market
conditions.

2.3.3 Simulation with Constant Intensity

Let $N(t) = \sum_{i=1}^{\infty} \mathbf{1}_{\{\tau_n \leq t\}}$, $t \geq 0$, be a counting process with a given constant intensity λ. We present two algorithms for simulating sample paths $(N(t))_{0 \leq t \leq T}$ and
the corresponding jump times $\tau_1, \ldots, \tau_{N(T)}$. The first method, called "*conditional
simulation*", simulates the number of jump times occurring on $[0, T]$ followed by
their exact location within the same interval. The second method, called "*countdown
simulation*", samples jump times in a sequential order, starting from the first one and
continuing until the time horizon T is reached.

Method I (Conditional simulation)   It can be shown that conditional to $N(t) = n$,
jump times τ_i of a homogeneous Poisson process are independent uniformly distributed r.v.'s on $[0, T]$. This fact leads to the following algorithm.

Algorithm

1. Simulate $N(T) \sim \text{Po}(\lambda T)$;
2. Simulate $N(T)$ independent uniform samples $\tau_1, \ldots, \tau_{N(T)}$ on $[0, T]$;
3. Return jump times $\tau_1, \ldots, \tau_{N(T)}$, and the Poisson realization $N(T)$.

Notice that, contrary to the construction (2.7), here $N(T)$ is sampled first and
jump times are obtained then. The Poisson random variable should be sampled by
any of the methods introduced in chapter "Static Monte Carlo".

Method II (Countdown simulation)   Since λ is constant, we know that N is a Poisson process. Therefore interarrival times are exponentially distributed with parameter λ and consecutive jump times on the interval $[0, T]$ can be simulated by the
following algorithm.

Algorithm

1. Let $\tau_0 = 0$, $i = 0$;
2. Set $i = i + 1$;
3. Generate $U \sim \mathcal{U}[0, 1]$;
3. Set $T_i = -\lambda^{-1} \ln U$ and $\tau_i = \tau_{i-1} + T_i$;
4. If $\tau_i \leq T$, then go to Step 2;
5. Return jump times $\tau_1, \ldots, \tau_{i-1}$, and the Poisson sample $N(T) = i - 1$.

This is in agreement with construction (2.7).

Example (Simulation of a Poisson random variable) Since $N(t)$ is a Poisson r.v. with parameter λt, $N(1)$ is a Poisson r.v. X with parameter λ. From expression (2.6), we see that:

$$N(1) = \sum_{n=1}^{\infty} n \mathbf{1}_{\{\tau_n \leq 1 < \tau_{n+1}\}} = \sum_{n=1}^{\infty} n \mathbf{1}_{\{T_1 + \cdots + T_n \leq 1 < T_1 + \cdots + T_{n+1}\}}.$$

This expression selects n in a way that the sum of n exponentially distributed samples $T_1, \ldots, T_n$ does not exceed 1 and the same number plus a further independent exponential sample T_{n+1} is greater than 1. This is just the algorithm illustrated above, with $T = 1$. The procedure is somehow inefficient. Indeed, each step involves the computation of a logarithm. Fortunately, using the equivalence

$$\{T_1 + \cdots + T_n \leq 1 < T_1 + \cdots + T_{n+1}\}$$
$$= \{U_1 \times \cdots \times U_{n+1} \leq e^{-\lambda} \leq U_1 \times \cdots \times U_n\},$$

we may avoid computing the logarithmic function. This observation leads to the following algorithm.

Algorithm

1. Set $P = 1, n = 0$ and $d = e^{-\lambda}$;
2. Generate $U \sim \mathcal{U}[0, 1]$;
3. Define $P = P \times U$ and $n = n + 1$;
4. If $P > d$, then go to Step 2;
5. Return n.

   The reason for setting d at the outset is to prevent the implementation code from repeatedly compute an exponential during the run.

2.3.4 Simulation with Deterministic Intensity

The two methods described in the previous paragraph can be adapted to simulate samples from inhomogeneous Poisson processes with given deterministic intensity $\lambda(t)$. A third method, called "*thinning*", is based on an acceptance–rejection scheme: jump times of a homogeneous Poisson process (=constant intensity) are sampled first; then, a test is performed to decide whether to accepted them or not as sample jump times of the inhomogeneous Poisson process.

Method I (Conditional simulation) The cumulated jump intensity over the time horizon $[0, T]$ is given by $\Lambda = \int_0^T \lambda(s)\, ds$. This leads to the following algorithm.

Algorithm

1. Generate $N(T) \sim \text{Po}(\Lambda)$;
2. Generate $N(T)$ independent samples $\tau_1, \ldots, \tau_{N(T)}$ with common distribution density $f_\tau(t) = \Lambda^{-1} \lambda(t)$;
3. Return jump times $\tau_1, \ldots, \tau_{N(T)}$ and sample number $N(T)$.

Method II (Countdown simulation)   It can be shown that for any pair of consecutive jump times τ_i and τ_{i+1}, the conditional distribution of the compensator $\int_{\tau_i}^{\tau_{i+1}} \lambda(s)\, ds$ given τ_i is an exponential variable $\mathcal{E}(1)$. Consequently, we may draw a sample e_{i+1} from this distribution and then set τ_{i+1} as the first time for which the integral above exceeds the threshold e_{i+1}.

Algorithm

1. Let $\tau_0 = 0$, $i = 0$;
2. Set $i = i + 1$;
3. Generate $e_{i+1} \sim \mathcal{E}(1)$;
4. Let $T_i = \inf\{t \geq \tau_i \colon \int_{\tau_i}^{t} \lambda(s)\, ds \geq e_{i+1}\}$;
5. Set $\tau_i = \tau_{i-1} + T_i$;
6. If $\tau_i \leq T$, go to Step 2;
7. Return jump times $\tau_1, \dots, \tau_{i-1}$ and Poisson sample $N(T) = i - 1$.

Method III (Thinning simulation)   The idea is to generate jump times from a Poisson process with a suitable constant intensity. This can be done by using any of the algorithms detailed hitherto. Then, we select some of these sample times according to the outcome of a random test. More precisely, if τ is a Poisson jump time, we draw an independent r.v. $U \sim \mathcal{U}[0, \lambda]$, where λ is an upper bound for $\lambda(t)$ over the simulation time domain $[0, T]$. If U falls into $[0, \lambda(\tau)]$, then τ is accepted as a sample from a Poisson process with intensity function $\lambda(t)$. Notice that the closer $\lambda(\tau)$ is to λ, the higher is the probability that U belongs to $[0, \lambda(\tau)]$ and that τ is then accepted. Intuitively, if $\lambda(\tau)$ is close to λ, then the inhomogeneous Poisson process with intensity function $\lambda(t)$ is close to a homogeneous Poisson process with intensity λ. The optimal level for λ is the one leading to the shortest average time period before acceptance. It can be proven that this number is $\lambda^* = \sup_{s \in [0,T]} \lambda(s)$. This method can be implemented through the following algorithm.

Algorithm

1. Set $\lambda^* = \sup_{s \in [0,T]} \lambda(s)$;
2. Sample N exponential r.v.'s $e_i \sim \mathcal{E}(\lambda^*)$, with $N = \min\{n \colon \sum_{i=1}^{n} e_i > T\}$;
3. Sample $N - 1$ uniform variables U_i on $[0, 1]$;
4. Define

$$I_j = \begin{cases} 1 & \text{if } U_j \leq \dfrac{\lambda(\sum_{i=1}^{j} e_i)}{\lambda^*}, \\ 0 & \text{otherwise.} \end{cases}$$

5. Select the subset of the index set from 1 to $N - 1$ consisting of indices j for which I_j equals 1: $J = \{j \colon I_j = 1\}$. An *order function* associates its position $R(j)$ to each element j in J;
6. Return jump times $\tau_i = \sum_{k=1}^{R^{-1}(i)} e_i$.

The mathematical proof of this method is developed in an appendix for the general case of random intensity.

2.4 Mixed-Jump Diffusions

2.4.1 Statement of the Problem

A mixed-jump diffusion process, or simply jump diffusion, is the sum of a continuous diffusion process and a pure jump process. We tackle the issue of simulating a mixed-jump diffusion satisfying a stochastic differential equation

$$dX(t) = \mu(t, X(t)) dt + \sigma(t, X(t)) dW(t) + \eta(t, X(t^-)) dJ(t), \qquad (2.9)$$
$$X(t_0) = x,$$

where J is a compound jump process $J(t) = \sum_{j=1}^{N(t)} Y_j$, $Y_j \overset{\text{i.i.d.}}{\sim} f_Y$, and $X(t^-) = \lim_{s \uparrow t} X(s)$. This process is specified by the following ingredients:

- x = Initial state;
- $\mu(t, x)$ = Drift function;
- $\sigma(t, x)$ = Brownian volatility function;
- $\eta(t, x)$ = Jump impact function;
- $\lambda(t, x)$ = Jump intensity, or frequency, function;
- Y_j = Jump size variable.

Regular coefficients are assumed in order to ensure the existence of a unique solution to the differential system above. Moreover, λ is superiorly bounded by a constant. The Euler discretization of equation (2.9) reads as

$$X_{i+1} = X_i + \mu(t_i, X_i)\Delta t + \sigma(t_i, X_i)\sqrt{\Delta t} \times g_i$$
$$+ \eta(t_i, X_i) \times \sum_{j=N(t_i)+1}^{N(t_{i+1})} Y_j, \qquad (2.10)$$

where the convention $\sum_{i=n+1}^{n} a_i = 0$ is adopted. Notice that all jumps occurring in the semi-closed time interval $(t_i, t_{i+1}]$ contribute to the evaluation of X_{i+1}. Three methods for simulating the discretized process (2.10) are detailed below.

Example (Black–Scholes dynamics with jumps) We generalize the Black–Scholes framework for a market where two assets are traded:

- The risk-free asset, or money market account, is given by $B(t) = e^{rt}$;
- The risky asset is driven by three mutually independent sources of noise: a standard Brownian motion W, a counting process $N(t)$, and a sequence of random jump sizes $(Y_i)_{i \geq 1}$.

These terms are combined according to the following guidelines:

(a) The price process S follows a standard geometric Brownian motion between two consecutive jump times:

$$t \in [\tau_j, \tau_{j+1}) \to dS(t) = S(t)(\mu(t) dt + \sigma(t) dW(t)),$$

(b) At jump time τ_j, S jumps by a proportion Y_j of its pre-jump value $S(\tau_j^-)$

$$t = \tau_j \rightarrow \Delta S(t) = S(\tau_j) - S(\tau_j^-) = S(\tau_j^-)Y_j.$$

These dynamics can be made explicit in terms of the driving noise. On $[0, \tau_1)$, we have a continuous diffusion:

$$S(t) = s_0 e^{(\mu - \sigma^2/2)t + \sigma W(t)}. \tag{2.11}$$

In particular, $S(\tau_1^-) = s_0 \exp((\mu - \sigma^2/2)\tau_1 + \sigma W(\tau_1))$. Note that the superscript "$-$" can be omitted in all expressions on the right-hand side because both drift and W are continuous functions of time. At the first jump time, price jumps by a proportion Y_1, that is:

$$S(\tau_1) = S(\tau_1^-)(1 + Y_1)$$
$$= s_0 e^{(\mu - \sigma^2/2)\tau_1 + \sigma W(\tau_1)} \times (1 + Y_1).$$

On the interval $[\tau_1, \tau_2)$, the process S behaves like a continuous diffusion starting at value $S(\tau_1)$ at time τ_1:

$$S(t) = S(\tau_1)e^{\int_{\tau_1}^{t} (\mu - \sigma^2/2)\,ds + \int_{\tau_1}^{t} \sigma\,dW(s)} = s_0 e^{(\mu - \sigma^2/2)t + \sigma W(t)} \times (1 + Y_1).$$

By induction, we arrive at the general term

$$S(t) = s_0 e^{(\mu - \sigma^2/2)t + \sigma W(t)} \times \prod_{i=1}^{N(t)} (1 + Y_i). \tag{2.12}$$

If $N(t) = 0$, i.e., no jump has ever occurred, then the product $\prod$ is set equal to 1 to recover formula (2.11). We may derive a unique expression for the differential of S by summing up differentials of continuous and discontinuous parts of S:

$$dS(t) = S(t)\big(\mu(t)\,dt + \sigma(t)\,dW(t)\big) + \sum_{v \in (t, t+dt]: \Delta S(v) \neq 0} \Delta S(v)$$

$$= S(t^-)\left[\mu(t)\,dt + \sigma(t)\,dW(t) + d_t\left[\sum_{j=1}^{N(t)} Y_j\right]\right],$$

because:

$$\sum_{\substack{v \in (t, t+dt]: \\ \Delta S(v) \neq 0}} \Delta S(v) = \sum_{j=N(t)+1}^{N(t+dt)} S(\tau_j^-)Y_j = d_t \sum_{j=1}^{N(t)} S(\tau_j^-)Y_j = S(t^-)\,d_t \sum_{j=1}^{N(t)} Y_j.$$

The last equality stems from the fact that $d_t(\sum_{j=1}^{N(t)} Y_j)$ assumes nonzero values at jump times τ_j ($j \geq 1$) only. Note that this formulation prevents the price process from assuming negative values as long as we require that $Y_i \in [-1, \infty)$ and that the three sources of random noise $W(t)$, $N(t)$ and Y_i ($i \geq 1$) are mutually independent. This implies that they must be generated by using transformations of independent uniform numbers.

2.4.2 Method I: Transition Probability

Jump diffusion sample paths can be simulated by combining methods developed for continuous diffusions and for pure jump processes. If the exact transition distributions are known, we may follow the scheme presented in Sect. 2.2.1.

Example (Poisson–Gaussian process) Consider a mean reverting process with Poisson jump component:

$$dX(t) = k(\theta - X(t)) dt + v \, dW(t) + \iota \, dN(t), \qquad (2.13)$$

where N is a Poisson process with intensity λ and parameters are all constant. The analytic expression of the transition density for this process can be approximated by the exact transition of a discrete-time approximation of the process above. We may consider the discrete-time approximation:

$$X_{i+1} = k(\theta - X_i)\Delta t + v \times \sqrt{\Delta t} \times \mathcal{N}(0, 1) + \iota \times Be(q),$$

where $Be(q)$ denotes a Bernoulli distribution with parameter $q = \lambda \Delta t$. It can be shown that this scheme converges in law to the solution of equation (2.13). The transition density of the discrete time process is given by

$$p(t_i, dx; t_{i-1}, X_{i-1}) = \frac{q}{\sqrt{2\pi(v^2 \Delta t + \gamma^2)}} \times e^{-\frac{(x - X_{i-1} - k(\theta - X_{i-1})\Delta t - \mu)^2}{2(v^2 \Delta t + \gamma^2)}}$$
$$+ \frac{1 - q}{\sqrt{2\pi v^2 \Delta t}} \times e^{-\frac{(x - X_{i-1} - k(\theta - X_{i-1})\Delta t)^2}{2v^2 \Delta t}}.$$

2.4.3 Method II: Exact Solution

If the solution to the s.d.e. (2.9) can be derived in a closed-form, we may simulate paths from a suitable discretization of this process. The following example illustrates this technique.

Example (Black–Scholes dynamics with jumps (continued)) The process (2.12) is one of the few instances for which simulation can be performed on the explicit solution of the s.d.e. This solution can be discretized by factorizing it on a partition of the interval $[0, T]$:

$$S(n\Delta t) = s_0 \times \frac{S(\Delta t)}{s_0} \times \cdots \times \frac{S(n\Delta t)}{S((n - 1)\Delta t)}.$$

All factors share a common distribution, namely:

$$\frac{S(i\Delta t)}{S((i - 1)\Delta t)} = \prod_{j=N(i\Delta t)+1}^{N((i+1)\Delta t)} (1 + Y_j) \times e^{(\mu - \sigma^2/2)\Delta t + \sigma(W((i+1)\Delta t) - W(i\Delta t))}$$
$$\stackrel{d}{=} \prod_{j=1}^{N(\Delta t)} (1 + Y_j) \times e^{(\mu - \sigma^2/2)\Delta t + \sigma(W(\Delta t) - W(0))},$$

where $N(\Delta t) \sim \text{Po}(\lambda \Delta t)$, $Y_j \overset{\text{i.i.d.}}{\sim} f_Y$ and $W(\Delta t) - W(0) \sim \mathcal{N}(0, \Delta t)$. This expression leads to the following:

Algorithm (Simulation of the exact solution)

1. Simulate a Poisson variable N with parameter λt;
2. Simulate random samples $Y_1, \ldots, Y_N$ with distribution f_Y;
3. Simulate a standard normal variable n;
4. Return $x \times Y_1 \times \cdots \times Y_N \times \exp((\mu - \sigma^2/2)\Delta t + \sigma \sqrt{\Delta t} \times n)$.

2.4.4 Method III.A: Approximate Dynamics with Deterministic Intensity

In most models, simulation is performed on the exact solution of an approximate dynamics stemming from the original s.d.e.: this is almost the only method for simulating a jump-diffusion with varying jump intensity. If intensity is deterministic, then the jump part of the process can be sampled by any of the algorithms detailed in Sect. 2.3. Once jump times $\tau_1, \ldots, \tau_m$ have been sampled, we generate the continuous part of the process by implementing a standard path simulation between consecutive pairs of jump times:

$$X_{i+1} = X_i + \mu(t_i, X_i)\Delta t + \sigma(t_i, X_i)\sqrt{\Delta t} \times g_i.$$

Then, for each jump time τ_i, we sample a random jump size $\widetilde{Y}_i$ from f_Y, and add $\eta(\tau_i, X(\tau_i))\widetilde{Y}_i$ to the last value of the continuous part of process. Of course, τ_i need not belong to the partition $\mathcal{T} = \{t_0, \ldots, t_N\}$ of the time horizon $[0, T]$. We may overcome this problem by substituting each τ_i with its closest element in $\mathcal{T}$, that is $\tau_i' = t_{\arg\min_j |t_j - \tau_i|}$. If we adopt the thinning method, the resulting procedure is given by the following algorithm.

Algorithm (Thinning method)

1. Generate approximate jump times $\tau_1', \ldots, \tau_m'$ as above;
2. $\widetilde{X}_0 = x$, $\tau_0' = t_0$;
3. Set i_k: $t_{i_k} = \tau_k'$, for $k = 0, \ldots, m$;
4. For $k = 0, \ldots, m-1$,
   4.1. For $i = i_k, \ldots, i_{k+1}$, use the Euler scheme:

   $$X_{i+1} = X_i + \mu(t_i, X_i)\Delta t + \sigma(t_i, X_i)\sqrt{\Delta t} \times g_i, \qquad (2.14)$$

   and get to $\widetilde{X}_{i_k}, \ldots, \widetilde{X}_{i_{k+1}-1}, \widetilde{X}_{i_{k+1}}$;
   4.2. Sample $\widetilde{Y}_{k+1}$ from f_Y;
   4.3. Set $\widetilde{X}_{i_{k+1}} = \widetilde{X}_{i_{k+1}} + \eta(t_{i_{k+1}}, \widetilde{X}_{i_{k+1}}) \times \widetilde{Y}_{k+1}$;
5. For $i = i_m, \ldots, N$, use the Euler scheme (2.14) and arrive at $\widetilde{X}_{i_m}, \ldots, \widetilde{X}_N$;
6. Return $\widetilde{X}_0, \ldots, \widetilde{X}_N$.

2.4.5 Method III.B: Approximate Dynamics with Random Intensity

Randomness in the jump intensity is usually introduced through a process D and a function $\lambda(t, D)$. The resulting jump intensity process $\lambda(t, D(t))$ gives rise to a jump component whose sampling technique differs according to whether D in independent of the underlying state variable process X or not. We examine two cases:

1. *Exogenous random intensity*: D is statistically independent of the Brownian motion W, the jump sizes Y_i, and the state variable process X.
2. *Endogenous random intensity*: D is statistically dependent on the Brownian motion W and/or jump sizes Y_i and/or the state variable process X.

The case of exogenous random intensity is the easiest to simulate. We first generate a sample $(D(t), 0 \leq t \leq T)$ and the corresponding intensity is the deterministic function $\lambda(t) = \lambda(t, D(t))$ defined for $0 \leq t \leq T$. Then, we employ any of the methods available for the case of deterministic intensity.

If we adopt a thinning algorithm for simulating jumps, the corresponding procedure is as follows:

Algorithm (**Thinning method under exogenous intensity**) Preliminary step. Simulate a sample path $(D(t))_{0 \leq t \leq T}$ and set $\lambda(t) = \lambda(t, D(t))$. Steps 1–5 as above (Method III, Sect. 2.3.4).

Sometimes it is possible to avoid simulating the whole path $(\lambda(t), 0 \leq t \leq T)$ and obtain a jump time in one movement. This is the case of an intensity process whose jump time distribution can be computed in a closed-form. Consider, for instance, a one-point process $N(t) = \mathbf{1}_{\{\tau \leq t\}}$ with jump intensity following a mean reverting square-root process $d\lambda(t) = \alpha(\beta - \lambda(t)) \, dt + \sigma \sqrt{\lambda(t)} \, dW(t)$, starting at $\lambda(0) > 0$. Provided that $\alpha, \beta > 0$ and $2\alpha\beta \geq \sigma^2$, the process stays always positive during the entire lifetime. By recalling the definition of intensity, the process $\mathbf{1}_{\{\tau \leq t\}} - \int_0^t \lambda(u) \, du$ is a martingale. Then, $\mathbb{P}(\tau \geq T) = \mathbb{E}(\exp(-\int_0^T \lambda(u) \, du))$ and, in the case above, this quantity can be calculated analytically. The jump time distribution function is given by

$$
\begin{aligned}
F_\tau(t) &= \mathbb{P}(\tau \leq t) \\
&= 1 - \mathbb{P}(\tau \geq t) \\
&= 1 - \exp\big(-A(t)\lambda(0) + C(t)\big),
\end{aligned}
$$

with:

$$
\begin{aligned}
A(t) &= 2\exp(\gamma t - 1)/\big[\alpha\gamma\big(\exp(\gamma t) - 1\big) + 2\gamma\big], \\
C(t) &= 2\alpha\beta\sigma^{-2} \ln \frac{2\gamma \exp((\alpha + \gamma)t/2)}{(\alpha + \gamma)(\exp(\gamma t) - 1) + 2\gamma}, \\
\gamma &= \sqrt{\alpha^2 + 2\sigma^2}.
\end{aligned}
$$

This computation, coupled with the inverse transformation method, leads to the following algorithm.

Algorithm (Exact sampling)

1. Compute F_τ as above and set F_τ^{-1} as the generalized inverse of F_τ;
2. Sample $U \sim \mathcal{U}[0, 1]$;
3. Return $\tau = F_\tau^{-1}(U)$.

The case of endogenous random intensity is slightly harder to address. Indeed, X determines D which in turn affects λ, and ultimately the state variable process X through its discontinuous part. Therefore, we need to simultaneously sample D and X via $\lambda(t, D(t))$. We consider the case where $D = X$ (i.e., $\lambda = \lambda(t, X(t))$) and leave the extension to the more general situation where $D \neq X$ (e.g., $D = g(X)$) as an exercise.

By adopting the thinning method, we come up to the following algorithm.

Algorithm (Thinning method under endogenous intensity)

1. Set initial state $X_0 = x$, initial time $\bar{\tau}_0 = 0$, time partition $t_i = i\Delta$ ($i = 1, \ldots, N$) and counters $i_0 = 0$, $k = -1$, $j = 0$;
2. Define an intensity upper bound λ: $\lambda(t, x) \leq \lambda$ for all (t, x) in $[0, T] \times \text{Im}(X)$;
3. Set $k = k + 1$;
4. Generate independent $e_k \sim \mathcal{E}(\lambda)$ and set $\bar{\tau}_k = \bar{\tau}_k + e_k$;
5. If $\bar{\tau}_k > T$, go to Step 11, otherwise set $\bar{\tau}'_k = t_{i_k}$ where $i_k = \arg\min_j |t_j - \bar{\tau}'_k|$;
6. For $i = i_k, \ldots, i_{k+1}$ use (2.14) and arrive at $X_{i_k}, \ldots, X_{i_{k+1}-1}, X_{i_{k+1}}$;
7. Generate independent $U_k \sim \mathcal{U}[0, 1]$;
8. If $\lambda U_k > \lambda(\bar{\tau}'_k, X_{i_{k+1}})$ go to Step 3 (rejected);
9. Set $j = j + 1$ and $\tau_j = \bar{\tau}'_k$;
10. Sample Y_{k+1} from f_Y;
11. Set $X_{i_{k+1}} = X_{i_{k+1}} + \eta(t_{i_{k+1}}, X_{i_{k+1}}) \times Y_{k+1}$ and go to Step 3;
12. For $i = i_k, \ldots, N$ use the Euler scheme (2.14) and arrive at $X_{i_k}, \ldots, X_N$;
13. Return $\tau_1, \ldots, \tau_j$ and $X_0, \ldots, X_N$.

The first two steps initialize variables and parameters. Steps 4 and 5 generate the next potential jump time within the discrete set of sample times. Step 6 simulates the continuous path until the potential jump time. Steps 7 and 8 perform an acceptance–rejection test on this sample time. If this is accepted, Steps 9–11 generate and add a sample jump on the last generated value of the continuous trajectory; if the test fails, the next potential jump time is sampled. Step 11 simulates the continuous path between the last potential, or actual, jump time and the time horizon. Step 13 returns sample jump times and path. The algorithm can be implemented through pseudo-code 2.2-2.3; unfortunately, it may be highly time consuming, depending on the number of rejected jump times, which, in turn, depends on the average distance between the jump intensity and the upper bound λ. This distance is random as long as it varies with the sample path. Consequently, the method is viable to the extent that the function $\lambda(t, x)$ has a relatively narrow image. It is therefore natural to search for more efficient methods to simulate the desired process. It turns out that the sequential method detailed in Sect. 2.3.4 can be extended to the case of an endogenous random intensity rather naturally.

Algorithm (Countdown method under endogenous intensity)

1. Set $\widetilde{X}_0 = x$, $\tau_0 = 0$, $i = k = 0$;
2. Generate $e_k \sim \mathcal{E}(1)$ independently of e_{k-1}, if any;
3. Set $i = i + 1$; if $i\Delta \geq T$, then go to Step 9;
4. Generate $g_i \sim \mathcal{N}(0, 1)$ independently of all previous samples;
5. Set $X_{i+1} = X_i + \mu(t_i, X_i)\Delta t + \sigma(t_i, X_i)\sqrt{\Delta t} \times g_i$;
6. If $\sum_{j=0}^{i} \lambda(j\Delta, X_j)\Delta < e_{i+1}$, then go to Step 3;
7. Set $k = k + 1$, $\tau_k = i\Delta$;
8. Sample a jump size J_k and set $X_{i+1} = X_{i+1} + J_k$; go to Step 2;
9. Return jump times $\tau_1, \ldots, \tau_k$ and the sample path $X_1, \ldots, X_{i-1}$.

This algorithm can be implemented using the pseudo-code 2.4.

2.5 Gaussian Processes

Let X be a Gaussian process with mean $\mu(t) = \mathbb{E}(X(t))$ and covariance $c(t, s) = \mathbb{E}((X(t) - \mu(t))(X(s) - \mu(s)))$. We examine two methods for simulating samples from this process following the series expansion approach.

The first approach applies to stationary processes. It relies on a sampling theorem for deterministic bounded functions. This results states that the value of a function $f : \mathbb{R} \to [-v, v]$ computed at a point t can be represented as the sum of a series whose nth term involves the values of f at finitely many points in the set $\{k\Delta, k = 0, \pm 1, \pm 2, \ldots\}$, with $\Delta = \pi/v$. More formally, $f(t) = \lim_{n \to \infty} \sum_{k=-n}^{n} f(k\Delta) \times \alpha_k(t)$, where $\alpha_k(t) = [(\Delta - k\pi)/t\pi]\sin(t\pi/\Delta - k\pi)$. This leads to the following algorithm.

Algorithm (Sampling theorem for stationary Gaussian processes)

1. Fix sample size $n \geq 1$, frequency v, and time interval $[0, T]$;
2. Set $\Delta = \pi/v$ and $M = [T/\Delta] + n + 1$;
3. Generate samples of

$$X(-n\Delta)$$
$$X((-n + 1)\Delta)|X(-n\Delta)$$
$$X((-n + 2)\Delta)|X(-n\Delta), X((-n + 1)\Delta)$$
$$\cdots$$
$$X(0)|X(-n\Delta), \ldots, X(-\Delta)$$
$$\cdots$$
$$X(M\Delta)|X(-n\Delta), \ldots, X((M - 1)\Delta).$$

by using the Gaussian property of conditional distributions;

```
(* INPUT PARAMETERS *)

steps = Value;                  (* number of steps *)
dt= Value;                      (* length of one step in years *)
x[0] = Value;


(* DIFFUSION COEFFICIENTS *)

Drift [t_, x_] := Function(t, x);
Volatility[t_, x_] := Function(t, x);
λ[t_, x_] := Function (t, x); (* jump intensity *)
jumpCoefficient[t_, x_, Δ_] := Function(t, x);


(* DISTRIBUTIONS *)

uniform = UniformDistribution[0, 1];
normal = NormalDistribution[0, 1];
jumpRandomComponent = Distribution[];


(* GENERATOR *)

λmax = Value;
tau = 0;
(* last potential jump time (in Years), *)
t = 0;
(* last potential jump time (in Steps), *)
potentialInterarrivalTime = -Log[Random[uniform]]/λmax;
tau= tau+potentialInterarrivalTime; (* potential jump time *)

While tau < horizon,
    ini = t + 1;
    t= [tau* /dt] +1 ;

    (* GENERATE CONTINUOUS TRAJECTORY UNTIL POTENTIAL JUMP TIME *)

    For[i = ini, i < t + 1, i++,
      (* Time steps between ' 1 + last potential jump time' and
          the ' next potential jump time' *)

      x[i] = x[i- 1] + Drift [i, x[i- 1] ] +
          Volatility [i, x[i - 1] ] * Sqrt [dt]
          * Random [normal] ;

    ] ;
```

Fig. 2.2. Thinning pseudo-code.

```
(* TEST WHETHER POTENTIAL
JUMP TIME IS TO BE ACCEPTTED OR REJECTED *)

ratio = λ[t * dt, xLeft[t]] /λmax;(* Acceptance Rejection Ratio *)

If [Random [uniform] < ratio,
       (* ACCEPTED => GEN.JUMP COMPONENT &
       SUM TO THE CONTINUOUS *)

       Δ = Random [jumpRandomComponent];
       x[t] += jumpCoefficient[t* dt, x[t], Δ];

       ];

   potent ialInterarrivalTime
      = Max [-Log [Random [uniform] ] / Δmax, dt];
        (* set next potential interarrival time ≥ dt *)

   tau = tau + potentialInterarrivalTime;
  (* next potenatial jump time *)

      ];

(* GENERATE CONTINUOUS PATH
   BETWEEN LAST POTENTIAL JUMP TIME AND horizon *)

ini = t + 1;

For[i=ini, i<steps+1, i++,

   x[i] = x[i-1] + Drift[i, x[i-1]] +
   Volatility[i, x[i - 1] ] * Sqrt[dt] * Random[normal];

   ];
```

Fig. 2.3. Thinning pseudo-code (*continued*).

4. Return

$$X^{(n)}(t) = \sum_{k=[\frac{t}{\Delta}]-n}^{[\frac{t}{\Delta}]+n+1} X(k\Delta)\alpha_k(t),$$

where $\alpha_k(t) = [(\Delta - k)/t] \sin(t/\Delta - k)$, for all $t \in [0, T]$.

The second approach applies to non-stationary Gaussian processes with zero mean and covariance function $c(t, s)$. It builds on the Fourier series expansion of sample paths. The procedure is detailed in the following algorithm.

Algorithm (Fourier method for non-stationary Gaussian processes)

1. Fix a sample size $n \geq 1$ and a time interval $[0, T]$;

```
(* INPUT PARAMETERS *)

steps = Value;                        (* number of steps *)
dt = Value;                           (* length of one step in years *)
x[0] = Value;


(* DIFFUSION COEFFICIENTS *)

Drift [t_, x_] := Function (t, x);
Volatility [t_, x_] := Function (t, x);
λ[t_, x_] := Function (t, x);         (* jump intensity *)
jumpCoefficient [t_, x_, Δ_] := Function (t, x);


(* DISTRIBUTIONS *)

uniform = UniformDistribution [0, 1];
normal = NormalDistribution [0, 1];
jumpRandomComponent = Distribution [];


(* GENERATOR *)

i = 1;
exponentialTime = -Log [Random [uniform ]];
cumulatedIntensity = 0;

While [i < steps ,
    t = i * dt;
    noise = Random [normal ];
    x[i] =
  x [i - 1 ] + Drift [t, x [i - 1 ]] *dt
    + Volatility [t, x [i - 1 ]] * Sqrt [dt ] * noise;
    cumulatedIntensity += (λ[ (i - 1) * dt, x[i - 1]]
                         + λ[i * dt, x[i]]) * dt/2;

    If[cumulatedIntensity > exponentialTime ,
        Δ = Random [ jumpRandomComponent ]
        x[i] += JumpCoefficient [t, x[i], Δ];
        cumulatedIntensity = 0;
        exponentialTime = -Log [Random [u]];
        ];

    i + +
    ];
```

Fig. 2.4. Countdown pseudo-code.

2. Define $I(n) = \mathbf{1}_{\mathbb{Z}_+}(n)$, $v_k = 2\pi k/T$ $(k = 1, \ldots, n)$, and

$$c_{kh} = \frac{4}{T} \int_0^T dx \int_0^T dy \big[c(x, y) \cos(v_k x)^{I(k)} \sin(v_k x)^{1-I(k)}$$
$$\times \cos(v_h y)^{I(h)} \sin(v_h y)^{1-I(h)} \big],$$

for $h, k = 0, \pm 1, \ldots, \pm n$;

3. Generate $2n + 1$ centered correlated normal variables Y_k ($k = \pm 1, \ldots, \pm n$) with covariance $\mathbb{E}(Y_k Y_h) = c_{kh}$;

4. Return

$$X^{(n)}(t) = \frac{Y_0}{2} + \sum_{k=1}^{n} [Y_k \cos(\nu_k t) + Y_{-k} \sin(\nu_k t)],$$

for $t \in [0, T]$.

This algorithm requires that the Fourier series of all c_{kh} converges to c_{kh}. This is the case, for instance, if c_{kh} are C^2 on the square $[0, T] \times [0, T]$.

2.6 Comments

General treatments of continuous and mixed-jump diffusion processes include Jacod and Shiryaev (1988), Kloeden and Platen (2000), Rogers and Williams (1987), and Shreve (2004). Approximations of continuous time jump processes are explored in Asmussen and Rosinski (2001), Bruti-Liberati, Nikitopoulos-Sklibosios and Platen (2006), Bruti-Liberati and Platen (2006, 2007), Duffie, Pan and Singleton (1998), Jacod and Protter (1998), Lewis and Shedler (1979), and Talay (1982, 1984, 1995). Theoretical aspects of path simulation in finance are examined in Acworth, Broadie and Glasserman (1998) (comparison of alternative simulation methods), Boyle and Tan (1997) (quasi Monte Carlo methods), Duan and Simonato (1998) (martingale simulation), Dupire and Savine (1998) (speed reduction). Abken (2000), Glasserman, Heidelberger and Shahabuddin (1999b, 2000), and Picoult (1999) apply random number generation to the computation of Value-at-Risk figures. Akesson and Lehoczky (2000) use simulation for the valuation of mortgage positions, while Andersen (1995), Carr and Yang (1998), Glasserman, Heidelberger and Shahabuddin (1999a), Glasserman and Zhao (1999), Miltersen (1999), and Rebonato (1998) perform Monte Carlo studies in the context of interest rate models. American option valuation naturally requires backward induction (see chapter "Dynamic Programming"). However, path sampling can be adopted for the purpose of improving traditional algorithms. These topics are explored by Carrière (1996), Broadie and Detemple (1997), Broadie and Glasserman (1997), Broadie, Glasserman and Jain (1997), Carr and Yang (1997), Carr (1998), Andersen and Broadie (2001), Longstaff and Schwartz (2001), Boyle and Kolkiewicz (2002), and Garcia (2003), among others. Douady (1998) and Avellaneda et al. (2001) apply Monte Carlo to the calibration of stochastic models. Avellaneda and Gamba (2000), Broadie and Glasserman (1996), Fournié et al. (1999), and Gobet and Munos (2002) are major references for hedging related applications. General valuation schemes using simulation are developed in Boyle, Evnine and Gibbs (1989), Clewlow and Carverhill (1994) and Carverhill and Pang (1998). Stochastic volatility models have been simulated in Clewlow and Strickland (1997) and Fournié, Lasry and Touzi (1997). Applications to the econometrics of continuous time processes can be found in Duffie and Glynn (1995), Duffie and Singleton (1993), and Gourieroux and Monfort (1996). An extensive analysis of price returns using simulation has been performed by Tompkins and D'Ecclesia (2005). Special simulation methods have been developed for Levy processes in Këllezi and Webber (2004), Ribeiro and Webber (2003, 2005), and Tankov (2005). Cont and Tankov (2004) review methods based on the notion of Levy copula as introduced in Kallsen and Tankov (2004).

3

Dynamic Programming for Stochastic Optimization

In this short chapter we introduce one of the most popular and powerful methods to solve optimal control problems for deterministic and stochastic dynamic systems in discrete time. This method is known as Dynamic Programming (DP) and has been proposed in the fifties of the last century by the American mathematician Richard Bellman (Bellman (1957)) for the case of deterministic dynamics. The main feature of DP is the reduction of a computational problem of complexity order m^n to n problems of complexity order m. If the resulting algorithm needs to be reduced further, then more sophisticated techniques must be employed. For a treatment of these methods, we may refer to Bertsekas (2005).

Our presentation is organized as follows. Section 3.1 presents the general setting of a discrete time optimization problem. Section 3.2 states the problem for both deterministic and stochastic systems. Section 3.3 introduces the Bellman principle of optimality. Section 3.4 develops the DP algorithm for the case of deterministic systems. The extension to stochastic systems is presented in Sect. 3.5. Section 3.6 illustrates two major applications of DP in quantitative finance. A final section concludes with comments on further references.

3.1 Controlled Dynamical Systems

Let us consider a system whose description at a given time s is captured by a variable $X(s)$. This quantity is the *state variable* of the system and may take values in any given set $\mathbb{X}$. A *controlled dynamic system* in discrete finite time horizon is defined by a state variable assigned for each time $t, \ldots, T$ according to a law of the form $X(s + 1) = f(s, X(s), u(s, X(s)))$. Given an initial condition $X(t) = x$, this rule says that the state of the system prevailing at time $s + 1$ is a known function of (1) the previous time s, (2) the state $X(s)$ prevailing then, and (3) the value $u(s, X(s))$ assumed by a control policy computed as a function of time s and state $X(s)$.

Example X may represent the market value of a security portfolio and u the selection made by an investor about the proportion of his wealth to invest in the set of tradeable assets in the market under consideration.

It is important to distinguish between the notions of control policy and control. A *control policy* is a rule $u(\cdot, \cdot)$ associating to each pair of time s and state y a number $u = u(s, y)$, the *control*, in a set $\mathbb{U}$ of admissible actions affecting the dynamic link between the states of the system at consecutive times.[1] Notice that the last control occurs one time step before the last date T. A control policy is denoted by a symbol $\mathbf{u}$. The overall evolution of the system depends on:

(a) the selected policy $\mathbf{u}$,
(b) the initial time t, and
(c) the starting value x of the system. Consequently we may denote by $X^{t,x,\mathbf{u}}$ the solution, which we assume to exist and to be unique, of the resulting dynamic system:

$$\begin{cases} X(s+1) = f\big(s, X(s), u\big(s, X(s)\big)\big), & s = t, \ldots, T-1, \\ X(t) = x, & \mathbf{u} = u(\cdot, \cdot). \end{cases}$$

Whenever the initial setting (t, x) is unambiguously understood, we adopt the lighter notation $X^{\mathbf{u}}$.

In most applications, we consider a state space $\mathbb{X} = \mathbb{R}^n$ and a control set $\mathbb{U} = \mathbb{R}^m$. Moreover, we assume that the pair (state, control) $= (x, u)$ belongs to a specified set $\mathbb{K}(s)$, which is allowed to vary over time, i.e., $(X(s), u(s, X(s))) \in \mathbb{K}(s)$ for all $s = t, \ldots, T-1$. In this case, we say that $\mathbf{u}$ is *admissible* and write $\mathbf{u} \in \mathcal{U}_{t,T}$. Despite $\mathbb{K}(s) \subset \mathbb{X} \times \mathbb{U}$, it need not be itself a rectangle $\mathbb{A} \times \mathbb{B}$, meaning that some specific control u may be admissible for a limited number of times only.[2]

Control $\mathbf{u}$ is evaluated according to a *performance measure* (or *index*) J of the corresponding dynamic system $(X^{\mathbf{u}}(s))_{s=t,\ldots,T}$. This index is supposed to be time additive, that is:

$$J_{t,x}(\mathbf{u}) = \sum_{s=t}^{T-1} F\big(s, X^{t,x,\mathbf{u}}(s), u\big(s, X^{t,x,\mathbf{u}}(s)\big)\big) + \Psi\big(X^{t,x,\mathbf{u}}(T)\big). \tag{3.1}$$

This means that the system generates a cost/return at each time. In our previous example, J may represent the investor's utility generated by the selected portfolio strategy $\mathbf{u}$. Table 3.1 summarizes notation and quantities introduced so far. Notice that J also depends on dynamics f, the control set $\mathbb{U}$, the last date T, though this is implicitly taken for granted. Under a slight abuse of notation, $J_{s,X^{t,x,\mathbf{u}}(s)}(\mathbf{u})$ denotes the performance of the same system as measured from an intermediate time $s > t$. We underline that the index above is separable in time and that the last summand on the right-hand side of expression (3.1) is affected from the control policy $\mathbf{u}$ through the state of the system only.

---

[1] In general, a control policy is a time dependent function stating the control selected by the user at any time t. In the case of stochastic systems, this quantity may be random. We restrict our definition to the so-called *Markovian control policies*, namely those that depend on time and state only.

[2] $A \times B$ is the Cartesian product between A and B as defined by $\{(x, y): x \in A, y \in B\}$.

Table 3.1. Notation

Symbol	Space	Symbol	Variable	Symbol	Function
$\mathbb{X}$	State	t	Initial time	$f(\cdot,\cdot)$	Dynamics
$\mathbb{U}$	Control	T	Final time	$u(\cdot,\cdot)$	Ctrl. policy
$\mathbb{K}(s)$	Ctrl. constraint	s	Current time	$\Psi(\cdot)$	Final reward
$\mathcal{U}_{t,T}$	Admiss.	$\mathbf{u}$	Control	$F(\cdot;\cdot)$	Going reward
	Ctrl. policies	x	State	$J(\cdot)$	Total reward

3.2 The Optimal Control Problem

Let a controlled dynamic system in discrete time and a performance index be given as above. A *deterministic optimal control problem* consists of determining for any starting point an admissible control policy maximizing the performance of the system, i.e.,

$$\max_{\mathbf{u}\in\mathcal{U}_{t,T}} J_{t,x}(\mathbf{u}),$$

for all $(t,x) \in \mathbb{K}(t)$. The function $\hat{\mathbf{u}} = \arg\max J$ is the *optimal control policy* and the best performance $V(t,x) = J_{t,x}(\hat{\mathbf{u}})$ is referred to as the *value function* of the problem. Minimization obtains by replacing J with $-J$.

A *stochastic optimal control problem* generalizes the notion of optimal control problem to the case of stochastic dynamic systems, namely those described by a stochastic process. Accordingly, we identify three major differences between deterministic and stochastic problems, which we now illustrate.

- The evolution law for the state variable X is random. More precisely, $X(s+1)$ is a random variable whose distribution p is random. It may also depend on time s, state $X(s)$, and the value $u(s)$ of the control at that time, that is:

$$X(s+1) \sim p\big(\mathrm{d}x; s, X(s), u(s), \omega\big). \tag{3.2}$$

More formally, for each 4-uple of time s, state y, control u, and sample event ω, there is a probability measure $p(\mathrm{d}x; s, y, u, \omega)$ on the state space $\mathbb{X}$. This quantity represents the probability that the state variable belongs to the interval $[x, x+\mathrm{d}x]$ at time $s+1$, conditional to event that the system was in state y at the previous time s and has been controlled by u.

Typically, the dependence on randomness is achieved through a noise term W. More precisely, a sequence of independent and identically distributed random shocks $W(t), \ldots, W(T-1)$ is given. Then, system dynamics are described as follows:

$$X(s+1) = f\big(s, X(s), u(s), W(s)\big), \tag{3.3}$$

for $s = t, \ldots, T-1$. This relation, coupled with the usual initial condition $X(t) = x$, determines a stochastic process $X^{t,x,\mathbf{u}} = (X^{t,x,\mathbf{u}}(s))_{s=t,\ldots,T}$.

Example A typical instance of these dynamics is provided by the Euler discretization of a continuously controlled stochastic differential system:

$$\begin{cases} dX^{\mathbf{u}}(s) = \mu\big(s, X^{\mathbf{u}}(s), u\big(s, X^{\mathbf{u}}(s)\big)\big)\,ds + \sigma\big(s, X^{\mathbf{u}}(s), u\big(s, X^{\mathbf{u}}(s)\big)\big)\,dW(s), \\ X^{\mathbf{u}}(t) = x. \end{cases}$$

In this case, the resulting discrete time dynamics read as:

$$\begin{cases} X^{\mathbf{u}}(s + \Delta s) = X^{\mathbf{u}}(s) + \mu\big(s, X^{\mathbf{u}}(s), u\big(s, X^{\mathbf{u}}(s)\big)\big) \times \Delta s \\ \qquad\qquad + \sigma\big(s, X^{\mathbf{u}}(s), u\big(s, X^{\mathbf{u}}(s)\big)\big) \times \sqrt{\Delta s} \times \mathcal{N}(0, 1), \\ X^{\mathbf{u}}(t) = x, \end{cases}$$

where $\mathcal{N}(0, 1)$ denotes a standard Gaussian variable. This expression assumes form (3.3).

- The control variable u is random too. Again, we consider Markov control policies, namely those whose randomness enters their value through the state variable only, i.e.,

$$u(s, \omega) = u\big(s, X(s, \omega)\big).$$

- Since there is one trajectory $(X(s, \omega), t \le s \le T)$ per sample event ω, we need to synthesize the corresponding performances into a single number. It is customary to consider the expected value of a functional of the random trajectory corresponding to a given control policy. This expectation is computed under the probability measure induced over the path-space by the transition densities above indicated and delivers a measure of the overall performance of the system as:

$$J_{t,x}(\mathbf{u}) = \mathbb{E}\left(\sum_{s=t}^{T-1} F\big(s, X(s), u\big(s, X(s)\big)\big) + \Psi\big(X(T)\big) \right),$$

for $t \le T$, provided that $\sum_{s=T}^{T-1}(\cdot) := 0$, i.e., $J_{T,x}(\mathbf{u}) = \Psi(x)$. Other functionals may be adopted in place of the expected value. For instance, we may consider higher-order moments (e.g., the mean square error) or distributional quantiles (e.g., Value-at-Risk) of the distribution of the random performance index.

Example (Optimal stopping) An important type of stochastic control problems is the *optimal stopping time*. In its simplest form, the system is allowed to evolve freely until a time τ is selected by the user. The problem reads as follows:

$$\max_{\tau \in \mathcal{T}} \left(\sum_{s=t}^{\tau-1} F\big(s, X^{t,x}(s)\big) + \Psi\big(X^{t,x}(\tau)\big) \right), \tag{3.4}$$

where the system evolves according to $X^{t,x}(s + 1) = f(s, X^{t,x}(s), W(s))$. Here, $\mathcal{T}$ is a set of random variables taking value in the time set $\{t, \ldots, T\}$.[3] In financial applications, τ represents the time at which an option holder decides to exercise his right to a given cash flow or to perform a certain action. In an appendix to this book, we show how this problem can be cast in the framework developed so far. It can be

---

[3] Proper measurability conditions are taken for granted in the present context. For more information on this point, see references cited at the end of this chapter.

proven that under suitable conditions, a Markov optimal control policy for an optimal stopping problem defines a curve $\gamma(s, x)$, known as the *free boundary* associated to the problem. This curve has the property that the optimally controlled system is stopped at time s, i.e., $\tau = s$, if and only if the state $X(s)$ hits (or overcome) the threshold $\gamma(s, X(s))$. Dynamic programming allows us to compute this curve point by point.

3.3 The Bellman Principle of Optimality

Both deterministic and stochastic optimization problems detailed above involve a maximization, or minimization, on $\mathbb{R}^{m \times (T-t)}$, where m is the number of control variables and $T-t$ is the length of the control time horizon. Dynamic programming is a method to transform this problem in $\mathbb{R}^{m \times (T-t)}$ into a sequence of $T - t$ simplified optimization problems on $\mathbb{R}^m$. This method is based on the *Bellman principle of optimality*, which we now introduce.

We say that a control policy $\mathbf{u}^1$ is dominated by another control policy $\mathbf{u}^2$ if $J_{t,x}(\mathbf{u}^1) \leq J_{t,x}(\mathbf{u}^2)$ for all pairs (t, x). Correspondingly, we write $\mathbf{u}^1 \leq \mathbf{u}^2$. We consider an arbitrary control policy $\mathbf{u} = (u(t), \ldots, u(s - 1), u(s), \ldots, u(T - 1))$ and then build another control policy $\mathbf{u}'$ by keeping the first $s - t$ components of $\mathbf{u}$ and modifying the last $T - s$ entries as follows. First, we apply the original policy $\mathbf{u}$ to the system X until time $s - 1$ included, so that the resulting state at time s is $X^{t,x,\mathbf{u}}(s)$. Then, we let the system evolve optimally: that is, we adopt the control

$$\hat{\mathbf{u}}^s = \left(\hat{u}^s(s), \ldots, \hat{u}^s(T - 1) \right) := \arg \max_{\mathbf{v} \in \mathcal{U}_{s,T}} J_{s, X^{t,x,\mathbf{u}}(s)}(\mathbf{v}).$$

The resulting control policy read as

$$\mathbf{u}' = \left(u(t), \ldots, u(s - 1), \hat{u}^s(s), \ldots, \hat{u}^s(T - 1) \right).$$

The Bellman principle of optimality states that:

Proposition (Bellman principle of optimality) *The control policy* $\mathbf{u}$ *is dominated by* $\mathbf{u}'$, *i.e.,*

$$\mathbf{u} \leq \mathbf{u}'.$$

Remark One should not confuse $\mathbf{u}'$ with the control policy:

$$\mathbf{u}'' = \left(u(t), \ldots, u(s - 1), \hat{u}^t(s), \ldots, \hat{u}^t(T - 1) \right),$$

which is obtained from $\mathbf{u}$ by substituting the $T - s$ components of the optimal policy $\hat{u}^t = \arg \max_{\mathcal{U}_{t,T}} J_{t,x}(\mathbf{u})$ as *computed at time t* and *under state x* to the last $T - s$ entries in $\mathbf{u}$.

Remark The intuition behind Bellman principle of optimality lays on the additivity of the performance functional J and the Markov property of the underlying system dynamics: given time s and state y, the optimal policy on the remaining time horizon

$s, \ldots, T - 1$ is independent of the way the system reached the state y at time s. This is linked to the following flow property of the kind of system dynamics we adopted:

$$X^{t,x}(s) = X^{X^{t,x}(r)}(s),$$

for any $t \le r \le s$. Put in simple terms, the time s state of a system X starting at a prior time t in a state x is the same as the state of the system X had it started on any intermediate time r at the state $X^{t,x}(r)$ that the same system would have reached after starting at x on date t. In the stochastic case, this property, which should be interpreted as an equality in distribution, stems from the Markovian nature of the underlying stochastic evolution. This informal reasoning is turns into a proof in an appendix reported at the end of this book.

In the rest of the chapter, we assume that any optimization problem under consideration admits a unique solution for all $(s, y) \in \{t, \ldots, T\} \times \mathbb{X}$. Sufficients conditions for this assumption to hold true are provided in Bertsekas (2005).

3.4 Dynamic Programming

We now state the dynamic programming algorithm which is grounded on the Bellman principle of optimality.

Proposition (Dynamic programming algorithm) *For all* $y \in \mathbb{X}$, *define recursively:*

- (Begin)

$$V(T, y) := \Psi(y),$$

- (Step 1)

$$V(T - 1, y) := \max_{\{u:(y,u)\in\mathbb{K}(T-1)\}} \left\{ F(T - 1, y, u) + V\left(T, f(T - 1, y, u)\right) \right\},$$

$$\hat{u}^{T-1}(T - 1, y) := \operatorname*{arg\,max}_{\{u:(y,u)\in\mathbb{K}(T-1)\}} \left\{ F(T - 1, y, u) + V\left(T, f(T - 1, y, u)\right) \right\},$$

- (Step $T - n$)

$$V(n, y) := \max_{\{u:(y,u)\in\mathbb{K}(n)\}} \left\{ F(n, y, u) + V\left(n + 1, f(n, y, u)\right) \right\},$$

$$\hat{u}^n(n, y) := \operatorname*{arg\,max}_{\{u:(y,u)\in\mathbb{K}(n)\}} \left\{ F(n, y, u) + V\left(n + 1, f(n, y, u)\right) \right\}.$$

Then:

$$u^B := \left(\hat{u}^t(t, \cdot), \ldots, \hat{u}^{T-2}(T - 2, \cdot), \hat{u}^{T-1}(T - 1, \cdot) \right)$$

is the optimal control policy $\hat{\mathbf{u}}^t$ *and* V *is the value function of the optimal control problem* $\mathcal{C}(t, x)$.

Remark The procedure is initialized at time T, and then moves backward. At time T, no control is possible, so the best reward is the final reward itself. At any intermediate time n, the controller selects the control u so that the pair (current state y, control u) is admissible, i.e., it belongs to $\mathbb{K}(n)$, and leads to the best reward. This latter is decomposed into two components: the first one is the time n reward stemming from the current state y of the system and the selected control u, that is $F(n, y, u)$; the second component is the optimal reward from time $n + 1$ on, that is the value function evaluated at time $n + 1$ under the state of the system then, namely $X(n+1) = f(n, X(n), u(n, X(n))) = f(n, y, u)$. A proof of this algorithm is detailed in an appendix.

Remark (Determination of optimal controls) The number $V(t, x)$ defines the performance for the optimally controlled system. The optimal control policy is the rule $\hat{u} : \{t, \dots, T - 1\} \times \mathbb{R}^n \to \mathbb{R}^m$ that fully determines a trajectory $(X^{t,x,\hat{u}}(s), t \le s \le T)$ of the controlled dynamic system according to the following procedure:

1. Start at (t, x);
2. If state y is reached at time s, then time $s + 1$ state is $f(s, y, \hat{u}(s, y))$.

In other words, the optimal control policy is a sequence of functions of the state of the system y, one function per time between t and $T - 1$. For instance, $\hat{u}(s, \cdot)$ states that the control to be applied to the system at time s is $\hat{u}(s, y)$ if the state is y at that time.

Let us show the way this method can be worked out step by step. Starting with state x at time t, one obtains:

$$X^{\hat{u}}(t + 1) = f\big(t, x, \hat{u}(t, x)\big),$$

and the optimal control at $t + 1$ is:

$$\hat{u}\big(t + 1, X^{\hat{u}}(t + 1)\big) = \hat{u}\big(t + 1, f\big(t, x, \hat{u}(t, x)\big)\big).$$

At time $t + 2$, the optimal trajectory defines a state:

$$X^{\hat{u}}(t + 2) = f\big(t + 1, X^{\hat{u}}(t + 1), \hat{u}\big(t + 1, X^{\hat{u}}(t + 1)\big)\big),$$

and the optimal control is:

$$
\begin{aligned}
\hat{u}\big(t + 2, X^{\hat{u}}(t + 2)\big) &= \hat{u}\big(t + 2, f\big(t + 1, X^{\hat{u}}(t + 1), \hat{u}\big(t + 1, X^{\hat{u}}(t + 1)\big)\big)\big) \\
&= \hat{u}\big(t + 2, f\big(t + 1, f\big(t, x, \hat{u}(t, x)\big), \\
&\qquad \hat{u}\big(t + 1, f\big(t, x, \hat{u}(t, x)\big)\big)\big)\big).
\end{aligned}
$$

By recursion, one computes all the individual optimal controls $\hat{u}(s, X^{\hat{u}}(s))$ for $s = t, \dots, T - 1$. The overall optimal control is thus the optimal policy computed on the optimally controlled trajectory $(X^{t,x,\hat{u}}(t), \dots, X^{t,x,\hat{u}}(T))$, that is the $(T - t)$-dimensional sequence of individual optimal controls in $\mathbb{R}^m$:

$$\hat{u} = \big(\hat{u}\big(t, X^{t,x,\hat{u}}(t)\big), \dots, \hat{u}\big(T - 1, X^{t,x,\hat{u}}(T - 1)\big)\big).$$

Note that this result is achieved by using the following items only:

- Initial condition: time t and state x;
- Optimal policy: $\hat{u}(t, \cdot), \ldots, \hat{u}(t-1, \cdot)$;
- Input dynamics: $f(\cdot, \cdot, \cdot)$.

3.5 Stochastic Dynamic Programming

Stochastic dynamic programming is an algorithm for determining the optimal control policy in the case of stochastic dynamic systems of the kind presented in the first section of this chapter. Here, we slightly abuse of our notation by writing $X^{n,y,u}(n+1)$ for the random state of the system at time $n+1$ given that it started from state y at the previous time and that a control u has been applied between times n and $n+1$. As usual, $\mathbb{P}(d\zeta; n+1, y, u)$ denotes the probability distribution of the state variable at time $n+1$ given that the system was at state y one time period earlier and that control u has been adopted. Symbol $\mathbb{P}_{W(n)}(dw)$ denotes the probability distribution of the random noise W affecting the system dynamics at time n.

Proposition (Stochastic dynamic programming algorithm) *For all $y \in \mathbb{X}$, define recursively:*

- (Begin)
$$V(T, y) := \Psi(y),$$

- (Step $T - n$)

$$V(n, y) := \max_{\{u:(y,u)\in\mathbb{K}(n)\}} \left\{ F(n, y, u) + \mathbb{E}\left[V\left(n+1, X^{n,y,u}(n+1)\right)\right]\right\}$$

$$= \max_{\{u:(y,u)\in\mathbb{K}(n)\}} \left\{ F(n, y, u) + \int_{\mathbb{X}} V(n+1, \zeta)\mathbb{P}(d\zeta; n+1, y, u)\right\}$$

$$= \max_{\{u:(y,u)\in\mathbb{K}(n)\}} \left\{ F(n, y, u) + \int_{\mathbb{X}} V(n+1, f(n, y, u, w))\mathbb{P}_{W(n)}(dw)\right\}$$

$$\hat{\mathbf{u}}(n, y) = \arg\max\{\cdots\}.$$

The sequence of functions $\hat{\mathbf{u}} := (\hat{u}(t, \cdot), \ldots, \hat{u}(T-1, \cdot))$ is the optimal control policy and V is the value function of the optimal control problem.

Remark The only difference with the determinist case is that the way the system is optimally controlled from time $n+1$ on as seen as from time t depends on the random state assumed by the system at time $n+1$. Therefore, controller's choice is driven by the best forecast he can do about this state. This reflects into assessing the expected best performance of the system from time $n+1$ on.

Remark (Determination of sample optimal controls) The numerical value $V(t, x)$ defines the performance for the optimally controlled system starting at (t, x). The optimal control policy determines the optimally controlled stochastic dynamics $(X^{t,x,\hat{\mathbf{u}}}(t))_{t=1,\ldots,T}$, that is a probability distribution for the optimal trajectory. For each sample ω, the corresponding sample optimal control is given by the $(T - t)$-dimensional sequence of individual optimal controls in $\mathbb{R}^m$:

$$\hat{\mathbf{u}} = \left(\hat{u}\left(t, X^{t,x,\hat{\mathbf{u}}}(1, \omega)\right), \ldots, \hat{u}\left(T - 1, X^{t,x,\hat{\mathbf{u}}}(T - 1, \omega)\right)\right).$$

In general, the optimal control policy gives rise to different sequences of control vectors, one sequence per realization ω in the sample space.

3.6 Applications

3.6.1 American Option Pricing

We consider an American put option written at time 0 on an asset S for a strike price K and expiration date T. If exercised at time $\tau \in [0, T]$, the option pay-off is

$$\psi(S_\tau) = \max(K - S_\tau, 0).$$

Arbitrage pricing theory allows us to compute the fair price for the contract under consideration as:

$$P_0 = \sup_{\tau \in \mathcal{T}_{0,T}} \mathbb{E}\left[e^{-r\tau}\psi(S_\tau)\right],$$

where $\mathbb{E}[\cdot]$ denotes the the risk-neutral expectation operator and $\mathcal{T}_{0,T}$ is the set of stopping times assuming values on $[0, T]$.[4]

We assume throughout that a discrete time pricing model is given. This model may be derived by discretizing continuous time asset price dynamics on a time refinement $\{0, \Delta t, \ldots, N\Delta t = T\}$. The choice of the optimal exercise time τ^* reduces to a comparison between the *intrinsic value* of the option and its *continuation value* at each time step j. The former is the pay-off $\psi(S_j)$ stemming from exercising the option, whereas the latter is the value of the option provided that exercise is postponed; that is the conditional expected and discounted value of the option $\mathbb{E}[e^{-r\Delta t}P_{j+1}|S_j]$ one time step later. The optimal exercise time is the first date and which the intrinsic value exceeds the continuation value.

This is clearly a dynamic programming procedure:

- (Begin)
$$P_N = \psi(S_N).$$

- (Step $N - j$)
$$P_j = \max\left(\psi(S_j), \mathbb{E}\left[e^{-r\Delta t}P_{j+1}|S_j\right]\right).$$

The optimal stopping time τ^* is thus:

$$\tau^* = \min\left(k \geq 0\colon \ \psi(S_k) \geq \mathbb{E}\left[e^{-r\Delta t}P_{k+1}|S_k\right]\right).$$

---

[4] A more precise formulation would consider the *essential* supremum over $\mathcal{T}_{0,T}$. For the sake of simplicity, we skip on this subtle point in our presentation and refer to works cited at the end of the chapter.

In a binomial model setting, prices can move from one period to the next by either a proportion $u > 1$ or $d < 1$ with probability p and $1 - p$ respectively. Three parameters need to be identified: (1) the up jump factor u, (2) the downward jump factor d, (3) the probability p of an upward movement.

Standard approaches exist for the purpose of selecting u, d, and p in a way that is compatible to a time discretization of lognormal dynamics underlying the Black–Scholes model. Cox, Ross and Rubinstein (1979) propose the following choice:

$$u = \exp\left(\left(r - \frac{1}{2}\sigma^2\right)\Delta t + \sigma\sqrt{\Delta t}\right),$$

$$d = \exp\left(\left(r - \frac{1}{2}\sigma^2\right)\Delta t - \sigma\sqrt{\Delta t}\right),$$

$$p = \frac{1}{2},$$

where r is the 1-period risk-free rate of interest, σ is the Black–Scholes volatility, and Δt is the time lag. Jarrow and Rudd (1982) consider equally large absolute movements. This choice leads to unequal probabilities for the upward and downward movements:

$$u = \exp\left(\sigma\sqrt{\Delta t}\right),$$

$$d = \exp\left(-\sigma\sqrt{\Delta t}\right),$$

$$p = \frac{1}{2} + \frac{r - \sigma^2/2}{2\sigma}\sqrt{\Delta t}.$$

A major problem with these two formulations is that the quality of the resulting approximation is acceptable only for a very thin time refinement. In order to overcome this issue, Trigeorgis (1991) proposes the following discretization. Let x denote the logarithmic price return of the risky asset S. We assume this quantity can increase to $x + \Delta x_u$ with a probability p_u or decrease to $x - \Delta x_d$ with a probability $p_d = 1 - p_u$. By matching both mean and variance of the continuous and the discrete time models, we obtain:

$$\mathbb{E}[\Delta x] = p_u \Delta x_u + p_d \Delta x_d = \left(r - \frac{1}{2}\sigma^2\right)\Delta t,$$

$$\mathbb{E}[\Delta x^2] = p_u \Delta x_u^2 + p_d \Delta x_d^2 = \sigma^2 \Delta t + \left(r - \frac{1}{2}\sigma^2\right)^2 \Delta t^2,$$

$$p_u + p_d = 1.$$

Equal jump sizes lead to the following solution:

$$\Delta x = \sqrt{\sigma^2 \Delta t + \left(r - \frac{1}{2}\sigma^2\right)^2 \Delta t^2},$$

$$p_u = \frac{1}{2} + \frac{(r - \sigma^2/2)\Delta t}{2\Delta x},$$

$$p_d = \frac{1}{2} - \frac{(r - \sigma^2/2)\Delta t}{2\Delta x}.$$

The dynamic programming algorithm can be implemented by computing a tree of values for both asset and option prices. Each node is identified by a double index (j, l), where $j = 0, \ldots, N$ represents the time step and l is the level of the state variable at that time. The asset price at node (j, l) is

$$S_{j,l} = \exp(x_{j,l}) = \exp\big(x + l\Delta x_u + (j - l)\Delta x_d\big).$$

The DP problem reads as:

$$P_{j,N} = \max(K - S_{N,l}, 0),$$
$$P_{j,l} = \max\big(\max(K - S_{j,l}, 0), e^{-r\Delta t}\big(p_u P_{j+1,l+1} + (1 - p_u)P_{j+1,l}\big)\big).$$

This algorithm is implemented in the code `BinomialCallPutPrice.m`.

3.6.2 Optimal Investment Problem

Consider a market where $N+1$ assets prices evolve according to stochastic processes $S_0(t), \ldots, S_N(t)$ in discrete timeset $t = 0, \ldots, T$. An investor is faced with the problem of allocating an initial wealth w_0 and adjusting the resulting portfolio at each point in time, $n-r$ way that utility stemming from the value of the standing portfolio at time T is maximized. We assume the investor's utility is a differentiable, concave and strictly increasing function U of the investor's.... . Let $h_i(t)$ denote the quantity of asset i ($i = 0, \ldots, N$) that is selected at time t and held until following time. An admissible strategy $\mathbf{H} = (\mathbf{h}(t))_{t=0,\ldots,T-1}$, with $\mathbf{h}(t) = (h_0(t), h_1(t), \ldots, h_N(t))$, is self-financing if the standing wealth at each time t is totally reallocated among the same set of $N + 1$ assets available for the trading market. In other words, wealth is neither drawn out of the market, say for consumption, nor increased by any cash inflow. Consequently, given an outstanding capital w, if N asset quantities are freely decided upon, say $h_1(t), \ldots, h_N(t)$, then the residual quantity $h_0(t)$ is constrained to either finance the resulting deficit (case: $\sum_{i=1}^{N} h_i(t)S_i(t) - w > 0$) or to support the standing surplus (case: $w - \sum_{i=1}^{N} h_i(t)S_i(t) > 0$), i.e.,

$$h_0(t) = \frac{w - \sum_{i=1}^{N} h_i(t)S_i(t)}{S_0(t)}. \tag{3.5}$$

This constraint reduces by one unit the degree of freedom in selecting the optimal allocation of wealth. We denote the class of self-financing trading strategies by $\mathbb{H}$. In view of expression (3.5), the time $t + 1$ wealth $W(t)$ generated by a trading strategy $\mathbf{H} = (\mathbf{h}(t))_{t=0,\ldots,T-1}$ investing an initial wealth $W(0) = w$ is given by:

$$
W^{\mathbf{h}}(t+1) = w_0 + \sum_{u=0}^{t}\sum_{i=0}^{N} h_i(u)[S_i(u+1) - S_i(u)]
$$

$$
= w_0 + \sum_{u=0}^{t}\left\{ \frac{W^{\mathbf{h}}(u) - \sum_{i=1}^{N} h_i(u)S_i(u)}{S_0(u)}[S_0(u+1) - S_0(u)]. \right.
$$

$$
\left. + \sum_{i=1}^{N} h_i(u)[S_i(u+1) - S_i(u)] \right\}
$$

$$
= W^{\mathbf{h}}(t) + \frac{W^{\mathbf{h}}(t) - \sum_{i=1}^{N} h_i(t)S_i(t)}{S_0(t)}[S_0(t+1) - S_0(t)]
$$

$$
+ \sum_{i=1}^{N} h_i(t)[S_i(t+1) - S_i(t)]. \tag{3.6}
$$

This expression says that investing $W^{\mathbf{h}}(t)$ at time t in a self-financing portfolio yields an amount at time $t+1$ that consists of two components: (1) the invested capital $W^{\mathbf{h}}(t)$; (2) the capital gains occurred between t and $t+1$, i.e., value variations in the portfolio exclusively due to price movements. By inspecting the same formula, we remark that the outstanding wealth at any time $t+1$ depends on the following items:

- The outstanding wealth $W^{\mathbf{h}}(t)$ at time t: . This is a state variable of the system.
- The asset prices prevailing at the both investment and reallocation times. These constitute N further state variables of the system.
- The allocation quantities $h_1(t), \ldots, h_N(t)$. These are control variables of the system.

The optimization problem can be cast as follows:

$$
\max_{\mathbf{h}\in\mathbb{H}} \mathbb{E}\big(U\big(W^{\mathbf{h}}(T)\big)\big|W^{\mathbf{h}}(0) = w^0, \mathbf{S}(0) = \mathbf{s}^0\big),
$$

where w^0 is the initial wealth, $\mathbf{S}(0) = (S_0(0), \ldots, S_N(0))$, and $\mathbf{s}^0 = (s_0^0, \ldots, s_N^0)$ is the vector of market prices of the investment assets at time 0.

This problem can be solve by using dynamic programming. The algorithm reads as follows:

- (Begin)

$$
V\big(T, (w, \mathbf{s})\big) = U(w).
$$

- (Step $T - t$)

$$
V\big(t, (w, \mathbf{s})\big) = \max_{\mathbf{h}\in\mathbb{R}^N}\big\{\mathbb{E}\big[V\big(t+1, \mathbf{S}(t+1)\big)|W(t) = w, \mathbf{S}(t) = \mathbf{s}\big]\big\}
$$

$$
= \max_{\mathbf{h}\in\mathbb{R}^N}\left\{ \mathbb{E}\left[w + \frac{w - \sum_{i=1}^{N} h_i s_i}{s_0}[S_0(t+1) - s_0(t)] \right.\right.
$$

$$
\left.\left. + \sum_{i=1}^{N} h_i[S_i(t+1) - s_i] \right\}, \right.
$$

$$\mathbf{h}^{*}\big(t, (w, \mathbf{s})\big) = \arg \max_{\mathbf{h} \in \mathbb{R}^{N}} \{\cdots\}$$

where $\mathbf{h} = (h_1, \ldots, h_N)$, the expected value is computed with respect to the conditional distribution of $\mathbf{S}(t+1) = (S_0(t+1), \ldots, S_N(t+1))$ given that $\mathbf{S}(t) = (s_0, \ldots, s_N)$.

The value function at time 0 is given by:

$$V\big(0, (w_0, \mathbf{s}_0)\big),$$

and the optimal control reads as:

$$\mathbf{h}^{*}\big(t, \big(W^{\mathbf{h}^{*}}(t), \mathbf{S}(t)\big)\big), \quad t = 0, \ldots, T-1.$$

By formula (3.6), the wealth $W^{\mathbf{h}^{*}}(t)$ depends on $\mathbf{h}^{*}(u)$ only for $u = 0, \ldots, t-1$; consequently the expression above for $\mathbf{h}^{*}$ is well defined for each sample path $\mathbf{S}(t), t = 0, \ldots, T-1$.

3.7 Comments

The most comprehensive treatment of dynamic programming available to date is Bertsekas (2005). Other comprehensive treatments of both deterministic and stochastic optimal control theory include Kushner (1967), Fleming and Rishel (1975), Gihman and Skorohod (1979), Krylov (1980), Oksendal (2003), and Oksendal and Sulem (2004). We refer to the wide bibliography cited in these books for a complete list of references. A concise introductory to stochastic control in continuous time is contained in Björk (2004). Shiryaev (1978) provides a complete account of the optimal stopping problem. Kushner and Dupuis (1992) describe a wide variety of numerical methods for control problems. Demange and Rochet (1997) provide an instructive introductory chapter on optimal control and its applications to the economic theory. More refined methods in economics can be found in Grüne and Semmler (2004) and references therein. The use of trees has been extensively adopted in quantitative finance since Cox, Ross and Rubinstein (1979). Self-contained introductions to lattice methodologies can be found in Baxter and Rennie (1996), Musiela and Rutkowski (1997), Björk (2004), Briys et al. (1998), James and Webber (2000), Hull (2005) and, to a much wider extent, Brigo and Mercurio (2006). Applications of tree-based methods to dynamic hedging and the pricing of exotic options can be found in Avellaneda and Paras (1994) and Avellaneda and Wu (1999). Li, Ritchken and Sankarasubramanian (1995) develop a clever device to overcome the problem of non-Markovianity in HJM models for the term structure of interest rates. Amin (1993), Baz and Das (1996), Das (1997a,1997b), and Këllezi and Webber (2004) develop random walks approximating continuous time jump diffusions. Jamshidian (1991a,1991b) proposes a discrete version of the Kolmogorov forward equation satisfied by the pricing kernel of certain diffusion processes and applied the method to price interest rate derivatives. Forward induction has also been explored for the pricing of American-style options in Carr and Hirsa (2003). Baccara, Battauz and Ortu (2005) adopt an event-tree approach to model a security market where bid-ask spreads affect daily quotations and then solve the resulting super-replication problem via linear programming techniques. Dumas and Luciano (1991) solve a dynamic portfolio choice problem in continuous time with transaction costs, while Roncoroni (1995) apply the optimal control scheme to model and solve an economic policy problem affecting developing countries.

4

Finite Difference Methods

4.1 Introduction

4.1.1 Security Pricing and Partial Differential Equations

We have seen in a previous chapter that the arbitrage-free price of a European-style contingent claim can be expressed as the time t conditional expected value of its discounted payoff under the risk-neutral probability measure $\mathbb{P}^*$:

$$V(t) = \mathbb{E}_t^* \left(e^{-\int_t^T r(s)\, ds} V(T) \right). \tag{4.1}$$

Here $r(t)$ represents the risk-free short rate of interest prevailing at time t for the period between t and $t + dt$. If both this rate and the pay-off $V(T)$ can be expressed as a function of a k-dimensional Itô process $\mathbf{X} = (X_1, \ldots, X_k)$ satisfying a vector-valued stochastic differential equation (SDE):

$$d\mathbf{X}(s) = \mu\big(s, \mathbf{X}(s)\big)\, ds + \Sigma\big(s, \mathbf{X}(s)\big)\, d\mathbf{W}(s),$$

with $\mathbf{X}(t) = \mathbf{x}$, then the derivative price $V(t)$ can be written as a function F of time t and state $\mathbf{X}(t)$.[1] According to the Feynman–Kaç theorem, the function $F(t, \mathbf{x})$ satisfies the following PDE:[2]

$$0 = \partial_t F(t, \mathbf{x}) + \nabla_x F(t, \mathbf{x}) \cdot \mu(t, \mathbf{x}) \tag{4.2}$$
$$+ \frac{1}{2} \mathrm{Tr}\big[\mathrm{He}[F(t, \mathbf{x})] \Sigma(t, \mathbf{x}) \Sigma(t, \mathbf{x})^\top \big] - r(t) F(t, \mathbf{x})$$

---

[1] More precisely, the right-hand side in expression (4.1) is a conditional expectation of the argument with respect to the σ-algebra $\mathcal{F}_t = \sigma(\mathbf{X}_s, 0 \leq s \leq t)$ generated by the process $\mathbf{X}$ as observed until time t. Since $\mathbf{X}$ is a Markov process, this quantity is $\sigma(\mathbf{X}_t)$-measurable and thus it can be expressed as a function of t and $\mathbf{X}_t$.

[2] Here $\partial_t F$ denotes the partial derivative of F with respect to t, $\nabla_x F$ is the gradient vector collecting all partial derivatives of F with respect to the components of the state variable $\mathbf{x}$, He[F] is the Hessian matrix of F, and Tr[·] is the trace operator of a square matrix, i.e., the sum of all diagonal elements.

on a domain $\mathcal{D}$, with terminal condition $F(T, \mathbf{x}) = h(\mathbf{x}) = V(T)$. This result holds under regularity conditions which basically amount to requiring that $F(t, \mathbf{x})$ belongs to the class of continuously differentiable functions $\mathcal{C}^{1,2}(\mathbb{R}^+ \times \mathbb{R})$.

Let us consider the case of a single asset ($k = 1$) and assume that the spot price evolves according to a Geometric Brownian Motion (GBM) under the risk-neutral probability measure, i.e., $dX(t) = (r - q)X(t)\,dt + \sigma X(t)\,dW(t)$, with $X(0) = x$ as a starting condition, r denoting the risk-free rate, and q representing a continuous "dividend yield". In this setting, the PDE (4.2) is exactly the Black–Scholes (BS) equation

$$\partial_t F(t, x) + (r - q)x\partial_x F(t, x) + \frac{1}{2}\sigma^2 x^2 \partial_{xx} F(t, x) - rF(t, x) = 0. \qquad (4.3)$$

This equation, coupled with the payoff condition $F(T, x) = \max(0, x - K)$, returns a solution that is the price of a standard call option written on X, with maturity T and strike K. If the asset pays a continuous cash flow $g(t, x)$, then this latter appears as an additional term in the PDE above. For example, if the cash flow is proportional to a stock price X, $g(t, x) = gx$ say, then a term gx must be added to the right-hand side of formula (4.3).

In this chapter, we present basic numerical methods for solving one-dimensional PDEs of the kind above. In particular, we introduce the Finite Difference Method (FDM) as a simple technique for generating an approximate solution to the pricing PDE (4.3). A considerable number of problems arise in the case of high-dimensional PDEs (e.g., stochastic volatility models, stochastic interest rates models, basket options) and path-dependent payoffs (e.g., Asian options). A few of these issues can be tackled through analytical tools such as integral transforms (e.g., Laplace and Fourier transforms). While we defer a treatment of these and other related issues to Chapter 7, the pricing of path-dependent contracts (such as Asian and lookback options) will be considered through a number of case studies in the second part of this book. Instead, we will not consider how to deal with multidimensional PDEs, arising with stochastic volatility or with basket options.

4.1.2 Classification of PDEs

A PDE is a functional equation containing a function and some of its derivatives. Whereas for Ordinary Differential Equations (ODEs) the unknown function depends on one variable only (e.g., $u'(x) = 2x + u(x)$, where x is the independent variable and u is the unknown function to be determined), in PDEs the unknown function depends on several variables. In financial applications, the relevant variables are time t and the underlying state variable x, which is usually identified with an asset price or any financial index representing the underlying of the contract under investigation. PDE's may be classified according to several criteria:

(a) The order of the PDE, namely the order of the highest partial derivative in the equation.
(b) The number of independent variables.

(c) The kind of relation (linear/nonlinear) combining the unknown function F with its partial derivatives.

For instance, the Black–Scholes PDE is a linear second-order equation involving two variables, that is time and the spot price of an asset. As a more general example, we may consider a second order linear partial differential equation with independent variables x and y:

$$A\partial_{xx}F + 2B\partial_{xy}F + C\partial_{yy}F + D\partial_x F + E\partial_y F + HF = G, \qquad (4.4)$$

where $F = F(y, x)$ is the unknown function to be determined and A, B, C, D, E, H and G are known real-valued functions of both y and x. If $G \equiv 0$ for all y and x, this equation is called *homogeneous*, otherwise *inhomogenous*. It is easy to see that PDE (4.3) can be obtained from (4.4) by setting $y = t$ and

$$A(t, x) = \frac{1}{2}\sigma^2 x^2, \qquad B(t, x) = 0, \qquad C(t, x) = 0,$$
$$D(t, x) = (r - q)x, \qquad E(t, x) = 1, \qquad H(t, x) = -r, \qquad G(t, x) = 0.$$

We may classify equations (4.4) as belonging to one of the following three groups within a region R where $A(y, x) \neq 0$:

$$
\begin{array}{ll}
\text{Hyperbolic,} & \text{if } B^2 - AC > 0, \\
\text{Parabolic,} & \text{if } B^2 - AC = 0, \\
\text{Elliptic,} & \text{if } B^2 - AC < 0.
\end{array}
$$

Since in general $B^2 - AC$ is a function of the independent variables, a PDE can change in nature depending on the part of the domain where it is considered. In the Black–Scholes equation, we have $B^2 - AC = 0$, so that (4.3) is a parabolic equation on $\{x > 0\}$. In quantitative finance, we usually come across parabolic equations:

$$\partial_y F(y, x) + A\partial_{xx}F(y, x) + D\partial_x F(y, x) + HF(y, x) = G,$$

where y represents the time variable and x represents a spatial variable. In terms of physical interpretation, $A\partial_{xx}F$ is a *diffusive* term, $D\partial_x F$ represents a *convective* term, HF describes a *conservative* term, and G is a *source* term. Let us comment on the meaning of these terms with reference to the BS equation (4.3) as an example.

- The term $\partial_t F(t, x)$ measures the effect of time flow on the option price: *ceteris paribus*, the option loses time value as time goes by. This is because the likelihood that the option ends up in-the-money decreases along time.
- The diffusive term $\sigma^2 x^2 \partial_{xx} F$ records the effect of noise on the stock price and consequently on the derivative price written on this asset.
- The convective term $rx\partial_x F$ takes into account the deterministic effect of time flow on the stock price. This is given by opportunity cost of investing in the risk-free asset and thus is reflected in an option price variation $rx\partial_x F$.
- The source term represents a continuous cash flow to be received by the holder of the derivative contract.

By Itô's lemma, the sum $\partial_t F(t, x) + \sigma^2 x^2 \partial_{xx} F + rx \partial_x F$ describes the expected capital gain on the derivative asset, i.e., $\mathbb{E}_t^*(dF(t, x))/dt$. By adding expected cash flow generated by the asset, i.e., a source term, we obtain the total return on the option contract. The sum of the expected capital gain and the expected cash flow ECF provides the expected total return of the option. In a risk-neutral world, all the assets have the same instantaneous expected total return. Consequently $\mathbb{E}_t^*(dF(t, x))/dt + ECF = rF$, where rF is the return on a risk-free investment. This is the famous Black–Scholes pricing equation.

A complete description of a financial valuation problem using PDEs requires:

1. A PDE describing the no-arbitrage relationship between the derivative price and the price of the underlying assets.
2. Boundary conditions describing the financial characteristics of the contract at boundary values of the underlying price process.
3. Terminal conditions, describing the payoff at maturity of the contract.

When the values of the state variable x are restricted to a bounded (or semi-infinite) interval, we require that the solution satisfies assigned boundary conditions which we assume linear in that they may involve either $F(t, x)$ (Dirichlet boundary condition), or its space derivative $\partial_x F(t, x)$ (Neumann boundary condition), or both (Robin boundary condition).

For example, if we have a left-hand boundary at $x = a$, the boundary condition at this point can assume the form

$$\gamma_0(t) F(t, a) - \gamma_1(t) \partial_x F(t, a) = \gamma_2(t),$$

where $\gamma_0(t) \geq 0$, $\gamma_1(t) \geq 0$ and $\gamma_0(t) + \gamma_1(t) > 0$ for all $t > 0$. If: (a) $\gamma_0(t) \neq 0$ and $\gamma_1(t) = 0$, we have Dirichlet boundary conditions; (b) $\gamma_0(t) = 0$ and $\gamma_1(t) \neq 0$ we have Neumann boundary conditions, (c) otherwise, we have Robin condition. At the right-hand boundary $x = b$, the boundary condition can be

$$\lambda_0(t) F(t, b) + \lambda_1(t) \partial_x F(t, b) = \lambda_2(t),$$

where $\lambda_0(t) \geq 0$, $\lambda_1(t) \geq 0$ and $\lambda_0(t) + \lambda_1(t) > 0$ for all t. When $\gamma_2(t)$ and $\lambda_2(t)$ are nonzero, the boundary conditions are of inhomogeneous type.

A PDE coupled with the prescription of its solution and/or its derivatives on the initial (resp. final) line $t = 0$ ($t = T$) and on the boundary lines defines an *initial* (resp. final) *and boundary value problem.*

We now examine the importance of the terminal and boundary conditions through some financial examples. Distinct terminal conditions lead to prices of different contracts. For example, the final condition $F(T, X(T)) = \max(0, K - X(T))$ returns a put option price. It is also important to specify the domain for x depending on the type of option one is evaluating.

Example (Vanilla call option) For a standard call option, the asset price can assume values in the range $[0, +\infty)$. From a computational viewpoint, this domain requires a truncation from above, $[0, M]$ say, and an indication about the solution behavior for both small and large value of x, i.e., for $x \to 0$ and $x \to M$. For a GBM process, we observe that if $x = 0$, then the process will be absorbed, i.e., $X(t) = 0$ for all $t > 0$. Consequently, there is no chance that the option matures in-the-money, i.e., $\mathbb{P}(X(T) > K \,|\, X(0) = 0) = 0$. The boundary condition at $x = 0$ is thus $F_{\text{call}}(t, x = 0) = 0, t \geq 0$. For $x = M$ (large), it is likely that the option will expire in-the-money and will then be exercised. Therefore, its current value can be approximated by $F_{\text{call}}(t, x = M) = M - K e^{-r(T-t)}$. Alternatively, we may assign the Neumann boundary condition $\partial_x F_{\text{call}}(t, x = M) = 1$. This reflects the observation that for large values of x, the option price tends to resemble the underlying stock price.

Example (Vanilla put option) For a standard put option, the infinite domain $[0, +\infty)$ needs be limited to a finite interval $[0, M]$. If $x = 0$, the put option is exercised with certainty since $X(T) = 0 < K$. The option value at $x = 0$ is then $F_{\text{put}}(t, x = 0) = K e^{-r(T-t)}$. Conversely, for large values of x, it is very unlikely that the put option is exercised. This suggests setting the upper boundary condition $F_{\text{put}}(t, x = M) = 0$. A Neumann boundary condition for large x might be $\partial_x F_{\text{put}}(t, x = M) = 0$.

Example (Barrier option) Barrier options are activated (knock-in feature) or terminated (knock-out feature) whenever a specific threshold is hit before the expiry date. In particular, a *down-and-out* call option is terminated if the asset price hits a low barrier $l < X(0)$ before time T. Then the option becomes worthless and the computational domain is $[l, M]$, for a large value M. Boundary conditions at $x = l, M$ become $F_{\text{doc}}(t, x = l) = 0$ and $F_{\text{doc}}(t, x = M) = M - K e^{-r(T-t)}$, respectively. A *down-and-in* call option is activated whenever the asset price hits a low barrier $l < X(0)$ before time T. If in this case, the holder receives a standard call option with a residual life equal to $T - t$. The computational domain is again $[l, M]$, however boundary conditions read as $F_{\text{dic}}(t, x = l) = F_{\text{call}}(t, l)$, and $F_{\text{dic}}(t, x = M) = M - K e^{-r(T-t)}$. A *knock-out double barrier* call is terminated whenever the asset price hits a low barrier $l < X(0)$ or a high barrier $u > X(0)$ before time T. The computational domain is $[l, u]$ and boundary conditions at $x = l$ and $x = u$ are $F_{\text{kodbcall}}(t, x = l) = 0$ and $F_{\text{kodbcall}}(t, x = u) = 0$, respectively.

Example (Discrete monitoring) In the previous example, threshold hitting has been checked with continuity over time. Discrete monitoring refers to the case where the triggering event is checked at a fixed number of times. In this setting, a knock-out (resp. knock-in) option becomes less (resp. more) expensive as long as the number of monitoring dates increases. No monitoring at all makes a knock-out option equal to a plain vanilla option. As for boundary conditions, we observe that out of monitoring dates, the asset price can freely span the range $[0, +\infty)$. For computational purposes, this domain must be truncated into $[0, M]$. For a double barrier option, we can then set boundary conditions:

$$F(t_i, x) = F(t_i^-, x) \mathbf{1}_{\{x \in [l, u]\}}, \tag{4.5}$$

where $\mathbf{1}_{\{x \in [l, u]\}}$ is the indicator function for the interval $[l, u]$ as defined by

$$\mathbf{1}_{\{x \in [l, u]\}} = \begin{cases} 1 & \text{if } x \in [l, u], \\ 0 & \text{if } x \notin [l, u]. \end{cases}$$

This expression checks whether either of the two barriers has been hit. If this latter condition occurs, the value of the option is set equal to zero. Otherwise, it is set equal to its value just before the monitoring date under consideration.

4.2 From Black–Scholes to the Heat Equation

Under the assumption of a GBM process, the Black–Scholes equation (4.3) can be simplified by simple variable transformations. First, we change the origin of time. Next, we work with undiscounted option prices. Then, we transform prices into price returns. Last, we reduce the BS equation to the much simpler heat equation. The first two steps are general and can be applied independently of the stock price dynamics. The remaining two steps are possible only under the GBM assumption. However, we will see how to cope with more general processes in Sects. 4.2.5 and 4.5.

4.2.1 Changing the Time Origin

In financial application, PDEs are usually characterized by a terminal condition expressing the security payoff. In physics, it is more natural to think in terms of initial conditions and then model the way information is gradually lost as time goes by. We instead use a terminal condition because dynamics represent the gradual updating of price information as the contract maturity approaches. In order to use numerical methods for PDEs, it is convenient to change the origin of time. In financial terms, this corresponds to deal with the time *to* maturity of the option instead of its time *of* expiry, or calendar time. This change of time simply requires introducing a variable τ defined as:

$$\tau = T - t,$$

and defining a function

$$F(t, x) = f(\tau, x), \quad \text{with } \tau = T - t.$$

We observe that

$$\partial_t F(t, x) = -\partial_\tau f(\tau, x),$$

and so f solves the PDE:

$$-\partial_\tau f(\tau, x) + rx\partial_x f(\tau, x) + \frac{1}{2}\sigma^2 x^2 \partial_{xx} f(\tau, x) - rf(\tau, x) = 0. \tag{4.6}$$

The terminal condition affecting F is then transformed into an initial condition involving f, namely $f(0, x) = F(T, x)$. Boundary conditions are unaffected by this simple transformation, except for replacing $T - t$ by τ in all formulas. For instance, a call option is characterized by conditions $f(\tau, 0) = 0$ and $f(\tau, M) = M - Ke^{-r\tau}$.

4.2.2 Undiscounted Prices

A second transformation removes the term rf from the BS equation. We just need to define $g(\tau, x) = e^{r\tau} f(\tau, x)$ and then obtain:

$$\partial_\tau g(\tau, x) = r e^{r\tau} f(\tau, x) + e^{r\tau} \partial_\tau f(\tau, x),$$
$$\partial_x g(\tau, x) = e^{r\tau} \partial_x f(\tau, x),$$
$$\partial_{xx} g(\tau, x) = e^{r\tau} \partial_{xx} f(\tau, x).$$

By replacing partial derivatives in (4.6), we obtain:

$$-\partial_\tau g(\tau, x) + r x \partial_x g(\tau, x) + \frac{1}{2}\sigma^2 x^2 \partial_{xx} g(\tau, x) = 0. \tag{4.7}$$

From a financial point of view, this transformation is equivalent to considering forward prices instead of spot prices.

4.2.3 From Prices to Returns

A third transformation allows us to obtain a PDE with constant coefficients. The method is strictly connected to the GBM assumption and, in general, cannot be extended to other processes.

Indeed, if $X(t)$ evolves according to a Geometric Brownian motion process $dX(t) = X(t)(r\,dt + \sigma\,dW)$, then, applying the Itô's lemma, $\ln X(t)$ has dynamics given by:

$$d \ln X(t) = \left(r - \frac{\sigma^2}{2}\right) dt + \sigma\,dW, \tag{4.8}$$

i.e. an SDE with constant coefficients.

If we define the new function $g(\tau, x) = G(\tau, z = \ln x)$, we have:

$$\partial_x g(\tau, x) = \frac{1}{x} \partial_z G(\tau, z),$$

$$\partial_{xx} g(\tau, x) = -\frac{1}{x^2} \partial_z G(\tau, z) + \frac{1}{x^2} \partial_{zz} G(\tau, z).$$

By inserting these derivatives into (4.7) after a few algebraic manipulations, we obtain a PDE with constant coefficients:

$$-\partial_\tau G(\tau, z) + \left(r - \frac{\sigma^2}{2}\right) \partial_z G(\tau, z) + \frac{1}{2}\sigma^2 \partial_{zz} G(\tau, z) = 0. \tag{4.9}$$

Notice that spot returns in this PDE replace spot prices in the previous equation. The coefficients in front of partial derivatives are $r - \sigma^2/2$ (convective term) and $\sigma^2/2$ (diffusive term).

By considering log-returns, the price domain $[0, +\infty)$ is transformed into the return domain $(-\infty, \infty)$. Consequently, we need to truncate the domain on both of its sides for the purpose of performing numerical computations. The advantage of working with returns is that the transformed PDE displays constant coefficients. As we shall see below, this is a good property for a better understanding of the theoretical properties of the numerical schemes to be used for solving (4.9).

4.2.4 Heat Equation

A further variable transformation allows us to obtain the standard heat equation. By setting $G(\tau, z) = e^{\alpha z + \beta \tau} u(\tau, z)$ we have:

$$\partial_z G(\tau, z) = \alpha e^{\alpha z + \beta \tau} u(\tau, z) + e^{\alpha z + \beta \tau} \partial_z u(\tau, z),$$
$$\partial_{zz} G(\tau, z) = \alpha^2 e^{\alpha z + \beta \tau} u(\tau, z) + 2\alpha e^{\alpha z + \beta \tau} \partial_z u(\tau, z) + e^{\alpha z + \beta \tau} \partial_{zz} u(\tau, z),$$
$$\partial_\tau G(\tau, z) = \beta e^{\alpha z + \beta \tau} u(\tau, z) + e^{\alpha z + \beta \tau} \partial_\tau u(\tau, z).$$

If we replace the partial derivatives in (4.9) and choose α and β in order to eliminate the terms $\partial_z u(\tau, z)$ and $u(\tau, z)$, where

$$\alpha = -\frac{1}{2}\left(r - \frac{\sigma^2}{2}\right), \qquad \beta = m\alpha + \frac{1}{2}\sigma^2\alpha^2, \quad m = r - \frac{\sigma^2}{2},$$

we come up with:

$$\partial_\tau u(\tau, z) + \frac{\sigma^2}{2}\partial_{zz}u(\tau, z) = 0. \tag{4.10}$$

The problem (4.10) is a problem of first order in time. The determination of the solution requires to fix an initial condition in time. On the boundary, various boundary conditions can be taken into account for determining completely the solution (Dirichlet, Neumann or Robin boundary conditions). Notice that the general form for a heat equation is $-\partial_\tau u(\tau, z) + c\partial_{zz}u(\tau, z) = 0$. It is said to be dimensionless whenever $c = 1$. Indeed, by recalling the time scaling property of the Brownian motion $W(\tau)$, i.e., $W(c\tau) = \sqrt{c}W(\tau)$, we can use the additional transformation $s = \sqrt{c}\tau$ and get to this form.

4.2.5 Extending Transformations to Other Processes

Let us examine how to extend the above transformation to processes exhibiting a diffusion coefficient more general than the one associated with the GBM process, i.e., $\sigma(x) \neq \sigma x$. Popular specifications in financial literature are: (1) $\sigma\sqrt{x}$ (square-root process), (2) σ (Gaussian process), (3) σx^α (constant elasticity of variance process, CEV).

The Black–Scholes PDE associated to a general volatility function $\sigma(x)$ is given by:

$$-\partial_\tau g(\tau, x) + rx\partial_x g(\tau, x) + \frac{1}{2}\sigma^2(x)\partial_{xx}g(\tau, x) = rg(\tau, x). \tag{4.11}$$

In general, the logarithmic transformation $z = \ln x$ does not work for the purpose of transforming the original PDE into a heat equation. However, we may define $z = f(x)$ and apply Itô's lemma:

$$dz(t) = \left(rxf'(x) + \frac{1}{2}\sigma^2(x)f''(x)\right)dt + \sigma(x)f'(x)\,dW(t).$$

If we set $f(x)$ in such a way that the new process shows a constant diffusion coefficient:

$$\sigma(x)f'(x) = c,$$

with $c \neq 0$, then

$$f(x) = \int_0^x \frac{c}{\sigma(u)} du, \qquad f'(x) = \frac{c}{\sigma(x)},$$

and $f''(x) = -c\sigma'(x)/\sigma^2(x).$[3] Consequently,

$$dz(t) = c\left(\frac{rx}{\sigma(x)} - \frac{1}{2}\sigma'(x)\right) dt + c\, dW(t).$$

By setting $g(\tau, x) = G(\tau, z = f(x))$, then $G(\tau, z)$ satisfies the PDE:

$$-\partial_\tau G(\tau, z) + c\left(\frac{rx}{\sigma(x)} - \frac{1}{2}\sigma'(x)\right)\Bigg|_{x=f^{-1}(z)} \partial_z G(\tau, z)$$

$$+ \frac{1}{2}c^2\partial_{zz}G(\tau, z) = 0, \tag{4.12}$$

where $f^{-1}(z)$ is the inverse function of $f(x)$. That is, a PDE with a constant diffusion coefficient and a variable convective term.

Unfortunately, the transformed PDE (4.12) does not correspond to the heat equation, as we showed to be possible in the GBM case. So in general, it is preferable to work directly with the numerical solution of (4.11) rather than with the transformed version (4.12). The numerical approximation is discussed in Sect. 4.5.

4.3 Discretization Setting

We now describe a numerical procedure allowing us to compute a numerical solution of the PDE:

$$-\partial_\tau u(\tau, z) + \frac{\sigma^2}{2}\partial_{zz}u(\tau, z) = 0, \tag{4.13}$$

with initial condition:

$$u(0, z) = \varphi(z),$$

and boundary conditions:

$$u(\tau, z_L) = \psi_L(\tau),$$
$$u(\tau, z_U) = \psi_U(\tau),$$

at points z_L and z_U, which may be infinite. We proceed with the following steps:

1. Approximating both spatial and time derivatives using finite differences;

---

[3] If $\sigma(x) = \sigma x$, then $f(x) = \ln x$ and we obtain the logarithmic transformation.

2. Creating time–space grid of points over which the approximate solution is computed;
3. Replacing partial derivatives appearing in the PDE by their finite differences computed at grid points;
4. Deriving a recursive procedure to compute the approximate solution;
5. Establishing the main properties of the resulting function.

4.3.1 Finite-Difference Approximations

When a function $f(z)$ is continuously differentiable an appropriate number of times, then, by Taylor's formula, we have:

$$f(z+h) = f(z) + hf'(z) + \frac{1}{2}h^2 f''(z) + \frac{1}{6}h^3 f'''(z) + O(h^4) \qquad (4.14)$$

and:

$$f(z-h) = f(z) - hf'(z) + \frac{1}{2}h^2 f''(z) - \frac{1}{6}h^3 f'''(z) + O(h^4), \qquad (4.15)$$

where $O(h^k)$ denotes terms that, when $h \to 0$, tend to zero as fast as or faster than the power function h^k. If we subtract (4.15) from (4.14), we have:

$$f(z+h) - f(z-h) = 2hf'(z) + \frac{1}{3}h^3 f'''(z) + O(h^4),$$

and therefore:

$$f'(z) = \frac{f(z+h) - f(z-h)}{2h} + O(h^2).$$

This suggests that we approximate the first derivative computed at z by

$$f'(z) \simeq \frac{f(z+h) - f(z-h)}{2h}. \qquad (4.16)$$

This approximation generates a leading error of order h^2. It is referred to as the "central difference approximation", because it is centered at z and it involves the symmetrical quantities $f(z+h)$ and $f(z-h)$. Geometrically, the derivative is approximated by the slope of the chord connecting points A and C in Fig. 4.1.

Using (4.14) and considering the Taylor polynomial up to the first term, we can approximate the derivative by the slope of the chord connecting points B and C in Fig. 4.1:

$$f'(z) \simeq \frac{f(z+h) - f(z)}{h}. \qquad (4.17)$$

This approximation, involving the points $z+h$ and z, has a leading error of order h and is called a "forward difference".

Similarly, we can alternatively approximate the derivative by using (4.15) through the slope of the chord connecting points A and B:

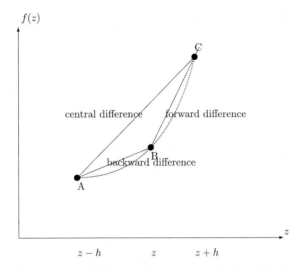

Fig. 4.1. Finite difference approximations to the first derivative.

$$f'(z) \simeq \frac{f(z) - f(z - h)}{h}. \tag{4.18}$$

This gives rise to a leading error of order h. The resulting approximation involves the points z and $z - h$ and is accordingly called "backward difference". Note that central difference is the most accurate among the three approximations. However, it requires an evaluation of the target function at two different points, namely $z - h$, and $z + h$, whereas the two alternatives require only one additional evaluation, i.e. at $z + h$ or at $z - h$. This simplicity comes at the price of a lower accuracy of these methods. Table 4.1 synthetizes the different approximations to the first derivative $f'(z)$.

Approximating the second-order derivative can be accomplished by summing (4.14) and (4.15):

$$f(z + h) + f(z - h) = 2f(z) + h^2 f''(z) + O(h^4)$$

and then solving by $f''(z)$:

$$f''(z) \simeq \frac{f(z + h) - 2f(z) + f(z - h)}{h^2}. \tag{4.19}$$

This formula leads to an error of order h^2.

Partial derivatives $\partial_\tau u(\tau, z)$ and $\partial_{zz} u(\tau, z)$ appearing in the heat equation (4.13) are approximated by expressions similar to those derived above.

The three main finite difference schemes (explicit, implicit, and Crank–Nicolson) are derived as a result of discretizing the time derivative $\partial_\tau u(\tau, z)$ according to the three methods illustrated above. This is summarized in Table 4.2.

Table 4.1. Finite difference approximations of the first partial derivative

Approximation	$f'(z)$	Leading error
Backward	$\frac{f(z)-f(z-h)}{h}$	$O(h)$
Central	$\frac{f(z+h)-f(z-h)}{2h}$	$O(h^2)$
Forward	$\frac{f(z+h)-f(z)}{h}$	$O(h)$

Table 4.2. Approximations of the time derivative and corresponding order of accuracy

Approximation for $\partial_\tau u(\tau, z)$	Method	Order of accuracy
Backward	Implicit	$O(d\tau)$
Central	Crank–Nicolson	$O(d\tau^2)$
Forward	Explicit	$O(d\tau)$

4.3.2 Grid

With reference to the time variable, we set $t = 0$, so that τ can range in the domain $(0, T)$. With reference to the spatial variable, we truncate the working domain at point (z_L, z_U). If the natural domain is $(-\infty, +\infty)$, as is the case for plain vanilla options, or $(-\infty, z_U)$ (or $(z_L, +\infty)$), as it is the case with a single barrier option, then choosing an appropriate working domain is crucial for determining the accuracy of the numerical solution. If we are interested in finding a solution in the interval (z_{-1}, z_1), both z_L and z_U should be selected far from z_{-1} and z_1 a multiple number n of times the standard deviation of the process. For most applications, a common choice consists in setting $n = 6$, so that $z_L = z_{-1} + 6\sigma\sqrt{T}$ and $z_U = z_{-1} + 6\sigma\sqrt{T}$.

For the sake of simplicity, we adopt a uniform grid and partition the z axis in $m-1$ equally spaced intervals of length δz, that is $\delta z = (z_U - z_L)/(m + 1)$. The τ axis is split into $n - 1$ equally spaced intervals of length $\delta\tau$, that is $\delta\tau = T/(n + 1)$. The resulting grid is illustrated in Fig. 4.2, where the black circles denote the initial values and the white circles underline values on the boundary of the domain. We build a mesh consisting of n points for the τ axis and m points for the z axis and denote the solution of the PDE at the mesh point $(i\delta\tau; z_L + j\delta z)$ by $u_{i,j}$, with $i = 1, \ldots, n+1$, and $j = 1, \ldots, m$:

$$u_{i,j} = u(i\delta\tau, z_L + j\delta z).$$

For $i = 0$, we have the initial value on the grid:

$$u_{0,j} = u(0, z_L + j\delta z).$$

For $j = 0$ and $j = m + 1$, we have boundary conditions at points z_L and z_U:

$$u_{i,0} = u(i\delta\tau, z_L) = \psi_L(i\delta\tau),$$
$$u_{i,m+1} = u(i\delta\tau, z_U) = \psi_U(i\delta\tau).$$

4.3.3 Explicit Scheme

We approximate the time derivative using the forward difference (4.17) and the spatial derivative using the finite difference (4.19):

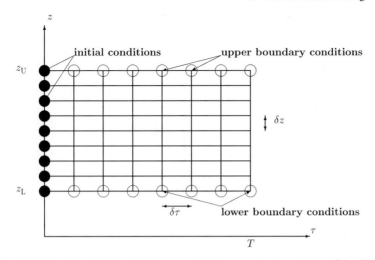

Fig. 4.2. The grid points at which to compute the approximate solution of the PDE.

$$\frac{\partial u}{\partial \tau} = \frac{u(\tau + \delta\tau, z) - u(\tau, z)}{\delta\tau} + O(\delta\tau),$$

$$\frac{\partial^2 u}{\partial z^2} = \frac{u(\tau, z + \delta z) - 2u(\tau, z) + u(\tau, z - \delta z)}{\delta z^2} + O(\delta z^2).$$

By replacing these expressions into (4.13), we have:

$$\frac{u(\tau + \delta\tau, z) - u(\tau, z)}{\delta\tau} + O(\delta\tau)$$

$$= \frac{\sigma^2}{2} \frac{u(\tau, z + \delta z) - 2u(\tau, z) + u(\tau, z - \delta z)}{\delta z^2} + O(\delta z^2). \qquad (4.20)$$

If we restrict our attention to the mesh points $(i\delta\tau, j\delta z)$, we can write:

$$\frac{u_{i+1,j} - u_{i,j}}{\delta\tau} + O(\delta\tau) = \frac{\sigma^2}{2} \frac{u_{i,j+1} - 2u_{i,j} + u_{i,j-1}}{\delta z^2} + O(\delta z^2).$$

Let us ignore all error terms and denote the approximate solution by $v_{i,j}$. The scheme above reads as:

$$\frac{v_{i+1,j} - v_{i,j}}{\delta\tau} = \frac{\sigma^2}{2} \frac{v_{i,j+1} - 2v_{i,j} + v_{i,j-1}}{\delta z^2}. \qquad (4.21)$$

Note that $u_{i,j}$ is the true solution of the PDE computed at grid points $(i\delta\tau, j\delta z)$, whereas $v_{i,j}$ indicates the approximate solution of the PDE, which is also the exact solution of the difference equation (4.21) computed at grid points $(i\delta\tau, j\delta z)$. We can solve the recursive relation (4.21) with respect to $v_{i+1,j}$ and obtain:

$$v_{i+1,j} = v_{i,j} + \frac{\delta\tau}{\delta z^2} \left(\frac{\sigma^2}{2} v_{i,j+1} - \sigma^2 v_{i,j} + \frac{\sigma^2}{2} v_{i,j-1} \right),$$

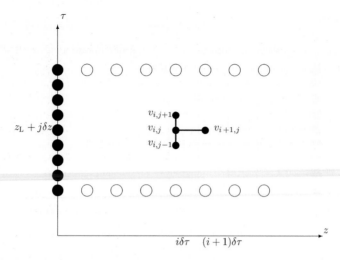

Fig. 4.3. Nodes involved in the updating of the solution in the Explicit scheme.

or, equivalently,

$$v_{i+1,j} = \frac{\sigma^2}{2}\alpha v_{i,j+1} + (1 - \sigma^2\alpha)v_{i,j} + \frac{\sigma^2}{2}\alpha v_{i,j-1}, \qquad (4.22)$$

where $\alpha = \delta\tau/\delta z^2$, $i = 1, 2, \ldots$, and $j = 1, \ldots, m$. Expression (4.22) suggests that the unknown value at node $(i + 1, j)$ can be computed in terms of known values at the previous time step, i.e., at nodes $(i, j + 1)$, (i, j) and $(i, j - 1)$, as is illustrated in Fig. 4.3. This compels us to use the attribute "explicit" in the resulting numerical scheme. For $i = 0$, we obtain the initial condition $v_{0,j} = u(0, z_L + j\delta\tau)$. For $j = 0$, or $j = m + 1$, we identify boundary conditions, $v_{i,0} = \psi_L(i\delta\tau)$ and $v_{i,m+1} = \psi_U(i\delta\tau)$, respectively.

Algorithm

```
EXPLICIT
Set vold(j) = initial condition evaluated
at jδz, j = 1,...,m;
For i = 1,...,n:
    compute v(0) = lowerboundary(iδτ)
    compute v(m) = upperboundary(iδτ)
    update the vector v using vold according to
    the iteration (4.22), j = 1,...,m;
    copy v(j) in vold(j)
End cycle
return v(j) for j = 1,...,m
```

Matrix form

It is convenient to use vector notations and rewrite the iteration of the explicit scheme in terms of matrix multiplications. Let us define the $(m \times 1)$ vector $\mathbf{v}_i$ $(i = 0, 1, 2, \ldots)$, the $(m \times m)$ matrix $\mathbf{A}$, and the $(m \times 1)$ vector $\mathbf{b}_i{}^4$ by:

$$\mathbf{v}_i = \begin{bmatrix} v_{i,m} \\ \vdots \\ v_{i,1} \end{bmatrix}; \qquad \mathbf{v}_0 = \begin{bmatrix} u(0, z_L + m\delta z) \\ u(0, z_L + (m-1)\delta z) \\ \vdots \\ u(0, z_L + 2\delta z) \\ u(0, z_L + \delta z) \end{bmatrix};$$

$$\mathbf{b}_i = \begin{bmatrix} v_{i,m+1} \\ 0 \\ \vdots \\ 0 \\ v_{i,0} \end{bmatrix}; \tag{4.23}$$

$$\mathbf{A} = \begin{bmatrix} (1-\sigma^2\alpha) & \frac{\sigma^2\alpha}{2} & 0 & 0 & 0 \\ \frac{\sigma^2\alpha}{2} & (1-\sigma^2\alpha) & \ddots & \ddots & 0 \\ 0 & \ddots & \ddots & \ddots & 0 \\ 0 & \ddots & \ddots & (1-\sigma^2\alpha) & \frac{\sigma^2\alpha}{2} \\ 0 & 0 & 0 & \frac{\sigma^2\alpha}{2} & (1-\sigma^2\alpha) \end{bmatrix}. \tag{4.24}$$

The recursion in (4.22) can be written using matrix algebra as follows:

$$\mathbf{v}_{i+1} = \mathbf{A}\mathbf{v}_i + \frac{\sigma^2\alpha}{2}\mathbf{b}_i, \quad i = 1, 2, \ldots .$$

This representation will be useful later on in order to understand the mathematical properties of this numerical scheme. Note that this matrix form is not convenient for programming the explicit scheme.

Example This example is borrowed from Smith (1985, p. 14). Let us consider the heat equation in the interval $[0, 1]$ with $\sigma = \sqrt{2}$:

$$-\partial_\tau u(\tau, z) + \partial_{zz} u(\tau, z) = 0, \tag{4.25}$$

initial condition:

$$u(0, z) = \begin{cases} 2z & 0 \le z < \frac{1}{2}, \\ 2(1-z) & \frac{1}{2} \le z \le 1, \end{cases} \tag{4.26}$$

and boundary conditions:

$$u(\tau, 1) = u(\tau, 0) = 0. \tag{4.27}$$

---

[4] Note that we have used the somewhat strange numbering of the vector components $m, \ldots, 1$ in order to conform to the graphic presentation.

The analytical solution of this initial value problem is:

$$u(\tau, z) = \frac{8}{\pi^2} \sum_{n=1}^{\infty} \frac{1}{n^2} \sin\left(\frac{n\pi}{2}\right) \sin(n\pi z) e^{-n^2\pi^2\tau}. \tag{4.28}$$

This example is interesting for at least four reasons. It allows us to:

(a) measure the effect of the initial condition with discontinuous first derivative on the numerical solution, a common phenomenon in option pricing problems;
(b) examine the interpolation procedure to be adopted for the purpose of computing solution values at points not included in the grid;
(c) assess the effect of the value of constant $\alpha = \delta z/\delta\tau^2$ on the numerical stability of the scheme;
(a) establish a correspondence with the pricing of a double barrier knock-out option, that is a contract extinguishing whenever the underlying asset hits either of two selected barriers before maturity.

We set $z_L = 0$, $z_U = 1$ and choose $m = 9$ nodes on the z-axis, so that $\delta z = 1/10 = 0.1$. We are interested in the evolution of the solution until time 0.01 under a time step $\delta\tau = 0.001$. This means we are considering 11 nodes $(0, 0.001, 0.002, \ldots, 0.01)$ on the time grid. The recursion has been implemented in Excel and is illustrated in Fig. 4.4.

Let us examine the behavior of the numerical solution at points $z = 0.2$ and $z = 0.5$. The latter has been chosen due to the presence of a discontinuity in the first derivative of the initial condition there. We obtain a very accurate solution at $z = 0.2$. However, the percentage error increases as long as we move forward in time, see Table 4.3. At $z = 0.5$, the numerical solution appears rather inaccurate due to the above mentioned discontinuity. Fortunately, the negative effect of the discontinuity

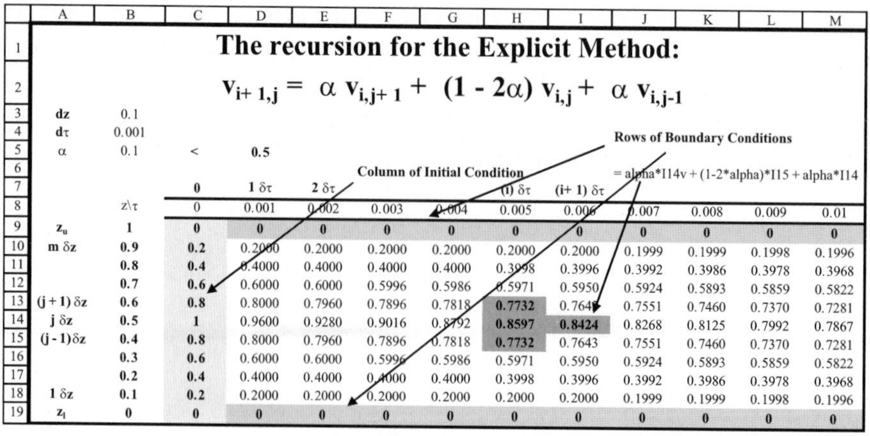
The recursion for the Explicit Method:

$$v_{i+1,j} = \alpha\, v_{i,j+1} + (1 - 2\alpha)\, v_{i,j} + \alpha\, v_{i,j-1}$$

	A	B	C	D	E	F	G	H	I	J	K	L	M
3	dz	0.1											
4	dτ	0.001								Rows of Boundary Conditions			
5	α	0.1	<	0.5									
6						Column of Initial Condition				= alpha*I14v + (1-2*alpha)*I15 + alpha*I14			
7			0	1 δτ	2 δτ			(i) δτ	(i+ 1) δτ				
8		z\τ	0	0.001	0.002	0.003	0.004	0.005	0.006	0.007	0.008	0.009	0.01
9	z_u	1	0	0	0	0	0	0	0	0	0	0	0
10	m δz	0.9	0.2	0.2000	0.2000	0.2000	0.2000	0.2000	0.2000	0.1999	0.1999	0.1998	0.1996
11		0.8	0.4	0.4000	0.4000	0.4000	0.4000	0.3998	0.3996	0.3992	0.3986	0.3978	0.3968
12		0.7	0.6	0.6000	0.6000	0.5996	0.5986	0.5971	0.5950	0.5924	0.5893	0.5859	0.5822
13	(j+1) δz	0.6	0.8	0.8000	0.7960	0.7896	0.7818	0.7732	0.764[illegible]	0.7551	0.7460	0.7370	0.7281
14	j δz	0.5	1	0.9600	0.9280	0.9016	0.8792	0.8597	0.8424	0.8268	0.8125	0.7992	0.7867
15	(j - 1)δz	0.4	0.8	0.8000	0.7960	0.7896	0.7818	0.7732	0.7643	0.7551	0.7460	0.7370	0.7281
16		0.3	0.6	0.6000	0.6000	0.5996	0.5986	0.5971	0.5950	0.5924	0.5893	0.5859	0.5822
17		0.2	0.4	0.4000	0.4000	0.4000	0.4000	0.3998	0.3996	0.3992	0.3986	0.3978	0.3968
18	1 δz	0.1	0.2	0.2000	0.2000	0.2000	0.2000	0.2000	0.2000	0.1999	0.1999	0.1998	0.1996
19	z_l	0	0	0	0	0	0	0	0	0	0	0	0

Fig. 4.4. Implementing the explicit recursion in Excel.

Table 4.3. Numerical and analytical solution when $z = 0.2$

t	Numerical	Analytical	% Difference
0.005	0.39983	0.39985	0.00%
0.01	0.39678	0.39656	0.06%
0.02	0.37808	0.3766	0.39%
0.1	0.17961	0.17756	1.15%
0.5	0.00351	0.00343	2.49%

Table 4.4. Numerical and analytical solution when $z = 0.5$

t	Numerical	Analytical	% Difference
0.005	0.85972	0.84042	2.30%
0.01	0.78674	0.77432	1.60%
0.02	0.68915	0.68085	1.22%
0.1	0.30562	0.30212	1.16%
0.5	0.00597	0.00583	2.49%

fades away as long as time increases and the error becomes comparable to the one reported at $z = 0.2$, see Table 4.4. The explicit method has an order of accuracy equal to $\delta\tau$, but, as the example above has clearly shown, it can be reduced by the presence of discontinuities. Smith (1985, pp. 16–17) reports that when the initial function and its first $p - 1$ derivatives are continuous and the pth-order derivative exhibit a discontinuity, then the difference between numerical and analytical solutions of the PDE approaches the order $(\delta\tau)^{(p+2)/(p+4)}$ for a "small" $\delta\tau$. Consequently, if all derivatives are continuous, i.e., $p \rightarrow \infty$, the error becomes of order $\delta\tau$. In the illustrated above example illustrated, $p = 1$ so that the error has order $(\delta\tau)^{3/5}$ in a neighborhood of point $z = 0.5$. This explains the inaccuracy of numerical results for small times τ.

It is worth mentioning that the scheme provides a numerical solution only at the nodes under consideration. For example, once we transform the Black–Scholes equation into a heat equation, this has been solved over an evenly spaced z-grid. It may therefore be possible that these values of z do not correspond to desired spot prices. To determine values over these regions, we need to adopt suitable interpolation methods. For instance, we may assume that v varies linearly from $v_{i,j}$ to $v_{i,j+1}$. The resulting linear interpolation is as accurate as the values on the grid. For any $z \in [j\delta z, (j + 1)\delta z]$, we obtain the approximation:

$$v(i\delta\tau, z) \simeq v(i\delta\tau, j\delta z) + \frac{v(i\delta\tau, (j + 1)\delta z) - v(i\delta\tau, j\delta z)}{(j + 1)\delta z - j\delta z}(z - j\delta z)$$

$$= v(i\delta\tau, j\delta z) + \frac{v(i\delta\tau, (j + 1)\delta z) - v(i\delta\tau, j\delta z)}{\delta z}(z - j\delta z).$$

It is interesting to examine the way α affects the numerical solution resulting from the proposed scheme. Figures 4.5 and 4.6 exhibit both analytical and numerical solutions for $\alpha = 0.48$ and $\alpha = 0.52$. We notice an oscillating behavior occurring in the latter case. A heuristic explanation for this effect may be as follows. Updated

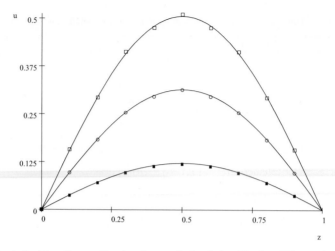

Fig. 4.5. Analytical (continuous lines) and numerical solution (dots) at different times (0.0520, 0.1040 and 0.2080) when $\delta z = 0.1$ and $\delta \tau = 0.0048$ ($\alpha = 0.48$).

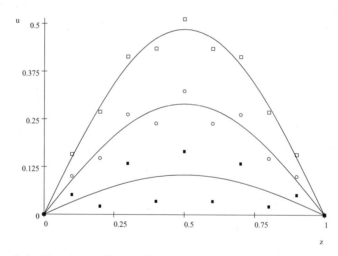

Fig. 4.6. Analytical (continuous lines) and numerical solution (dots) at different times (0.0520, 0.1040 and 0.2080) when $\delta z = 0.1$ and $\delta \tau = 0.0052$ ($\alpha = 0.52$).

values $v_{i+1,j}$ in (4.22) can be thought of as the expected value of the three preceding values, $v_{i,j+1}$, $v_{i,j}$, and $v_{i,j-1}$, much like is a trinomial tree. This probabilistic interpretation makes sense provided all coefficients are positive and that they sum up to 1. This amounts requiring that $\alpha < 1/2$. If this condition is not fulfilled, the round-off error generated by summing up terms with opposite sign progressively deteriorates the quality of the numerical solution as time goes by. Refining both time and spatial grids, while keeping α fixed, does not improve the situation. This behavior is actually due to an instability problem, which can be eliminated by taking appropriate values for α, that is by setting $\delta\tau$ to a very small value compared to δz. This choice ensures that the difference between numerical and exact solutions remains bounded as the number of time steps diverges to infinity. This will be discussed in detail in Sect. 4.4.

4.3.4 Implicit Scheme

If we approximate the time derivative using the backward difference (4.18) and the spatial derivative using the finite difference (4.19), we have:

$$\frac{u(\tau, z) - u(\tau - \delta\tau, z)}{\delta\tau} + O(\delta\tau)$$

$$= \frac{\sigma^2}{2} \frac{u(\tau, z + \delta z) - 2u(\tau, z) + u(\tau, z - \delta z)}{\delta z^2} + O(\delta z^2).$$

Computed on the grid points $\tau = i\delta\tau$, $z = j\delta z$, this expression provides:

$$\frac{u_{i,j} - u_{i-1,j}}{\delta\tau} + O(\delta\tau) = \frac{\sigma^2}{2} \frac{u_{i,j+1} - 2u_{i,j} + u_{i,j-1}}{\delta z^2} + O(\delta z^2).$$

If we ignore the error terms and denote the approximate solution by $v_{i,j}$, we obtain:

$$\frac{v_{i,j} - v_{i-1,j}}{\delta\tau} = \frac{\sigma^2}{2} \frac{v_{i,j+1} - 2v_{i,j} + v_{i,j-1}}{\delta z^2}.$$

By rearranging terms, we come up with a recursive relation:

$$-\frac{\sigma^2}{2}\alpha v_{i,j-1} + \left(1 + \sigma^2\alpha\right)v_{i,j} - \frac{\sigma^2}{2}\alpha v_{i,j+1} = v_{i-1,j},$$

for $i = 1, 2, \ldots$ and $j = 1, \ldots, m$, where

$$\alpha = \frac{\delta\tau}{\delta z^2}. \tag{4.29}$$

If $i = 0$, we have the initial condition. For $j = 0$ and $j = m + 1$, boundary conditions are obtained. By proceeding forward in time from the initial condition, we have an equation connecting three unknowns ($v_{i,j+1}$, $v_{i,j}$, and $v_{i,j-1}$) to a single known value ($v_{i-1,j}$). Thus, in order to find the updated values of the numerical solution, we need to solve a linear system, as detailed in the next subsection. The scheme is illustrated in Fig. 4.7. The order of accuracy of this scheme is $O(\delta\tau, \delta z^2)$. This means that no improvement in accuracy is granted by this scheme compared to explicit scheme. However, the implicit scheme is unconditionally stable in that no restriction on α is required. The price to pay is represented by the need to solve a linear system at each time step.

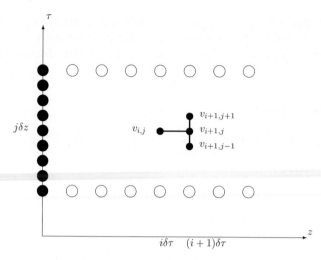

Fig. 4.7. The recursion for the implicit scheme.

Algorithm

```
IMPLICIT
Set v(j) = initial condition evaluated at jδz , j = 1,...,m;
For i = 1,...,n:
    compute the right-hand side in (4.31);
    solve the linear system (4.31);
    update the solution v;
End cycle
return v(j) for j = 1,...,m;
```

Matrix form

By using a vector notation, we may write the iterations involved in the implicit scheme in terms of matrix products. More precisely, let $\mathbf{v}_i$ and $\mathbf{b}_i$ be the $(m \times 1)$ vectors introduced in (4.23). Let also $\mathbf{B}$ denote the $m \times m$ matrix defined as:

$$
\mathbf{B} = \begin{bmatrix}
1+\sigma^2\alpha & -\frac{\sigma^2\alpha}{2} & 0 & 0 & 0 \\
-\frac{\sigma^2\alpha}{2} & 1+\sigma^2\alpha & \ddots & \ddots & 0 \\
0 & \ddots & \ddots & \ddots & 0 \\
0 & \ddots & \ddots & 1+\sigma^2\alpha & -\frac{\sigma^2\alpha}{2} \\
0 & 0 & 0 & -\frac{\sigma^2\alpha}{2} & 1+\sigma^2\alpha
\end{bmatrix} .
\qquad (4.30)
$$

The recursive relation (4.22) can be then written as:

$$\mathbf{B}\mathbf{v}_{i+1} = \mathbf{v}_i + \frac{\sigma^2 \alpha}{2}\mathbf{b}_i, \quad i = 1, 2, \ldots. \tag{4.31}$$

This expression clearly shows the need to solve a linear system at every time step. Fortunately, this system of equations is tridiagonal, i.e. the matrix $\mathbf{B}$ has non-zero elements only on the diagonal and in the positions immediately to the left and to the right of the diagonal. This implies that the computational cost is directly proportional to the number of grid points. In fact, the cost per time step for this method is approximately twice that of the explicit method. In principle, to solve (4.31) we could compute the inverse of matrix $\mathbf{B}$, provided it exists, and proceed forward according to the following relation:

$$\mathbf{v}_{i+1} = \mathbf{B}^{-1}\left(\mathbf{v}_i + \frac{\sigma^2 \alpha}{2}\mathbf{b}_i\right),$$

starting with the initial condition mentioned above. The resulting algorithm resembles an explicit recurrence, where the matrix $\mathbf{A}$ appearing in the explicit method has now been replaced by the matrix $\mathbf{B}^{-1}$. However, in the explicit method, $\mathbf{A}$ is tridiagonal and to update the solution, we just need to store the elements of the three diagonals. On the other hand, the matrix $\mathbf{B}^{-1}$ has all non-zero entries, so that its computational cost and storage can be expensive especially when a large number of grid points is considered. The $m \times m$ matrix $\mathbf{B}^{-1}$ requires m^2 elements to be stored, whereas the tridiagonal matrix $\mathbf{B}$ needs only $3m - 2$ numbers to be recorded. As m is typically around 1,000, one method leads to 1,000,000 elements while the other only 2,998 figures. In addition, if the coefficients are time dependent, then we need to compute the inverse of matrix $\mathbf{B}$ at each time step.

Example Let us reconsider the example developed for illustrating the explicit method. We take $\delta z = 0.1$ and $\delta \tau = 0.001$, so that $\alpha = 0.1$. Vector $\mathbf{b}$ has all zero entries and is independent of the index i. Matrix $\mathbf{B}$ has main diagonal elements equal to 1.2, whereas entries in the upper and lower diagonal are all equal to -0.1. The recursion is illustrated in Fig. 4.8, where we report an Excel spreadsheet implementing an implicit scheme. For illustrative purposes, the recursion has been performed using the inverse of matrix $\mathbf{B}$.[5] Note that $\mathbf{B}^{-1}$ has all nonzero entries, so that storage related limitations are likely to arise whenever the number of spatial nodes is increased.

Table 4.5 shows a comparison between the implicit solution at time 0.01 to the one obtained using an explicit method and the analytical solution.

Figure 4.9 shows the root mean square error of the implicit method versus the time step. We can see that this relationship is linear as expected: the time derivative has been indeed computed using a finite difference scheme accurate to order $d\tau$.

4.3.5 Crank–Nicolson Scheme

This scheme is based on the idea of approximating the PDE at points $(i\delta\tau + \frac{\delta\tau}{2}, j\delta z)$.

---

[5] In Excel this recursion can be done using the functions MInverse and MProduct.

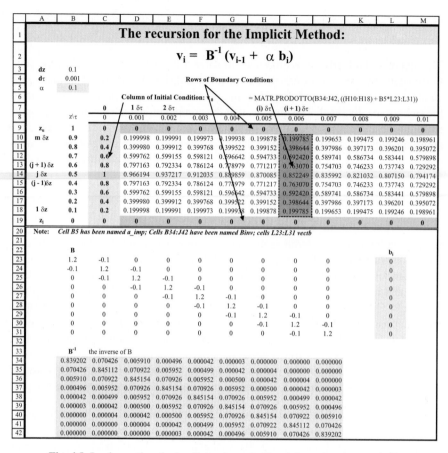

Fig. 4.8. Implementing the implicit scheme in Excel, inverting the matrix **B**.

Table 4.5. Accuracy of the explicit and implicit schemes

z	Analytical, A	Explicit, E	$(E-A)/A$	Implicit, I	$(I-A)/A$
0.9	0.19961	0.19958	−0.0002	0.19896	−0.0033
0.8	0.39655	0.39678	0.0006	0.39507	−0.0037
0.7	0.57990	0.58221	0.0040	0.57990	0.0000
0.6	0.72014	0.72811	0.0111	0.72929	0.0127
0.5	0.77432	0.78674	0.0160	0.79417	0.0256
0.4	0.72014	0.72811	0.0111	0.72929	0.0127
0.3	0.57990	0.58221	0.0040	0.57990	0.0000
0.2	0.39655	0.39678	0.0006	0.39507	−0.0037
0.1	0.19961	0.19958	−0.0002	0.19896	−0.0033

- For the time derivative, we use a central Taylor series expansion around $(i\delta\tau + \frac{\delta\tau}{2}, j\delta z)$:

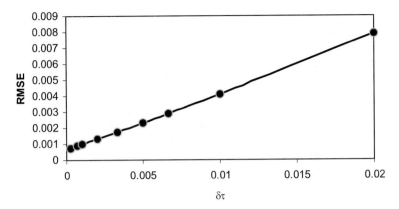

Fig. 4.9. Implicit scheme: root mean square error versus time step.

$$\left.\frac{\partial u(\tau, z)}{\partial \tau}\right|_{\tau=i\delta\tau+\frac{\delta\tau}{2}, z=z_L+j\delta z} \simeq \frac{u_{i+1,j} - u_{i,j}}{\delta\tau}.$$

This provides an order of accuracy equal to $O(\delta\tau^2)$.

- For the second-order spatial derivative, we adopt the average of finite differences computed at time steps $(i\delta\tau, j\delta z)$ and $((i+1)\delta\tau, j\delta z)$:

$$\left.\frac{\partial^2 u}{\partial z^2}(\tau, z)\right|_{\tau=(i+1/2)\delta\tau, z=z_L+j\delta z}$$
$$\simeq \frac{1}{2}\left\{\frac{u_{i+1,j+1} - 2u_{i+1,j} + u_{i+1,j-1}}{\delta z^2}\right\} + \frac{1}{2}\left\{\frac{u_{i,j+1} - 2u_{i,j} + u_{i,j-1}}{\delta z^2}\right\}.$$

The accuracy here is $O(\delta z^2)$.

By inserting these differences into the heat equation, we obtain the Crank–Nicolson (CN) scheme:

$$\frac{u_{i+1,j} - u_{i,j}}{\delta\tau} + O(\delta\tau^2)$$
$$= \frac{\sigma^2}{2}\left(\frac{1}{2}\left\{\frac{u_{i+1,j+1} - 2u_{i+1,j} + u_{i+1,j-1}}{\delta z^2}\right\} + \frac{1}{2}\left\{\frac{u_{i,j+1} - 2u_{i,j} + u_{i,j-1}}{\delta z^2}\right\}\right)$$
$$+ O(\delta z^2).$$

Although the Taylor expansion has been developed at point $i\delta\tau + \frac{\delta\tau}{2}$, only values of u at the grid points appear in the expression above. This leads to an approximated recursive relation:

$$\frac{v_{i+1,j} - v_{i,j}}{\delta\tau}$$
$$= \frac{\sigma^2}{4}\left(\left(\frac{v_{i+1,j+1} - 2v_{i+1,j} + v_{i+1,j-1}}{\delta z^2}\right) + \left(\frac{v_{i,j+1} - 2v_{i,j} + v_{i,j-1}}{\delta z^2}\right)\right),$$

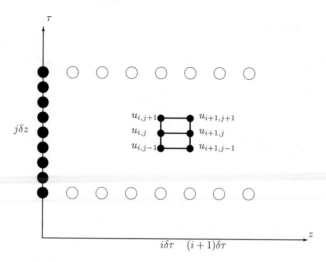

Fig. 4.10. The recursion for the Crank–Nicolson scheme.

or, equivalently,

$$-\frac{\sigma^2\alpha}{4}v_{i+1,j-1} + \left(1 + \frac{\sigma^2\alpha}{2}\right)v_{i+1,j} - \frac{\sigma^2\alpha}{4}v_{i+1,j+1}$$

$$= \frac{\sigma^2\alpha}{4}v_{i,j-1} + \left(1 - \frac{\sigma^2\alpha}{2}\right)v_{i,j} + \frac{\sigma^2\alpha}{4}v_{i,j+1}, \qquad (4.32)$$

where $\alpha = \delta\tau/\delta z^2$. At each time step, the resulting scheme relates six points on the grid as is shown in Fig. 4.10. As in the case of a fully implicit method, here no relevant limitation on the size of the time step is required for the method to converge. The approximation error is $O(\delta\tau^2, \delta z^2)$ and allows us to say the Crank–Nicolson recursion is more accurate than both explicit and fully implicit methods. Another similar feature to the implicit scheme is that we need to solve a tridiagonal system of linear equations at each time step. Consequently, the cost becomes proportional to the number of grid points.

Matrix form

We may write the CN scheme using a matrix notation. Let us define ($m \times m$) matrices **C** and **D** as

$$\mathbf{C} = \begin{bmatrix} 1 + \frac{\sigma^2\alpha}{2} & -\frac{\sigma^2\alpha}{4} & 0 & 0 & 0 \\ -\frac{\sigma^2\alpha}{4} & 1 + \frac{\sigma^2\alpha}{2} & \ddots & \ddots & 0 \\ 0 & \ddots & \ddots & \ddots & 0 \\ 0 & \ddots & \ddots & 1 + \frac{\sigma^2\alpha}{2} & -\frac{\sigma^2\alpha}{4} \\ 0 & 0 & 0 & -\frac{\sigma^2\alpha}{4} & 1 + \frac{\sigma^2\alpha}{2} \end{bmatrix}, \qquad (4.33)$$

$$\mathbf{D} = \begin{bmatrix} 1 - \frac{\sigma^2\alpha}{2} & \frac{\sigma^2\alpha}{4} & 0 & 0 & 0 \\ \frac{\sigma^2\alpha}{4} & 1 - \frac{\sigma^2\alpha}{2} & \ddots & \ddots & 0 \\ 0 & \ddots & \ddots & \ddots & 0 \\ 0 & & \ddots & 1 - \frac{\sigma^2\alpha}{2} & \frac{\sigma^2\alpha}{4} \\ 0 & 0 & 0 & \frac{\sigma^2\alpha}{4} & 1 - \frac{\sigma^2\alpha}{2} \end{bmatrix}, \qquad (4.34)$$

and an $(m \times 1)$ vector $\mathbf{b}_i$ as

$$\mathbf{b}_i^T = \begin{bmatrix} \frac{v_{i,m+1}+v_{i+1,m+1}}{2} & 0 & \cdots & 0 & \frac{v_{i,0}+v_{i+1,0}}{2} \end{bmatrix}.$$

The recursive relation (4.32) can be written as:

$$\mathbf{C}v_{i+1} = \mathbf{D}v_i + \frac{\sigma^2\alpha}{2}\mathbf{b}_i, \quad i = 1, 2, \ldots. \qquad (4.35)$$

The solution of the linear system above can efficiently exploit the tridiagonal form of matrix $\mathbf{C}$.

Algorithm

```
CRANK-NICOLSON
Set v(j) = initialcondition(jδz), j = 1,...,m
For i = 1,...,n:
    compute the right-hand side in (4.35)
    solve the linear system in (4.35)
End cycle
return v(j), j = 1,...,m;
```

Example We reconsider the prototypical example previously illustrated. We now set $\delta z = 0.1$ and $\delta \tau = 0.001$, so that $\alpha = 0.1$. Vector $\mathbf{b}$ has all zero entries and does not depend on i. Matrix $\mathbf{C}$ has main diagonal elements equal to 1.1. Entries in the upper and lower diagonals are all equal to -0.05. Diagonal elements of $\mathbf{D}$ are all set to 0.9. Elements in the upper and lower diagonals of the same matrix are all equal to 0.05. Figure 4.11 shows a spreadsheet solving the linear system using the inverse of matrix $\mathbf{C}$. Efficient methods for solving linear systems are presented in the next chapter.

Figure 4.12 shows the root mean square error of the Crank–Nicolson scheme versus the square of the time step. We can see that this relationship is linear as expected: the time derivative has been indeed computed using finite differences accurate to order $d\tau^2$.

Although the Crank–Nicolson method can be proven to be stable for all values of α, large values of this quantity may negatively affect the actual performance of the scheme due to unexpected and spurious oscillations in the numerical solution. This phenomenon is illustrated in figure where we use different values of α. Indeed, if we

The recursion for the Crank Nicolson Method:

$$C\, v(i+1) = D\, v(i) + (\alpha/2)*b(i)$$

$$v(i+1) = C^{-1}\,[D\, v(i) + (\alpha/2)*b(i)]$$

	A	B	C	D	E	F	G	H	I	J	K	L	M
3	dx	0.1											
4	$d\tau$	0.001											
5	α	0.1						{= MATR.PRODOTTO(invC;H21:H29)}					
7			z\τ	0	0.001	0.002	0.003	0.004	0.005	0.006	0.007	0.008	0.009	0.01
8	z_u	0	0	0	0	0	0	0	0	0	0	0	0
9	m δz	0.1	0.2	0.20000	0.2000	0.2000	0.2000	0.1999	0.199895	0.1998	0.1997	0.1995	0.19926
10		0.2	0.4	0.40000	0.4000	0.3999	0.3998	0.3995	0.399101	0.3985	0.3978	0.3969	0.39588
11		0.3	0.6	0.59992	0.5996	0.5988	0.5976	0.5958	0.593599	0.5910	0.5880	0.5846	0.58101
12	(j+1) δz	0.4	0.8	0.79834	0.7939	0.7877	0.7803	0.7722	0.763716	0.7550	0.7462	0.7375	0.72881
13	j δz	0.5	1	0.96349	0.9331	0.9072	0.8848	0.8651	0.847406	0.8314	0.8167	0.8031	0.79038
14	(j-1) δz	0.6	0.8	0.79834	0.7939	0.7877	0.7803	0.7722	0.763716	0.7550	0.7462	0.7375	0.72881
15		0.7	0.6	0.59992	0.5996	0.5988	0.5976	0.5958	0.593599	0.5910	0.5880	0.5846	0.58101
16		0.8	0.4	0.40000	0.4000	0.3999	0.3998	0.3995	0.399101	0.3985	0.3978	0.3969	0.39588
17	1 δz	0.9	0.2	0.20000	0.2000	0.2000	0.2000	0.1999	0.199895	0.1998	0.1997	0.1995	0.19926
18	z_1	1	0	0	0	0	0	0	0	0	0	0	0

Note: *invC refers to a range of cells containing the matrix C^{-1}*

D v(i)+ (α/2)*b(i)

0.2	0.199999686	0.19999745	0.199989404	0.199969303	0.199929199	0.199860301	0.199753768	0.19960133	0.199395719
0.4	0.399993099	0.399956449	0.399856343	0.39965904	0.399335881	0.39886517	0.398232316	0.397429122	0.396452739
0.6	0.599848486	0.599320485	0.598316127	0.596809862	0.594818295	0.592380093	0.589543723	0.586360173	0.582878839
0.8	0.796673582	0.791154805	0.784218455	0.776392855	0.768037181	0.759394512	0.750628241	0.741847103	0.733122384
0.98	0.946970326	0.919141938	0.895240973	0.874343642	0.855774802	0.83903681	0.823759458	0.809664588	0.79654098
0.8	0.796673582	0.791154805	0.784218455	0.776392855	0.768037181	0.759394512	0.750628241	0.741847103	0.733122384
0.6	0.599848486	0.599320485	0.598316127	0.596809862	0.594818295	0.592380093	0.589543723	0.586360173	0.582878839
0.4	0.399993099	0.399956449	0.399856343	0.39965904	0.399335881	0.39886517	0.398232316	0.397429122	0.396452739
0.2	0.199999686	0.19999745	0.199989404	0.199969303	0.199929199	0.199860301	0.199753768	0.19960133	0.199395719

Fig. 4.11. Implementing the Crank–Nicolson recursion in Excel.

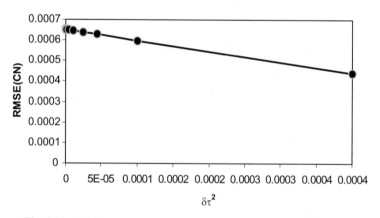

Fig. 4.12. CN scheme: root mean square error versus squared time step.

take $d\tau$ too large the relationship between error and $d\tau^2$, as illustrated in Fig. 4.12, is lost. Indeed the presence of the kink in the initial condition introduces spurious oscillations, as illustrated in Fig. 4.13. These oscillations disappear only reducing α, i.e. or reducing the time step dt or increasing the space step dx.

This bias slowly disappears as i increases. Moreover, it usually occurs in a neighborhood of points of discontinuity in the initial values or between initial values and boundary values. This effect can be observed in Table 4.6, where the error near $z = 0.5$ oscillates between positive and negative values as α increases from 0.1

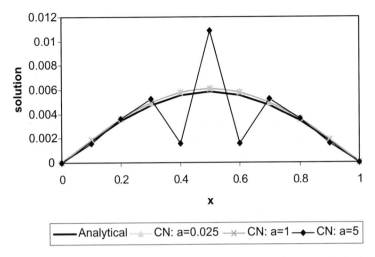

Fig. 4.13. CN: oscillations in the numerical solution due to the discontinuity in the first derivative of the initial condition.

Table 4.6. Analytical and numerical solution of the problem at $\tau = 0.01$

z	Analytical, A	CN ($\alpha = 1$)	$\frac{(CN-A)}{A}$	CN ($\alpha = 0.1$)	$(CN-A)/A$
0.9	0.19961	0.19890	−0.0036	0.19926	−0.0017
0.8	0.39655	0.39558	−0.0024	0.39588	−0.0017
0.7	0.57990	0.58343	0.0061	0.58101	0.0019
0.6	0.72014	0.73812	0.0250	0.72881	0.0120
0.5	0.77432	0.76906	−0.0068	0.79038	0.0207
0.4	0.72014	0.73812	0.0250	0.72811	0.0120
0.3	0.57990	0.58343	0.0061	0.58101	0.0019
0.2	0.39655	0.39558	−0.0024	0.39588	−0.0017
0.1	0.19961	0.19890	−0.00360	0.19926	−0.0017

to 1. These oscillations can be eliminated only by constraining the time-step of the Crank–Nicolson method, that is by decreasing the value of α.[6]

4.3.6 Computing the Greeks

Financial "Greeks" are the partial derivatives of the option price with respect to the underlying price, time to maturity and model parameters. These quantities can be computed using finite difference approximations. We consider approximations for the option delta (Δ), gamma (Γ) and theta (Θ) defined as:

$$\Delta_{i,j} = \left.\frac{\partial u(\tau, z)}{\partial z}\right|_{\tau=i\delta\tau, z=j\delta z} \simeq \frac{v_{i,j+1} - v_{i,j-1}}{2\delta z},$$

---

[6] A discussion on this point can be found in Smith (1985, p. 122).

$$\Gamma_{i,j} = \left.\frac{\partial^2 u(\tau,z)}{\partial z^2}\right|_{\tau=i\delta\tau,z=j\delta z} \simeq \frac{v_{i,j+1} - 2v_{i,j} + v_{i,j-1}}{(\delta z)^2},$$

$$\Theta_{i,j} = \left.\frac{\partial u(\tau,z)}{\partial \tau}\right|_{\tau=i\delta\tau,z=j\delta z} \simeq \frac{v_{i+1,j} - v_{i,j}}{\delta\tau}.$$

Since the schemes are accurate to the order $O(\delta z^2)$, it follows that delta is accurate to the order $O(\delta z)$, while gamma is merely $O(1)$. Theta is accurate to the order $O(\delta\tau)$ for the CN scheme and $O(1)$ for the Implicit and Explicit schemes. Fortunately, numerical Greeks do not display great inaccuracies in most instances. However, some bias shows up for small diffusion coefficients.

Computing Greeks relative to model parameters (e.g., Rho, Kappa, and Vega) can be performed using finite difference approximations. For example, Rho can be computed with an accuracy of order two by (1) solving the pricing PDE with $r + \delta r$ and then with $r - \delta r$; (2) calculating the appropriate finite difference. However, this requires solving two additional PDEs.

4.4 Consistency, Convergence and Stability

As Figs 4.5 and 4.6 clearly show, the explicit method suffers from an instability effect. Also the Crank–Nicolson scheme may generate spurious oscillations near discontinuity points of the boundary condition. This section tries to clarify this point through the examination of convergence properties owned by the proposed schemes. We start with preliminary definitions.

Let us write the heat equation (4.13) as follows:

$$L(u) = -\partial_\tau u(\tau,z) + \mathcal{L}u(\tau,z) = 0,$$

where $\mathcal{L}$ is the differential operator $\mathcal{L}u(\tau,z) = \frac{\sigma^2}{2}\partial_{zz}u(\tau,z)$. Let F be the finite difference approximation of L at grid points (i,j), v be the exact solution of the difference equation, i.e., $F(v) = 0$, and u be the exact solution of the PDE, namely $L(u) = 0$.

Definition (Truncation error) For any continuous function ϕ of τ and z, we define the *truncation error* at a grid point (i,j) by

$$T_{i,j}(\phi) = F(\phi_{ij}) - L(\phi_{i,j}),$$

where $\phi_{ij} = \phi(i\delta\tau, z_L + j\delta z)$. By setting $\phi = u$ and using $L(u) = 0$, we have

$$T_{i,j}(u) = F(u_{ij}) - L(u_{i,j}) = F(u_{ij}). \tag{4.36}$$

The value $F(u_{ij})$ in expression (4.36) is the *local truncation error* at grid point (i,j).

In other words, $T_{i,j}(\phi)$ represents an estimate of the error generated by replacing $L(\phi_{i,j})$ with $F(\phi_{i,j})$. Let C be a positive constant independent on $\delta\tau$ and on δz. If p and q are the largest positive integers for which

$$|T_{i,j}(u)| \le C\big((\delta\tau)^p + (\delta z)^q\big)$$

as $\delta\tau \to 0$ and $\delta z \to 0$, the scheme is said to have order of accuracy p in $\delta\tau$ and q in δz. For instance, the local truncation error in the explicit scheme as applied to equation (4.13) has order $p = 1$ and $q = 2$, indeed it is given by $O(\delta\tau) + O(\delta z^2)$. In the Crank–Nicolson scheme we have $p = 2$ and $q = 2$.

Definition (Consistency) The approximating difference equation is said to be *consistent* with the original PDE provided that $F(u_{i,j}) \to 0$ as $\delta\tau \to 0$ and $\delta z \to 0$.

Since the local truncation error in the explicit scheme vanishes as $\delta\tau \to 0$ and $\delta z \to 0$, it follows that this scheme is consistent with the original PDE. The same statement holds for both implicit and CN schemes.

Definition (Convergence) A finite difference scheme is *convergent* according to a given norm $\| \cdot \|$ provided that its exact solution converges to the exact solution of the original PDE

$$\max_{i,j} \| v_{i,j} - u_{i,j} \| \to 0,$$

as $\delta\tau \to 0$ and $\delta z \to 0$.

Let us introduce the discretization error that quantifies the accuracy of the solution at the grid point (i, j):

$$\varepsilon_{i,j} = v_{i,j} - u_{i,j}.$$

We observe that truncation error is a local concept. For the purpose of assessing the proximity of the solution stemming from the difference operator F to the one resulting from the differential operator L, the error clearly depends on $\delta\tau$ and δz. On the contrary, the proposed definition of convergence is a global assessment. More precisely, we are interested the way the difference between true and numerical solutions behave at a generic point τ, z as long as the grid becomes more and more refined.

To prove convergence, it is convenient to use the maximum norm given by:

$$E_i = \|\varepsilon_i\| = \max_{j=1,\dots,m} |\varepsilon_{ij}|.$$

We aim at investigating $\lim_{\delta z \to 0} E_i$, where τ is arbitrary and $\delta z \to 0$ with $\delta\tau = \alpha\delta z$ for some constant α.

In general the problem of convergence is dealt with using the Lax equivalence theorem. Given a well-posed linear initial-value problem[7] and a corresponding linear finite-difference approximation satisfying the consistency condition, this result states that stability is a necessary and sufficient condition for convergence.

Stability refers to the fact that small perturbations introduced through numerical rounding at any stage do not grow and dominate the solution. In other words, if we could use exact arithmetic, the whole error should be represented by a truncation error.

---

[7] Broadly speaking, a problem is well-posed if: (a) a solution always exists for initial data that is arbitrarily close to initial data for which no solution exists; (b) it is unique; (c) it depends on initial data with continuity.

Definition (Lax–Richtmyer stability) A scheme is *Lax–Richtmyer stable* if the solution of the finite difference equation at a fixed time level remains bounded as $\delta z \to 0$. It is asymptotically stable provided that boundedness holds true as $i \to \infty$, $\delta \tau$ fixed.

Given the definition of stability, we can now state the Lax Equivalence Theorem. This theorem studies the relation between consistency, stability, and convergence of the approximations of linear initial value problems by finite difference equations.

Theorem (Lax Equivalence Theorem) *For a consistent difference approximation to a well-posed linear initial-value problem, the stability of the scheme is necessary and sufficient for convergence.*

The implication of this theorem is that, once consistency has been established, we need to verify the condition for stability. Stability analysis can be conducted by using the matrix form of the time recursion. Let $\mathbf{F}$ be the matrix arising from the second-order centered difference approximation of $\mathcal{L}u(\tau, z)$, i.e., $\mathbf{F}/\delta z^2 = \text{tridiag}\{1, -2, 1\}/\delta z^2$. In the following, we will use the symbol $\text{tridiag}\{a, b, c\}$ to indicate a tridiagonal matrix, where the lower (resp. main and upper) diagonal contains a (resp. b and c) entries only:

$$\text{tridiag}\{a, b, c\} = \begin{bmatrix} b & c & & & \\ a & b & c & & \\ & a & \ddots & \ddots & \\ & & \ddots & b & c \\ & & & a & b \end{bmatrix}. \tag{4.37}$$

We have seen that the three schemes can be written as:

$$\mathbf{v}_{i+1} = \mathbf{A}\mathbf{v}_i + \mathbf{b}_i, \tag{4.38}$$

where vector $\mathbf{b}_i$ keeps track of boundary conditions and may vary over time, whilst matrix $\mathbf{A}$ is set according to the following scheme:

Explicit $\quad \dfrac{\mathbf{v}_{i+1} - \mathbf{v}_i}{\delta \tau} = \dfrac{\sigma^2}{2} \dfrac{\mathbf{F}}{\delta z^2} \mathbf{v}_i \implies \mathbf{A} = \mathbf{I} + \alpha \dfrac{\sigma^2}{2} \mathbf{F},$

Implicit $\quad \dfrac{\mathbf{v}_i - \mathbf{v}_{i-1}}{\delta \tau} = \dfrac{\sigma^2}{2} \dfrac{\mathbf{F}}{\delta z^2} \mathbf{v}_i \implies \mathbf{A} = \left(\mathbf{I} - \alpha \dfrac{\sigma^2}{2} \mathbf{F}\right)^{-1},$

CN $\quad \dfrac{\mathbf{v}_i - \mathbf{v}_{i-1}}{\delta \tau} = \dfrac{\sigma^2}{2} \dfrac{\mathbf{F}\mathbf{v}_{i+1} + \mathbf{F}\mathbf{v}_i}{2\delta z^2}$

$\quad\quad \implies \mathbf{A} = \left(\mathbf{I} - \sigma^2 \dfrac{\alpha}{4} \mathbf{F}\right)^{-1} \left(\mathbf{I} + \sigma^2 \dfrac{\alpha}{4} \mathbf{F}\right),$

and where $\alpha = \delta \tau / \delta z^2$.

Lax–Richtmyer stability requires that the solution remains bounded as $i \to \infty$. If $\mathbf{v}_0$ represents a stated initial condition, an iterative application of (4.37) leads to

$$\mathbf{v}_i = \mathbf{A}^i \mathbf{v}_0 + \mathbf{A}^{i-1}\mathbf{b}_0 + \mathbf{A}^{i-2}\mathbf{b}_1 + \cdots + \mathbf{b}_{i-1}.$$

Let us now consider the way a perturbation $\mathbf{e}_0$ (e.g., a numerical rounding) affecting initial condition $\mathbf{v}_0$ propagates on the solution. The biased initial condition is now $\mathbf{v}_0^* = \mathbf{v}_0 + \mathbf{e}_0$ and the corresponding solution at time step i reads as:

$$\mathbf{v}_i^* = \mathbf{A}^i \mathbf{v}_0^* + \mathbf{A}^{i-1}\mathbf{b}_0 + \mathbf{A}^{i-2}\mathbf{b}_1 + \cdots + \mathbf{b}_{i-1}.$$

The error vector can be computed as:

$$\mathbf{e}_i = \mathbf{v}_i^* - \mathbf{v}_i = \mathbf{A}^i (\mathbf{v}_0^* - \mathbf{v}_0) = \mathbf{A}^i \mathbf{e}_0.$$

We need to establish the extent error $\mathbf{e}_i$ stays bounded as i diverges. In other words, we look for a constant $M > 0$ (independent of $\delta\tau$ and δz) such that:

$$\|\mathbf{e}_i\| = \|\mathbf{A}^i \mathbf{e}_0\| \le \|\mathbf{A}^i\| \|\mathbf{e}_0\| \le M \|\mathbf{e}_0\|,$$

for compatible matrix and vector norms.[8] In this case,

$$\|\mathbf{A}^i\| = \|\mathbf{A}\mathbf{A}^{i-1}\| \le \|\mathbf{A}\|\|\mathbf{A}^{i-1}\| \le \cdots \le \|\mathbf{A}\|^i,$$

and the stability condition is satisfied provided that $\|\mathbf{A}\| \le 1$.

Possible matrix norms are:

1. The 1-norm defined as the greatest among the sums of absolute values of column entries of $\mathbf{A}$:

$$\|\mathbf{A}\|_1 = \max_j \sum_{i=1}^{m} |a_{ij}|;$$

2. The ∞-norm defined as the greatest among the sums of absolute values of raw entries of $\mathbf{A}$:

$$\|\mathbf{A}\|_\infty = \max_i \sum_{j=1}^{m} |a_{ij}|;$$

3. The 2-norm defined as the spectral radius $\rho(\mathbf{A})$ of matrix $\mathbf{A}$, that is the largest absolute eigenvalue of $\mathbf{A}$:

$$\|\mathbf{A}\|_2 = \rho(\mathbf{A}),$$

provided that $\mathbf{A}$ is real and symmetric.

We now verify the cited stability condition for each of the three schemes mentioned above.

Explicit scheme   Let us consider norm $\|\cdot\|_\infty$. Matrix $\mathbf{A}$ is given by:

$$\mathbf{I} + \alpha \frac{\sigma^2}{2} \mathbf{F} = \text{tridiag}\left\{ \alpha \frac{\sigma^2}{2}, 1 - \alpha\sigma^2, \alpha \frac{\sigma^2}{2} \right\},$$

---

[8] The *norm of a matrix* $\mathbf{A}$ is a real positive number giving a measure of the size of the matrix. It satisfies the following axioms: (1) $\|\mathbf{A}\| > 0$ if $\mathbf{A} \ne \mathbf{0}$ and $\|\mathbf{A}\| = 0$ if $\mathbf{A} = \mathbf{0}$, (2) $\|c\mathbf{A}\| = |c|\|\mathbf{A}\|$ for a real or complex scalar c, (3) $\|\mathbf{A} + \mathbf{B}\| \le \|\mathbf{A}\| + \|\mathbf{B}\|$, (4) $\|\mathbf{AB}\| \le \|\mathbf{A}\|\|\mathbf{B}\|$. Matrix and vector norms are said to be *compatible*, or consistent, provided that $\|\mathbf{Ax}\| \le \|\mathbf{A}\|\|\mathbf{x}\|$, $\mathbf{x} \ne \mathbf{0}$.

and then

$$\|\mathbf{A}\|_\infty = \alpha\frac{\sigma^2}{2} + \left|1 - \alpha\sigma^2\right| + \alpha\frac{\sigma^2}{2} = \alpha\sigma^2 + \left|1 - \alpha\sigma^2\right|$$

$$= \begin{cases} 1 & \text{if } 1 - \alpha\sigma^2 \geq 0, \\ 2\alpha\sigma^2 - 1 > 0 & \text{if } 1 - \alpha\sigma^2 < 0. \end{cases}$$

Therefore, the explicit scheme is stable provided that

$$\alpha \leq \frac{1}{\sigma^2}.$$

In a standard heat equation, we have $\sigma^2 = 2$, so that the stability condition amounts to requiring that $\alpha \leq 1/2$. This condition is actually violated in the experiment exhibited in Fig. 4.6, where the numerical solution displays large oscillations. The explicit scheme turns out to be *conditionally* stable and consistent. By the Lax equivalence theorem, the scheme is also conditionally convergent.

Implicit scheme   Let us consider norm $\|\cdot\|_2$. Matrix $\mathbf{A}$ is given by $(\mathbf{I} - \sigma^2\frac{\alpha}{4}\mathbf{F})^{-1}$. $\mathbf{F}$ has known eigenvalues given by $-4\sin^2(s\pi/(2m))$ (see Smith (1985), pp. 58–59 and 154–156). The matrix $\mathbf{A}$ is symmetric and its eigenvalues are given by:

$$\lambda_s = \frac{1}{1 + \sigma^2\alpha\sin^2(\frac{s\pi}{2m})}, \quad s = 1, \ldots, m-1.$$

Consequently,

$$\|\mathbf{A}\|_2 = \rho(\mathbf{A}) = \max_s\left|\frac{1}{1 + \sigma^2\alpha\sin^2(\frac{s\pi}{2m})}\right| < 1, \quad \forall \alpha > 0,$$

proving that the implicit scheme is *unconditionally* stable. The method is also consistent and, by the Lax equivalence theorem, it is convergent.

Crank–Nicolson scheme   Let us consider norm $\|\cdot\|_2$. Matrix $\mathbf{A}$ is given by $(\mathbf{I} - \sigma^2\frac{\alpha}{4}\mathbf{F})^{-1}(\mathbf{I} + \sigma^2\frac{\alpha}{4}\mathbf{F})$. It can be shown that $\mathbf{A}$ is symmetric and has eigenvalues:[9]

$$\lambda_s = \frac{2 - \sigma^2\alpha\sin^2(\frac{s\pi}{2m})}{2 + \sigma^2\alpha\sin^2(\frac{s\pi}{2m})},$$

for $s = 1, \ldots, m$. Consequently,

$$\|\mathbf{A}\|_2 = \rho(\mathbf{A}) = \max_s\left|\frac{2 - \sigma^2\alpha\sin^2(\frac{s\pi}{2m})}{2 + \sigma^2\alpha\sin^2(\frac{s\pi}{2m})}\right| < 1,$$

for any $\alpha > 0$. This proves that the CN scheme is *unconditionally* stable. The method is also consistent and, by the Lax equivalence theorem, it is convergent.

---

[9] We exploit the fact that if two $m \times m$ symmetric matrices $\mathbf{B}$ and $\mathbf{C}$ commute (i.e., $\mathbf{BC} = \mathbf{CB}$), then $\mathbf{B}^{-1}\mathbf{C} = \mathbf{BC}^{-1}$ and $\mathbf{B}^{-1}\mathbf{C}^{-1}$ are symmetric too. In the present case, $\mathbf{F}$ is symmetric, and then $(\mathbf{I} \pm \sigma^2\frac{\alpha}{4}\mathbf{F})$ are also symmetric. Moreover, they commute. Hence, $\mathbf{A}$ is symmetric too. The eigenvalues of $\mathbf{F}$ are $-4\sin^2(s\pi/(2m))$, where $s = 1, \ldots, m$. Henceforth, the claimed result about the eigenvalues of $\mathbf{A}$ follows.

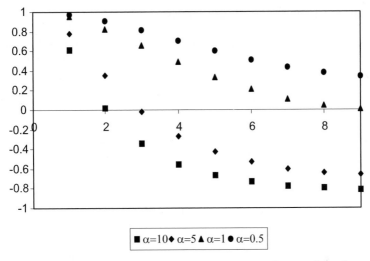

Fig. 4.14. Eigenvalues of the CN iteration matrix for different values of α.

The discussion above explains the presence of exploding oscillations in the explicit scheme whenever α is larger than $1/2$. It does not however justify the presence of spurious oscillations in the Crank–Nicolson scheme near discontinuity points. For this specific issue, we need to distinguish A_0-stability (i.e., all eigenvalues lay in $[-1, 1]$), from L_0-stability (i.e., all eigenvalues lay in $[0, 1]$). (See, again, Smith (1985, pp. 120–121).) The implicit scheme is unconditionally L_0-stable, whilst the CN scheme is conditionally L_0-stable, that is the iteration matrix can have negative eigenvalues. Although stability guarantees that the eigenvalues of **A** belong to the interval $[-1, 1]$, it is possible to find eigenvalues close to -1 for some value of α. As a consequence, the numerical solution will be affected by spurious oscillations. The phenomenon can be particularly pronounced in the neighborhood of points of discontinuity of the initial condition. There, oscillations are damped out only very slowly. For instance, while pricing a discrete barrier option, a discontinuity in the price function is introduced on every monitoring date (e.g., we set the option value equal to zero outside the barriers). Spurious oscillations can be avoided only by reducing the time-step of the grid. With reference to the example previously discussed, Fig. 4.14 shows the eigenvalues of the iteration matrix resulting from alternative values for the constant α and for $\delta z = 0.1$. As long as α decreases, the resulting eigenvalues all become positive. Of course, this fact implies a binding restriction on the time step $\delta\tau$. For example, let us set $\delta\tau = 0.1$, $\alpha = 10$, and a time to maturity 1. The resulting algorithm requires 10 time steps only. However, the corresponding numerical solution can be affected by spurious oscillations. By setting $\delta\tau = 0.005$ and $\alpha = 0.5$, we can prevent from this problem to occur. However, the number of time steps needs be increased to 200.

4.5 General Linear Parabolic PDEs

In the previous sections, we considered finite difference solutions to the standard heat equation. In general, we cannot reduce a linear parabolic PDE to this kind of equation. We now discuss how to numerically solve a general linear parabolic PDE:

$$-\partial_\tau u(\tau, z) + a(\tau, z)\partial_{zz}u + b(\tau, z)\partial_z u + c(\tau, z)u + d(\tau, z) = 0, \qquad (4.39)$$

where the function $a(\tau, z)$ is assumed to be strictly positive. We associate an initial condition at $\tau = 0$ and boundary conditions at states z_L and z_U to the above PDE. Here, $z_L = 0$ and $z_U = +\infty$ are allowed assignments.

Example (Double knock out option) Assuming the underlying asset evolves according to a square-root process, $dz = rz\,dt + \sigma\sqrt{z}\,dW_t$, then $a(\tau, z) = \sigma^2 z/2$, $b(\tau, z) = rz$, $c(\tau, z) = -r$, $d(\tau, z) = 0$ and the pricing PDE is

$$-\partial_\tau u(\tau, z) + \frac{1}{2}\sigma^2 z\partial_{zz}u + rz\partial_z u - ru = 0.$$

For a double knock out option, the initial condition reads as $u(0, z) = (z - k)_+$ and boundary conditions are $u(\tau, z_L) = u(\tau, z_u) = 0$.

The application of finite difference schemes requires an approximation of the partial derivative $\partial_z u$. In order to preserve an $O(\delta z^2)$ accuracy, it is natural to use a central difference approximation:

$$b(\tau, z)\partial_z u = b_{i,j}\frac{u_{i,j+1} - u_{i,j-1}}{2\delta z} + O(\delta z^2).$$

We now examine three benchmark schemes.

4.5.1 Explicit Scheme

The explicit scheme reads as:

$$-\frac{v_{i+1,j} - v_{i,j}}{\delta\tau} + a_{i,j}\frac{v_{i,j+1} - 2v_{i,j} + v_{i,j-1}}{\delta z^2}$$

$$+ b_{i,j}\frac{v_{i,j+1} - v_{i,j-1}}{2\delta z} + c_{i,j}v_{i,j} + d_{i,j} = 0,$$

leading to a recursion:

$$v_{i+1,j} = v_{i,j} + \frac{\delta\tau}{\delta z^2}a_{i,j}(v_{i,j+1} - 2v_{i,j} + v_{i,j-1})$$

$$+ \frac{\delta\tau}{2\delta z}b_{i,j}(v_{i,j+1} - v_{i,j-1}) + c_{i,j}v_{i,j} + d_{i,j}$$

$$= \left(\alpha a_{i,j} - \frac{\beta}{2}b_{i,j}\right)v_{i,j-1} + (1 - 2\alpha a_{i,j} + \delta\tau c_{i,j})v_{i,j}$$

$$+ \left(\alpha a_{i,j} + \frac{\beta}{2}b_{i,j}\right)v_{i,j+1} + \delta\tau d_{i,j},$$

where

$$\alpha = \frac{\delta\tau}{\delta z^2} \quad \text{and} \quad \beta = \frac{\delta\tau}{\delta z}.$$

Using a matrix notation, we have:

$$\mathbf{v}_{i+1} = \mathbf{A}_i \mathbf{v}_i + \mathbf{b}_i, \quad i = 1, 2, \ldots,$$

where $\mathbf{A}_i$ is the $m \times m$ tridiagonal matrix defined by

$$\mathbf{A}_i = \text{tridiag}\left\{ \alpha a_{i,j} - \frac{\beta}{2} b_{i,j}, 1 - 2\alpha a_{i,j} + \delta\tau c_{i,j}, \alpha a_{i,j} + \frac{\beta}{2} b_{i,j} \right\},$$

and $\mathbf{b}_i$ is the $m \times 1$ vector

$$\mathbf{b}_i^\mathsf{T} = \left[\left(\alpha a_{i,j} - \frac{\beta}{2} b_{i,j} \right) v_{i,0} \quad 0 \quad \cdots \quad 0 \quad \left(\alpha a_{i,j} + \frac{\beta}{2} b_{i,j} \right) v_{i,m+1} \right].$$

Here, $v_{i,0}$ and $v_{i,m+1}$ have been set according to the boundary conditions at states $z = z_\mathrm{L}$ and $z = z_\mathrm{U}$. Stability conditions now require that coefficients are all nonnegative and sum up to a number smaller than 1. This gives conditions:

$$(2\alpha a_{i,j} - c_{i,j}\delta\tau) < 1 \quad \text{and} \quad \frac{\beta}{2}|b_{i,j}| < \alpha a_{i,j},$$

which lead to restrictions on both space and time steps:

$$\delta z < 2 \frac{a_{i,j}}{|b_{i,j}|} \quad \text{and} \quad \delta\tau < \left(\frac{2}{\delta z^2} a_{i,j} - c_{i,j} \right)^{-1}.$$

The restriction can become relevant when the diffusion coefficient $a(\tau, z)$ is much smaller than the drift coefficient $b(\tau, z)$.

4.5.2 Implicit Scheme

Implicit discretization leads to:

$$-\left(\alpha a_{i,j} - \beta \frac{b_{i,j}}{2} \right) v_{i,j-1} + (1 + 2\alpha a_{i,j} - \delta\tau c_{i,j}) v_{i,j} - \left(\alpha a_{i,j} + \beta \frac{b_{i,j}}{2} \right) v_{i,j+1}$$
$$= v_{i-1,j},$$

for $j = 1, \ldots, m$ and $i = 1, \ldots, n$. At each time step, we need to solve the linear system:

$$\mathbf{B}\mathbf{v}_{i+1} = \mathbf{v}_i + \mathbf{b}_i, \quad i = 1, 2, \ldots,$$

where $\mathbf{B}_i$ is the tridiagonal matrix

$$\mathbf{B}_i = \text{tridiag}\left\{ \alpha a_{i,j} - \beta \frac{b_{i,j}}{2}, -1 + \frac{\delta\tau}{\delta z^2} a_{i,j} + \delta\tau c_{i,j}, \alpha a_{i,j} + \beta \frac{b_{i,j}}{2} \right\},$$

and

$$\mathbf{b}_i^\top = \left[-\left(\alpha a_{i,j} - \beta \frac{b_{i,j}}{2}\right) v_{i+1,0} \quad 0 \quad \cdots \quad 0 \quad -\left(\alpha a_{i,j} + \beta \frac{b_{i,j}}{2}\right) v_{i+1,m+1} \right].$$

Here, $v_{i+1,0}$ and $v_{i+1,m+1}$ are set according to the boundary conditions at $z = z_\mathrm{L}$ and $z = z_\mathrm{U}$.

Differently from discretizing a standard heat equation, $\mathbf{B}_i$ need not be a symmetric matrix.

4.5.3 Crank–Nicolson Scheme

We use a Taylor series expansion around point $(i\delta\tau + \frac{\delta\tau}{2}, z_\mathrm{L} + j\delta z)$. After performing tedious algebraic calculations, we arrive at a system of difference equations for $v_{i,j}$:

$$\left(\alpha \frac{a^*}{2} - \beta \frac{b^*}{4}\right) v_{i+1,j-1} - \left(1 + \alpha a^* - \frac{c^*}{2}\delta\tau\right) v_{i+1,j} + \left(\alpha \frac{a^*}{2} + \beta \frac{b^*}{4}\right) v_{i+1,j+1}$$
$$= \left(-\alpha \frac{a^*}{2} + \beta \frac{b^*}{4}\right) v_{i,j-1} - \left(1 - \alpha a^* + \frac{c^*}{2}\delta\tau\right) v_{i,j}$$
$$- \left(\alpha \frac{a^*}{2} + \beta \frac{b^*}{4}\right) v_{i,j+1} - d^*\delta\tau,$$

where:

$$a^* = a\left(i\delta z + \frac{\delta\tau}{2}, z_\mathrm{L} + j\delta z\right),$$
$$b^* = b\left(i\delta\tau + \frac{\delta\tau}{2}, z_\mathrm{L} + j\delta z\right),$$
$$c^* = r,$$
$$d^* = 0.$$

This iteration can be written in a shorter form by using a matrix algebra notation:

$$\mathbf{C}v_{i+1} = \mathbf{D}v_i + \mathbf{b}_i, \quad i = 1, 2, \ldots,$$

with matrices $\mathbf{C}$ and $\mathbf{D}$ defined by:

$$\mathbf{C} = \mathrm{tridiag}\left\{\alpha \frac{a^*}{2} - \beta \frac{b^*}{4}, -1 - \alpha a^* + \frac{c^*}{2}\delta\tau, \alpha \frac{a^*}{2} + \beta \frac{b^*}{4}\right\},$$
$$\mathbf{D} = \mathrm{tridiag}\left\{-\alpha \frac{a^*}{2} + \beta \frac{b^*}{4}, -1 + \alpha a^* - \frac{c^*}{2}\delta\tau, -\alpha \frac{a^*}{2} - \beta \frac{b^*}{4}\right\},$$

and vectors $\mathbf{b}_i$ defined as:

$$\mathbf{b}_i = \begin{bmatrix} -\left(\alpha \frac{a^*}{2} - \beta \frac{b^*}{4}\right) v_{i+1,0} + \left(-\alpha \frac{a^*}{2} + \beta \frac{b^*}{4}\right) v_{i,0} - d^*\delta\tau \\ -d^*\delta\tau \\ \cdots \\ -d^*\delta\tau \\ -\left(\alpha \frac{a^*}{2} + \beta \frac{b^*}{4}\right) v_{i+1,m+1} - \left(\alpha \frac{a^*}{2} + \beta \frac{b^*}{4}\right) v_{i,m+1} - d^*\delta\tau \end{bmatrix}.$$

Table 4.7. List of main VBA® functions for numerically solving (4.39)

VBA® function	Description	Default value
payoff	Returns the option payoff	$(z - K)_+$
upperbc	Returns the option value at the upper boundary	$z - e^{-r\tau} K$
lowerbc	Returns the option value at the lower boundary	0
PDEfunctionA	Returns the function $a(\tau, z)$ in (4.39)	$\frac{1}{2}\sigma^2 z^2$
PDEfunctionB	Returns the function $b(\tau, z)$ in (4.39)	rz
PDEImplicit	Solves (4.39) using implicit scheme and LU dec.	–
PDECN	Solves (4.39) using CN scheme and LU dec.	–
PDEExplicit	Solves (4.39) using explicit scheme and LU dec.	–
PDECNSOR	Solves (4.39) using CN scheme and SOR iteration	–

4.6 A VBA® Code for Solving General Linear Parabolic PDEs

A VBA® code for solving a general linear parabolic PDE has been implemented according to the description reported in Table 4.7. We assume the underlying asset follows a geometric Brownian motion and we aim at evaluating the arbitrage-free price a European call option. The user can easily modify the VBA® routine for the purpose of pricing other contracts. One needs only to modify the payoff function as expressed through the upper and lower boundary conditions appearing in the code. Under alternative driving processes, a change in the VBA® function PDEfunctionA is required.

4.7 Comments

The literature on PDE is vast. An introduction to analytical methods for PDEs can be found in Strauss (1992) and Zauderer (2006). The most readable introductory texts to numerical solution of PDEs are Smith (1985), Morton and Mayers (1994), Mitchell and Griffiths (1980). The relationship between stability and convergence (Lax Equivalence Theorem) was brought into organized form by Lax and Richtmyer (1956). Interested readers can find a proof in Richtmyer and Morton (1967), pp. 34–46. The different concepts of stability are discussed in Lambert (1991). More general references to numerical methods are Atkinson (1989) and Press et al. (1992). In particular, Smith also discusses the problem of spurious oscillations in the CN method. Morton and Mayers discuss how a small diffusion coefficient can alter the solution. The use of PDE in finance has been introduced by Brennan and Schwartz, Brennan and Schwartz (1977, 1978), Courtadon (1982) and Hull and White (1990). Nowadays, standard references in finance are Wilmott, Dewynne and Howison (1993), Tavella and Randall (2000) and James and Webber (2000). The PDE approach to solve problems in two or more dimensions is also discussed in these references. Contributions on specific topics are by Boyle and Tian (1998), Carr (2000), Fusai and Tagliani (2001), Pacelli, Recchioni and Zirilli (1999), Zvan, Forsyth and Vetzal (1998b). Extensions to jump processes are considered in D'Halluin, Forsyth and

Vetzal (2003), D'Halluin, Forsyth and Labahn (2005), Hirsa and Madan (2003) and Zhang (1997). Several papers, among the others we recall Pooley and Forsyth (2002) and Fusai, Sanfelici and Tagliani (2002), have studied how to cope with the oscillations that can affect the Crank–Nicolson solution. A finite element approach to PDEs arising in finance is given in Forsyth, Vetzal and Zvan (1999) and in Topper (2005). Sanfelici (2004) introduces the use in finance of the infinite element method, a simple and efficient modification of the more common finite element method. In the setting of diffusion models for price evolution. Corielli (2006) suggests an easily implementable approximate evaluation formula for measuring errors arising in option pricing and hedging due to volatility misspecification.

5

Numerical Solution of Linear Systems

In this chapter we present several methods for solving linear systems of the form $\mathbf{A}\mathbf{x} = \mathbf{b}$. Here $\mathbf{A}$ is a $(m \times m)$ matrix and both $\mathbf{x}$ and $\mathbf{b}$ are m-dimensional vectors. Our interest in linear systems is related to the solution of PDEs and therefore we will consider the case where the coefficient matrix $\mathbf{A}$ is tridiagonal, i.e.,

$$\mathbf{A} = \begin{bmatrix} a_1 & c_1 & 0 & 0 & 0 \\ b_2 & a_2 & \ddots & \ddots & 0 \\ 0 & \ddots & \ddots & \ddots & 0 \\ 0 & \ddots & \ddots & a_{m-1} & c_{m-1} \\ 0 & 0 & 0 & b_m & a_m \end{bmatrix}.$$

In the following we will adopt the notation $\operatorname{tridiag}(\{b_i\}_{i=2,\dots,n}, \{a_i\}_{i=1,\dots,n}, \{c_i\}_{i=1,\dots,n})$ to denote a tridiagonal matrix with diagonal entries given respectively by $\{b_i\}_{i=2,\dots,n}, \{a_i\}_{i=1,\dots,n}, \{c_i\}_{i=1,\dots,n}$. As a shorthand notation, we will also write $\operatorname{tridiag}(b_i, a_i, c_i)$. If the entries do not depend on the index i, we just write $\operatorname{tridiag}(b, a, c)$. In the previous chapter, we saw that when we deal with the implicit or the Crank–Nicolson discretization of a PDE, we have to solve a system of linear equations at each time step. In principle, if we compute the inverse $\mathbf{A}^{-1}$, assuming that it exists, then the solution of the linear system can be obtained by matrix vector multiplication, $\mathbf{x} = \mathbf{A}^{-1}\mathbf{b}$. Indeed, this would require a number $O(m^3)$ of operations, to which we add further $O(m^2)$ arithmetic operations for multiplying the resulting inverse matrix by vector $\mathbf{b}$. The problem with matrix inversion is that we lose the tridiagonal structure of $\mathbf{A}$, see Fig. 5.1. Therefore, storage requirements of $\mathbf{A}$ for large enough values of m can also become an issue.

Fortunately, when the matrix $\mathbf{A}$ has a tridiagonal structure, we have highly efficient algorithms exploiting this particular structure in the solution of the linear system at our disposal. Therefore, in this chapter, we will focus our attention on the solution of linear systems where the coefficient matrix $\mathbf{A}$ is tridiagonal. We will distinguish between direct and iterative methods. The former provide a solution in a finite number of steps, i.e. they will come up to an exact solution of the linear system in as many as $O(m)$ operations: a striking improvement with respect to matrix

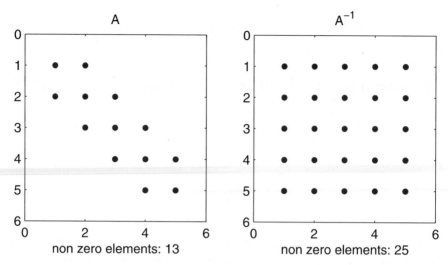

Fig. 5.1. The structure of a (5×5) tridiagonal matrix and the non-sparseness of its inverse.

inversion. Iterative methods begin with an initial vector $\mathbf{x}^{(0)}$ and generate a sequence of vectors $\mathbf{x}^{(1)}, \ldots, \mathbf{x}^{(k)}, \ldots$ which converge toward the desired solution as $k \to \infty$. Therefore, a stopping criteria, such as the norm of the difference between $\mathbf{x}^{(k)}$ and $\mathbf{x}^{(k-1)}$, is introduced. The main feature of iterative methods is the fact that an individual iteration requires an amount of work which is comparable to the multiplication of $\mathbf{A}$ with a vector, a very modest amount if $\mathbf{A}$ is tridiagonal. For this reason, one can, with a reasonable amount of work, still carry out a relatively large number of iterations. This is necessary, if for no other reason than the fact that these methods converge only linearly, and very slowly at that. Iterative methods are therefore usually inferior to the elimination methods if $\mathbf{A}$ is a small matrix (a 100×100 matrix is small in this sense) or not a sparse matrix. However, they prove their relevance when we deal with American options, for which direct methods are inappropriate.

In Sect. 5.1 we present one of the most famous direct methods: LU decomposition, a simplification of the Gaussian elimination procedure. Section 5.2 illustrates iterative methods, such as Jacobi, Gauss–Seidel, Successive over-relaxation (SOR) and Conjugate Gradient (CG) methods; we will also discuss the convergence properties of the different methods. Sections 5.3.1 and 5.3.2 present VBA® and MATLAB® codes. Finally, we conclude with Sect. 5.4 presenting several applications of PDEs to pricing problems in finance.

5.1 Direct Methods: The LU Decomposition

The most appropriate algorithm for tridiagonal linear systems is based on the Gaussian elimination procedure. A linear system where the matrix of coefficients is lower triangular, i.e.,

$$\begin{bmatrix} 1 & 0 & 0 \\ l_{21} & 1 & 0 \\ l_{22} & l_{23} & 1 \end{bmatrix} \begin{bmatrix} x_1 \\ x_2 \\ x_3 \end{bmatrix} = \begin{bmatrix} b_1 \\ b_2 \\ b_3 \end{bmatrix},$$

can be solved by forward substitution, starting from the first equation and proceeding upward to the last one. Similarly, a linear system with an upper triangular matrix of coefficients, i.e.,

$$\begin{bmatrix} u_{11} & u_{12} & u_{13} \\ 0 & u_{22} & u_{23} \\ 0 & 0 & u_{33} \end{bmatrix} \begin{bmatrix} x_1 \\ x_2 \\ x_3 \end{bmatrix} = \begin{bmatrix} b_1 \\ b_2 \\ b_3 \end{bmatrix}$$

can be solved by backward substitution starting from last equation and proceeding backward to the first one.

Let us now consider the linear system $\mathbf{Ax} = \mathbf{b}$, where $\mathbf{A}$ is a $(m \times m)$ matrix and both $\mathbf{x}$ and $\mathbf{b}$ are m-dimensional vectors. The algorithm can be formalized by factorizing the matrix $\mathbf{A}$ as:

$$\mathbf{A} = \mathbf{LU},$$

where $\mathbf{L}$ is lower triangular and $\mathbf{U}$ is upper triangular. For example, if $\mathbf{A}$ is a (3×3) matrix, we would write

$$\mathbf{A} = \begin{bmatrix} 1 & 0 & 0 \\ l_{21} & 1 & 0 \\ l_{22} & l_{23} & 1 \end{bmatrix} \begin{bmatrix} u_{11} & u_{12} & u_{13} \\ 0 & u_{22} & u_{23} \\ 0 & 0 & u_{33} \end{bmatrix}.$$

Therefore, the linear system $\mathbf{Ax} = \mathbf{b}$ can be written as $\mathbf{L(Ux)} = \mathbf{b}$. If we set $\mathbf{Ux} = \mathbf{y}$, then we have $\mathbf{Ly} = \mathbf{b}$. In other words, once the $\mathbf{LU}$ factorization has been done, the original linear system can be solved in two steps:

1. Solve the linear system $\mathbf{Ly} = \mathbf{b}$ for the unknown vector $\mathbf{y}$ by forward substitution, i.e.,

$$\begin{cases} y_1 = b_1, \\ y_2 = b_2 - l_{21}y_1, \\ y_3 = b_3 - l_{22}y_1 - l_{23}y_2. \end{cases}$$

2. Given $\mathbf{y}$, solve $\mathbf{Ux} = \mathbf{y}$ for the vector $\mathbf{x}$ by backward substitution, i.e.,

$$\begin{cases} x_3 = \frac{y_3}{u_{33}}, \\ x_2 = \frac{y_2 - u_{23}x_3}{u_{22}}, \\ x_1 = \frac{y_1 - l_{22}y_1 - l_{23}y_2}{u_{11}}. \end{cases}$$

When matrix $\mathbf{A}$ is tridiagonal, one can immediately find the decomposition $\mathbf{A} = \mathbf{LU}$. Indeed, most elements of $\mathbf{L}$ and $\mathbf{U}$ are equal to zero. This decomposition leads to the following factorization:

$$\begin{bmatrix} a_1 & c_1 & 0 & 0 & 0 \\ b_2 & a_2 & \ddots & \ddots & 0 \\ 0 & \ddots & \ddots & \ddots & 0 \\ 0 & \ddots & \ddots & a_{m-1} & c_{m-1} \\ 0 & 0 & 0 & b_m & a_m \end{bmatrix}$$

$$
= \begin{bmatrix} 1 & 0 & 0 & 0 & 0 \\ l_2 & 1 & 0 & \ddots & 0 \\ 0 & \ddots & \ddots & \ddots & 0 \\ 0 & \ddots & \ddots & 1 & 0 \\ 0 & 0 & 0 & l_m & 1 \end{bmatrix} \begin{bmatrix} d_1 & u_1 & 0 & 0 & 0 \\ 0 & d_2 & u_2 & \ddots & 0 \\ 0 & \ddots & \ddots & \ddots & 0 \\ 0 & \ddots & \ddots & d_{m-1} & u_{m-1} \\ 0 & 0 & 0 & 0 & d_m \end{bmatrix}.
$$

If we multiply $\mathbf{L}$ and $\mathbf{U}$ we obtain a recursive procedure to compute $d_1, u_1, l_2, d_2, \ldots, d_m$:

Algorithm (LU Decomposition)
```
Assign elements a(j), b(j), c(j) of the three main diag-
onals
d(1) = a(1)
For j = 2,...,m
   u(j-1) = c(j-1)
   l(j) = b(j)/d(i-1)
   d(j) = a(j)-l(j)*c(j-1)
Next j
Return vectors l, d, u.
```

A numerical example is provided in Fig. 5.2.

Note that, once we have factorized $\mathbf{A}$, we do not store the whole matrices $\mathbf{L}$ and $\mathbf{U}$, but just the $3m - 2$ entries of vectors $[l_2, \ldots, l_m]^\top$, $[d_1, \ldots, d_m]^\top$ and $[u_1, \ldots, u_{m-1}]^\top$. These are the only quantities entering the linear systems arising from implicit and Crank–Nicolson schemes. The following pseudo-code details the procedure that combines LU factorization with forward and backward substitution and return the solution of the tridiagonal system.

Algorithm (Tridiagonal linear system)
```
Assign the three main diagonals and the constant vector
vecb.
auxvar = diag(1)
```

	A	B	C	D	E	F	G	H	I	J	K	L	M
1						**LU Decomposition:**							
2				Find L and U such that A=L U, where A is tridiagonal and L (U) is a lower (upper) triangular matrix									
3													
4	j	lower diagonal (b)	main diagonal (a)	upper diagonal (c)	l	d	u	l	d	u	Algorithm: LU Decomposition		
5	1		1.4	-0.2		1.4	-0.2			= E7			
6	2	-0.2	1.4	-0.2	-0.14286	1.37143	-0.2	= C8/H7	= D8-G8*E7	= E8	d(1) = a(1)		
7	3	-0.2	1.4	-0.2	-0.14583	1.37083	-0.2	= C9/H8	= D9-G9*E8	= E9	j = 2...m		
8	4	-0.2	1.4	-0.2	-0.14590	1.37082	-0.2	= C10/H9	= D10-G10*E9	= E10	u(j-1) = c(j-1)		
9	5	-0.2	1.4	-0.2	-0.14590	1.37082	-0.2	= C11/H10	= D11-G11*E10	= E11	l(j) = b(j)/d(j-1)		
10	6	-0.2	1.4	-0.2	-0.14590	1.37082	-0.2	= C12/H11	= D12-G12*E11	= E12	d(j) = a(j)-l(j)*c(j-1)		
11	7	-0.2	1.4	-0.2	-0.14590	1.37082	-0.2	= C13/H12	= D13-G13*E12	= E13	Next j		
12	8	-0.2	1.4	-0.2	-0.14590	1.37082	-0.2	= C14/H13	= D14-G14*E13	= E14			
13	9	-0.2	1.4		-0.14590	1.37082		= C15/H14	= D15-G15*E14				
14													

Fig. 5.2. An example of LU decomposition.

```
solution(1) = vecb(1)/auxvar
// Forward substitution
For j = 2 To m
  gam(j) = updiag(j-1)/auxvar
  auxvar = diag(j)-lowdiag(j)*gam(j)
  solution(j) = (vecb(j)-lowdiag(j)*solution(j-1))/aux-
var
Next j
// Backward substitution
For j = m - 1 To 1 Step -1
  solution(j)=solution(j)-gam(j+1)*solution(j+1)
Next j
End
Return solution.
```

A spreadsheet-based illustration of this algorithm is shown in Fig. 5.3. Figure 5.4 shows a chart detailing the implementation of a solution algorithm, using the user-defined VBA® function tridag, for the example

$$-\partial_\tau u(\tau, z) + \partial_{zz} u(\tau, z) = 0, \tag{5.1}$$

initial condition:

$$u(0, z) = \begin{cases} 2z & 0 \le z \le \frac{1}{2}, \\ 2(1-z) & \frac{1}{2} \le z \le 1, \end{cases} \tag{5.2}$$

and boundary conditions:

$$u(\tau, 1) = u(\tau, 0) = 0. \tag{5.3}$$

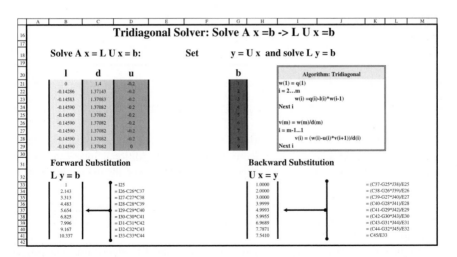

Fig. 5.3. Implementing the tridiagonal algorithm using forward and backward substitution.

126 5 Numerical Solution of Linear Systems

The recursion for the C-N Method using LU decomposition

C v(i+1) = D v(i)+(α/2)*b(i)

	A	B	C	D	E
3	dx	0.1	This cell has been named dxi		
4	dτ	0.001	This cell has been named dtc		
5	α	0.1	This cell has been named alphac		

	lower	diag	upper
7		1.1	-0.05
8	-0.05	1.1	-0.05
9	-0.05	1.1	-0.05
10	-0.05	1.1	-0.05
11	-0.05	1.1	-0.05
12	-0.05	1.1	-0.05
13	-0.05	1.1	-0.05
14	-0.05	1.1	-0.05
15	-0.05	1.1	0

=tridag($B$7:$B$8,$C$7:$C$8,$D$7:$D$8,G32:G40,9)

v(m+1) is found by LU decomposition of the matrix C

z\τ	0	0.001	0.002	0.003	0.004	0.005	0.006	0.007	0.008	0.009	0.01
0	0	0	0	0	0	0	0	0	0	0	0
0.1	0.2	0.2	0.199999	0.199993	0.199979	0.199949	0.199895	0.199807	0.199678	0.199499	0.199263
0.2	0.4	0.399997	0.399975	0.399906	0.399758	0.399497	0.399101	0.398549	0.397831	0.396941	0.395879
0.3	0.6	0.599924	0.599584	0.598818	0.597563	0.595814	0.593599	0.590962	0.587952	0.58462	0.581012
0.4	0.8	0.798337	0.793914	0.787687	0.780306	0.772215	0.763716	0.755011	0.746238	0.737485	0.728811
0.5	1	0.963485	0.933056	0.907191	0.884792	0.865059	0.847406	0.831398	0.816712	0.803103	0.790384
0.6	0.8	0.798337	0.793914	0.787687	0.780306	0.772215	0.763716	0.755011	0.746238	0.737485	0.728811
0.7	0.6	0.599924	0.599584	0.598818	0.597563	0.595814	0.593599	0.590962	0.587952	0.58462	0.581012
0.8	0.4	0.399997	0.399975	0.399906	0.399758	0.399497	0.399101	0.398549	0.397831	0.396941	0.395879
0.9	0.2	0.2	0.199999	0.199993	0.199979	0.199949	0.199895	0.199807	0.199678	0.199499	0.199263
1	2.22E-16	0	0	0	0	0	0	0	0	0	0

Z(m)= D v(m)-(α/2)*h(m)

0.2	0.199999686	0.19999745	0.199989404	0.199969303	0.199929199	0.199860301	0.199753768	0.19960133	0.199395719
0.4	0.399993099	0.399956449	0.399856343	0.39996590	0.399335881	0.39886517	0.398232316	0.397429122	0.396452739
0.6	0.599048486	0.599320485	0.593316127	0.596809862	0.594818295	0.592380093	0.589543723	0.586360173	0.582078039
0.8	0.796673582	0.791154805	0.784218455	0.776392855	0.768037181	0.759304512	0.750628241	0.741847103	0.733122384
0.98	0.945970326	0.919141938	0.895240973	0.874343642	0.855774802	0.83903681	0.823739458	0.809664588	0.79654098
0.8	0.796673582	0.791154805	0.784218455	0.776392855	0.768037181	0.759304512	0.750628241	0.741847103	0.733122384
0.6	0.599048486	0.599320485	0.593316127	0.596809862	0.594818295	0.592380093	0.589543723	0.586360173	0.582878839
0.4	0.399993099	0.399956449	0.399856343	0.39996590	0.399335881	0.39886517	0.398232316	0.397429122	0.396452739
0.2	0.199999686	0.19999745	0.199989404	0.199969303	0.199929199	0.199860301	0.199753768	0.19960133	0.199395719

h(m)

0	0	0	0	0	0	0	0	0
0	0	0	0	0	0	0	0	0
0	0	0	0	0	0	0	0	0
0	0	0	0	0	0	0	0	0
0	0	0	0	0	0	0	0	0
0	0	0	0	0	0	0	0	0
0	0	0	0	0	0	0	0	0
0	0	0	0	0	0	0	0	0
0	0	0	0	0	0	0	0	0

Fig. 5.4. Implementing the Crank–Nicolson scheme using the tridiagonal solver at each time step.

The total number of basic operations, i.e., multiplications and divisions, needed to calculate L and U is $2m-2$; in order to solve $Ax = b$, additional $3m-2$ operations are required. The amount of work is therefore $O(m)$. However, the main advantage of the procedure described above is that a limited amount of numbers, that is $l_2, \ldots, l_m$, $u_1, \ldots, u_m$, and $d_1, \ldots, d_m$, needs to be stored, just $3m - 2$ elements. Inverting matrix A requires storing m^2 numbers and the overall computation requires $O(m^3)$ operations.

Atkinson (1989) performs an error analysis of the LU algorithm, focusing in particular on the stability of the solution under small perturbations affecting vector b. In particular, a measure for the degree of stability in this respect is given by the *condition number* v of matrix $\mathbf{A}$ as defined by the ratio between the largest and the smallest absolute eigenvalue of $\mathbf{A}$:

$$v = \frac{\max_{\lambda \in \Lambda(\mathbf{A})} |\lambda|}{\min_{\lambda \in \Lambda(\mathbf{A})} |\lambda|}.$$

Here, $\Lambda(\mathbf{A})$ denotes the spectrum of $\mathbf{A}$, i.e. the set of the eigenvalues of $\mathbf{A}$. If this ratio is nearly 1, the system is well-conditioned. As this ratio rises, the system becomes more and more ill-conditioned, meaning that a small perturbation of $\mathbf{b}$ produces a large variation in the solution of the corresponding linear system. This can be of great concern for the final user due to the finiteness of the computer machine accuracy.[1] Stability is guaranteed by the following conditions: (a) $l_i > 0$, $d_i > 0$, and $u_i > 0$, (b) $d_i > l_{i+1} + u_{i-1}$, for $i = 1, \ldots, m - 1$, with $l_m = u_0 = 0$, (c) $d_i > l_i + c_i$, with $l_1 = u_{m-1} = 0$. Namely, conditions (a) and (b), or (a) and (c), ensure that forward (backward) substitution is stable. These conditions are satisfied when LU decomposition is applied to matrices obtained by discretizing the heat equation according to both implicit and CN schemes.

5.2 Iterative Methods

Several pricing problems require highly accurate solutions. These can be achieved by thoroughly refining the space grid. In all these cases, the LU decomposition can be excessively time consuming. Iterative methods may constitute a valid alternative in these instances. The basic idea is to provide a recursive procedure that keeps on improving the degree of approximation until a confidence threshold is achieved. As iterative methods operate on nonzero elements only, the tridiagonal form of matrix $\mathbf{A}$ in the above examples is effectively exploited for the purpose of decreasing the number of computations. Moreover, no matrix inversion is required, hence the sparsity of the original linear system is preserved. Finally, iterative methods can be used to price American options, while direct methods are not applicable in that context. This actually is the main reason for using iterative methods for solving PDE.

In this section we consider some of the most common and simple iterative methods for the solution of a linear system $\mathbf{A}\mathbf{x} = \mathbf{b}$. In particular, we describe four methods: (1) Jacobi, (2) Gauss–Seidel, (3) SOR, and (4) Conjugate Gradient Method.

Let us write the matrix $\mathbf{A}$ as a sum of three matrices, i.e., $\mathbf{A} = \mathbf{D} + \mathbf{L} + \mathbf{U}$, where $\mathbf{D}$ is a diagonal matrix and $\mathbf{L}$ and $\mathbf{U}$ are a lower and an upper triangular matrix with zeros on the main diagonal. The linear system $\mathbf{A}\mathbf{x} = \mathbf{b}$ then becomes

$$(\mathbf{D} + \mathbf{L} + \mathbf{U})\mathbf{x} = \mathbf{b},$$

so that $\mathbf{D}\mathbf{x} = -(\mathbf{L} + \mathbf{U})\mathbf{x} + \mathbf{b}$. By assuming that diagonal elements of $\mathbf{A}$ are all nonzero,[2] the last expression gives

---

[1] This figure usually amounts to either 16 or 32 digits.
[2] If $\mathbf{A}$ is nonsingular and has some diagonal element equal to zero, we can always interchange rows and colums and obtain a nonsingular matrix $\mathbf{D}$.

$$\mathbf{x} = -\mathbf{D}^{-1}(\mathbf{L} + \mathbf{U})\mathbf{x} + \mathbf{D}^{-1}\mathbf{b}. \tag{5.4}$$

This equation represents the starting point for the first three iterative methods which will be developed below. Then the idea of iterative methods is to transform the above equation into an iterative procedure. Given the result of the $(k-1)$th iteration (the right-hand side in (5.4)), a new vector is generated (the left-hand side in (5.4)) for the kth step.

5.2.1 Jacobi Iteration: Simultaneous Displacements

In this iterative method, if we know the approximate solution $\mathbf{x}^{(k-1)}$ at the $(k-1)$th iteration, we use (5.4) to update the solution according to the iteration

$$\mathbf{x}^{(k)} = -\mathbf{D}^{-1}(\mathbf{L} + \mathbf{U})\mathbf{x}^{(k-1)} + \mathbf{D}^{-1}\mathbf{b}.$$

By setting $\mathbf{H} = -\mathbf{D}^{-1}(\mathbf{L} + \mathbf{U})$, we have the general iteration

$$\mathbf{x}^{(k)} = \mathbf{H}\mathbf{x}^{(k-1)} + \mathbf{D}^{-1}\mathbf{b}.$$

The starting condition $\mathbf{x}^{(0)}$ is somehow arbitrary. If the linear system arises in the solution of a PDE, a possible choice is to use the solution of the PDE computed at the previous time step i or just set $\mathbf{x}^{(0)} = \mathbf{0}$.

When $\mathbf{A}$ is tridiagonal, $\mathbf{A} = \text{tridiag}(l_i, d_i, u_i)$, the iterations stemming from the Jacobi method read as follows:

$$
\begin{bmatrix} x_1^{(k)} \\ x_i^{(k)} \\ x_m^{(k)} \end{bmatrix} = -
\begin{bmatrix} \frac{1}{d_1} & & 0 & 0 \\ & \frac{1}{d_i} & & \\ 0 & & 0 & \\ 0 & & 0 & \frac{1}{d_m} \end{bmatrix}
$$

$$
\times \left(\begin{bmatrix} 0 & & 0 & 0 \\ l_2 & & & \\ & 0 & & \\ 0 & l_{i+1} & & 0 \\ 0 & & l_m & 0 \end{bmatrix} + \begin{bmatrix} 0 & u_1 & 0 & 0 \\ & 0 & & \\ 0 & & u_{m-1} & \\ 0 & & 0 & 0 \end{bmatrix} \right)
$$

$$
\times \begin{bmatrix} x_1^{(k)} \\ x_i^{(k)} \\ x_m^{(k)} \end{bmatrix} + \begin{bmatrix} \frac{b_1}{d_1} \\ \frac{b_i}{d_i} \\ \frac{b_m}{d_m} \end{bmatrix}.
$$

Then, we have a simple recursive formula:

$$x_i^{(k)} = \frac{1}{d_i}\left(b_i - l_{i-1}x_{i-1}^{(k-1)} - u_i x_{i+1}^{(k-1)}\right), \quad i = 1, \ldots, m. \tag{5.5}$$

where we have set $l_0 = u_m = 0$. The order in which the equations are processed is irrelevant: the Jacobi method deals with them independently. Hence, we have the expression "simultaneous displacements" meaning that updates could in principle be done simultaneously.

Example Let us consider the following linear system:

$$
\begin{bmatrix} 3 & 1 & 0 \\ -1 & 4 & 2 \\ 0 & 1 & 2 \end{bmatrix} \begin{bmatrix} x_1 \\ x_2 \\ x_3 \end{bmatrix} = \begin{bmatrix} 1 \\ 2 \\ 0 \end{bmatrix},
$$

with a starting value $\mathbf{x}^{(0)} = \mathbf{0}$. The first three iterations read as:

$k = 1$:

$$x_1^{(1)} = \frac{1}{3}(1 - 1 \times 0) = 0.333\,33,$$

$$x_2^{(1)} = \frac{1}{4}\big(2 - (-1) \times 0 - 2 \times 0\big) = 0.5,$$

$$x_3^{(1)} = \frac{1}{2}(0 - 1 \times 0) = 0;$$

$k = 2$:

$$x_1^{(2)} = \frac{1}{3}(1 - 1 \times 0.5) = 0.166\,67,$$

$$x_2^{(2)} = \frac{1}{4}\big(2 - (-1) \times (0.333\,33) - 2 \times 0\big) = 0.583\,33,$$

$$x_3^{(2)} = \frac{1}{2}(0 - 1 \times (0.5)) = -0.25;$$

$k = 3$:

$$x_1^{(3)} = \frac{1}{3}\big(1 - 1 \times (0.583\,33)\big) = 0.138\,89,$$

$$x_2^{(3)} = \frac{1}{4}\big(2 - (-1) \times 0.166\,67 - 2 \times (-0.25)\big) = 0.666\,67,$$

$$x_3^{(3)} = \frac{1}{2}\big(0 - 1 \times (0.583\,33)\big) = -0.291\,67.$$

The next iterations are reported in the Table 5.1. In last column, we indicate the Euclidean distance between vectors $\mathbf{x}^{(k)}$ and $\mathbf{x}^{(k-1)}$. This figure provides a criterion to decide upon terminating the recursive procedure. The iteration is broken as soon as the distance $\|\mathbf{x}^{(k)} - \mathbf{x}^{(k-1)}\| \leq \varepsilon$, where ε represents a tolerance threshold that is

Table 5.1. Iterations of the Jacobi method

k	$x_1^{(k)}$	$x_2^{(k)}$	$x_3^{(k)}$	$\|\mathbf{x}^{(k)} - \mathbf{x}^{(k-1)}\|$
4	0.1111	0.6806	−0.3333	0.0027
5	0.1065	0.6944	−0.3403	0.00026
6	0.1019	0.6968	−0.3472	0.00008
7	0.1011	0.6991	−0.3484	0.00001
8	0.1003	0.6995	−0.3495	2.1×10^{-6}
9	0.1002	0.6998	−0.3497	2.0×10^{-7}
10	0.1001	0.6999	−0.3499	5.8×10^{-8}
11	0.1000	0.7000	−0.3500	5.6×10^{-9}

selected exogenously. A small value for ε determines a large number of iterations. In our example, we set $\varepsilon = 10^{-8}$, so that a break occurs after eleven iterations. The exact solution is $[\frac{1}{10} \quad \frac{7}{10} \quad -\frac{7}{20}]$.

5.2.2 Gauss–Seidel Iteration (Successive Displacements)

In this iterative scheme, each new component of $\mathbf{x}^{(k)}$ is immediately used in the computation of the following component. The attribute "successive" underlines the dependence of the iterates on the order in which equations are processed. This kind of updating process is also convenient for reasons related to memory storage. The new value can immediately be stored in the memory location of the previous value, thus minimizing the overall storage cost. Compared to the Jacobi method, the storage requirement for vector $\mathbf{x}$ is reduced by half.

Let us consider again the system (5.4). We may do the following transformation:

$$\mathbf{x} = -\mathbf{D}^{-1}\mathbf{L}\mathbf{x} - \mathbf{D}^{-1}\mathbf{U}\mathbf{x} + \mathbf{D}^{-1}\mathbf{b}.$$

If we need to compute the ith element of the vector $\mathbf{x}$ in the left-hand side, we observe that in the right-hand side (a) the elements x_j, with $1 \leq j < i$, are involved in the product $\mathbf{D}^{-1}\mathbf{L}\mathbf{x}$, (b) in the product $\mathbf{D}^{-1}\mathbf{U}\mathbf{x}$ we need the elements x_j, with $i < j \leq m$. Therefore, to accelerate the convergence in doing the multiplication $\mathbf{D}^{-1}\mathbf{L}\mathbf{x}$ we can use the already updated components. Then the iteration $\mathbf{x}^{(k)} = -\mathbf{D}^{-1}\mathbf{L}\mathbf{x}^{(k-1)} - \mathbf{D}^{-1}\mathbf{U}\mathbf{x}^{(k-1)} + \mathbf{D}^{-1}\mathbf{b}$ is replaced by

$$\mathbf{x}^{(k)} = -\mathbf{D}^{-1}\mathbf{L}\mathbf{x}^{(k)} - \mathbf{D}^{-1}\mathbf{U}\mathbf{x}^{(k-1)} + \mathbf{D}^{-1}\mathbf{b}, \qquad (5.6)$$

which gives the matrix-form of the Gauss–Seidel iteration:

$$\mathbf{x}^{(k)} = \left(\mathbf{I} + \mathbf{D}^{-1}\mathbf{L}\right)^{-1}\mathbf{D}^{-1}\left(\mathbf{b} - \mathbf{U}\mathbf{x}^{(k-1)}\right).$$

When $\mathbf{A}$ is tridiagonal, the iterative relation (5.6) collapses to the following algorithm:

$$x_i^{(k)} = \frac{1}{d_i}\left(b_i - l_{i-1} \underbrace{x_{i-1}^{(k)}}_{\text{updated value}} - u_i x_{i+1}^{(k-1)}\right), \quad i = 1, \ldots, m, \qquad (5.7)$$

where we have set $l_0 = u_m = 0$ once again.

Example With reference to the previous example, we have the following iterations, where the bold figures represent the updated values that are used in the iterations:

$k = 1$:

$$x_1^{(1)} = \frac{1}{3}(1 - 1 \times 0) = 0.333\,33,$$

$$x_2^{(1)} = \frac{1}{4}\left(2 - (-1) \times \mathbf{0.333\,33} - 2 \times 0\right) = 0.583\,33,$$

$$x_3^{(1)} = \frac{1}{2}(0 - 1 \times \mathbf{0.583\,33}) = -0.291\,67;$$

Table 5.2. Iterations of the Gauss–Seidel method

k	$x_1^{(k)}$	$x_2^{(k)}$	$x_3^{(k)}$	$\|\mathbf{x}^{(k)} - \mathbf{x}^{(k-1)}\|$
4	0.10108	0.69946	−0.34973	3.8×10^{-5}
5	0.10018	0.69991	−0.34995	1.1×10^{-6}
6	0.10003	0.69998	−0.34999	3.0×10^{-8}
7	0.10001	0.70000	−0.35000	8.2×10^{-10}

$k = 2$:

$$x_1^{(2)} = \frac{1}{3}(1 - 1 \times 0.583\,33) = 0.138\,89,$$

$$x_2^{(2)} = \frac{1}{4}\big(2 - (-1) \times (\mathbf{0.138\,89}) - 2 \times (-0.291\,67)\big) = 0.680\,56,$$

$$x_3^{(2)} = \frac{1}{2}\big(0 - 1 \times (\mathbf{0.680\,56})\big) = -0.340\,28;$$

$k = 3$:

$$x_1^{(3)} = \frac{1}{3}\big(1 - 1 \times (0.680\,56)\big) = 0.106\,48,$$

$$x_2^{(3)} = \frac{1}{4}\big(2 - (-1) \times \mathbf{0.106\,48} - 2 \times (-0.340\,28)\big) = 0.696\,76,$$

$$x_3^{(3)} = \frac{1}{2}\big(0 - 1 \times (\mathbf{0.696\,76})\big) = -0.348\,38.$$

The next iterations are given in the Table 5.2. If the stopping criterion is defined by a tolerance threshold $\varepsilon = 10^{-8}$, then exactly seven iterations are required to break the loop, whereas the Jacobi method leads to eleven steps.

5.2.3 SOR (Successive Over-Relaxation Method)

Generally speaking, iterative methods show a regularly decreasing pattern in the approximation error. This property can be exploited in order to speed up the convergence of the Gauss–Seidel method. At each step in the procedure, the SOR method consists of looking for an optimal linear combination between the previous approximate solution and the Gauss–Seidel iterate. To clarify this point, let us consider the system $\mathbf{Dx} = -(\mathbf{L} + \mathbf{U})\mathbf{x} + \mathbf{b}$ in the form:

$$(\mathbf{D} + \mathbf{L})\mathbf{x} = -\mathbf{Ux} + \mathbf{b}.$$

We may solve for $\mathbf{x}$ and obtain the following iterative procedure:

$$\mathbf{x}^{(k)} = (\mathbf{D} + \mathbf{L})^{-1}\big(-\mathbf{Ux}^{(k-1)} + \mathbf{b}\big).$$

By adding and subtracting $\mathbf{x}^{(k-1)}$ on the right-hand side, we get to:

$$\mathbf{x}^{(k)} = \underbrace{\mathbf{x}^{(k-1)} - (\mathbf{D} + \mathbf{L})^{-1}(\mathbf{D} + \mathbf{L})\mathbf{x}^{(k-1)}}_{=0} + (\mathbf{D} + \mathbf{L})^{-1}\big(-\mathbf{Ux}^{(k-1)} + \mathbf{b}\big)$$

$$= \mathbf{x}^{(k-1)} - (\mathbf{D} + \mathbf{L})^{-1}\big((\mathbf{L} + \mathbf{D} + \mathbf{U})\mathbf{x}^{(k-1)} - \mathbf{b}\big).$$

Notice that $\xi^{(k-1)} = (L + D + U)x^{(k-1)} - b$ represents the error at the $(k-1)$th iteration. We may then write:

$$x^{(k)} = x^{(k-1)} - (D + L)^{-1}\xi^{(k-1)}.$$

It would be interesting to examine if we can accelerate convergence by giving a larger correction to the previous iterate. This can make sense if successive corrections display a regular pattern, e.g., they exhibit a common sign.

The idea underlying SOR is indeed to improve the convergence speed of the algorithm by introducing an *over-relaxation* parameter ω as follows:

$$x^{(k)} = x^{(k-1)} - \omega(D + L)^{-1}\xi^{(k-1)}. \tag{5.8}$$

This iterative relation is called the SOR method (Successive Over-Relaxation). The case $\omega = 1$ recovers the Gauss–Seidel method. An alternative way of interpreting formula (5.8) is to represent it as a linear combination of the previous solution and the regular Gauss–Seidel iterate, that is

$$= (1 - \omega) \underbrace{x^{(k-1)}}_{\text{previous iterate}} + \omega \underbrace{\left(x^{(k-1)} - (D + L)^{-1}\xi^{(k-1)}\right)}_{\text{Gauss–Seidel iterate}}$$

$$= x^{(k-1)} - \omega(D + L)^{-1}\xi^{(k-1)}.$$

When matrix A is tridiagonal, and setting $l_0 = u_m = 0$, iteration (5.8) becomes

$$x_i^{(k)} = (1 - \omega)x_i^{(k-1)} + \frac{\omega}{d_i}\left(b_i - l_{i-1}x_{i-1}^{(k)} - u_i x_{i+1}^{(k-1)}\right), \quad i = 1, \ldots, m. \tag{5.9}$$

Example With reference to the previous example, Table 5.3 provides the first few SOR iterations, for an optimal value of ω chosen according to criterion (5.19) to be introduced in the next section. This figure is 1.045549. Table 5.4 provides us with the SOR iteration when ω has been set equal to 1.2. Choosing a nonoptimal value for ω increases the number of iterations necessary to achieve a given tolerance.

The three algorithms can be implemented using the following code.

Table 5.3. SOR iterations when $\omega = 1.045549$

k	$x_1^{(k)}$	$x_2^{(k)}$	$x_3^{(k)}$	$\|x^{(k)} - x^{(k-1)}\|$
1	0.348516	0.613872	−0.320917	0.60129
2	0.118697	0.693606	−0.347982	0.05991
3	0.101377	0.699596	−0.349881	3.4×10^{-4}
4	0.100078	0.699976	−0.349993	1.8×10^{-6}

Table 5.4. SOR iterations when $\omega = 1.2$

k	$x_1^{(k)}$	$x_2^{(k)}$	$x_3^{(k)}$	$\|x^{(k)} - x^{(k-1)}\|$
1	0.400000	0.720000	−0.432000	0.86502
2	0.032000	0.724800	−0.348480	0.14242
3	0.103680	0.695232	−0.347443	6.01×10^{-3}
4	0.101171	0.699771	−0.350374	3.55×10^{-5}

Algorithm (Jacobi/Gauss–Seidel/SOR method)

```
Assign the maximum number of allowed iterations (MAXITS)
Choose a termination scalar ε and an initial point x⁽⁰⁾
Set l(i) = u(m) = 0
While err > ε
//iteration: select the algorithm
Case JACOBI: temp(i) = Apply formula (5.5)
Case GAUSS-SEIDEL: temp(i) = Apply formula (5.7)
Case SOR: temp(i) = Apply formula (5.9)
err = 0
//Compute the error norm
For i = 1 To m
err = err + (temp(i) - v(i)) ^2
v(i) = temp(i)
Next i
Numberiterations = Numberiterations + 1
//Return a error number if MAXITS has been
If Numberiterations > MAXITS Then
Msg("Too many iterations in JACOBI/G-S/SOR")
Exit Function
End if
Wend
Return v
```

5.2.4 Conjugate Gradient Method (CGM)

This approach has be proposed by Hestenes and Stiefel (1952). Nowadays, CGM is largely known as the algorithm for solving unconstrained optimization problems. Indeed it has been proved to be extremely effective in dealing with general objective functions. CGM is the oldest and best known method belonging to the class of *non-stationary iterative methods*. It differs from the techniques presented above in that computations involve information that changes at each iteration.

If we consider the problem of minimizing a quadratic function $f(\mathbf{x}) = \frac{1}{2}\mathbf{x}^T\mathbf{A}\mathbf{x} - \mathbf{b}\mathbf{x}$, where matrix $\mathbf{A}$ is symmetric positive definite, the optimality conditions become $\mathbf{A}\mathbf{x} = \mathbf{b}$. In this case, CGM can be very effective. Here, we illustrate the CGM when $\mathbf{A}$ is symmetric positive definite.

The idea of the CGM is to accelerate the typically slow convergence associated with the *steepest descent method*. This is defined by the iterative algorithm $\mathbf{x}^{(k+1)} = \mathbf{x}^{(k)} - \alpha_k \mathbf{g}^{(k)}$, where $\mathbf{g}^{(k)}$ is the gradient of f and α_k is the step length minimizing f along the direction of the negative gradient. Each step in the CGM is at least as good as the steepest descent step starting at the same point. The first directional vector in CGM is the unit gradient.

The basic procedure in CGM is as follows:

(a) generate a sequence of iterates according to the following rule

$$\mathbf{x}^{(k+1)} = \mathbf{x}^{(k)} + \alpha_k \mathbf{d}^{(k)},$$

where $\mathbf{d}^{(k)}$ is the search direction and α_k is the step length which minimizes $f(\mathbf{x} + \alpha \mathbf{d}^{(k)})$ along $\mathbf{d}^{(k)}$ starting from point $\mathbf{x}^{(k)}$;

$$\alpha_k = \frac{-(\mathbf{g}^{(k)})^\top \mathbf{d}^{(k)}}{(\mathbf{d}^{(k)})^\top \mathbf{A} \mathbf{d}^{(k)}}, \tag{5.10}$$

(b) update directions $\mathbf{d}^{(k)}$ by using formula:

$$\mathbf{d}^{(k+1)} = -\mathbf{g}^{(k+1)} + \beta_k \mathbf{d}^{(k)},$$

where β_k is a scalar given by formula (5.11).

The pseudo-code illustrated below refers to $m \times m$ symmetric and positive definite matrices and terminates within m steps at most.[3,4]

Algorithm (CGM)

1. Starting at $\mathbf{x}^{(0)}$ compute $\mathbf{g}^{(0)} = \mathbf{A}\mathbf{x}^{(0)} - \mathbf{b}$ and set $\mathbf{d}^{(0)} = -\mathbf{g}^{(0)}$.
2. For $k = 0, 1, \ldots, m - 1$:
   (a) Compute α_k according to (5.10) and $\mathbf{x}^{(k+1)} = \mathbf{x}^{(k)} + \alpha_k \mathbf{d}^{(k)}$.
   (b) Compute the gradient of the function f

$$\mathbf{g}^{(k+1)} = \mathbf{A}\mathbf{x}^{(k+1)} - \mathbf{b}.$$

   (c) Unless $k = m - 1$, set

$$\mathbf{d}^{(k+1)} = -\mathbf{g}^{(k+1)} + \beta_k \mathbf{d}^{(k)},$$

   where

$$\beta_k = \frac{(\mathbf{g}^{(k+1)})^\top \mathbf{A} \mathbf{d}^{(k)}}{(\mathbf{d}^{(k)})^\top \mathbf{A} \mathbf{d}^{(k)}} \tag{5.11}$$

   and repeat (a).
3. Return $x^{(m)}$.

Note that if $\mathbf{A} = \mathrm{tridiag}(l_i, d_i, u_i)$, the (i, i) entry of the product $\mathbf{A}\mathbf{x}$ is:

$$(\mathbf{A}\mathbf{x})_{ii} = l_i x_{i-1} + d_i x_i + u_i x_{i+1}.$$

Consequently, we just need to store the three main diagonals of $\mathbf{A}$.

---

[3] If matrix $\mathbf{A}$ is nonsymmetric and possibly indefinite, we can apply the CGM to the related symmetric positive definite system $\mathbf{A}^\top \mathbf{A}\mathbf{x} = \mathbf{A}^\top \mathbf{b}$.

[4] Due to rounding errors, in practice this is not the case, so we need a stopping criteria. In addition, if the function is not quadratic, the CGM may not terminate in m iterations.

5.2.5 Convergence of Iterative Methods

We explore the convergence of iterative methods together with their rate of convergence. The discussion focuses on the Jacobi and the SOR methods. Indeed, the Gauss–Seidel method is merely a special case of the SOR method, namely the one corresponding to $\omega = 1$. A brief discussion of the CGM convergence properties is also done.

All the iterative methods introduced so far solve $\mathbf{Ax} = \mathbf{b}$, where $\mathbf{A} = \mathbf{L} + \mathbf{D} + \mathbf{U}$. The iterations can be represented as follows:

$$\mathbf{x}^{(k)} = \mathbf{Hx}^{(k-1)} + \mathbf{c}, \tag{5.12}$$

where $\mathbf{H}$ is the *iteration matrix* corresponding to the selected method. Table 5.5 reports matrix $\mathbf{H}$ and vector $\mathbf{c}$ for the examined techniques. The solution of any iteration satisfies

$$\mathbf{x} = \mathbf{Hx} + \mathbf{c},$$

and the error $\varepsilon^{(k)} = \mathbf{x} - \mathbf{x}^{(k)}$ generated at the k-iteration satisfies

$$\varepsilon^{(k)} = \mathbf{H}\varepsilon^{(k-1)} = \mathbf{H}^n\varepsilon^{(0)}. \tag{5.13}$$

Convergence of the iteration (5.12), i.e., $\mathbf{x}^{(k)} \to \mathbf{x}$ as $k \to \infty$ or, equivalently, $\varepsilon^{(k)} \to \mathbf{0}$ as $k \to \infty$, is ensured provided that all the eigenvalues of $\mathbf{H}$ lie within the unit circle. This means that $\rho(\mathbf{H}) = |\lambda_{max}| = \max_{\lambda \in \Lambda(\mathbf{H})} |\lambda(\mathbf{H})| < 1$, where $\Lambda(\mathbf{H})$ denotes the set of all eigenvalues of $\mathbf{H}$.

The *rate of convergence* indicates the number of iterations necessary to achieve an assigned level of accuracy. In particular, this number provides an indication of the number of decimal digits by which the error is decreased by each convergent iteration. For a large k, the iteration (5.13) is mainly driven by the greatest eigenvalue of $\mathbf{H}$, i.e., λ_{max}. Therefore:

$$\varepsilon^{(k)} \simeq \lambda_{max}\varepsilon^{(k-1)} \simeq (\lambda_{max})^k \varepsilon^{(0)}.$$

The number p of iterations required to reduce the error size by 10^{-q} is the smallest value p for which

$$\frac{\|\varepsilon^{(p)}\|}{\|\varepsilon^{(0)}\|} = (\rho(\mathbf{H}))^p \le 10^{-q}.$$

This inequality can also be written as $p \log_{10} \rho(\mathbf{H}) \le -q$, i.e.,

Table 5.5. Iteration matrices for the Jacobi, GS and SOR methods

Iter. meth.	Matrix $\mathbf{H}$	Vector $\mathbf{c}$
Jacobi	$-\mathbf{D}^{-1}(\mathbf{L} + \mathbf{U})$	$\mathbf{D}^{-1}\mathbf{b}$
Gauss–Seidel	$-(\mathbf{I} + \mathbf{D}^{-1}\mathbf{L})^{-1}\mathbf{D}^{-1}\mathbf{U}$	$(\mathbf{I} + \mathbf{D}^{-1}\mathbf{L})^{-1}\mathbf{D}^{-1}\mathbf{b}$
SOR	$(\mathbf{I} + \omega(\mathbf{D}^{-1} + \mathbf{L}))^{-1}((1 - \omega)\mathbf{I} - \omega\mathbf{D}^{-1}\mathbf{U})$	$\omega(\mathbf{I} + \omega(\mathbf{D}^{-1} + \mathbf{L}))^{-1}\mathbf{D}^{-1}\mathbf{b}$

$$p \geq \frac{q}{-\ln_{10} \rho(\mathbf{H})} = \frac{q}{\ln_{10}(1/\rho(\mathbf{H}))} = \frac{q}{-\ln \rho(\mathbf{H})},$$

where $\log_{10} \rho(\mathbf{H}) < 0$. Consequently, we may use $r = -\ln \rho(\mathbf{H})$ to measure the *asymptotic* rate of convergence. The attribute "asymptotic" underlines the significance of r as a rate of convergence for large values of k.

We now examine the cases of Jacobi and SOR methods when $\mathbf{A}$ is tridiagonal.

Jacobi method

Let us consider the row norm μ of matrix $\mathbf{H}$ as defined by

$$\mu(\mathbf{H}) = \max_{1 \leq i \leq m} \sum_{i=1}^{m} |h_{ij}|.$$

Condition $\mu(\mathbf{H}) < 1$ implies $\rho(\mathbf{H}) < 1$. When matrix $\mathbf{A}$ is tridiagonal, $\mathbf{H}$ is tridiagonal as well

$$\mathbf{H} = \mathrm{tridiag}\left(\frac{l_i}{d_i}, 0, \frac{u_i}{d_i}\right),$$

where, as usual, l_i, d_i, and u_i denote the elements appearing on the three main diagonals of $\mathbf{A}$. Condition $\mu(\mathbf{H}) < 1$ amounts to requiring that the tridiagonal matrix $\mathbf{A}$ be *strictly diagonally dominant*, in that its entries satisfy:

$$|l_i| + |u_i| < |d_i|, \quad i = 1, \ldots, m.$$

In other words, the sum of the moduli of the diagonal elements must exceed the sum of the moduli of the off-diagonal entries. Let us verify that this is the case for the finite difference approximation to the heat equation. We do not consider the explicit scheme, because it does not involve the solution of a linear system.[5]

- In the implicit scheme, matrix $\mathbf{A}$ is given by (5.14)

$$\mathbf{A} = \mathrm{tridiag}\left(-\frac{\sigma^2 \alpha}{2}, 1 + \sigma^2 \alpha, -\frac{\sigma^2 \alpha}{2}\right). \tag{5.14}$$

The diagonally dominant condition reads as:

$$\left| -\frac{\sigma^2 \alpha}{2} \right| + \left| -\frac{\sigma^2 \alpha}{2} \right| < |1 + \sigma^2 \alpha|,$$

i.e., $\sigma^2 \alpha < 1 + \sigma^2 \alpha$. This inequality is always satisfied.

---

[5] However, note that in the explicit scheme the updated solution is given by $\mathbf{v}_{i+1} = \mathbf{A}\mathbf{v}_i + \frac{\sigma^2 \alpha}{2}\mathbf{b}_i$, where $\mathbf{A} = \mathrm{tridiag}(\frac{\sigma^2 \alpha}{2}, (1 - \sigma^2 \alpha), \frac{\sigma^2 \alpha}{2})$. This corresponds to the Jacobi iteration, where the iteration matrix $\mathbf{H}$ is exactly $\mathbf{A}$. Therefore the convergence condition becomes $|\frac{\sigma^2 \alpha}{2}| + |\frac{\sigma^2 \alpha}{2}| < |1 - \sigma^2 \alpha|$, i.e., $\sigma^2 \alpha < 1$. This is the stability condition already obtained for the explicit scheme.

- In the Crank–Nicolson scheme, matrix $\mathbf{A}$ is given by (5.15)

$$\mathbf{A} = \text{tridiag}\left(-\frac{\sigma^2\alpha}{4}, 1 + \frac{\sigma^2\alpha}{2}, -\frac{\sigma^2\alpha}{4}\right). \tag{5.15}$$

The convergent condition reads as:

$$\left|-\frac{\sigma^2\alpha}{4}\right| + \left|-\frac{\sigma^2\alpha}{4}\right| < \left|1 + \frac{\sigma^2\alpha}{2}\right|,$$

i.e., $\sigma^2\alpha/2 < 1 + \sigma^2\alpha/2$. This is always satisfied.

To examine the rate of convergence, we need the eigenvalues of $\mathbf{H}$. If matrix $\mathbf{A}$ is obtained by discretizing the heat equation, e.g., it has a structure (5.14) or (5.15), we can use a result stating that the eigenvalues of a tridiagonal matrix $\text{tridiag}\{c, a, b\}$ of order m are given by:[6]

$$\lambda_s = a + 2\sqrt{bc}\cos\left(\frac{s\pi}{m+1}\right), \quad s = 1, \ldots, m.$$

(See Smith (1985), pp. 154–155.) In particular, if $\mathbf{A} = \mathbf{D} + \mathbf{L} + \mathbf{U}$ is tridiagonal, $\mathbf{H} = -\mathbf{D}^{-1}(\mathbf{L} + \mathbf{U})$ is tridiagonal too and we have the following two cases:

- In the implicit scheme, the eigenvalues of $\mathbf{H}$ are given by

$$\lambda_s = \frac{\sigma^2\alpha}{(1 + \sigma^2\alpha)}\cos\left(\frac{s\pi}{m+1}\right), \quad s = 1, \ldots, m. \tag{5.16}$$

- In the CN scheme, the eigenvalues of $\mathbf{H}$ are given by

$$\lambda_s = \frac{\sigma^2\alpha}{(2 + \sigma^2\alpha)}\cos\left(\frac{s\pi}{m+1}\right), \quad s = 1, \ldots, m. \tag{5.17}$$

Table 5.6 reports the spectral radius of the Jacobi iteration as applied to both difference schemes. We can see that there is a trade-off between increasing spatial accuracy (i.e., reducing $\delta z = (z_u - z_l)/(m+1)$) and augmenting the asymptotic rate of convergence. The spectral radius tends to 1 as $m \to \infty$ (see rows 2 and 4 in Table 5.6), i.e. as the grid size is refined. Therefore, unless the matrix is strongly diagonally dominant, the Jacobi method is not practical, since for a reasonable grid size the iteration converges very slowly. To improve the things, as we increase the number m of grid points, we can try to reduce the time spacing $\delta\tau$, so that we keep $\alpha = \delta\tau/\delta z^2$ constant (see rows 3 and 5 in Table 5.6). But this again increases the computational cost, because, for pricing an option with a given time to maturity, more time steps will be necessary.

---

[6] If instead the spectral radius of the Jacobi matrix is not known, it can be estimated for large k by using (see Smith (1985), p. 273)

$$\rho(\mathbf{H}) \simeq \frac{\|\mathbf{x}^{(k+1)} - \mathbf{x}^{(k)}\|}{\|\mathbf{x}^{(k)} - \mathbf{x}^{(k-1)}\|},$$

where a possible vector norm is $\|\mathbf{x}\| = \max_{1 \leq i \leq m} |x_i|$.

Table 5.6. Behavior of the spectral radius in the Jacobi iteration when **A** comes from a finite difference scheme. n is the number of time steps, m the number of space points. The time to maturity has been fixed equal to 1. The spatial domain is the interval $[0, 1]$

$\alpha = \delta\tau/\delta x^2$	Scheme	Spectral radius $\rho(\mathbf{H})$		
		$m = 10$	$m = 100$	$m = 1000$
$0.001/\frac{1}{(m+1)^2}$	Implicit	0.186954	0.952814	0.999496
0.1	Implicit	0.15991	0.166586	0.1666
		$(n = 1210)$	$(n = 102010)$	$(n = 1.002 \times 10^7)$
$0.001/\frac{1}{(m+1)^2}$	CN	0.103567	0.910282	0.998998
0.1	CN	0.0872266	0.0908651	0.0909086
		$(n = 1210)$	$(n = 102010)$	$(n = 1.002 \times 10^7)$

SOR method

The following results illustrate the role of parameter ω in determining the rate of convergence of the SOR method (Press et al. (1992)).

- Necessary condition for the convergence of the SOR method is that $0 < \omega < 2$. If the iteration matrix is symmetric and definite positive, the condition is also sufficient.
- Under constraints that are usually satisfied by matrices arising from a finite difference scheme, only over-relaxation (i.e., $1 < \omega < 2$) ensures a quicker convergence than the one provided by the Gauss–Seidel method. If $0 < \omega < 1$, we speak of under-relaxation.
- The eigenvalues λ of the SOR iteration matrix are related to the eigenvalues μ of the Jacobi iteration matrix by the equation

$$(\lambda + \omega - 1)^2 = \lambda\omega^2\mu^2. \tag{5.18}$$

- From the previous result it can be proved that if ρ_{Jacobi} is the spectral radius of the Jacobi iteration, then the optimal choice for ω is given by

$$\omega = \frac{2}{1 + \sqrt{1 - \rho_{\text{Jacobi}}^2}}. \tag{5.19}$$

- For the optimal choice (5.19), the spectral radius of SOR is:

$$\rho_{\text{SOR}} = \left(\frac{\rho_{\text{Jacobi}}}{1 + \sqrt{1 - \rho_{\text{Jacobi}}^2}}\right)^2. \tag{5.20}$$

- The SOR method is more efficient than direct methods for solving finite difference schemes provided that ω is set equal to the optimal value given in (5.19).
- If we set $\omega = 1$ in (5.18), we can also find the spectral radius of the Gauss–Seidel method in terms of ρ_{Jacobi}:

Table 5.7. Optimal value of ω and spectral radius in the SOR iteration when $\mathbf{A}$ stems from a finite difference scheme. n denotes the number of time steps. Time-to-maturity has been fixed equal to 1. The spatial domain is the whole interval [0,1]

$\alpha = \delta\tau/\delta x^2$	Scheme		$m = 10$	$m = 100$	$m = 1000$
$0.001/\frac{1}{(m+1)^2}$	Implicit	ω	1.00889	1.53427	1.93848
		ρ_{SOR}	0.0088941	0.534267	0.938483
0.1	Implicit	ω	1.00648	1.00704	1.00704
		ρ_{SOR}	0.0064763	0.0070357	0.0070425
$0.001/\frac{1}{(m+1)^2}$	CN	ω	1.0027	1.41444	1.91433
		ρ_{SOR}	0.0026960	0.414438	0.914328
0.1	CN	ω	1.00191	1.00207	1.00207
		ρ_{SOR}	0.0019094	0.0020727	0.0020747

$$\rho_{GS} = \rho_{Jacobi}^2,$$

that is an improvement by a factor of 2 in the number of iterations over the Jacobi method. This improvement is still modest, so this method is also slowly convergent and not very useful to accelerate the convergence of the Jacobi method.

Formulae (5.19), (5.16) and (5.17) allow us to find the optimal value of ω for both implicit and CN schemes applied to the heat equation $-\partial_\tau u(\tau, z) + \partial_{zz} u\ (\tau, z) = 0$. A numerical example is illustrated in Table 5.7. For more general problems than the heat equation, the main difficulty with the SOR method is the estimation of the optimal value ω. Indeed, the benefits of the SOR method are visible in a narrow neighborhood of the optimal value ω. Figure 5.5 illustrates this problem. We consider a 9×9 tridiagonal matrix with $l_i = l = -0.2$, $d_i = d = 1.4$, $u_i = u = -0.2$. The convergence criterion has been set equal to $\varepsilon = 10^{-9}$. Using the optimal value of ω, that is 1.02129, we achieve convergence in 8 iterations. As we move away from this optimal value, the number of iterations required to obtain convergence increases quite rapidly, as is illustrated in Fig. 5.5. If we have to solve similar problems several times, we can perform sample experiments starting with alternative values for ω and then select the figure that minimizes the number of iterations. For example, if we need to solve linear systems arising from discretization of PDEs, we have to solve a linear system at each time step. We may start with $\omega = 1$; then, we may slightly increase ω on the second time step and record the number of iterations necessary to achieve convergence. After a few time steps, we can choose the value ω providing the smallest number of iterations. Sometimes the following heuristic estimate is used: $\omega = 2 - O(\delta z)$, where δz is the mesh spacing of the discretization of the underlying PDE. By inspecting the second and third rows in Table 5.7, we argue that this prescription appears to be quite reasonable.

The error of CGM can be bounded in terms of the condition number ν defined as the ratio of the largest and the smallest eigenvalue of the matrix $\mathbf{A}$. If $\mathbf{x}$ is the exact solution of the linear system and $\mathbf{A}$ is a symmetric positive definite matrix, then it can be shown that

$$\left\| \mathbf{x}^{(k)} - \mathbf{x} \right\|_A \le 2\gamma^k \left\| \mathbf{x}^{(0)} - \mathbf{x} \right\|_A,$$

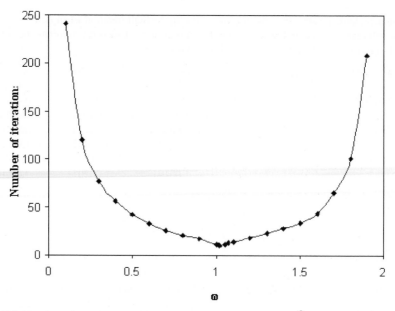

Fig. 5.5. Number of iterations required to achieve a tolerance of 10^{-8} as seen as a function of the over-relaxation parameter ω.

where $\gamma = (\sqrt{\nu} - 1)/(\sqrt{\nu} + 1)$ and $\|\mathbf{y}\|_A^2 = \mathbf{y}^\top \mathbf{A} \mathbf{y}$ and ν is the condition number of **A** defined previously. This is a conservative bound, implying poor convergence for ill-conditioned problems. To increase the rate of convergence, we can transform the problem $\mathbf{Ax} = \mathbf{b}$ to an equivalent with a smaller condition number, so that the iterates will converge more rapidly. The transformed problem is written as $(\mathbf{MA})\mathbf{x} = \mathbf{Mb}$, where **M** is called a preconditioner and needs to be chosen carefully. A discussion about the choice of a suitable preconditioner can be found in Barrett et al. (1994).

5.3 Code for the Solution of Linear Systems

5.3.1 VBA® Code

In the Excel file available on the book web site, we provide Visual Basic codes for all five methods presented in this chapter for the purpose of numerically solving a linear system $\mathbf{Ax} = \mathbf{q}$, with a tridiagonal matrix **A**. The CG method applies provided that **A** is a symmetric and positive definite matrix.

The VBA® functions are named `tridag`, `Jacobi`, `GaussSeidel`, `SOR` and `CGM`:

- `tridag(subdiag As variant, diag As variant, superdiag As variant, q As variant, NumberElements As Integer)`
  implements the LU decomposition, forward, and backward substitution to solve the two triangular systems;

- `Jacobi(subdiag As variant, diag As variant, superdiag As variant, q As variant, x0 As Object, NumberElements As Integer, tol As Double, ByRef nits As Integer)` implements the Jacobi iteration;
- `GaussSeidel(subdiag As variant, diag As variant, superdiag As variant, q As variant, x0 As Object, NumberElements As Integer, tol As Double, ByRef nits As Integer)` implements the Gauss–Seidel iteration;
- `SOR(subdiag As variant, diag As variant, superdiag As variant, q As variant, x0 As Object, NumberElements As Integer, tol As Double, ByRef nits As Integer)` implements the SOR iteration;
- `CGM(subdiag As variant, diag As variant, superdiag As variant, q As variant, x0 As Object, NumberElements As Integer, tol As Double, ByRef nits As Integer)` implements the CGM method.

The arguments of these functions are as follows:

(a) `subdiag`, `diag`, and `superdiag` represent the three main diagonals of the matrix **A**;
(b) `q` is the constant vector in the linear system;
(c) `x0` the starting point for the iterative procedures;
(d) `NumberElements` is an integer number representing the number of rows (or columns) of the matrix **A**;
(e) omega is the parameter ω required in the SOR iteration;
(f) `tol` is the assigned error margin (e.g., 10^{-9}).
(g) `nits` is passed to VBA® as a reference variable, so that it can be changed inside the code. Indeed, when the iterative method terminates, `nits` contains the number of iterations.

Vectors `subdiag`, `diag`, `superdiag`, `q`, `x0`, have to be passed to the function as column vectors. The functions return the solution as a vector with NumberElements components. The iterative methods also change the input parameter `nits`. When the tolerance criterion has been satisfied and the linear system is solved, `nits` is updated and contains the number of iterations necessary to achieve the required tolerance.

5.3.2 MATLAB® Code

MATLAB® has built in functions to solve linear systems, either using direct methods or iterative methods. We briefly present them. In comparison to VBA®, MATLAB® exploits its array structure, so that linear system of large dimensions can be solved in a fraction of second. For example, VBA® cannot invert 1000×1000 matrices. This is not a problem for MATLAB®.

Solution of a linear system

The most direct method to solve a linear system in MATLAB® is to use the matrix division operator \. For example A\q returns the solution of the linear system $\mathbf{Ax} = \mathbf{q}$. If $\mathbf{A}$ is triangular, MATLAB® implements a triangular solver.

Matrix inversion

To get the solution of the linear system, we can also compute inv(A)*q, where with inv(A) we compute the inverse of A.

LU factorization

LU factorization can be performed using the lu function, that expresses a matrix $\mathbf{A}$ as the product of two essentially triangular matrices, one of them a permutation of a lower triangular matrix and the other an upper triangular matrix. [L,U] = lu(A) returns an upper triangular matrix in U and a lower triangular matrix in L, so that $\mathbf{A} = \mathbf{L} * \mathbf{U}$. For example, if we write in the command window:

```
A = [ 1 2 3; 4 5 6; 7 8 0 ];
```

and check for its LU factorization, we may call lu with two output arguments, i.e.,

```
[L,U] = lu(A)
```

As a result, we obtain

```
L =
0.1429 1.0000 0
0.5714 0.5000 1.0000
1.0000 0 0
U =
7.0000 8.0000 0
0 0.8571 3.0000
0 0 4.5000
```

Notice that $\mathbf{L}$ is a permutation of a lower triangular matrix that has 1s on the permuted diagonal, and that U is upper triangular. To check that the factorization does its job, compute the product $\mathbf{L} * \mathbf{U}$, which should return the original $\mathbf{A}$. The solution of linear system using LU decomposition and forward and backward substitution is depicted in the following lines:

LU factorization	[L,U] = lu(A);
Forward substitution	y = L\q;
Backward substitution	x = U\y

The procedure in the second and third line can be replaced by the one shot operation $x = U \backslash (L \backslash q)$. In practice, this operation is equivalent to the `tridag` algorithm implemented in VBA®.

Tridiagonal solver

Several algorithms can be found on the web to solve the linear system when $\mathbf{A}$ is tridiagonal. For example, the function:

```
function x = tridiag(a, b, c, q)
```

can be downloaded from the web site:

    http://www.math.toronto.edu/almgren/tridiag/tridiag1.m

Here a, b, c are the three diagonals of the matrix $\mathbf{A} = \text{tridiag}(a_i, b_i, c_i)$ whilst $\mathbf{q}$ is the fixed term in the linear system $\mathbf{Ax} = \mathbf{q}$. Here $\mathbf{q}$ is a vector; m is determined from its length, a, b, c must be vectors of lengths at least $m - 1$, m, and $m - 1$ respectively. Similar programs can be downloaded from web sites:

    http://www.columbia.edu/itc/applied/e3101/
    http://www.dms.uaf.edu/~bueler/tri.m.

Iterative solvers

MATLAB® code for iterative methods such as Jacobi and SOR can be downloaded from the web site:

    http://www.netlib.org/templates/matlab/

They are called using the commands

```
[x, error, iter, flag] = jacobi(A, x, q, max_it, tol)
[x, error, iter, flag] = sor(A, x, q, w, max_it, tol)
```

where x as input is the initial guess and as output is the solution of the linear system, q is the constant vector, w is the relaxation parameter. Iterations are repeated until the number of iterations is larger than `maxit` or until norm(xnew-xold)/norm(xnew) is less than `tol`.

The conjugate gradient method is implemented as a built in function in MATLAB® and can be used only for symmetric positive definite matrices. The function name is `pcg` and can be used as

```
x = pcg(A,q,tol,maxit)
```

This function performs the CG method with initial guess the zero vector and stops when the 2-norm of the residual vector is less than `tol` or the number of iterations is larger than `maxit`. The $m \times m$ coefficient matrix $\mathbf{A}$ must be symmetric and positive and the right-hand side column vector q must have length m. `tol` specifies the tolerance of the method. If `tol` is [] then PCG uses the default, 1e-6. `maxit` specifies the maximum number of iterations. If `maxit` is [] then PCG uses the default, $\min(m, 20)$. MATLAB® makes also available other functions that implement more advanced iterative methods for sparse linear systems. All of them can make use of preconditioners. More detailed information is available in the MATLAB® help.

5.4 Illustrative Examples

In this section, we present some applications of the presented numerical methods to pricing financial contracts. We consider different processes and different products. The first example is related to the classical Black–Scholes model for pricing call options. Then we present the pricing of the same contract, but assuming that the dynamics of the underlying is described by the Constant Elasticity of Variance (CEV) process. Then we price American options under both the GBM and the CEV process and we introduce the PSOR method, an extension of the SOR algorithm that allows to consider the possibility of early exercise. These examples have been implemented using the VBA® functions described in Sect. 5.3.1. The last two examples are related to pricing double barrier options under the GBM process and options on Coupon Bond in the Cox, Ingersoll and Ross (1985) model. These examples have been implemented in MATLAB® using the pdepe function discussed in the Appendix.

5.4.1 Pricing a Plain Vanilla Call in the Black–Scholes Model (VBA®)

As a first example we compare the implicit and the CN method. We consider the problem of pricing a plain vanilla call option and computing its Greeks in the standard Black–Scholes model. Therefore we numerically solve

$$-\partial_\tau u(\tau, x) + \frac{\sigma^2}{2} x^2 \partial_{xx} u + rx\,\partial_x u + ru = 0,$$
$$u(0, x) = (K - x)_+,$$
$$u(\tau, x_{\max}) = 0,$$
$$u(\tau, 0) = Ke^{-r\tau}.$$
$$(5.21)$$

To this aim we use the VBA® functions:

- PDEImplicit(phi As Integer, spot As Double, strike As Double, t As Double, rf As Double, sg As Double, numspacestep As Integer, numtimestep As Integer, Smin As Double, Smax As Double)
- PDECN(phi As Integer, spot As Double, strike As Double, t As Double, rf As Double, sg As Double, numspacestep As Integer, numtimestep As Integer, Smin As Double, Smax As Double)
- PDECNSOR(phi As Integer, spot As Double, strike As Double, t As Double, rf As Double, sg As Double, numspacestep As Integer, numtimestep As Integer, Smin As Double, Smax As Double, omega As Double, tol As Double)

These functions, presented in the Chapter on PDE's, allow one to solve the above problem using LU decomposition (PDEImplicit and PDECN) or using the SOR method (PDECNSOR).

Table 5.8. Numerical results of different finite difference schemes for the Black–Scholes equation

Spot price	Implicit	Crank–Nicolson		BS formula
		LU	SOR ($\omega = 1.3$)	
0.7	0.049569	0.049570	0.049605	0.049600
0.8	0.088912	0.088923	0.088962	0.088965
0.9	0.140579	0.140598	0.140636	0.140645
1	0.203115	0.203139	0.203173	0.203185
1.1	0.274675	0.274699	0.274727	0.274740
1.2	0.353412	0.353434	0.353456	0.353469
1.3	0.437690	0.437709	0.437724	0.437736

Table 5.9. Crank–Nicolson method implemented with the SOR iterative method and different values of ω

	$\omega = 1.1$	$\omega = 1.2$	$\omega = 1.3$	$\omega = 1.5$
MAE	0.000096	0.0000582	0.000010	0.000107925
SSE	0.0006016	0.0003254	0.000058	0.00060338

Model parameters have been set to $r = 0.1$, $\sigma = 0.4$, $K = 1$, $T = 1$. We have used a 500×1000 grid, so that $\delta x = 10/500$ and $\delta \tau = 1/1000$. The SOR method has been implemented with $\omega = 1.3$ and a stopping criterion of 10^{-9}. The Crank–Nicolson (CN) implemented with SOR method seems to perform better respect to others methods, Implicit and Crank–Nicolson with LU decomposition, see Table 5.8. However, in the SOR method the choice of ω, can make a difference. Indeed, in general it is very hard to estimate its optimal value, so we have to try different values of ω and compare the results with alternative methods before choosing a "good value". For the example chosen, the best choice for ω appears to be around 1.3, as illustrated in Table 5.9 where we report the mean absolute error (MAE) and the sum of squared errors (SSE) of the SOR method implemented with different values of ω (1.1, 1.2, 1.3 and 1.5). Figures 5.6 and 5.7 represent the difference between the Black–Scholes price and delta versus the corresponding numerically computed quantities. In particular, we observe that the greatest error occurs near the strike price, due to the non-differentiability of the payoff condition. Moreover, as expected the error is larger for the delta. The computational cost has been approximately 1 second. We do not report the results relative to the explicit method, for which a very fine time step is necessary to avoid stability problems.

5.4.2 Pricing a Plain Vanilla Call in the Square-Root Model (VBA®)

We now examine the problem of pricing a call option given alternative assumptions about the stock price dynamics. This example is also useful to illustrate the flexibility of the PDE approach. Indeed, very little change in the VBA® code is necessary. In particular, we compare the GBM process and the CEV (constant elasticity of variance) process. For this process the price dynamics are given by:

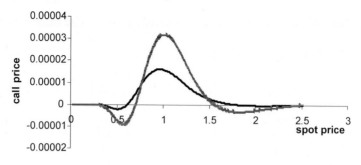

Fig. 5.6. Call option price varying the spot price (Implicit and Crank–Nicolson schemes).

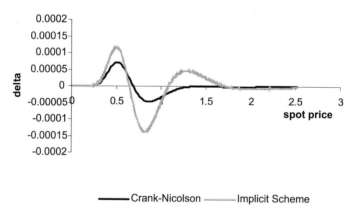

Fig. 5.7. Call option delta varying the spot price (Implicit and Crank–Nicolson schemes).

$$dX(t) = rX(t)\,dt + \sigma\left(X(t)\right)^{\gamma/2} dW(t).$$

In particular, we examine the three cases: $\gamma = 2$ (lognormal process), $\gamma = 1$ (square-root process) and $\gamma = 3$ and we examine at first the prices of plain vanilla call options. The VBA® implementation of this model does not require particular effort. In the module mPDEIngredients, we then need to modify the code of the VBA® function PDEfunctionA(spot As Double, rf As Double, sg As Double) that is used in the code in the following way. Lines

```
Dim gamma As Double
gamma = 2 'lognormal model
PDEfunctionA = 0.5 * sg * sg * spot^gamma
```
should be modified as follows:
```
Dim gamma As Double
gamma = 1 'square root process
PDEfunctionA = 0.5 * sg * sg * spot^gamma
```
for considering the square-root process, or into

Table 5.10. Call option prices for different models: Square-root ($\gamma = 1$), GBM ($\gamma = 2$), CEV ($\gamma = 3$)

Strike	$\gamma = 1$	$\gamma = 2$	$\gamma = 3$
0.7	0.381200	0.386449	0.392787
0.80	0.348966	0.354652	0.361020
0.9	0.263868	0.267052	0.270700
1	0.203386	0.203139	0.203385
1.1	0.156680	0.152836	0.149616
1.2	0.126610	0.119314	0.112866
1.3	0.096124	0.086599	0.078166

```
Dim gamma As Double
gamma = 3 'CEV model
PDEfunctionA = 0.5 * sg * sg * spot^gamma
```
for considering the CEV process.

In Table 5.10 we report the prices of call options for different strikes assuming different values for the parameter γ. In order to have comparable models, we chose volatility parameters such as

$$\sigma_{\gamma=1}^2 X = \sigma_{\gamma=2}^2 X^2 = \sigma_{\gamma=3}^2 X^3,$$

and, given $X = 1$, we have simply $\sigma_{\gamma=1} = \sigma_{\gamma=2} = \sigma_{\gamma=3}$. In the example, we set $X = 1$, $r = 0.1$, $\sigma = 0.4$, $T = 1$ and we let K vary. Table 5.10 shows that as we increase γ, calls in-the-money (and therefore puts out-of-the-money) increase in value. Vice versa, for calls out-of-the money (puts in-the-money). Therefore, choosing $\gamma > 2$ we can generate an implied volatility curve with a skewed shape, as is usually observed for index options.

5.4.3 Pricing American Options with the CN Scheme (VBA®)

Let us consider a put option written at an initial time 0 with a strike price K and maturity date T on a nondividend paying stock. Suppose the option may be exercised early. As discussed in the chapter on Dynamic Programming the value of the option is the greater of the value on immediate exercise and that from holding the option. The latter equals the risk-neutral expectation of its value at the next possible exercise date discounted at the risk-free rate r

$$u(x, \tau) = \max\left(K - e^x, e^{-r\delta\tau} \mathbb{E}_{x,\tau}^* \left(u\left(xe^\xi, \tau + \delta\tau\right)\right)\right), \quad n = 1, 2, \dots$$

and where ξ is $\mathcal{N}((r - \sigma^2/2)\tau, \sigma^2\tau)$, at maturity $v(x, 0) = (K - x)_+$. The largest stock price at which the put option value equals its exercise value is the exercise boundary, $b(\tau)$, which satisfies $K - xe^{b(\tau)} = u(b(\tau), \tau)$. Therefore, $u(x, \tau) = K - x$ when $x \leq b(\tau)$ (*stopping region*). When the asset price is above the critical price, $x > b(\tau)$, it is convenient to keep the option alive (rather than exercising it, it would be better to sell it). Therefore, in this region (*continuation region*) u satisfies the Black–Scholes equation

$$-\partial_\tau u(\tau, x) + \frac{\sigma^2}{2}x^2\partial_{xx}u + rx\partial_x u + ru = 0, \quad x > b(\tau).$$

We can combine the two conditions into the following single equation

$$\left(-\partial_\tau u(\tau, x) + \frac{\sigma^2}{2}x^2\partial_{xx}u + rx\partial_x u + ru\right)(u(\tau, x) - (K - x)) = 0, x > 0, \quad (5.22)$$

and together the two inequalities and (5.22) constitute the so called *linear complementarity formulation* of the American option pricing problem. Using finite differences we can discretize (5.22) and obtain the following linear system of inequalities:

$$(\mathbf{A}\mathbf{v}_{i+1} - \mathbf{b}_i)(\mathbf{v}_{i+1} - \mathbf{g}_{i+1}) = 0, \quad\quad (5.23)$$
$$(\mathbf{v}_{i+1} - \mathbf{g}_{i+1}) \geq 0, \quad\quad (5.24)$$
$$(\mathbf{A}\mathbf{v}_{i+1} - \mathbf{b}_i) \geq 0, \quad\quad (5.25)$$

where $\mathbf{v}_i$ is the vector containing the solution at time step i, $\mathbf{A}$ and $\mathbf{b}$ depend on the finite difference scheme that we have adopted, and $\mathbf{g}_i$ is the payoff function at time step i. Cryer (1971) has suggested a numerical solution to it using the Projected SOR, PSOR. In this algorithm, the early exercise condition is included in the SOR algorithm with the additional line of code

```
temp(i) = maximum(temp(i), earlypayoff(i)).
```

This line compares the result of the i-th iteration with the early exercise condition and takes the largest of the two. Note that direct methods cannot incorporate the early condition in such a simple way. Indeed, using LU we first compute the updated solution and then we compare it with the payoff condition. This is suboptimal. The Projected SOR method is implemented in the VBA® function PSOR(earlypayoff As Variant, subdiag As Variant, diag As Variant, superdiag As Variant, q As Variant, x0 As Variant, NumberElements As Integer, omega As Double, tol As Double). In this function, we have included a column vector named earlypayoff and containing the early exercise value of the option, e.g. $(K - x)_+$ for a put option, as first input.

As a numerical example, we consider the evaluation of an American put option when $r = 0.05$, $\sigma = 0.4$, $X = 1$ and we let the strike price vary. In Table 5.11 we examine the Black–Scholes model and we compare the European Black–Scholes price (BS) with the CN solution implemented with the PSOR algorithm[7] with a binomial tree (Bin) and the numerical approximation due to Barone-Adesi and Whaley (BA–W) (1987). The three methods confirm each other quite well showing a correspondence to the third digit.

---

[7] The CN method has been implemented using a 500×1000 grid, so that $\delta x = 10/500$ and $\delta \tau = 1/1000$. The PSOR method has been implemented with $\omega = 1.2$ and a stopping criterion of 10^{-9}.

In Table 5.12, we compare American option prices using three different dynamics: square-root, GBM and CEV processes. The PSOR algorithm has been implemented setting $\omega = 1.2$. Table 5.12 confirms that puts out-of-the money ($K = 1.1$, 1.2 and 1.3) increase in value with γ.

Table 5.11. American (A) option prices in the Black–Scholes model using different numerical approximations: Crank–Nicolson (CN), Barone-Adesi and Whaley formula (BA–W) and a binomial tree with 1000 steps (BIN)

Strike	BS	CN	BA–W	BIN
0.7	0.026007	0.026717	0.02712	0.026658
0.80	0.050748	0.052303	0.05275	0.052277
0.9	0.085954	0.088954	0.08932	0.088904
1	0.131459	0.136692	0.13681	0.136691
1.1	0.186395	0.194825	0.19453	0.194813
1.2	0.249535	0.262279	0.26146	0.262322
1.3	0.319548	0.337851	0.33650	0.337895

5.4.4 Pricing a Double Barrier Call in the BS Model (MATLAB® and VBA®)

Let us consider a double barrier down-and-out call option in the BS model. The two barriers are fixed at the price level U and L. The option strike is K, with $L < K < U$. Let us consider the BS PDE in the log-return form

$$\partial_\tau G(\tau, z) = \frac{1}{2}\sigma^2 \partial^2_{zz} G(\tau, z) + \left(r - \frac{\sigma^2}{2}\right)\partial_z G(\tau, z) - rG(\tau, z),$$

and, in order to use the MATLAB® PDE solver, let us write it in the following form

$$\partial_\tau G(\tau, z) = +\frac{1}{2}\sigma^2 \partial_z\left(\partial_z G(\tau, z)\right) + \left(r - \frac{\sigma^2}{2}\right)\partial_z G(\tau, z) - rG(\tau, z), \quad (5.26)$$

Table 5.12. European (E) and American (A) option prices for different models: square-root ($\gamma = 1$), GBM ($\gamma = 2$), CEV ($\gamma = 3$)

Strike	$\gamma = 1$		$\gamma = 2$		$\gamma = 3$	
	E	A	E	A	E	A
0.7	0.033484	0.034127	0.026059	0.026717	0.019815	0.020448
0.80	0.057184	0.058544	0.050791	0.052303	0.045042	0.046661
0.9	0.089773	0.092386	0.085978	0.088954	0.082635	0.085932
1	0.131713	0.136360	0.131456	0.136692	0.131670	0.137443
1.1	0.182880	0.190639	0.186364	0.194825	0.190445	0.199531
1.2	0.242665	0.254961	0.249480	0.262279	0.257113	0.270347
1.3	0.310122	0.328763	0.319472	0.337851	0.330016	0.337851

Table 5.13. Functions to be used as arguments to PDEp to solve the problem (5.26)

```
pdefun      c = 1      f = 0.5*DuDx*sg^2  s = (r-sg^2/2)*DuDx-r*u
pdex1ic     (x>=log(strike)).*(exp(x)-strike)
pdex1bcfun  pl = ul; ql = 0;             pr = ur; qr = 0;
```

Table 5.14. Prices of double knock-out call options, with barriers at L and U. Results are from Kunitomo and Ikeda (KI) and from the numerical solution of the PDE. Parameters setting: $T - t = 0.5$, $X = 1000$, $K = 1000$, $r = 0.05$

				PDE ($m \times n$)		
σ	L	U	KI (1992)	grid	MATLAB®	VBA® (CN)
0.2	500	1500	66.12866	(300 × 300)	66.126651	66.12262
0.2	800	1200	22.08201	(300 × 300)	22.081897	22.08121
0.2	950	1050	0.00066	(30 × 300)	0.00056	0.00057
0.4	500	1500	53.34555	(300 × 300)	53.34755	53.34348
0.4	800	1200	3.13712	(300 × 300)	3.137155	3.13556
0.4	950	1050	0.00098	(30 × 300)	0.00000	0.00000

$$G(0, z) = \left(e^z - K\right)_+,$$
$$G(\tau, \ln U) = 0,$$
$$G(\tau, \ln L) = 0.$$

As described in the Appendix on the MATLAB® solver, we need to define functions describing (a) the PDE, (b) the initial condition, (c) the boundary conditions. This has been done in the M-file pdeBSDoubleBarrierExample.m, where the function pdepe is called using the structure

```
pdepe(m,@pdefun,@pdex1ic,@pdex1bcfun,x,t,[],riskfree,
sigma,strike).
```

Notice the presence of the parameters (riskfree, sigma, strike) as additional inputs in the function pdepe. The functions pdefun, pdex1ic and pdex1bcfun define the PDE, the initial condition and the boundary conditions; see Table 5.13.

The complete code can be found in the Matlab® module

```
function [solpoints, UOUT, DUOUTDX]
    = pdeBSDoubleBarrierExample
```

In Table 5.14 we report the prices of a double knock-out option taken from Table 3.1 in Kunitomo and Ikeda (1992), which provides an analytical formula in the GBM case. In Table 5.14, for comparison, we give numerical results obtained using the VBA® code (Crank–Nicolson with LU factorization).[8] The numerical solution

---

[8] In VBA® we have solved the BS PDE expressed in prices and not in returns. Moreover, we have considered a standardized problem with barriers set at U/K and L/K and the spot

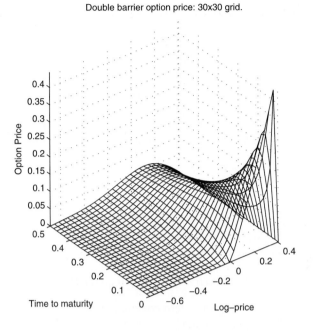

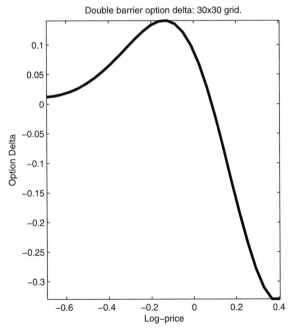

Fig. 5.8. Numerical pricing of a double barrier down-and-out call option. Parameters are set as follows: $\sigma = 0.2$, $r = 0.05$, $K = 1$, $L = 0.5$, $U = 1.5$, $\tau = 0.5$.

(price and delta) is also illustrated in Fig. 5.8. Notice that at the starting time, we have an inconsistency between the initial condition evaluated at $z = \ln U$ and the upper boundary condition. This inconsistency produces inaccurate solutions for very short maturities. However, as time progresses, this problem fades away, except when the ratio of the time step to the square of the space step, i.e. $\delta \tau / \delta z^2$, is too large. In such cases, as discussed in the chapter on PDEs, the Crank–Nicolson solution generates spurious oscillations. For this reason, in Table 5.14 we set the number of grid points equal to (30×300) when the barriers are 950 and 1050. We suggest that the reader modifies the MATLAB® code and prices a double barrier option using the CEV model.

5.4.5 Pricing an Option on a Coupon Bond in the Cox–Ingersoll–Ross Model (MATLAB®)

Let us consider the Cox, Ingersoll and Ross (1985) model that assigns the following risk-neutral dynamics to the short rate r:

$$dr(t) = \alpha\big(\mu - r(t)\big)\, dt + \sigma \sqrt{r(t)}\, dW(t).$$

At time t, the price $P(t, T)$ of a zero-coupon expiring in T can be obtained by evaluating the following expectation

$$P\big(t, T; r(t)\big) \equiv P(t, T) = \mathbb{E}_t^*\big(e^{-\int_t^T r(s)\, ds}\big),$$

that admits the closed-form expression (see Hull (2005))

$$P(t, T) = A(t, T)e^{-B(t,T)r(t)},$$

where

$$A(t, t + \tau) = \left(\frac{2\phi_1 e^{\phi_2 \tau / 2}}{\phi_2(e^{\phi_1 \tau} - 1) + 2\phi_1}\right)^{\phi_3}, \tag{5.27}$$

$$B(t, t + \tau) = 2\frac{e^{\phi_1 \tau} - 1}{\phi_2 + 2\phi_1}, \tag{5.28}$$

$$\phi_1 = \sqrt{\alpha^2 + 2\sigma^2}, \qquad \phi_2 = \phi_1 + \alpha, \qquad \phi_3 = \frac{2\alpha\mu}{\sigma^2}.$$

The zero coupon bond price can be also obtained numerically solving the pricing PDE

---

price at X/K. We reobtain the correct solution multiplying the numerical one by K. The VBA® implementation of the double barrier model simply requires a modification of the code of the VBA® functions upperbc and lowerbc that must return 0. In particular, the function upperbc has to be modified as follows. The line of code

```
upperbc = spot - strike * Exp(-rf * t)  'bc for a call
option
```

has to be commented out and we need to activate the line of code

```
upperbc = 0  'bc for a up-out barrier option
```

$$-\partial_\tau u(\tau, r) + \frac{\sigma^2}{2} r \partial_{rr} u(\tau, r) + \alpha(\mu - r)\partial_r u(\tau, r) - ru(\tau, r) = 0, \qquad (5.29)$$

with initial condition

$$u(0, r) = 1,$$

and boundary conditions at $r = r_{max}$ and at $r = r_{min}$:

$$u(0, r_{max}) = 0, \qquad u(0, r_{min}) = 1.$$

Our aim is now to price, using PDEs, an option on a coupon bond. The option will expire at time T. If we let c_i to be the coupon that is paid at time T_i, $T_i > T$, $i = 1, \ldots, n$, with c_n inclusive of the notional, the option payoff with strike K is

$$\left(\phi \left(\sum_{i=1}^{n} P(T, T_i; r(T))c_i - K \right) \right)_+ = \left(\phi \left(\sum_{i=1}^{n} A(T, T_i)e^{-B(T,T_i)r(T)}c_i - K \right) \right)_+,$$

where the parameter ϕ has been introduced to distinguish between call option ($\phi = 1$) and put option ($\phi = -1$). We need to compute

$$u(\tau, r) = \mathbb{E}_t^* \left(e^{-\int_t^{t+\tau} r(s)\,ds} \left(\phi \left(\sum_{i=1}^{n} A(t + \tau, T_i)e^{-B(t+\tau, T_i)r(t+\tau)}c_i - K \right) \right)_+ \right),$$

or equivalently, we can solve the pricing PDE

$$-\partial_\tau u(\tau, r) + \frac{\sigma^2}{2} r \partial_{rr} u(\tau, r) + \alpha(\mu - r)\partial_r u(\tau, r) - ru(\tau, r) = 0,$$

with initial condition

$$u(0, r) = \left(\phi \left(\sum_{i=1}^{n} A(T, T_i)e^{-B(T,T_i)r(T)}c_i - K \right) \right)_+,$$

and boundary conditions at $r = r_{max}$ and at $r = r_{min}$:

$$u(0, r_{max}) = \begin{cases} 0, & \phi = 1, \\ -\sum_{i=1}^{n} A(t, T_i)e^{-B(t,T_i)r_{max}}c_i + A(t, T)e^{-B(t,T)r_{max}}K, & \phi = -1, \end{cases}$$

$$u(0, r_{min}) = \begin{cases} \sum_{i=1}^{n} A(t, T_i)e^{-B(t,T_i)r_{min}}c_i - A(t, T)e^{-B(t,T)r_{min}}K, & \phi = 1, \\ 0, & \phi = -1. \end{cases}$$

The two boundary conditions state that if we have a call option (put option) and r is sufficiently large (small), the price of the coupon bond will be so low that the exercise will be unlikely (likely).

The above problem can again be solved using the MATLAB® function pdepe. This is done in the MATLAB® script pdeCIRoptiononazcb. In order to use the function pdepe we need to assign the PDE to be solved (pdefun), the initial condition (pdex1ic) and the boundary conditions (pdex1bcfun). Table 5.15 illustrates how to construct these functions. In particular, we have introduced a quantity named r_0 that represents the interest rate level for which the coupon bond price equals the exercise price

$$\sum_{i=1}^{n} A(T, T_i)e^{-B(T,T_i)r_0}c_i = K.$$

The coupon bond price is a decreasing function of r, so that if $r(T) < r_0$, the call option will be exercised. Vice versa, for the put option. In Table 5.16 we compare the prices of an option on a zero coupon bond given by: (a) the numerical solution of the PDE as described above, (b) Monte Carlo simulation with 500,000 runs, (c) the analytical solution given in the CIR paper. However, note that in order to implement the CIR formula for options on a zero coupon bond we need the cumulative distribution of a non-central chi-square distribution. For this we have used the approximation due to Sankaran (1963).[9] In the Monte Carlo simulation we generate the short rate directly at the option maturity, according to the algorithm described in the Monte Carlo chapter. The different methods agree very well. In Table 5.17 we price an option on a coupon bond. The coupon bond pays a notional at maturity and a 5% annual coupon every year for four years, starting one year later the option expiry. The price of the option on the coupon bond can be computed using the so-called "Jamshidian trick", Jamshidian (1996), that exploit the monotonicity of the coupon bond with respect to the short rate and is able to express the option on the coupon bond as an appropriate portfolio of options on zero-coupon bonds. Each option can then be still priced using the Sankaran approximation.[10] Unfortunately, this approximation results falls outside the 95% MC interval most of the time. This is not the case for the numerical solution of the PDE.

Exercise Write a MATLAB® code using the function pdepe to price a zero-coupon bond when the dynamics of the short rate are specified as

$$dr(t) = \alpha(\mu - r(t))\, dt + \sigma(r(t))^\gamma\, dW(t),$$

where $\gamma > 0$. Use the solution to price a zero-coupon bond and a call option on it.

---

[9] The Sankaran approximation is implemented in the Matlab® module Chi2RipSankaran.m. The corresponding option pricing formula is in the Matlab® module ZCBOption_CIR.m.

[10] The Jamshidian trick is implemented in the Matlab® module BondOption_CIR.m.

Table 5.15. Functions to be used as arguments to pdepe to solve the problem (5.26)

pdefun	c = 1
	f = $\frac{\sigma^2}{2}$*x*DuDx;
	s = α*(μ-x-$\frac{\sigma^2}{2\alpha}$)*DuDx-x*u;
pdex1ic	$(\phi*(x-r0)<=0).*(\sum_{i=1}^{n} A(t, T_i)e^{-B(t,T_i)x}c_i-K)*\phi$
pdex1bcfun	p1 = ul-phi*$(\sum_{i=1}^{n} A(s, T_i)e^{-B(s,T_i)x}c_i+A(s, T)e^{-B(s,T)x}*K)$;
	ql = 0; pr = ur; qr = 0;

Table 5.16. Prices of options on zcb in the CIR model. Parameters $\alpha = 0.1$, $\mu = 0.1$, $\sigma = 0.1$, $r_{\min} = 0$, $r_{\max} = 0.5$. Strikes have been set equal to the forward price of the zero-coupon bond

Option maturity	Zcb maturity	Strike	PDE ($m = 500, n = 1000$)	MC (s.e. $\times 10^3$) 500,000 runs	Sankaran
0.25	0.5	98.7132	0.106508	0.10645 (0.0021)	0.106475
0.25	5	75.3712	1.20600	1.20440 (0.0242)	1.205606
1	2	94.5136	0.731992	0.73481 (0.0142)	0.734048
1	5	78.4225	2.046608	2.04619 (0.0406)	2.051997
5	6	93.5933	1.0694595	1.07078 (0.0236)	1.080315
5	10	71.027	3.143115	3.15348 (0.0732)	3.175804

Table 5.17. Prices of options on a coupon bond in the CIR model. Parameters $\alpha = 0.1$, $\mu = 0.1$, $\sigma = 0.1$, $r_{\min} = 0$, $r_{\max} = 0.5$. Strikes have been set equal to the forward price of the coupon bond

Option maturity	Strike	PDE ($m = 500, n = 1000$)	MC (s.e.) 500,000 runs	Sankaran
1	95.703393	2.3386243	2.3405161 (0.0036606)	2.343645
2	94.825867	2.9534488	2.9505254 (0.0046964)	2.968641
3	94.129268	3.2217921	3.2113425 (0.0052998)	3.245655
4	93.576725	3.3090647	3.3185218 (0.0057044)	3.338724
5	93.138299	3.2897274	3.2953126 (0.0059728)	3.322710

5.5 Comments

The literature on solving the linear systems arising from the numerical solution of PDEs is very large. Standard reference texts are Mitchell and Griffiths (1980), Morton and Mayers (1994) and Smith (1985). Hageman and Young (1981) is devoted to iterative methods applied to linear systems arising from PDEs. Introductory textbooks to numerical analysis are Atkinson (1989) and Burlisch and Stoer (1992). Quarteroni, Sacco and Saleri (2000) provide theoretical insights combined with examples and counterexamples implemented in MATLAB®. A general introduction to numerical linear algebra is given in Stewart (1973) and Golub and Van Loan (1996). The best codes for the direct solution of linear systems are given in the package LINPACK, see Dongarra et al. (1979). A detailed treatment of iterative methods can be found in the essential book of Varga (1962), and also in Young (1971). For a presentation and templates for the solution of linear systems, we suggest Barrett et al. (1994), available at the web address:

    http://www.netlib.org/templates

From the same website, MATLAB® codes can be downloaded free of charge. For generalizations of the Conjugate Gradient Method see Luenberger (1989), Bazaraa, Sherali and Shetty (1993) for optimization problems, and Barrett et al. (1994) for methods with asymmetric matrices. A discussion about the choice of a suitable preconditioner can be found in the same text. See also Golub and Van Loan (1996, Sects. 10.2 and 10.3).

A detailed treatment of exotic option pricing using PDEs is given in Sydel (2006), Tavella and Randall (2000), and Wilmott, Dewynne and Howison (1993). Other useful references are Kwok (1998), Zhu, Wu and Chern (2005). They present the application of PDEs to pricing multiasset products as well as more advanced methods.

Analytical formulae for barrier options have first been obtained by Geman and Yor (1996) and Kunitomo and Ikeda (1992). The square-root process has been introduced by Feller (1951), who studied the existence and uniqueness of the solution. Cox, Ingersoll and Ross (1985) used this process for the description of the dynamics of term structure of interest rates. Chen and Scott (1995) present a multifactor extension. A detailed analytical treatment can be found in Lamberton and Lapeyre (1996) and in Cairns (2004). Vetzal (1998) discusses how to set appropriate boundary conditions when we apply finite difference methods to the CIR model. The Constant Elasticity of Variance Option Pricing Model has been introduced in finance by Cox (1996), and studied in Schroder (1989). Webber and Kuan (2003) investigate the pricing of barrier options in one-factor interest rate models.

Since the appearance of the cornerstone papers by Samuelson (1967), McKean (1967), the fast computation of American puts became practical with the works of MacMillan (1986) and Barone-Adesi and Whaley (1987). Since then the literature on numerical methods for American options has grown enormously. Barone-Adesi (2005) reviews the most important references. Detemple (2005) and Salopek (1997) provide a detailed overview from both theoretical and computational approaches. Brennan and Schwartz (1977) propose the first model for valuing American puts using finite difference (implicit method). Dempster and Hutton (1999) investigate the use of a finite difference scheme for American option pricing using a direct numerical solver based on the simplex method. Zvan, Forsyth and Vetzal (1998a) (1998b) introduce the penalty method for American options. Broadie and Detemple (1996) perform an extensive study comparing the performance of various methods. Villeneuve and Zanette (2002) investigate the pricing problem of American options with the payoff depending on two assets. Battauz (2002) investigates the pricing of American options written on two assets using a change of numéraire technique.

Numerical methods for PDEs with jump diffusion processes are studied in D'Halluin, Forsyth and Vetzal (2003), Hirsa and Madan (2003), Zhang (1997).

6

Quadrature Methods

Quadrature methods allow for numerical computations of integrals. In quantitative finance, these methods directly evaluate conditional expected values representing derivative prices. This task can be achieved whenever the distribution of the underlying variable is available in closed form. The resulting method turns out to be very effective for low-dimensional problems.

Let us consider a one-dimensional diffusion describing the evolution of an underlying asset price under the risk-neutral probability measure $\mathbb{P}^*$:

$$dX(s) = \mu(s, X(s)) \, ds + \sigma(s, X(s)) \, dW(s), \quad X(t) = x.$$

The arbitrage-free price of a derivative with payoff F is given by the conditional expectation:

$$F(t, X(t)) = \mathbb{E}_t^* (e^{-\int_t^T r(s) \, ds} F(T, X(T))), \tag{6.1}$$

where $r(t)$ is the risk-free instantaneous rate for interest prevailing at time t. Let $f(s, y; t, x)$ denote the transition density from state x at time t to state y at time $s > t$. The analytic expression of this quantity is available in a few instances. For example, if X is a geometric Brownian motion, i.e., $\mu(s, x) = rx$, $\sigma(s, x) = \sigma x$, then

$$f(s, y; t, x) = \frac{1}{y\sqrt{2\pi\sigma^2(s-t)}} \exp\left(-\frac{1}{2}\left(\frac{\ln y - \ln x - (r - \sigma^2/2)(s-t)}{\sigma\sqrt{s-t}}\right)^2\right).$$

By assuming deterministic interest rates and defining the discount factor as $P(t, T) = \exp(-\int_t^T r(s) \, ds)$, we can write (6.1) as follows:

$$\mathbb{E}_t^* (e^{-\int_t^T r(s) \, ds} F(T, X(T))) = P(t, T) \int_0^{+\infty} F(T, y) f(T, y; t, x) \, dy. \tag{6.2}$$

Quadrature methods allow one to numerically evaluate this integral.[1]

---

[1] If the integrand is continuous on a compact domain, then integral (6.2) is finite. (Sometimes, this condition is not easy to verify.) From now on, we assume this is the case.

This chapter is organized as follows. Section 6.1 introduces the general idea underlying quadrature rules. Section 6.2 presents the Newton–Cotes formulas and details rectangle, trapezoid, Romberg and Simpson rules. Section 6.3 illustrates Gaussian rules which improve the computing accuracy compared to the Newton–Cotes formulas. VBA® code is detailed in Sect. 6.4. Adaptive quadratures and the corresponding Matlab® functions are shown in Sect. 6.5. Section 6.6 presents the derivative pricing problem using the Fourier transform with VBA® and Matlab® implementation. Section 6.7 presents numerical examples. Finally we conclude with a few comments and bibliographic suggestions.

6.1 Quadrature Rules

Consider an integral

$$I(f) = \int_A f(x)\,dx,$$

where f is a real-valued function defined on a closed and bounded subset A of the real line. This quantity is traditionally computed by using an approximating finite sums corresponding to a suitable partition of the set A. A *quadrature rule* of order n assumes the form

$$I_n(f) = \sum_{i=1}^{n} w_i f(x_i), \tag{6.3}$$

where the w_i's are called weights and the x_i are called abscissas or quadrature nodes. These quantities depend on n, however we suppress the explicit indication of this fact whenever it does not produce any ambiguity. Alternative quadrature methods arise from different ways of building the set

$$R_n = \{(w_i, x_i): i = 1, \ldots, n\}$$

such that $I_n(f) = I(f)$ for some class of functions f.

Each of the rules presented below is coupled with theoretical results showing that the resulting $I_n(f)$ converges to $I(f)$ as $n \to \infty$ for any integrable function f. As a consequence, we can improve our estimate by increasing the number n, although this fact cannot be practically guaranteed. Indeed, the finite precision of computer arithmetic may lead to round-off errors that can deteriorate the estimate of I when n increases. For instance, this problem occurs whenever weights are negative, so that the rule becomes unstable and round-off errors dominate discretization errors.

The basic strategy for computing I_n is as follows:

- Approximate function f by an interpolating polynomial p_n of order n in a way that $p_n(x_i) = f(x_i)$ for all i's;
- Integrate p_n and return $I(p_n)$ as approximation to $I(f)$.

We consider Newton–Cotes formulae, for which abscissas x_i are evenly spaced, and Gaussian quadrature formulae, where both weights and abscissas are selected in a way to maximize the order of the interpolating polynomial.

6.2 Newton–Cotes Formulae

The ith Lagrange polynomial of degree $n-1$ with respect to $\{x_1,\ldots,x_n\}$ is defined as:

$$L_i(x) = \frac{(x-x_1)\cdots(x-x_{i-1})(x-x_{i+1})\cdots(x-x_n)}{(x_i-x_1)\cdots(x_i-x_{i-1})(x_i-x_{i+1})\cdots(x_i-x_n)}$$

$$= \prod_{k=1,k\neq i}^{n} \frac{(x-x_k)}{(x_i-x_k)},$$

for $i = 1,\ldots,n$. These functions satisfy $L_i(x_i) = 1$ and $L_i(x_k) = 0$ for $k \neq i$. The interpolating polynomial p_{n-1} of a function f on the set of points $\{x_1,\ldots,x_n\}$ is:

$$p_{n-1}(x) = \sum_{i=1}^{n} f(x_i)L_i(x). \tag{6.4}$$

This expression is known as the *Lagrange interpolation formula*. It can be shown that:

- Functions $L_1(x),\ldots,L_n(x)$ form a basis for the linear space of polynomials of degree up to $n-1$;
- Expression (6.4) is the unique representation of the polynomial $p_{n-1}(x)$ with respect to this basis.

Example Let us suppose that the function f assumes the values 2, 4 and 7 at $x_1 = 1$, $x_2 = 3$ and $x_3 = 5$, respectively. We look for a polynomial of order 2 such that $p_2(x_i) = f(x_i)$, for $i = 1, 2, 3$. Then:

$$L_1(x) = \frac{(x-3)(x-5)}{(1-3)(1-5)} = \frac{1}{8}x^2 - x + \frac{15}{8},$$

$$L_2(x) = \frac{(x-1)(x-5)}{(3-1)(3-5)} = \frac{3}{2}x - \frac{1}{4}x^2 - \frac{5}{4},$$

$$L_3(x) = \frac{(x-1)(x-3)}{(5-1)(5-3)} = \frac{1}{8}x^2 - \frac{1}{2}x + \frac{3}{8},$$

so that:

$$p_2(x) = 2L_1(x) + 4L_2(x) + 7L_3(x)$$
$$= 2\left(\frac{1}{8}x^2 - x + \frac{15}{8}\right) + 4\left(\frac{3}{2}x - \frac{1}{4}x^2 - \frac{5}{4}\right) + 7\left(\frac{1}{8}x^2 - \frac{1}{2}x + \frac{3}{8}\right)$$
$$= \frac{1}{2}x + \frac{1}{8}x^2 + \frac{11}{8}.$$

It is natural to wonder how large the error $f(x) - p_{n-1}(x)$ can be for a fixed n, whenever $x \neq x_i$ ranges between x_1 and x_n. It is possible to show that if $f \in C^n([a,b])$, then there is a point ξ depending on x, with $x_1 < \xi < x_n$, such that

$$f(x) - p_{n-1}(x) = \frac{f^{(n)}(\xi)}{n!} \prod_{i=1}^{n} (x - x_i), \quad \forall x \in A, \tag{6.5}$$

for all $x \in A$. Here $f^{(n)}(\xi) = \partial^n f(x)/\partial x^n$. Unfortunately, the problem with the application of this formula is that the point ξ is not known a priori. However, expression (6.5) will still be useful in evaluating the error in quadrature methods.

We now move to integration of the interpolating function. Let $A = [a, b]$ and consider a uniform partition such that

$$x_i = a + (i - 1)h,$$

for $i = 1, \ldots, n-1$ and $n \geq 2$. Therefore, $x_1 = a$, $x_n = b$ and $h = (b-a)/(n-1)$. This partition gives rise to the so-called "closed formulae", i.e., those for which the integrand function is also computed at the endpoints a and b. Conversely, "open formulae" are obtained by setting $h = (b-a)/(n+1)$ and $x_i = a+ih, i = 1, \ldots, n$, ($x_1 = a + h$, $x_n = b - h$). These latter are useful whenever the integrand function is not defined at the endpoints a and b. In what follows, we explicitly deal with closed formulae. However, the results can be extended to open formulae as well.

By integrating the interpolation formula (6.4), we may approximate $I(f) := \int_a^b f(x)\,dx$ by

$$I_n(f) = \int_a^b p_{n-1}(x)\,dx = \sum_{i=1}^{n} f(x_i) \int_a^b L_i(x)\,dx.$$

In order to simplify the right-hand side of this expression, we consider the change of variable $x = x_1 + (t - 1)h$, so that:

$$\begin{aligned}
\frac{x - x_k}{x_i - x_k} &= \frac{x_1 + (t - 1)h - (x_1 + (k - 1)h)}{x_1 + (i - 1)h - (x_1 + (k - 1)h)} \\
&= \frac{(t - k)}{(i - k)}.
\end{aligned}$$

Consequently, $L_i(x) = \prod_{k=1, k \neq i}^{n} (t-k)/(i-k) = \phi_i(t)$, $\int_a^b L_i(x)\,dx = \int_1^n \phi_i(t)h\,dt$ and

$$I_n(f) = \int_a^b p_{n-1}(x)\,dx = \sum_{i=1}^{n} f(x_i) \int_1^n \phi_i(t)h\,dt$$

$$= h \sum_{i=1}^{n} f(x_i)w_i,$$

with $w_i = \int_1^n \phi_i(t)\,dt$. We remark that weights w_i do not depend on the interval $[a, b]$, but only on the order n. This allows to tabulate quantities $\alpha_i = \int_1^n \phi_i(t)\,dt$ and use their values for integrating other functions on various intervals.

Example   Let us consider the case $n = 2$. This corresponds to interpolating f with a polynomial of order 1 and computing the function at two points only. As $h = b - a$,

$$w_1 = \int_1^2 \phi_1(t)\,dt = \int_1^2 \prod_{k=1, k\neq 1}^2 \frac{(t-k)}{(1-k)}\,dt = \int_1^2 \frac{(t-2)}{(1-2)}\,dt = \frac{1}{2},$$

$$w_2 = \int_1^2 \phi_2(t)\,dt = \int_1^2 \prod_{k=1, k\neq 2}^2 \frac{(t-k)}{(2-k)}\,dt = \int_1^2 \frac{(t-1)}{(2-1)}\,dt = \frac{1}{2}.$$

As $h = b - a$, then $I(f)$ is approximated by

$$I_2(f) = \int_a^b f(x)\,dx = h\big(w_1 f(a) + w_2 f(b)\big) = \frac{(b-a)}{2}\big(f(a) + f(b)\big),$$

that is the average of the function f at the interval endpoints.

Combining interpolation using Lagrange polynomials and integration performed as illustrated above gives rise to the so-called closed Newton–Cotes formulae:

$$\int_a^b p_n(x)\,dx = h\sum_{i=1}^n f_i w_i,$$

where $f_i = f(a + (i-1)h)$, $h = (b-a)/(n-1)$ for closed formulae and $f_i = f(a + ih)$, $h = (b-a)/(n-1)$ for open formulae providing an approximate value for $\int_a^b f(x)\,dx$. The coefficients (weights) w_i are rational numbers satisfying $\sum_{i=1}^n w_i = n - 1$ for closed formulae and $n + 1$ for open formulae. They are presented in Table 6.1 for closed formulae and Table 6.2 for open formulae. As n increases, some of the values w_i become negative and the corresponding formulae become unsuitable for numerical purposes as long as cancellations tend to occur in the sum $h\sum_{i=1}^n f_i w_i$. The approximation error in using Newton–Cotes formulae can be computed through expression (6.5):

$$I(f) - I_n(f) = \begin{cases} e_n h^{n+1} f^{(n)}(\xi) & f \in C^{(n)}([a,b]), n \text{ even,} \\ o_n h^{n+2} f^{(n+1)}(\xi) & f \in C^{(n+1)}([a,b]), n \text{ odd.} \end{cases}$$

Table 6.1. Weights of closed Newton–Cotes formulae and corresponding error

Rule	$n-1$	w_1	w_2	w_3	w_4	w_5	w_6	w_7	|Error|
Trapezoid	1	$\frac{1}{2}$	$\frac{1}{2}$	0	0	0	0	0	$\frac{h^3}{12}f^{(2)}(\xi)$
Simpson	2	$\frac{1}{3}$	$\frac{4}{3}$	$\frac{1}{3}$	0	0	0	0	$\frac{h^5}{90}f^{(4)}(\xi)$
3/8-rule	3	$\frac{3}{8}$	$\frac{9}{8}$	$\frac{9}{8}$	$\frac{3}{8}$	0	0	0	$\frac{3}{80}h^5 f^{(4)}(\xi)$
Milne	4	$\frac{14}{45}$	$\frac{64}{45}$	$\frac{24}{45}$	$\frac{64}{45}$	$\frac{14}{45}$	0	0	$\frac{8}{945}h^7 f^{(6)}(\xi)$
–	5	$\frac{95}{288}$	$\frac{375}{288}$	$\frac{250}{288}$	$\frac{250}{288}$	$\frac{375}{288}$	$\frac{95}{288}$	0	$\frac{275}{12096}h^7 f^{(6)}(\xi)$
Weddle	6	$\frac{41}{140}$	$\frac{216}{140}$	$\frac{27}{140}$	$\frac{272}{140}$	$\frac{27}{140}$	$\frac{216}{140}$	$\frac{41}{140}$	$\frac{9}{1400}h^9 f^{(8)}(\xi)$

Table 6.2. Weights of open Newton–Cotes formulae and corresponding error

$n-1$	w_1	w_2	w_3	w_4	w_5	w_6	w_7	|Error|
0	1	0	0	0	0	0	0	$\frac{h^3}{24}f^{(2)}(\xi)$
1	$\frac{3}{2}$	$\frac{3}{2}$	0	0	0	0	0	$\frac{h^3}{4}f^{(2)}(\xi)$
2	$\frac{8}{3}$	$-\frac{4}{3}$	$\frac{8}{3}$	0	0	0	0	$\frac{28}{90}h^5 f^{(4)}(\xi)$
3	$\frac{55}{24}$	$\frac{5}{24}$	$\frac{5}{24}$	$\frac{55}{24}$	0	0	0	$\frac{95}{144}h^5 f^{(4)}(\xi)$
4	$\frac{66}{20}$	$-\frac{84}{20}$	$\frac{156}{20}$	$-\frac{84}{20}$	$\frac{66}{24}$	0	0	$\frac{41}{140}h^7 f^{(6)}(\xi)$

Here ξ is a suitable number in $[a, b]$, and both e_n and o_n depend on n but not on the integrand function f. Their values are given in the last column in Tables 6.1 and 6.2. We conclude with a couple of remarks.

- If n is even (odd) and f is a polynomial of degree $n - 1$ (n), then $f^{(n)}(x) = 0$ ($f^{(n+1)}(x) = 0$) and the integration rule is exact in that we can exactly integrate polynomials of degree $n - 1$ (n);
- As $n \to \infty$, then $h \to 0$ and the error bound vanishes. However, from a numerical point of view, using n larger than 6 can generate round-off errors. This fact compels one to consider alternative formulae such as the ones presented below.

6.2.1 Composite Newton–Cotes Formula

Newton–Cotes formulae are usually not applied to the entire interval of integration $[a, b]$. A popular practice consists of splitting this interval into $m - 1$ evenly spaced subintervals $[x_j, x_{j+1}]$, $j = 1, \ldots, m - 1$, i.e., $x_j = a + (j - 1)H$ with $H = (b - a)/(m - 1)$. The required integral is given by the sum of the integrals computed on each subinterval, each one being calculated using a Newton–Cotes formula with $n+1$ equally spaced nodes. The resulting procedure is known as a *composite* Newton–Cotes formula and can be described as follows. The integral

$$I(f) = \int_a^b f(x)\,dx = \sum_{j=1}^{m-1} \int_{x_j}^{x_{j+1}} f(x)\,dx$$

is approximated by $I_{n,m}(f)$ defined as

$$I_{n,m}(f) = h \sum_{j=1}^{m-1} \sum_{i=1}^{n} f_{i,j} w_i,$$

where $f_{i,j} = f(x_i^j)$ and the x_i^j's are nodes refining subinterval $[x_j, x_{j+1}]$ and where $h = H/(n - 1)$ for close formula and $h = H/(n + 1)$ for open formula. We remark that integers $m - 1$ and $n - 1$ respectively refer to the number of subintervals and to the order of the interpolating polynomial.

The error associated to a composite Newton–Cotes formula is given by

$$I(f) - I_{n,m}(f)$$

$$= \begin{cases} \frac{b-a}{(n)!} \frac{e_n}{(n-1)^{n+1}} H^n f^{(n)}(\xi) & f \in C^{(n)}([a,b]) \text{ and } n \text{ even,} \\ \frac{b-a}{(n+1)!} \frac{o_n}{(n+1)^{n+2}} H^{n+1} f^{(n+1)}(\xi) & f \in C^{(n+1)}([a,b]) \text{ and } n \text{ odd,} \end{cases} \quad (6.6)$$

where $\xi \in [a,b]$. This figure says that approximation $I_{n,m}(f)$ is exact up to the order $n-1$ (resp. n) if n is even (resp. odd). In other words, it integrates exactly all polynomials of degree $n-1$ (resp. n). For a fixed value n, this error vanishes to zero as $m \to \infty$ (i.e. $H \to 0$).

From expression (6.6), it is clear that estimating an approximation error requires derivatives of f. If this is not the case, we may proceed as follows. We let n be an even number and then compute the error corresponding to a half of the space length h:

$$I(f) - I_{n,m}(f) = \frac{b-a}{n!} \frac{e_n}{(n-1)^{n+1}} H^n f^{(n)}(\xi), \quad (6.7)$$

$$I(f) - I_{n,2m-1}(f) = \frac{b-a}{n!} \frac{e_n}{(n-1)^{n+1}} \left(\frac{H}{2}\right)^n f^{(n)}(\xi), \quad (6.8)$$

$$|I_{n,m} - I_{n,2m-1}(f)| = \frac{b-a}{n!} \frac{|e_n|}{(n-1)^{n+1}} \left| f^{(n)}(\xi) \left(H^n - \frac{H^n}{2^n}\right) \right|$$

$$= \frac{b-a}{n!} \frac{|e_n|}{(n-1)^{n+1}} \left(\frac{H}{2}\right)^n |f^{(n)}(\xi)|(2^n - 1)$$

$$= |I(f) - I_{n,2m-1}(f)|(2^n - 1).$$

Consequently, the error estimate of the Newton–Cotes formula reads as:

$$|I(f) - I_{n,2m-1}(f)| = |I_{n,m} - I_{n,2m-1}(f)| \left(\frac{1}{2^n - 1}\right) \quad (6.9)$$

for n even. Similarly, for n odd we get

$$|I(f) - I_{n,2m-1}(f)| = |I_{n,m} - I_{n,2m-1}(f)| \left(\frac{1}{2^{n+1} - 1}\right). \quad (6.10)$$

These formula can be used for estimating the error. We now detail composite formulae corresponding to $n = 1, 2$, and 3. Higher values are usually not desirable due to cancellation errors produced by abnormal weights arising in these instances.

Rectangle and Midpoint Rules ($n = 1$)

These are the simplest approximations. They work rather well whenever function f is smooth and the space between points x_i is adequately small (i.e., m is quite large). In the Rectangle rule,[2] the function f on each interval is approximated by a piecewise constant function using the value attained at one of the two vertices of

---

[2] To be precise this is not a Newton–Cotes formula.

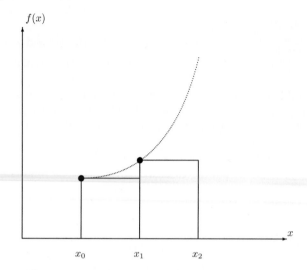

Fig. 6.1. Approximation of the integral using rectangles.

Table 6.3. Pseudo-code for implementing the rectangle rule

```
Input a, b, m and f.
Set H = (b - a)/(m - 1)
Initialize sum = 0
For i = 0 to m - 1
   Compute f at abscissas xi = a+i*H
   Update the sum according to sum = sum+H*f(xi)
Next i
Return sum
```

each subinterval. For example, in Fig. 6.1, we use the points x_0 and x_1. All weights are equal to 1 and the approximated formula reads as:

$$I_{0,m}(f) = h \sum_{j=1}^{m-1} f_j.$$

Table 6.3 provides the pseudo-code for implementing the rectangle rule.

In the *composite midpoint formula* the function on each subinterval is approximated by the value attained at the subinterval midpoint, that is $(x_{i+1} - x_i) f((x_{i+1} + x_i)/2)$, see Fig. 6.2. The composite formula reads as

$$I_{1,m}(f) = h \sum_{j=1}^{m-1} f_{j+\frac{1}{2}},$$

where $f_{i+\frac{1}{2}} := f((x_{i+1} + x_i)/2)$.

Example Let us consider the computation of the following integral

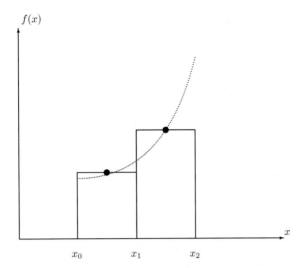

Fig. 6.2. Approximation of the integral using the midpoint formula.

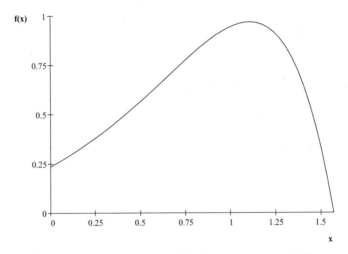

Fig. 6.3. Graph of the integrand function in expression (6.11).

$$\frac{5}{e^\pi - 2} \int_0^{\pi/2} e^{2x} \cos(x) = 1. \tag{6.11}$$

The integrand function is shown in Fig. 6.3. Table 6.4 shows the behavior of approximated value $I_{0,m}$, error, $E_{n,m} = I(f) - I_{n,m}(f)$, and ratio $R_{n,m} = |I(f) - I_{n,m}(f)|/|I(f) - I_{n,2m-1}(f)|$ across alternative values for m. Following expressions (6.9) and (6.10), this ratio is expected to converge to 2 for large values of m as long as $n = 1$. This means a reduction of the absolute error by a factor of 2 following a transition from m to $2m - 1$ subintervals, i.e., reducing length h of each subinterval

Table 6.4. Example of the rectangle rule

m	$I_{0,m}(f)$	$E_{0,m(h)} = (I(f) - I_{0,m}(f))$	$E_{0,m(h)}/E_{0,2m(n)-1}$
2	0.371510	6.285E–01	
3	0.817605	1.824E–01	3.446
5	0.972004	2.800E–02	6.515
9	1.004241	−4.241E–03	−6.601
17	1.006842	−6.842E–03	0.620
33	1.004611	−4.611E–03	1.484
65	1.002604	−2.604E–03	1.771
129	1.001377	−1.377E–03	1.892
257	1.000707	−7.070E–04	1.947
513	1.000358	−3.581E–04	1.974
1025	1.000180	−1.802E–04	1.987
2049	1.000090	−9.041E–05	1.994

by a half. If m exceeds the threshold 2049, the discretization error becomes smaller than the round-off error and the ratio does not tend to approach level 2 anymore.

Trapezoid Rule ($n = 2$)

This rule approximates the function on each subinterval using linear interpolation as illustrated in Fig. 6.4, that is by replacing $\int_{x_j}^{x_{j+1}} f(x)\,dx$ with $(f(x_j) + f(x_{j+1}))h/2$:

$$I_{1,m}(f) = \frac{h}{2} \sum_{j=1}^{m-1} (f(x_j) + f(x_{j+1}))$$

$$= h\left(\frac{1}{2}f(x_1) + f(x_2) + \cdots + f(x_{m-1}) + \frac{1}{2}f(x_m)\right).$$

The resulting error can be computed by summing up all error terms for subintervals and observing that $h = H$. If $f \in C^2([a, b])$, we have:

$$|I(f) - I_{1,m}(f)| = \sum_{j=1}^{m-1} \frac{h^3}{12} f^{(2)}(\xi_j) = \frac{h^2}{12} \frac{b-a}{m-1} \sum_{j=1}^{m-1} f^{(2)}(\xi_j),$$

where $\xi_j \in [x_j, x_{j+1}]$. Since $\min_j f^{(2)}(\xi_j) \leq \sum_{j=1}^{m-1} f^{(2)}(\xi_j)/(m-1) \leq \max_j f^{(2)}(\xi_j)$ and $f^{(2)}(\xi)$ is continuous, a number $\xi \in [\min_j \xi_j, \max_j \xi_j] \subset [a, b]$ exists, with $f^{(2)}(\xi) = \sum_{j=1}^{m-1} f^{(2)}(\xi_j)/(m-1)$. Therefore:

$$|I(f) - I_{1,m}(f)| = \frac{h^2}{12}(b - a)f^{(2)}(\xi),$$

with $\xi \in [a, b]$, showing that the error stemming from trapezoid rule approaches zero at the same rate as h^2. This rule is said to be of order 2 and integrates exactly first-order polynomials.

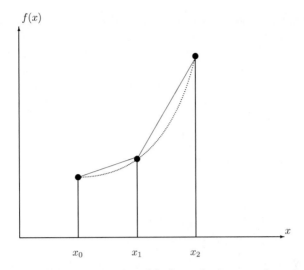

Fig. 6.4. Approximation of the integral using trapezium.

An interesting remark relative to the accuracy of the trapezoid rule can be made by exploiting the so-called Euler–MacLaurin formula. This formula states that

$$I_{1,m}(f) - \int_a^b f(x)\,dx = \sum_{l=1}^{k} h^{2l} \frac{B_{2l}}{(2l)!} \left(f^{(2l-1)}(b) - f^{(2l-1)}(a) \right)$$

$$+ h^{2k+2} \frac{B_{2k+2}}{(2k+2)!} (b-a) \left(f^{(2k+2)}(\xi) \right), \qquad (6.12)$$

for a suitable $\xi \in [a, b]$ and all $f \in C^{2k+2}([a, b])$. Here, B_{2l} are *Bernoulli numbers:*[3]

$$B_2 = \frac{1}{6}, \qquad B_4 = -\frac{1}{30}, \qquad B_6 = \frac{1}{42}, \qquad B_8 = -\frac{1}{30}, \dots$$

In particular, if $f \in C^3([a, b])$ and satisfies $f^{(1)}(b) = f^{(1)}(a)$, we obtain

$$I_{1,m}(f) - \int_a^b f(x)\,dx \simeq h^4 \frac{B_2}{(2)!} \left(f^{(3)}(b) - f^{(3)}(a) \right).$$

In this case, the trapezoid rule exhibits an error of order h^4. If $f \in C^5([a, b])$, then $f^{(1)}(b) = f^{(1)}(a)$ and $f^{(2)}(b) = f^{(2)}(a)$ and the error is of order h^6. Therefore, when f is a highly periodic smooth function, that is $f^{(j)}(b) = f^{(j)}(a)$ holds true up

---

[3] Bernoulli numbers are the values assumed by Bernoulli polynomials for $x = 0$. The kth Bernoulli polynomial $B_k(x)$ satisfies $(-1)^k B_k(1 - x) = B_k(x)$. The first five Bernoulli polynomials are: $B_0(x) = 1$, $B_1(x) = x - \frac{1}{2}$, $B_2(x) = x^2 - x + \frac{1}{6}$, $B_3(x) = x^3 - \frac{3}{2}x^2 + \frac{1}{2}x$, $B_4(x) = x^4 - 2x^3 + 2x^2 - \frac{1}{30}$. Bernoulli numbers of an odd index $k > 1$ are all equal to zero.

to a large value for j, the trapezoid rule turns out to be a very accurate integration formula. This fact is particularly useful for the numerical inversion of Fourier and Laplace transforms, that involve the numerical integration of oscillating functions.

Again, the Euler–MacLaurin formula can be used for the purpose of improving the trapezoid rule whenever both $f^{(1)}(b)$ and $f^{(1)}(a)$ are available. A formula of order h^4 reads as

$$I_{2,m}^*(f) = I_{2,m}(f) + \frac{h^2}{12}\big(f^{(1)}(a) - f^{(1)}(b)\big).$$

The corresponding error $I(f) - I_{2,m}^*(f)$ is $(b-a)h^4 f^{(4)}(\xi)/720$, provided the $f \in C^4([a, b])$. The order of trapezoid formula improves by a factor of 2 with a simple computation of the first derivative at the boundaries.

If we replace derivatives with finite differences of suitably high order, we obtain the so-called *end corrections* of a trapezoidal sum. For example, the following variant of the trapezoid rule has an error of order 3:

$$\widehat{I}_{2,m}(f) = h\bigg(\frac{5}{12}f(a) + \frac{13}{12}f(a+h) + f(a+2h)$$

$$+ \cdots + f(b-2h) + \frac{13}{12}f(b-h) + \frac{5}{12}f(b)\bigg). \qquad (6.13)$$

Example   Let us consider the integral (6.11)

$$\frac{5}{e^\pi - 2} \int_0^{\pi/2} e^{2x}\cos(x) = 1. \qquad (6.14)$$

Table 6.5 shows the behavior of approximated values $I_{2,m}$, errors $E_{n,m} = I(f) - I_{n,m}(f)$, and ratios $R_{n,m} = |I(f) - I_{n,m}(f)|/|I(f) - I_{n,2m-1}(f)|$ across alternative values for m. Following expressions (6.9) and (6.10), this ratio is expected to converge to 4 (resp., 8) for large values of m as long as $n = 2$ – Trapezoid rule (resp., $n = 3$ – Extended Trapezoid rule).

The Romberg Extrapolation

The Trapezoid Rule can be further improved using Romberg integration.[4] This procedure consists of using two estimates of the integral, say $I_{2,m_1}(f)$ and $I_{2,m_2}(f)$, in order to extrapolate a better approximation to $I(f)$. The estimates use different grid spacing. Romberg integration starts with the Euler–MacLaurin formula and represents $I_{2,m}(f)$ in terms of step length $h = (b-a)/(m-1)$, that is:

$$I_{2,m}(f) = I(f) + \gamma_1 h^2 + \gamma_2 h^4 + \cdots + \gamma_k h^{2k} + C(h)h^{2k+2}, \qquad (6.15)$$

---

[4] Romberg integration is a particular case of the more general Richardson extrapolation as applied to the trapezoid rule. This method exploits representation (6.15) of the error of the trapezoid formula in terms of power of h^2.

Table 6.5. Example of trapezoid and extended trapezoid rules

m	Trapezoid $I_{2,m}(f)$	$E_{2,m(h)}$		Extended trapezoid $\widehat{I}_{2,m}(f)$		
2	0.185755	8.142E–01				
3	0.724727	2.753E–01	2.958			
5	0.925565	7.443E–02	3.698	0.964759	3.524E–02	0
9	0.981022	1.898E–02	3.922	0.994570	5.430E–03	6.490
17	0.995232	4.768E–03	3.980	0.999247	7.531E–04	7.210
33	0.998807	1.193E–03	3.995	0.999901	9.914E–05	7.596
65	0.999702	2.985E–04	3.999	0.999987	1.272E–05	7.796
129	0.999925	7.462E–05	4.000	0.999998	1.610E–06	7.897
257	0.999981	1.866E–05	4.000	1.000000	2.026E–07	7.949
513	0.999995	4.664E–06	4.000	1.000000	2.541E–08	7.974
1025	0.999999	1.166E–06	4.000	1.000000	3.181E–09	7.987
2049	1.000000	2.915E–07	4.000	1.000000	3.981E–10	7.990

Table 6.6. Pseudo-code for implementing the trapezoid rule

```
Input a, b, m and f
Set h = (b - a)/(m - 1)
Initialize sum = 0.5 * h * (f(a) + f(b))
For i = 1 to m-2
    Compute f at abscissas ai = a+i*h
    Update the sum according to sum = sum+h*f(ai)
Next i
Return sum
```

where the expressions for $\gamma_1, \ldots, \gamma_k$ and $C(h)$ derive from (6.12). Therefore, we have devised for $I_{2,m}(f)$ a polynomial expansion in powers of h^2. The Romberg extrapolation is based on the following idea. We halve the integration range and compute $I_{2,m}(f)$ and $I_{2,2m-1}(f)$, see Fig. 6.5. According to formula (6.15), we have:

$$I_{2,m}(f) = I(f) + \gamma_1 h^2 + \gamma_2 h^4 + \cdots + \gamma_k h^{2k} + C(h)h^{2k+2},$$
$$I_{2,2m-1}(f) = I(f) + \gamma_1 4h^2 + \gamma_2(2h)^4 + \cdots + \gamma_k(2h)^{2k} + C(h)(2h)^{2k+2}.$$

We multiply the first of these expressions by 4 and then subtract the second one from it. By solving with respect to $I(f)$, we obtain:

$$I(f) = I_{2,m}(f) + \frac{1}{3}\left(I_{2,m}(f) - I_{2,2m-1}(f)\right) + \gamma_2^* h^4$$
$$+ \cdots + \gamma_k^*(2h)^{2k} + C^*(h)(2h)^{2k+2},$$

that is an estimate of the integral with accuracy to the order of h^4. If we approximate the integral above using formula

$$I_{2,m}(f) + \frac{1}{3}\left(I_{2,m}(f) - I_{2,2m-1}(f)\right),$$

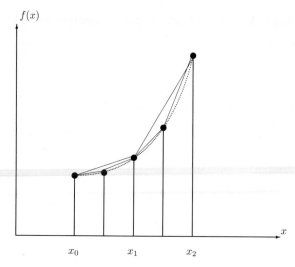

Fig. 6.5. Approximation of the integral using trapezium and Romberg extrapolation.

then we can exactly compute integrals of polynomials up to the third order.
    We consider the following sequence of interval refinements:

$$h_0 = b - a, \qquad h_1 = \frac{h_0}{2}, \qquad \dots, \qquad h_i = \frac{h_{i-1}}{2}, \quad i = 2, 3, \dots \qquad (6.16)$$

and the corresponding trapezoid formula

$$T_{i,0} = I_{1, m(h_i)}(f),$$

where $m(h_i) = (b - a)/h_i + 1$. We can extrapolate the value of the integral by using
the recursion

$$T_{i,s+1} = \frac{4^{s+1} T_{i,s} - T_{i-1,s}}{4^{s+1} - 1}, \qquad s = 0, \dots, n - 1; i = s + 1, \dots, n, \qquad (6.17)$$

which can be represented by the diagram reported in Table 6.7. Starting on the upper-
west case and then proceeding downward in this table amounts to increasing the
number of subintervals used in the trapezoid rule, whereas moving towards the right
end refers to augmenting the order of the integration rule. Furthermore, as $n \to \infty$,
the values on each column, as well as those on the diagonal, converge to the definite
integral. It can be proved that the error $I(f) - T_{i,n}$ goes to zero as fast as $h^{2(n+1)}$
does. The extrapolating procedure can be speeded up by using the fact that under
refinement (6.16), half of the function values required to compute the trapezoidal
sum $T_{i+1,0}$ stem from the calculation of $T_{i,0}$. This is illustrated in Fig. 6.6. In order
to avoid recomputing the function values at these points, we can use the following
updating rule:

$$T_{0,0} = \frac{h_0}{2} \big(f(a) + f(b) \big), \qquad (6.18)$$

Table 6.7. Romberg extrapolation

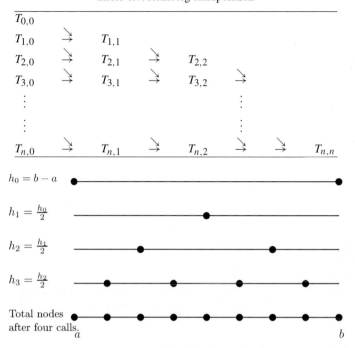

Fig. 6.6. New points necessary to refine the grid in the trapezoidal rule, starting with $h_0 = b-a$ and then setting $h_i = h_{i-1}/2$.

$$T_{1,0} = \frac{T_{0,0}}{2} + h_1 f(a + h_1),$$

$$\vdots$$

$$T_{i+1,0} = \frac{T_{i,0}}{2} + h_{i+1}\big(f(a + jh_{i+1}) + f(a + 3h_{i+1})$$
$$+ f(a + 5h_{i+1}) + \cdots + f(b - h_{i+1})\big).$$

The Romberg algorithm provides no clue about the proper size of parameter i defining the number of elements in the first column of diagram (6.7). In practice, one only computes a few columns, seven or eight for example, and then interrupts the calculation as soon as $|T_{i+1,6} - T_{i,6}|$ becomes "reasonably small".[5] However, the stopping criteria related to $|T_{i+1,6} - T_{i,6}|$ sometimes leads to an early stopping. Therefore, more refined stopping rules can be used. Table 6.8 provides the pseudo-code for Romberg extrapolation.

---

[5] This criterion has been implemented in the VBA® function `romberg(a, b, mmax)`. Here, `mmax` is the maximum number of elements in the first column in Table 6.7. Alternative stopping rules are discussed in Stoer and Bulirsch (1980), p. 149.

Table 6.8. Pseudo-code for implementing the Romberg extrapolation

```
Input a, b, maxiter, EPSILON and f
Set h = b - a; t(1) = 0.5*h*(f(a)+f(b));
For i = 1 to maxiter-1
  h = 0.5*h //halve the integration length
  compute t[i+1,1] using (6.18)
  For j = 2 To i + 1
    extrapolate up to t[i+1,j] using (6.17);
  next j
  err = Abs(t[i+1,i+1]-t[i,i])
  If err < EPSILON return {t[i+1,i+1], err, i+1}
next i
return {t[maxiter, maxiter], err, maxiter}
```

Table 6.9. Example of the Romberg extrapolation

i	h	$m(h)/s$	0	1	2	3	4
1	1.570796	2	0.185755				
2	0.785398	3	0.724727	0.904385			
3	0.392699	5	0.925565	0.992510	0.998386		
4	0.196350	9	0.981022	0.999507	0.999974	0.999999	
5	0.098175	17	0.995232	0.999969	0.99999958	1.000000	1.000000

Example We aim at computing integral (6.19)

$$\frac{5}{e^\pi - 2} \int_0^{\pi/2} e^{2x} \cos(x) = 1 \tag{6.19}$$

by using the Romberg extrapolation. The first rows in the iteration matrix indicated in Table 6.7 are reported in Table 6.9. Numbers shown in the column "0" result from applying the trapezoidal approximation and then halving the interval length h. The remaining columns report extrapolated values obtained by using recursion (6.17). We see that the level of accuracy is $|T_{4,4} - T_{3,3}| = 1.23 \times 10^{-6}$. Notice that the trapezoidal rule would require as much as $m = 1,025$ subintervals in order to produce a result with comparable accuracy.

Simpson Rule ($n = 3$)

The Simpson formula is based on a local quadratic approximation of the integrand using nodes a, $(a + b)/2$ and b. From Table 6.1, we obtain:

$$\int_a^b f(x)\,dx = \frac{h}{3}\left(f(a) + 4f\left(\frac{a+b}{2}\right) + f(b)\right).$$

If $f \in C^4([a, b])$, then the error $I(f) - I_n(f)$ is $-h^5 f^{(4)}(\xi)/90$. For any even m, the composite formula is obtained by applying the Simpson rule to consecutive nonoverlapping pairs of subintervals $[x_{2j+1}, x_{2j+2}, x_{2j+3}]$, for $j = 0, 1, \ldots, m/2 - 3$, individually, yielding the approximation $(f(x_{2j+1}) + 4f(x_{2j+2}) + f(x_{2j+3}))h/3$. Summing these contribution yields to the composite estimate

$$I_{3,m}(f) = \frac{h}{3} \sum_{j=0}^{m/2-3} \left(f(x_{2j+1}) + 4f(x_{2j+2}) + f(x_{2j+3}) \right)$$

$$= \frac{h}{3}(f_1 + 4f_2 + 2f_3 + 4f_4 + \cdots + 2f_{m-2} + 4f_{m-1} + f_m).$$

It can be shown that the error in the composite Simpson formula is given by

$$I(f) - I_{3,m}(f) = -\frac{b-a}{180} h^4 f^{(4)}(\xi), \quad \xi \in [a, b],$$

provided that $f \in C^4([a, b])$. Table 6.10 provides the pseudo-code for the Simpson rule.

Example Table 6.11 reports values obtained for integrating

$$\frac{5}{e^\pi - 2} \int_0^{\pi/2} e^{2x} \cos(x) = 1$$

Table 6.10. Pseudo-code for implementing the Simpson rule

```
Input a, b, m and f
Set h = (b - a)/(m - 1)
sum = h * (f(a) + f(b)) / 3
For i = 1 To m - 2 Step 2
  sum = sum + h * f(a + i * h) * 4 / 3
Next i
For i = 2 To m - 2 Step 2
  sum = sum + h * f(a + i * h) * 2 / 3
Next i
return sum
```

Table 6.11. Integration of (6.11) using composite Simpson quadrature under alternative numbers of subintervals

m	$I_{3,m}(f)$	$E_m = I(f) - I_{3,m}(f)$	E_m/E_{2m+1}
3	0.904385	9.562E–02	0.000
5	0.992511	7.489E–03	12.767
9	0.999507	4.928E–04	15.196
17	0.999969	3.119E–05	15.803
33	0.999998	1.955E–06	15.951
65	1.000000	1.223E–07	15.988
129	1.000000	7.645E–09	15.996
257	1.000000	4.780E–10	15.992
513	1.000000	3.009E–11	15.887

by using the Simpson rule. We can see that the ratio between absolute errors approaches 16. This figure is in accordance with expression (6.9) as applied to the Simpson rule ($n = 2$).

6.3 Gaussian Quadrature Formulae

Newton–Cotes formulae use a fixed number of equally spaced abscissas and choose the weights to achieve the highest order of accuracy. Moreover, they are usually implemented in a composite form: the interval of integration is divided into subintervals of equal size, then the rule is applied to each subinterval, and finally the sum of resulting numbers is taken as an estimate of the overall integral. In this framework, convergence to the actual value of the integral can be rather slow. Consequently, an accurate estimation may require a excessively large number of subintervals and corresponding evaluations of the integrand function.

Gaussian quadrature rules select both the n abscissas and the n weights to produce a rule of order $2n - 1$. This way we can exactly integrate polynomials of degree $2n - 1$, that is the highest degree for which a polynomial can be integrated using n points. Therefore, compared to the Newton–Cotes formulas, we may freely select the abscissas at which the integrand function is evaluated.

Example We build a quadrature rule for integrating exactly all polynomials up to the third order, i.e., combinations of 1, x, x^2, and x^3, over the closed interval $[-1, 1]$ by using 2 points only ($n = 2$). By setting $\{w_i, x_i\}_{i=1,2}$ so that:

$$I(1) = \int_{-1}^{1} 1 \, dx = 2 = w_1 f(x_1) + w_2 f(x_2) = w_1 + w_2,$$

$$I(x) = \int_{-1}^{1} x \, dx = 0 = w_1 f(x_1) + w_2 f(x_2) = w_1 x_1 + w_2 x_2,$$

$$I(x^2) = \int_{-1}^{1} x^2 \, dx = \frac{2}{3} = w_1 f(x_1) + w_2 f(x_2) = w_1 x_1^2 + w_2 x_2^2,$$

$$I(x^3) = \int_{-1}^{1} x^3 \, dx = 0 = w_1 f(x_1) + w_2 f(x_2) = w_1 x_1^3 + w_2 x_2^3,$$

we derive a set of nonlinear equations:

$$\begin{cases} w_1 + w_2 = 2 \\ w_1 x_1 + w_2 x_2 = 0 \\ w_1 x_1^2 + w_2 x_2^2 = \frac{2}{3} \\ w_1 x_1^3 + w_2 x_2^3 = 0. \end{cases}$$

The solution for this system is:

$$w_1 = 1, \qquad w_2 = 1, \qquad x_1 = -\frac{1}{\sqrt{3}}, \qquad x_2 = \frac{1}{\sqrt{3}}.$$

Consequently, the 2-point quadrature rule for a function f reads as:

$$I(f) = \int_{-1}^{1} f(x)\,dx \simeq 1 f\left(-\frac{1}{\sqrt{3}}\right) + 1 f\left(\frac{1}{\sqrt{3}}\right).$$

If we are interested in integrating the same function in the interval $[a, b]$, we define $\alpha = (a + b)/2$ and $\beta = (b - a)/2$, and compute

$$\int_{a}^{b} f(x)\,dx = \int_{-1}^{1} f(\alpha + \beta x)\,dx \simeq 1 f\left(\alpha - \beta\frac{1}{\sqrt{3}}\right) + 1 f\left(\alpha + \beta\frac{1}{\sqrt{3}}\right).$$

We see that weights remain unchanged, while the two abscissas selected inside the interval $[a, b]$ are $\alpha - \beta\frac{1}{\sqrt{3}}$ and $\alpha + \beta\frac{1}{\sqrt{3}}$, namely those resulting from applying a linear transformation of the abscissas obtained in case of integration on the interval $[-1, 1]$. As a sample case, we consider $f(x) = 1/(0.2x + 3)$ on $[-1, 1]$. The exact value is

$$\int_{-1}^{1} \frac{1}{0.2x + 3}\,dx = 5\ln|3 + 0.2x|\big|_{-1}^{1} = 5\ln\left(\frac{8}{7}\right) = 0.6676569631.$$

By using a two-point Gaussian quadrature rule, we get to an approximated value

$$1\frac{1}{0.2(-1/\sqrt{3}) + 3} + 1\frac{1}{0.2(1/\sqrt{3}) + 3} = 0.6676557864,$$

with a bias evaluated as 1.1768×10^{-6}.

The quadrature formula illustrated above can be extended to integrate the product of polynomials with any function ϕ. We need to select weights w_i and abscissas x_i in a way that the approximated integral

$$\int_{a}^{b} \phi(x) f(x)\,dx \simeq \sum_{i=1}^{n} w_i f(x_i)$$

is exact provided that f is a polynomial of degree $2n - 1$.

The error in the quadrature rule, when $\phi(x) \equiv 1$, can be evaluated by the following estimate. See Davis and Rabinowitz (1975)

$$\int_{a}^{b} f(x)\,dx - \sum_{i=1}^{n} w_i f(x_i) = \frac{(b - a)^{2n+1}(n!)^4}{(2n + 1)(2n!)^3} f^{(2n)}(\xi), \qquad (6.20)$$

where $\xi \in (a, b)$, provided that $f^{(2n)}(\xi)$ is continuous on $[a, b]$.

At first glance, we may think that computing abscissas and weights requires solving a nonlinear system with $2n$ unknown variables. However, it can be shown that the abscissas are the zeros of a polynomial with degree $n + 1$, which in turn belong to a sequence of polynomials that are orthogonal on $[a, b]$. (See Davis and Rabinowitz (1975), pp. 95–100.) We now briefly illustrate this construction. The integral

$\int_a^b \phi(x) f(x) g(x) \, dx$ defines an inner product of functions g and f over $[a, b]$ with respect to the weighing function ϕ satisfying appropriate regularity conditions (i.e., positive, integrable and continuous on $[a, b]$). Following standard notation, we denote this quantity by (f, g). Two functions are said to be "orthogonal" on $[a, b]$ with respect to ϕ whenever $(f, g) = 0$. For a given weighing function ϕ, we can build a sequence of pairwise orthogonal polynomials $p_0, p_1, \ldots, p_n$ of degree $0, 1, \ldots, n$, respectively, that is:

$$\int_a^b \phi(x) p_n(x) p_k(x) \, dx = 0 \quad \text{for } k \neq n.$$

Given such a sequence, we can obtain a new one $(p_n^*)_{n \geq 1}$ where each element has a unit norm, namely:

$$\left(p_n^*, p_k^*\right) = \int_a^b \phi(x) p_n^*(x) p_k^*(x) \, dx = \begin{cases} 0 & \text{if } k \neq n, \\ 1 & \text{if } k = n. \end{cases}$$

This is referred to as an "orthonormal system". Notice, in particular, that $p_n^*(x) = p_n(x)/(p_n, p_n)^{1/2}$. From now on, we assume inner products are all computed with respect to a given weighing function ϕ defined on $[a, b]$.

We now report three results relating the zeros of orthogonal polynomials to the abscissas in quadrature rules.

Result 1 (Properties of zeros of orthogonal polynomials) The zeros of (real) orthogonal polynomials are real, simple (i.e. not multiple), and located in the interior of $[a, b]$.

Result 2 (Construction algorithm for orthonormal polynomials) Orthogonal polynomials satisfy the following iteration:

$$p_{n+1}(x) = (x - a_n) p_n(x) - b_n p_{n-1}(x), \tag{6.21}$$
$$p_{-1}(x) = 0, \ p_0(x) = 1,$$

where

$$a_n = \frac{(x p_n, p_n)}{(p_n, p_n)}, \tag{6.22}$$

$$b_n = \frac{(p_n, p_n)}{(p_{n-1}, p_{n-1})}.$$

This leads to an iterative rule for building orthonormal polynomials, that is:

$$p_n^*(x) = \frac{p_n(x)}{(p_n, p_n)^{1/2}}.$$

Result 3 (Relationship between polynomial zeros and quadrature abscissas) Let the zeros of $p_n^*(x)$ be denoted by $x_1, \ldots, x_n$, where $a < x_1 < x_2 < \cdots < x_n < b$. Then, we can find positive constants $w_1, w_2, \ldots, w_n$ such that:

$$\int_a^b \phi(x) f(x) \, dx = \sum_{i=1}^n w_i f(x_i), \tag{6.23}$$

whenever f is a polynomial of order $2n - 1$. Moreover, weights w_k can be explicitly represented as:

$$w_k = \frac{(p_{n-1}, p_{n-1})}{p_{n-1}(x_k) p'_{n-1}(x_k)},$$

where $p'_{n-1}(x_k)$ is the first-order derivative of the orthogonal polynomial at the zero point x_j.

A *Gaussian quadrature rule* is one whose abscissas and weights have been determined according to the prescriptions in Result 3. In summary, this rule requires:

(a) generate orthogonal polynomials $p_0, \dots, p_n$ using iterations (6.21) and (6.22);
(b) determine the zeros of $p_n(x)$;
(c) compute the corresponding weights;
(d) return (6.23) as a result.

Nowadays, these steps are well understood for a quite large class of weighing functions and integration ranges. Also, numerical routines are widely available. The most important weighing functions and their quadrature rules are given in Table 6.12. The error term of the Gauss–Legendre rule is given in expression (6.20). If we define $M_n = \max_{-1 \le x \le 1} |f^{(n)}(x)|/n$, this error is superiorly bounded:

$$\left| \int_{-1}^{1} f(x)\, dx - \sum_{i=1}^{n} w_i f(x_i) \right| \le \frac{(2)^{2n+1}(n!)^4}{(2n+1)(2n!)^2} M_{2n} = e_n n_{2n},$$

so that as $n \to \infty$, $e_n \simeq \pi/4^n$ and the error term is bounded by $\pi M_{2n}/4^n$, i.e., by an exponential rate of decrease as a function of n. The composite trapezoidal and the composite Simpson rules have only polynomial rates of decrease, i.e. $1/m^2$ and $1/m^4$. Gaussian quadrature results to be always better than the trapezoidal rule, except in the case of periodic integrands.

A pseudo-code for implementing the Gaussian quadrature rule is given in Table 6.13. The algorithm is straightforward provided we have at our disposal a numerical routine to compute abscissas and weights. For example, Chebyshev polynomials have analytical abscissas and weights given by

$$x_i = \cos\left(\frac{\pi(i - 0.5)}{n}\right),$$

$$w_i = \frac{\pi}{n}.$$

Table 6.12. Principal Gaussian quadrature formulas

Rule	$\phi(x)$	Range	Recursion
Gauss–Legendre	1	$-1 \le x \le 1$	$p_{j+1} = \frac{2j+1}{j+1} x p_j - \frac{j}{j+1} p_{j-1}$
Gauss–Chebyshev	$\frac{1}{\sqrt{1-x^2}}$	$-1 \le x \le 1$	$p_{j+1} = 2x p_j - p_{j-1}$
Gauss–Laguerre	$x^\alpha e^{-x}$	$0 \le x < +\infty$	$p_{j+1} = \frac{-x+2j+\alpha+1}{j+1} p_j - \frac{(j+\alpha)}{j+1} p_{j-1}$
Gauss–Hermite	e^{-x^2}	$-\infty < x < +\infty$	$p_{j+1} = 2x p_j - 2j p_{j-1}$

Table 6.13. Pseudo-code for implementing the Gaussian quadrature

```
Input a, b, m and f.
Compute the abscissas ai and the weights wi
Initialize sum = 0
For i = 1 to m
  Compute f at abscissas ai
  Update the sum according to sum = sum+wi*f(ai)
Next i
Return sum
```

VBA® functions:

```
function gauleg(x1 As Double, x2 As Double,
                n As Integer) as Variant
```

and

```
function gaulag(n As Integer, alfa As Double) As Variant
```

return a set $\{w_i, x_i\}_{i=1,...,n}$ for Gauss–Legendre and Gauss–Laguerre rules, respectively. These functions are presented in Sect. 6.5.

Example We compute integral (6.11)

$$\frac{5}{e^\pi - 2} \int_0^{\pi/2} e^{2x} \cos(x) = 1$$

by using the 10-point Gauss–Legendre rule. The set $\{w_i, x_i\}_{i=1,...,10}$ is reported in Table 6.14. The first column contains the abscissas in $[0, \pi/2]$, the second column exhibits the weights, while the third column indicates the quantities $f(x_i)w_i$.
The last row contains the approximated value $\sum_{i=1}^{10} w_i f(x_i)$. In this example, the absolute error is 8.90066×10^{-13}.

Example We evaluate $\int_{-1}^1 1/(1+x^2)\,dx$ by using the 10-point Gauss–Chebyshev rule. The exact value of this integral is $\pi/2 = 1.570796327$. The weighing function according to this rule is $(1-x^2)^{-1/2}$, which we consider on the interval $[-1, 1]$. Then:

$$\int_{-1}^1 \frac{1}{(1+x^2)}\,dx = \int_{-1}^1 \frac{1}{(1-x^2)^{1/2}} \frac{(1-x^2)^{1/2}}{(1+x^2)}\,dx \simeq \sum_{i=1}^{10} w_i f(x_i),$$

where the function f is given by $(1-x^2)^{1/2}/(1+x^2)$. This point is illustrated in Table 6.15. In particular, we have $\sum_{i=1}^{10} w_i f(x_i) = 1.57488489$ and the absolute error equals 0.00408856.

An important property of the Gaussian quadrature rule is that the all weights are positive, so that Gaussian formulas display nice round-off properties even for large

Table 6.14. Example of Gauss–Legendre quadrature

x_i	w_i	$f(x_i)w_i$
0.02049376	0.052363551	0.012899989
0.10597898	0.117378815	0.034123158
0.25179114	0.172070027	0.065214501
0.44501022	0.211481587	0.109939222
0.66847253	0.232104183	0.164022574
0.9023238	0.232104183	0.206788039
1.12578611	0.211481587	0.204601451
1.31900519	0.172070027	0.141798329
1.46481734	0.117378815	0.054975751
1.55030256	0.052363551	0.005636986
	$\sum_{i=1}^{10} w_i f(x_i) =$	1

Table 6.15. Example of Gauss–Chebyshev quadrature

$x_i = \cos(\pi(j - 1/2)/n)$	$w_i = \pi/n$	$f(x_i) \times w_i$
0.987688	0.314159265	0.024877061
0.891007	0.314159265	0.079506053
0.707107	0.314159265	0.148096098
0.45399	0.314159265	0.232083777
0.156434	0.314159265	0.302879456
−0.156434	0.314159265	0.302879456
−0.45399	0.314159265	0.232083777
−0.707107	0.314159265	0.148096098
−0.891007	0.314159265	0.079506053
−0.987688	0.314159265	0.024877061
	$\sum_{i=1}^{10} w_i f(x_i) =$	1.57488489

values of n. This does not hold true for Newton–Cotes formula for which the weights associated to high-order quadratures can be negative, round-off errors are magnified, and the rule fails to converge. We can also use a composite Gaussian rule, that is split the interval into several bins and apply the formula over each bin independently of the other. In contrast with the Newton–Cotes formula where old points can still be used, here the new abscissas are all different from those computed one step before and a new computation is required. This problem also occurs whenever we try to increase the accuracy of the Gaussian rule. In this case, all information obtained while computing the lower-order rule is discarded at the following step as long as weights and abscissas of rules of distinct orders are different from each other. This problem has been partially solved by the Gauss–Kronrod formulas which enable us to add new abscissas and return a new rule with higher order. In practice, the weights need to be recomputed whereas to the original n abscissas $n + 1$ new abscissas are added, so that the rule becomes exact for all polynomials of degree less than or equal to $3n + 1$, when n is even, or of degree $3n + 2$ when n is odd. Finally, note that integration formulas of Gaussian type with a certain number abscissas assigned by the rule at the outset exist. Radau and Lobatto rules are among the most popular integration methods. These rules prescribe abscissas at the endpoints of the intervals and use a unit weight $\phi = 1$.

Note that integrals over infinite or semi-infinite intervals can be computed through several approaches. For instance, we can replace an infinite limit of integration by appropriately selecting finite values and then using a Gauss–Legendre rule. Another option consists of transforming the variable of integration so that the new interval is finite. Needless to say, some care is necessary in order to avoid introducing singularities in the new system of variables. Another possibility is to use quadrature rules designed for infinite intervals, such as Gauss–Laguerre and Gauss–Hermite.

6.4 Matlab® Code

The Matlab® implementation of Newton–Cotes rules is quite straigthforward. For example, the following trap() integrates the function fun in the interval $[a, b]$ using the trapezoid rule.

6.4.1 Trapezoidal Rule

```
function result = trap(a,b,m,fun)
   h = (b-a)/(m-1); x = [a:h:b]; y = eval(fun);
   result = h*(0.5*y(1)+sum(y(2:end-1))+0.5*y(end));
```

Example Let us write in the command window

```
>> trap(0,pi/2,513,'exp(2*x).*cos(x)*5/(exp(pi)-2)')
```

we get 0.99999535430049.

6.4.2 Simpson Rule

```
function result = simpson (a,b,m,fun)
   m = m-1; h = (b-a)/m; x = [a:h/2:b]; y = eval(fun);
   result = (h/6)*(y(1)+2*sum(y(3:2:2*m-1))
                   +4*sum(y(2:2:2*m))+y(2*m+1)));
```

Example Let us write in the command window

```
>> simpson(0,pi/2,125,'exp(2*x).*cos(x)*5/(exp(pi)-2)')
```

we get 0.99999999947464.

6.4.3 Romberg Extrapolation

```
function [T] = romberg(a,b,n,fun);
  %generate the first column of the Romberg matrix
  h = (b-a) ;
  for i = 1: n+1, T(i,1) = trap(a,b,(b-a)/h+1,fun);
  h = h/2; end;
  %start the extrapolation
  for s = 0:n-1, for i = s+1:n
    ss = s+1; ii = i+1;
  T(ii,ss+1) = (4^(s+1)*T(ii,ss)
                  -T(ii-1,ss))/(4^(s+1)-1);end ;end
```

Let us write in the command window

```
>> romberg(0,pi/2,3,'exp(2*x).*cos(x)*5/(exp(pi)-2)')
```

and we obtain the Table 6.9.

6.5 VBA® Code

The Newton–Cotes and Gauss–Legendre rules have been implemented in VBA®, in the modules associated to the Excel file Quadrature.xls. These functions take the extremes of the integration range, a and b, and the number of subintervals m for the composite Newton–Cotes or the number of points n for the Gaussian rules as inputs.

The integrating function must have the form

```
function f(x As Double, Optional Parameters as Variant)
as Double
```

where `Parameters` is a row vector containing the parameters that eventually enter into the definition of the function. Examples of integrand are given in the VBA® module `mf_x`.

The Newton–Cotes functions, VBA® module `mNewtonCotes`, return a number representing the estimated value of the integral (see Table 6.16). The function `romberg` is an exception. Instead of requiring the number of subintervals m, it asks for the maximum number (mmax) of subdivisions of the integration range that can be used for extrapolating the final estimate, provided the stopping criterion is not yet met. mmax cannot exceed 16. The function `romberg` returns a row vector containing the estimated value of the integral, the difference between the last two diagonal elements, i.e. $T_{i,i}$ and $T_{i-1,i-1}$ and the number of iterations. The functions for the Gaussian quadrature are `gauleg`, `gq` (in the VBA® module `mLegendre`) and `gaulag` (in the VBA® module `mLaguerre`) (see Table 6.17). The function `gaulag` requires the parameter `alfa` as input that enters in the definition of the

Table 6.16. Newton–Cotes functions in VBA® module mNewtonCotes

```
function rectangular(a As Double, b As Double,
                    m As Integer) as Double.
function trap(a As Double, b As Double,
             m As Integer) as Double.
function ExtendedTrap(a As Double, b As Double,
                     m As Integer) as Double.
function simpson(a As Double, b As Double,
                m As Integer) as Double.
function romberg(a As Double, b As Double,
                mmax As Integer) as Variant.
```

Table 6.17. Gaussian quadrature rules in VBA® modules mLaguerre and mLegendre

```
function gauleg(a As Double, b As Double,
               n As Integer) as Variant.
function gq(a As Double, b As Double, n As Integer) as Double.
function gaulag(n As Integer, alfa As Double) As Variant.
```

	A	B	C	D	E
1	a	0			
2	b	1.570796327			
3	m	1024			
4					
5	Exact	1.000000000			Ass(I(f)-I$_{n,m}$(f))
6	Rectangle	1.000180411	rectangular(B1,B2,B3)		1.80E-04
7	Trapezium	0.999998832	trap(B1,B2,B3)		1.17E-06
8	Ext. Trap	0.999999997	ExtendedTrap(B1,B2,B3)		3.19E-09
9	Simpson	1.000000000	Simpson(B1,B2,B3)		2.09E-12
10	GQ	1.000000000	gq(B1,B2,B3)		1.14E-11
11	Romberg	1.000000000	romberg(B1,B2,B3)		3.03E-12

Fig. 6.7. Example of usage of the VBA® integration routines.

weighing function (see Table 6.12). The functions `gauleg` and `gaulag` return an $n \times 2$ array containing the abscissas in the first column and the weights in the second column. The function `gq` applies the Gauss–Legendre quadrature. Figure 6.7 illustrates how to use the above VBA® functions and compares the different numerical routines in the computation of (6.11).

6.6 Adaptive Quadrature

This method is based on the following idea. We split the integration range $[a, b]$ and then keep on refining it until the composite quadrature formula produces the required level of accuracy. First, we integrate $f(x)$ by using two numerical methods and come up with approximations I_1 and I_2. With no loss of generality, we may

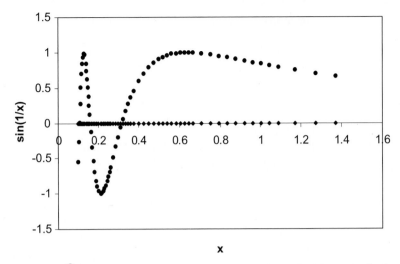

Fig. 6.8. Matlab® adaptive routine quad is used to integrate $\sin(1/x)$ over the interval $[0.1, \pi/2]$. Sample values are denoted by dots on the x-axis. The corresponding values of the integrand function are indicated by dots.

assume that I_1 is more accurate than I_2. If the difference of the two approximations is smaller than some prescribed tolerance, one accepts I_1 as the value of the integral. Otherwise, the interval $[a, b]$ is split into $[a, c]$ and $[c, b]$, with $c = (a + b)/2$ and the corresponding two integrals are computed independently of each other. Splitting intervals into subintervals continues until the stopping criterion

$$\frac{|I_1 - I_2|}{|I_1|} < tol$$

is met for each integral. Here, I_1 and I_2 are two estimates for the integral computed on the subinterval under consideration and *tol* is a predefined tolerance. In order to prevent the procedure from generating an excessive number of subdivisions, the suggested criterion is usually implemented with additional conditions.

Matlab® functions quad and quadl are based on an adaptive recursive Simpson's rule and on a recursive Lobatto quadrature, respectively. Figure 6.8 illustrates the way function quad adaptively splits the integration range $[0.1, \pi/2]$ for the purpose of integrating $\sin(1/x)$. As it is clear from the graph, the integrand is more thoroughly sampled in those portions of the domain where integration is more challenging, namely the ones where the integrand function exhibits an oscillatory behavior. A typical problem arising from the use of adaptive routines is encountered while dealing with discontinuous functions. In this case, a large number of successive evaluations may be produced by the algorithm. In order to circumvent this problem, it may be useful to apply an adaptive quadrature routine on each of the two sides neighboring the point of discontinuity.

The syntax of function quadl is reported in Table 6.18. Function quadl(fun,a,b) computes the integral of function fun between a and b by

Table 6.18. Syntax of the MATLAB® function

Syntax quadl
[I,n] = quadl(fun,a,b)
[I,n] = quadl(fun,a,b,tol)
[I,n] = quadl(fun,a,b,tol,trace)

using a recursive adaptive Lobatto quadrature with a default tolerance equal to 10^{-6}. Here, fun is a function handle through either an M-file function or an anonymous function. An anonymous function can be defined directly in the command window. For example,

```
>> fun = @(x) exp(x).*cos(x);
```

is a function handle that defines $f(x) = e^x \cos(x)$. Routine quadl returns the estimated value I of the integral and the number n of function evaluations. quadl(fun,a,b,tol) allows the user to specify the tolerance level to be adopted in the stopping criterion, whereas quadl(fun,a,b,tol,trace) generates a table $[n|a|b-a|I]$, where n is the number of functional evaluations (first column), a is the abscissas at which the integrand is computed (second column), $b - a$ is the size of each subinterval (third column), and I is the value of the integral in each subinterval (fourth column). quadl may issue one of the following warnings: "Minimum step size reached" indicates that the recursive interval subdivision has produced a subinterval whose length is on the order of round-off error in the length of the original interval. A nonintegrable singularity is possible. "Maximum function count exceeded" indicates that the integrand has been evaluated more than 10,000 times. A nonintegrable singularity is likely. "Infinite or Not-a-Number function value encountered" indicates a floating point overflow or division by zero during the evaluation of the integrand in the interior of the interval.

Example   Let us reconsider the integral (6.11)

$$\frac{5}{e^\pi - 2} \int_0^{\pi/2} e^{2x} \cos(x) = 1.$$

We want to compute this value by using quad and quadl. First, we define fun as an anonymous function.[6] In the Matlab® command window we write:

```
>> c = 5/(exp(pi)-2);
>> fun = @(x) c.*exp(2.*x).*cos(x);
```

---

[6] This is possible in Matlab® 7. In Matlab® 6 we need to use a different syntax:
```
>> fun = inline(' c.*exp(2.*x).*cos(x)','x','c');
>> c = 5/(exp(pi)-2);
>> quad(fun,0,pi/2,10^-6,'trace on',c)
```

Funct. Evals.	x_i	$x_{i+1}-x_i$	I	
9	0.000000	0.426597	0.153483	
11	0.000000	0.213298	0.062389	0.153483
13	0.213298	0.213298	0.091095	
15	0.426597	0.717603	0.562404	
17	0.426597	0.358801	0.234610	
19	0.426597	0.179401	0.103727	0.234610
21	0.605998	0.179401	0.130883	
23	0.785398	0.358801	0.327800	
25	0.785398	0.179401	0.156145	0.327800
27	0.964799	0.179401	0.171655	
29	1.144199	0.426597	0.284106	
31	1.144199	0.213298	0.190276	0.190276
33	1.357498	0.213298	0.093830	
35	1.357498	0.106649	0.066821	0.093830
37	1.464147	0.106649	0.027009	
			$I_n(f)$	**1.000000**

Fig. 6.9. Subdivision of the integration range for computing the integral in (6.11) by using the MATLAB® function quad.

Then, we apply the quadrature rule:

```
>>[I,n] = quad(fun,0,pi/2,10^-6)
```

and get to $I = 0.99999999502151$ and $n = 37$. If instead we use the command:

```
>>[I,n] = quadl(fun,0,pi/2,10^-6)
```

we obtain $I = 1.00000003176782$ and $n = 18$. Another input may be:

```
>>[I,n] = quad(fun,0,pi/2,10^-6,'trace on')
```

which returns results reported in Fig. 6.9.

There, the first column states the number of function evaluations; the second column indicates the abscissas at which the integrand function is computed; the third column contains the size of each subinterval; the fourth column reports the value of the integral on each subinterval; the last column underlines the subintervals that contribute to the computation of the final integral.

6.7 Examples

In this section we consider several examples. First, we apply quadrature methods to the pricing of options in the Black–Scholes model. Next, we consider the same problem under the assumption of square-root dynamics for the underlying asset. Then, we consider the problem of pricing options on coupon bonds in the Cox–Ingersoll–Ross model. Also, we examine the pricing of barrier options under discrete monitoring of the hitting condition. Finally, we implement the FFT algorithm to price derivatives in the case of underlying assets driven by Lévy processes.

6.7.1 Vanilla Options in the Black–Scholes Model

Let us consider the pricing problem of a plain vanilla option in the Black–Scholes setting. The option price requires computing the following integral:

$$c = e^{-r\tau} \int_{\ln K}^{+\infty} \left(e^{\xi} - K\right) h(\xi, \tau; z) \, d\xi,$$

where

$$h(\xi, \tau; z) = \frac{1}{\sqrt{2\pi\sigma^2\tau}} \exp\left(-\frac{1}{2\sigma^2\tau}\left(\xi - z - \left(r - \frac{\sigma^2}{2}\right)\tau\right)^2\right),$$

$z = \ln x$, and x denotes the spot price. In practice, we consider a range of possible log-prices at maturity, starting from $\ln K$ and moving up to z_{max}, a value selected such that $(e^{z_{max}} - K)h(z_{max}, \tau; z)$ turns out to be a small quantity, e.g., no greater that 10^{-8}. Numerical quadrature approximates the integral under consideration as $\sum_{i=1}^{n-1} w_i (e^{\xi_i} - K)h(\xi_i, \tau; z)$. Here, spacing between ξ_i's equals the constant $h = (\ln x_{max} - \ln K)/(m - 1)$, where $x_{max} = e^{z_{max}}$, in composite Newton–Cotes methods, whereas it varies in the Gauss–Legendre quadrature. The resulting assessment can be compared to the well-known Black–Scholes formula. We consider a call option with strike price $K = 100$ and time to maturity $\tau = 1$. We also assume an instantaneous volatility $\sigma = 0.2$, a risk-free rate of interest $r = 0.1$, and a standing market price of the underlying asset $x = 100$. The Black–Scholes price is 13.26967658466089. We set $z_{max} = \ln 400$. Table 6.19 reports prices for four alternative quadrature methods along successive halvings of h, taking $h = (\ln x_{max} - \ln K)/8$, $(\ln x_{max} - \ln K)/16$, $(\ln x_{max} - \ln K)/32$, and so on. Gaussian quadrature provides a result that is accurate to the sixth digit with $m = 33$. As theory suggests, among Newton–Cotes formulas, Simpson's is the preferred one, immediately followed by the Extended Trapezoidal formula. Table 6.20 shows the behavior of the absolute error $E_{n,m} = |I(f) - I_{n,m}(f)|$ and of the ratio $R_{n,m} = |I(f) - I_{n,m}(f)|/|I(f) - I_{n,2m-1}(f)|$ for varying values of m. In particular, $R_{n,m}$ corresponding to the Newton–Cotes formulae converge to 2^{n+1} for n odd (e.g., 4 for the Trapezoidal rule, $n = 2$), and to 2^{n+2} for n even (e.g., 16 for the Simpson rule, $n = 3$). The order of the extended Trapezoidal formula (6.13) is h^3, so that halving h results into a reduction of the error by a factor 8. For the Gauss–Legendre formula, m stands for the number of points in the quadrature rule. Table 6.21 reports results relative to the Romberg integration. After just 9 iterations, i.e., by extrapolating results out of the Trapezoidal formula with $m = 2, 3, 5, 9, 17, 33, 65, 129, 257$ and 513, we get to an option price equal to 13.26967658, that is an absolute error equal to 1.672×10^{-11}.

Finally, we have used the built-in Matlab® integration routine quadl. We have constructed a *.m file containing the function to be integrated as functionGBM. The calling command is

```
integrand = functionGBM(x,phi,spot,strike,
                        maturity,rf,volatility)
```

Table 6.19. Pricing a call option in the Black–Scholes model using numerical quadrature: T (Trapezoid), ET (Extended Trapezoid), S (Simpson), GL (Gauss–Legendre)

m	T	ET (6.13)	S	GL
9	12.713444	13.254415	13.547893	13.690968
17	13.134627	13.286321	13.275021	13.269557
33	13.236146	13.272851	13.269985	13.269677
65	13.261308	13.270135	13.269696	13.269677
129	13.267585	13.269737	13.269678	13.269677
257	13.269154	13.269684	13.269677	13.269677
513	13.269546	13.269678	13.269677	13.269677

Table 6.20. Absolute error and relative error in pricing call options using different quadrature methods: T (Trapezoid), ET (Extended Trapezoid), S (Simpson), GL (Gauss–Legendre)

	T		ET		S		GL
m	$E_{n,m}$	$R_{n,m}$	$E_{n,m}$	$R_{n,m}$	$E_{n,m}$	$R_{n,m}$	$E_{n,m}$
9	5.562E–01	5.501	1.526E–02	145.643	−2.782E–01	0.832	−4.213E–01
17	1.350E–01	4.119	−1.664E–02	−0.917	−5.345E–03	52.055	1.195E–04
33	3.353E–02	4.028	−3.174E–03	5.244	−3.086E–04	17.319	4.165E–10
65	8.369E–03	4.007	−4.581E–04	6.929	−1.893E–05	16.305	3.921E–11
129	2.091E–03	4.002	−6.081E–05	7.533	−1.177E–06	16.075	2.771E–11
257	5.228E–04	4.000	−7.813E–06	7.783	−7.349E–08	16.022	3.553E–10
513	1.307E–04	4.000	−9.896E–07	7.895	−4.576E–09	16.059	4.862E–10

Table 6.21. Pricing a call option in the Black–Scholes model using Romberg extrapolation

Price	Iterations	Error
13.5379776782	4	2.683E–01
13.2510395962	5	−1.864E–02
13.2699407831	6	2.642E–04
13.2696756268	7	−9.578E–07
13.2696765856	8	8.943E–10
13.2696765846	9	−1.672E–11

If we restrict the infinite integration interval to [0, 6], we can type (notice that we scale the spot price dividing it by the strike)

```
>>[price numeval] = quadl(@functionGBM,0.000001,6,
                         10^-6,'trace off',1,100/100,
                         1,1,0.1,0.2);
```

in the command window and obtain

```
                    price*100 = 13.26969508049296
```

and

```
                    numeval=168
```

Table 6.22. Pricing a call option in the Black–Scholes model using the MATLAB® adaptive quadrature routine

Tolerance	Price	Function evaluations (s)
10^{-6}	13.26969517231181	169
10^{-7}	13.26967472206082	199
10^{-8}	13.26967662879438	349
10^{-9}	13.26967662880944	409
10^{-10}	13.26967662880941	439
10^{-11}	13.26967658467796	589

In Table 6.22 we report prices obtained by running function `quadl` under varying tolerance levels. Results appear to be highly accurate.

6.7.2 Vanilla Options in the Square-Root Model

Most of the research in option pricing assumes that the underlying asset follows a simple GBM. However, it is well known that this model has several drawbacks, e.g., the volatility smile effect. In this section we reconsider the constant elasticity of variance (CEV) already presented in the context of pricing using PDEs. In this model, the dynamics of the underlying price are given by:

$$dx(t) = rx \, dt + \sigma x^{\lambda/2} \, dW(t),$$

where $0 \leq \lambda < 2$. We obtain the lognormal model as a special case by setting $\lambda = 2$. The transition density of the process can be expressed in terms of a noncentral chi-squared distribution. As an illustrative example, we consider the case $\lambda = 1$.[7] The (time-homogenous) transition density is given by

$$w(y, \tau_n; x) = \sqrt{\frac{xe^{r\tau}}{y}} \, \gamma e^{-\gamma(xe^{r\tau}+y)} I_1\left(2\gamma\sqrt{xe^{r\tau}y}\right), \tag{6.24}$$

where $\gamma = 2r/(\sigma^2(e^{r\tau} - 1))$ and $I_1(z)$ is the modified Bessel function of order 1

$$I_1(x) = \frac{1}{2}x \sum_{k=0}^{\infty} \frac{(x^2/4)^k}{k! \Gamma(k+2)},$$

---

[7] Notice that by using the transformation $y = x^{2-\lambda}$ and then applying Itô's lemma we obtain:

$$dy(t) = \left(rxx^{1-\lambda} + \frac{1}{2}\sigma^2 x^{\lambda}x^{-\lambda}\right) dt + \sigma x^{1-\lambda}x^{\lambda/2} \, dW(t)$$

$$= \left(\frac{1}{2}\sigma^2 + ry\right) dt + \sigma \sqrt{y} \, dW(t),$$

i.e., y is a square-root process. Therefore, the numerical example we are discussing with regard to the square-root process can be easily extended to processes with λ different from 1.

Table 6.23. Comparison of different quadrature methods, T (Trapezoid), ET (Extended Trapezoid), R (Romberg), S (Simpson), GL (Gauss–Legendre), AL (Adaptive Lobatto), in pricing options under the square-root process. RSSE stands for the root of the sum squared errors, taking as benchmark the GL rule

Strike	T	ET	R	S	GL	AL
80	24.703785	24.704346	24.704262	24.704259	24.704261	24.704251 (828)
90	15.905726	15.907228	15.907108	15.907108	15.907108	15.907102 (618)
95	11.994622	11.996561	11.996465	11.996466	11.996465	11.996460 (618)
100	8.601314	8.603483	8.603437	8.603439	8.603437	8.603433 (618)
105	5.832221	5.834345	5.834354	5.834356	5.834354	5.834352 (558)
110	3.722808	3.724647	3.724697	3.724699	3.724697	3.724696 (558)
120	1.251160	1.252142	1.252208	1.252208	1.252208	1.252207 (468)
RSSE	0.004388	0.000200	0.000001	0.000004	–	0.0000133

and where $\Gamma(z)$ is the Gamma function

$$\Gamma(z) = \int_0^{+\infty} t^{z-1} e^{-t} \, dt.$$

If we set:

$$\tau(t) = \frac{(1 - e^{-rt})}{r}; \qquad v = 1; \qquad \gamma = \frac{4x}{\sigma^2 \tau(t)},$$

then the random variable $4e^{-rt} y/(\sigma^2 \tau(t))$ has a non-central chi-squared density with four degrees of freedom and noncentrality parameter γ.

In Table 6.23 we compare the different quadrature methods for pricing a plain vanilla option. The quadrature methods have been implemented setting $m = 200$. The Romberg (R) extrapolation quadrature has been implemented by using 8 recursions, i.e., by extrapolating the trapezoidal rule starting from m equal to 2, 3, 5, 9, 17, 33, 65 and 129 according to the sequence $h_{i+1} = h_i/2$. Last column refers to the Adaptive Lobatto rule (AL) implemented in Matlab® by the function quadl.[8] Model parameters are: $r = 5.91\%$, $\sigma = 1.353885$, $\tau = 1$ year, and $x(0) = 100$.

In Table 6.24 we still use the Adaptive Lobatto rule to price a call option using the quadl Matlab® function. The tolerance has been set equal to 10^{-11}. In the example, we set $X = 1$, $r = 0.1$, $\sigma = 0.4$, $T = 1$ and we let K vary.

---

[8] In particular, the numbers in the last column have been obtained with the commands:

```
k = [80,90,95,100,105,110,120]
for i = 1:7
[op(i) neval(i)]
   = optionpricegbmcev(1,2,100,k(i),1,0.0591,1.353885,300,
                       10^-11)
end
op';neval'
```
This function makes use of the function
```
prob = densityGBMCEV(spotT,spot,maturity,rf,
                     volatility,model)
```
that returns the density of the GBM and square-root process.

Table 6.24. Call option prices for different models: square-root ($\lambda = 1$), GBM ($\lambda = 2$), CEV ($\lambda = 3$)

Strike	$\lambda = 1$	$\lambda = 2$
0.7	0.392836	0.386449
0.8	0.321699	0.316001
0.9	0.258349	0.254849
1	0.203433	0.203185
1.1	0.157091	0.160499
1.2	0.118996	0.125858
1.3	0.088464	0.098142

This example has been done using the function

```
function [op, neval]
    = optionpricegbmcev(phi,model,spot,strike,maturity,
                        rf,volatility,upperlimit,tol)
```

6.7.3 Bond Options in the Cox–Ingersoll–Ross Model

The Cox, Ingersoll and Ross (1985) model (CIR) assumes short rate risk-neutral dynamics:

$$dr(t) = \alpha\big(\mu - r(t)\big)\, dt + \sigma\sqrt{r(t)}\, dW(t).$$

The time t price of a zero-coupon expiring at time T is:

$$P\big(t, T; r(t)\big) \equiv P(t, T) = \mathbb{E}_t^*\big(e^{-\int_t^T r(s)\, ds}\big|r(t)\big).$$

This expression admits a closed-form:

$$P(t, T) = A(t, T)e^{-B(t,T)r(t)}, \tag{6.25}$$

where

$$A(t, t + \tau) = \left(\frac{2\phi_1 e^{\phi_2\tau/2}}{\phi_2(e^{\phi_1\tau} - 1) + 2\phi_1}\right)^{\phi_3}, \tag{6.26}$$

$$B(t, t + \tau) = 2\frac{e^{\phi_1\tau} - 1}{\phi_2 + 2\phi_1}, \tag{6.27}$$

$$\phi_1 = \sqrt{\alpha^2 + 2\sigma^2}, \qquad \phi_2 = \phi_1 + \alpha, \qquad \phi_3 = \frac{2\alpha\mu}{\sigma^2}.$$

We aim at pricing coupon bearing bond options by using quadrature methods. This problem can also be solved by using the PDE approach as illustrated in the chapter on the numerical solution of linear systems.

Let the option expire at time T and define c_i as the coupon paid off at time T_i, where $T_i > T$ for all $i = 1, \ldots, n$ (c_n includes the bond notional). The option pay-off with strike price K is given by:

$$\text{Pay-off} = \left(\phi \left(\sum_{i=1}^{n} P\big(T, T_i; r(T)\big) c_i - K \right) \right)_+$$

$$= \left(\phi \left(\sum_{i=1}^{n} A(T, T_i) e^{-B(T,T_i)r(T)} c_i - K \right) \right)_+ ,$$

where parameter ϕ has been introduced to distinguish a call option ($\phi = 1$) from a put option ($\phi = -1$). We aim at computing the arbitrage free price:

$$u(\tau, r) = \mathbb{E}_t^* \left(e^{-\int_t^{t+\tau} r(s)\, ds} \left(\phi \left(\sum_{i=1}^{n} A(t+\tau, T_i) e^{-B(t+\tau,T_i)r(t+\tau)} c_i - K \right) \right)^+ \right),$$

where the time to maturity is defined by $\tau = T - t$. The pricing problem can be considerably simplified by using an appropriate change of numéraire. Precisely, let $\mathbb{P}^{t+\tau}$ denote the equivalent probability measure that makes prices martingales once discounted by the zero coupon bond maturing at $t + \tau$. This is the so-called $t + \tau$ forward measure. The pricing problem requires computing

$$u(\tau, r) = P(t, t+\tau) \mathbb{E}_t^{t+\tau} \left(\left(\phi \left(\sum_{i=1}^{n} A(t+\tau, T_i) e^{-B(t+\tau,T_i)r(t+\tau)} c_i - K \right) \right)^+ \right),$$

where expectation is taken with respect to rate dynamics under the forward measure $\mathbb{P}^{t+\tau}$.

We observe that the coupon bond price is a strictly decreasing function of the instantaneous rate. Therefore, the call (resp. put) option is exercised provided that $r(t + \tau) < r_0$ (resp. $r(t + \tau) > r_0$), where r_0 is the unique solution of equation

$$\sum_{i=1}^{n} A(T, T_i) e^{-B(T,T_i)r_0} c_i = K. \tag{6.28}$$

The option price can be computed as follows:

$$V(t) = \mathbb{E}_t^{t+\tau} \left(\left(\phi \left(\sum_{i=1}^{n} A(t+\tau, T_i) e^{-B(t+\tau,T_i)r(t+\tau)} c_i - K \right) \right)^+ \right) \tag{6.29}$$

$$= \begin{cases} \int_0^{r_0} \pi(r) f(r)\, dr, & \phi = 1, \\ \int_{r_0}^{+\infty} \pi(r) f(r)\, dr, & \phi = -1, \end{cases} \tag{6.30}$$

where $\pi(r) = \phi(\sum_{i=1}^{n} A(t+\tau, T_i) e^{-B(t+\tau,T_i)r(t+\tau)} c_i - K)$ and $f(r)$ is the density function of the instantaneous rate $r(t + \tau)$ under the forward measure $\mathbb{P}^{t+\tau}$. This latter is given by

$$f(r) = \frac{1}{k_2} \chi^2 \left(\frac{x}{k_2}, d, \lambda_2 \right),$$

with

$$k_2 = \frac{\sigma^2(e^{\phi_1(t+\tau)} - 1)}{2(2\phi_1 + (\phi_1 + \alpha)(e^{\phi_1(t+\tau)}) - 1)},$$

$$\lambda_2 = \frac{8\phi_1^2 e^{\phi_1(t+\tau)} r(t)}{\sigma^2(e^{\phi_1(t+\tau)} - 1)(2\phi_1 + (\phi_1 + \alpha)(e^{\phi_1(t+\tau)} - 1))},$$

and $\chi^2(x, d, \lambda)$ representing the density of a non-central chi-square variable with d degrees of freedom and non-centrality parameter λ, namely:

$$\chi^2(x, d, \lambda) = e^{-\lambda/2} e^{-x/2} \left(\frac{x}{\lambda}\right)^{d/4 - 1/2} I_{d/2-1}(\sqrt{x\lambda}).$$

Here $I_d(x)$ is the modified Bessel function of the first kind as defined by

$$I_d(x) = \left(\frac{x}{2}\right)^d \sum_{k=0}^{+\infty} \frac{(x^2/4)^k}{k! \Gamma(d + k + 1)}.$$

In Table 6.25, we price an option on a coupon bond. The coupon bond pays a yearly coupon of 5% *per annum* for four years, beginning just one year after the option maturity. We compare the results obtained by numerical integration to those stemming from either solving the pricing PDE or running Monte Carlo simulations. Both numerical solution of the pricing PDE and the quadrature formula often provide estimates within the 95% MC confidence interval. Numbers reported in last column of this table have been obtained using a Matlab® code. In particular, the integral in (6.30) has been computed using function quadl with a tolerance level set equal to 10^{-9}. The MATLAB® function to be called is:

```
[op, nf]
  = OptionCouponBond_CIR(phi,strike,optionexpiry,
                         vCouponDates,vCouponAmount,
                         shortratet,speed,mu,volatility)
```

This routine implements the pseudo code given in Table 6.26 and returns option price and number of function evaluations required to compute the integral by using the quadl adaptive routine. Notice that this implementation requires a routine to compute the noncentral chi-squared density and the price of a zero-coupon in the CIR model through formula (6.25). As an example, 2.3386119 (408) in the last column have been obtained by writing in the command window:

```
>> [op,nf]
   = OptionCouponBond_CIR(1,95.703393,1,[1, 2, 3, 4],
                          [5 5 5 105],0.05,0.1,0.1,0.1)
```

6.7.4 Discretely Monitored Barrier Options

Let us consider a double barrier knock-out option. That is a call option that expires worthless provided that one of two given barriers has been hit at any monitoring date

Table 6.25. Prices of options on a coupon bond in the CIR model. Parameters $\alpha = 0.1$, $\mu = 0.1$, $\sigma = 0.1$, $r_{\min} = 0$, $r_{\max} = 0.5$. Strikes have been set equal to the forward price of the coupon bond

Option maturity	Strike	MC (s.e.) 500,000 runs	PDE ($m = 500, n = 1{,}000$)	Adaptive quadrature (s.e.)
1	95.703393	2.3405161 (0.0036606)	2.3386243	2.3386119 (408)
2	94.825867	2.9505254 (0.0046964)	2.9534488	2.9535513 (348)
3	94.129268	3.2113425 (0.0052998)	3.2217921	3.2220002 (348)
4	93.576725	3.3185218 (0.0057044)	3.3090647	3.3093217 (348)
5	93.138299	3.2953126 (0.0059728)	3.2897274	3.2898931 (348)

Table 6.26. Pseudo-code for pricing an option on a coupon bond in the CIR model

```
Inputs:
   model parameters: α, μ, σ, r,
   vector of payment dates and coupon amounts
   option expiry and payoff (call or put)
Solve (6.28)
Define the integrand π(r)f(r) in (6.30)
Compute using a quadrature method the integral (6.30)
```

during the contract lifetime. Let $0 = t_0 < t_1 < \cdots < t_p < \cdots < t_n = T$ be the monitoring dates and define l (resp. u) the lower (resp. upper) barrier active at time t_p. By setting $l = 0$ or $u = +\infty$ we recover the payoff of a more traditional single barrier option. To simplify our exposition we assume constant barriers and a time-homogenous process, i.e. the transition density does not depend on t. Therefore, we may denote by $p(y, \tau; x)$ the conditional transition density from state x at time t to state y at time $t + \tau$. Both GBM and the square root process are examples of time-homogeneous processes. Let us denote the price of the barrier option when $t > t_n$ by $v(x, t, n) \equiv v(x, t, n; l, u)$. By setting $\tau = t - t_n$, the option price satisfies the recursion relation:

$$v(x, t_n + \tau, n) = e^{-r\tau} \mathbf{1}_{\{x \in [l,u]\}} \int_{-\infty}^{+\infty} p(\xi, \tau; x) v(\xi, t_n, n - 1) \, d\xi, \qquad (6.31)$$

where $\mathbf{1}_{\{x \in [l,u]\}}$ is the indicator function for the interval $[l, u]$ as defined as:

$$\mathbf{1}_{\{x \in [l,u]\}} = \begin{cases} 1 & \text{if } x \in [l, u], \\ 0 & \text{if } x \notin [l, u]. \end{cases}$$

This function has been introduced to take into account the possibility of hitting the barrier at the monitoring date t_n. By using the fact that $v(\xi, t_n, n-1)$ is zero whenever $\xi \notin [l, u]$, we have:

$$v(x, t_n + \tau, n) = e^{-r\tau} \int_l^u p(\xi, \tau; x) v(\xi, t_n, n - 1) \, d\xi \quad \text{for } l < x < u.$$

The option price is given by a recursive univariate integration consisting of a single integral for every monitoring date. In order to exploit the recursive structure of the problem, we compute the option price at the monitoring dates $t_1, t_2, \ldots, t_n$ only. Then we use (6.31) at the intermediate date $t_n + \tau$. In other words, we want to compute $v(x, t_n, n)$, for $n \geq 0$. Actually, this is the main advantage of the present methodology compared to lattices and finite-difference techniques. More precisely, here we need not consider intermediate time steps and can move to any value of the underlying at each monitoring date. This procedure simply requires knowledge of the transition density in closed form.

As an example, let us consider the GBM process and then let us consider the log-return $z = \ln x$ and define $g(z, t_n, n) = v(e^z, t_n, n)$. The recursion formula becomes

$$g(z, t_n + \tau, n) = e^{-r\tau} \int_{\ln l}^{\ln u} h(\xi, \tau; z) g(\xi, t_n, n - 1) \, d\xi, \quad \ln l < z < \ln u$$

where

$$h(\xi, \tau; z) = \frac{1}{\sqrt{2\pi\sigma^2\tau}} \exp\left(-\frac{1}{2\sigma^2\tau} \left(\xi - z - \left(r - \frac{\sigma^2}{2} \right) \tau \right)^2 \right).$$

We compute the above integral using a quadrature with weighing coefficients w_j and abscissas $y_j \in [l, u]$:

$$g(z, t_n + \tau, n) = \sum_{j=1}^m w_j h(y_j, \tau; z) g(y_j, t_n, n - 1).$$

If we consider the option price computed at the m points $\mathbf{z} = (z_1, \ldots, z_m)$, we can write:

$$\mathbf{g}(\mathbf{z}, n + 1) = \mathbf{K}\mathbf{g}(\mathbf{z}, n),$$

where $\mathbf{K}$ is an $m \times m$ matrix whose element (i, j) is given by $w_j h(y_j, \tau; z_i) e^{-r\tau}$, and $\mathbf{g}(\mathbf{z}, n)$ is a vector with entries $g(z_i, t_n + \tau, n)$. The pricing procedure is described in Table 6.27. The computational cost is linear in the number of monitoring dates and requires the construction of the matrix $\mathbf{K}$, which involves m^2 operations, plus an additional m operations to compute the product $\mathbf{K}\mathbf{g}$. Therefore the resulting cost is in the order of $\mathcal{O}(m^2 + nm)$. Let us consider a double barrier option with lower threshold $l = 95$, upper threshold $u = 110$, risk-free rate $r = 10\%$, instantaneous volatility $\sigma = 20\%$, a number of 25 monitoring dates, 0.5 years to go and strike price $K = 100$. In Table 6.28, we compare different quadrature methods for $m = 301$. Note the excellent agreement (within 5 digits!) between the results from Extended Trapezoid, Simpson and Gauss–Legendre methods, whereas the Rectangular method appears inaccurate. Figure 6.10 shows the price, delta and gamma of a discrete barrier option as function of the spot price. Figures 6.11–6.13 illustrate the way the number of monitoring dates affects option prices (that have been computed by using

Table 6.27. Pseudo-code for pricing discrete barrier options

```
Set g(z,0)
Build iteration matrix K
i = 1
While i<n Do g = Kg, i = i+1.
```

Table 6.28. Prices of discrete barrier options for different quadrature methods ($m = 301$). Romberg extrapolation has been conducted using $m = 2, 3, 5, 9, 17, 33, 65, 129, 257$, for a total of eight iterations. Parameters $r = 0.1$, $\sigma = 0.2$, $K = 100$, $l = 95$, $u = 110$, $t = 0.5$, $n = 25$

Spot	Rectangular	T	R	ET	S	GL
95.00022	0.069795	0.068124	0.068132	0.068131	0.068131	0.068132
97	0.117652	0.115172	0.115186	0.115184	0.115184	0.115185
99	0.154918	0.151925	0.151942	0.15194	0.151939	0.151939
99.5	0.161189	0.158119	0.158139	0.158135	0.158134	0.158134
100	0.166097	0.162971	0.162987	0.162986	0.162985	0.162984
100.5	0.169626	0.166463	0.166479	0.166479	0.166478	0.166476
102	0.172133	0.168965	0.168983	0.168981	0.16898	0.168981
105	0.145087	0.142324	0.142340	0.142337	0.142337	0.142338
107	0.109937	0.107695	0.107707	0.107706	0.107706	0.107707
109.9997	0.052086	0.050837	0.050843	0.050843	0.050842	0.050843

the Gauss–Legendre quadrature method). Values can vary from 0.46455 with 5 monitoring dates up to 0.08255 with 100 monitoring dates. The continuous monitoring formula, developed by Kunitomo and Ikeda (1992), returns 0.02939. These price differences support the need for a numerical method to take into account monitoring in discrete time. Figure 6.11 illustrates the way delta and gamma may vary with the number of monitoring dates.

All numerical results in this section have been obtained by using the VBA® function:

```
Function DiscreteBarrier(quadrature As Integer,
phi as Integer, EuAm as Integer, spot As Double,
strike As Double, rf As Double, sg As Double, dt
As Double, ndates As Integer, lowbarrier As Dou-
ble, upbarrier As Double, npoints As Integer) As
Variant
```

The first argument (quadrature) allows the user to select the quadrature rule ($1 =$ "Rectangular", $2 =$ "Trapezoidal", $3 =$ "Extended Trapezoidal", $4 =$ "Simpson", $5 =$ "Gauss-Legendre"). The second argument (phi) allows to distinguish between call (phi = 1) and put options (phi = -1). The third argument (EuAm) allows to distinguish between the European (EuAm = 1)

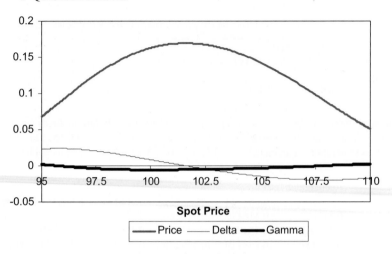

Fig. 6.10. Price, Delta and Gamma of a discrete barrier option vs. spot price. Parameters: $r = 0.1, \sigma = 0.2, t = 0.5, K = 100, n = 25, l = 95, u = 110$.

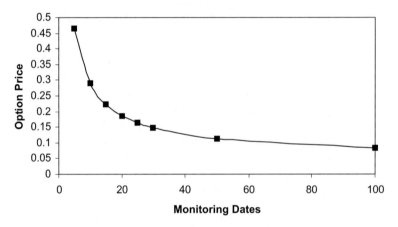

Fig. 6.11. Price of a discrete barrier option and monitoring dates. Parameters: $r = 0.1, \sigma = 0.2, t = 0.5, K = 100, x = 100.0194, l = 95, u = 110$.

and the American version (EuAm = 0).[9] The last parameter (npoints) represents the number of abscissas in the selected quadrature rule. Remaining parameters are easily understood. The function returns an array with two columns, one reporting the abscissas, the other indicating the corresponding option value.

---

[9] In this case, we compare at each monitoring date and node by node, the intrinsic value of the option to the residual value stemming from holding the option at each step of the iteration.

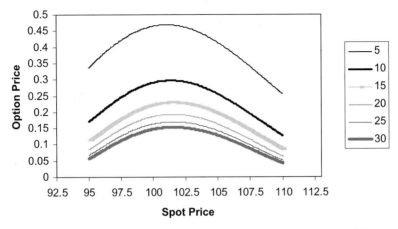

Fig. 6.12. Price of a discrete barrier option as function of the spot price, for different monitoring dates (5, 10, 15, 20, 25, 30). Parameters: $r = 0.1$, $\sigma = 0.2$, $t = 0.5$, $K = 100$, $x = 100.0194$, $l = 95$, $u = 110$.

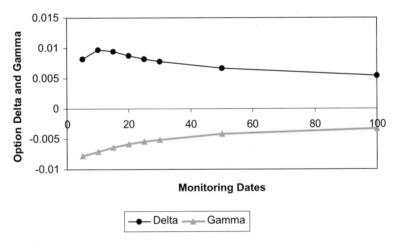

Fig. 6.13. Delta and Gamma of a discrete barrier option and monitoring dates. Parameters: $r = 0.1$, $\sigma = 0.2$, $t = 0.5$, $K = 100$, $x = 100.0194$, $l = 95$, $u = 110$.

6.8 Pricing Using Characteristic Functions

Several option pricing models generate analytical expressions for the characteristic function of the underlying variables. This allows us to price derivative contracts by using numerical approximations of the required probability distributions as obtained by Fourier inversion. In this section, we define the characteristic function of a r.v. and we show how it can be used for option pricing purposes. Then, we illustrate the way quadrature methods can be used to make usable this important tool. In particular, we present the Fast Fourier Transform (FFT) algorithm.

The characteristic function $\varphi(\gamma)$ of a random variable Z is defined as

$$\varphi(\gamma) = \mathbb{E}_t\left(e^{i\gamma Z}\right),$$

where $i := \sqrt{-1}$ is the imaginary unit and γ is a complex number. Observe that φ is a complex-valued function, except when z has a symmetric distribution, in which case it is real-valued. If Z has density function $\mathbb{P}_t(Z \in dz)$, we can equivalently write

$$\varphi(\gamma) = \int_{-\infty}^{+\infty} e^{i\gamma z} \mathbb{P}_t(Z \in dz).$$

It can be shown (e.g., Kendall (1994)) that if $\int_{\mathbb{R}} |\varphi(\gamma)| \, d\gamma < \infty$, then Z has continuous probability density function that can be recovered by the following Fourier inversion formula:

$$\mathbb{P}_t(Z \in dz) = \frac{1}{\pi} \int_0^{+\infty} \text{Re}\left(e^{-i\gamma z}\varphi(\gamma)\right) d\gamma, \tag{6.32}$$

where $\text{Re}(\cdot)$ stands for the real part of its argument.[10] Expression (6.32) shows that the characteristic function identifies the distribution of Z. More generally, it can be shown that characteristic functions unequivocally identify distribution functions. Elementary properties of the characteristic function are given in Table 6.29.

Example If $Z \sim \mathcal{N}(0, 1)$, then

$$\varphi(\gamma) = e^{-\gamma^2/2}.$$

Moreover, if $Y = \mu + \sigma Z \sim \mathcal{N}(\mu, \sigma^2)$, using property (d) in Table 6.29, we get

$$\mathbb{E}_t\left(e^{i\gamma Y}\right) = e^{i\gamma\mu - \gamma^2\sigma^2/2}.$$

Table 6.29. Elementary properties of the characteristic function

(a)	$\varphi(0) = 1.$
(b)	$|\varphi(\gamma)| \leq 1.$
(c)	If $Y = -Z$, then $\mathbb{E}_t(e^{i\gamma Y}) = \bar{\varphi}(\gamma)$, where $\bar{\varphi}(\gamma)$ is the complex conjugate of $\varphi(\gamma)$.
(d)	If $Y = a + bZ$, then $\mathbb{E}_t(e^{i\gamma Y}) = e^{i\gamma a}\varphi(b\gamma).$
(e)	$\mathbb{E}_t(Z^n) = (i)^n \frac{\partial^n \varphi(\gamma)}{\partial \gamma^n}\Big|_{\gamma=0}, \quad n \in \mathbb{N}.$
(f)	If Z_1 and Z_2 are independent r.v.s with c.f. $\varphi_1(\gamma)$ and $\varphi_2(\gamma)$ then, $Y = Z_1 + Z_2$ has c.f. given by $\varphi_1(\gamma)\varphi_2(\gamma).$

[10] Sometimes the inversion formula is given by $\frac{1}{2\pi}\int_{-\infty}^{+\infty} e^{-i\gamma z}\varphi(\gamma, T) \, d\gamma$. The two expressions coincide when the original function is a real function.

In order to understand the importance of the characteristic function in option pricing, we start with a simple example. Recall that the arbitrage-free price of a contingent claim can be expressed as a conditional expectation of its discounted payoff. If we consider a standard call option, then the price reads as:

$$c(K, T) = \mathbb{E}_t^* \left(e^{-r(T-t)} (X(T) - K)^+ \right)$$
$$= e^{-r(T-t)} \left\{ \mathbb{E}_t^* \left((X(T) \mathbf{1}_{\{X(T)>K\}}) \right) - K \mathbb{E}_t^* (\mathbf{1}_{\{X(T)>K\}}) \right\}.$$

If we define $z(T) = \ln X(T)$ and $k = \ln K$, we can write

$$c(e^k, T) = e^{-r(T-t)} \left\{ \mathbb{E}_t^* \left(e^{z(T)} \mathbf{1}_{\{z(T)>k\}} \right) - e^k \mathbb{E}_t^* (\mathbf{1}_{\{z(T)>k\}}) \right\}$$
$$= X(t) \Pi_1 - e^k e^{-r(T-t)} \Pi_2.$$

Quantities Π_1 and Π_2 can be interpreted as "stock-adjusted" and "money-market adjusted" probabilities, i.e., these probabilities have been computed by using respectively the stock and the money market account as numéraires. Π_1 and Π_2 both represent the probability of ending up in-the-money at the option expiry. However, they are computed under martingale measures for two different numéraires: Π_1 uses as a numéraire the stock itself, whereas Π_2 uses the money market account:

$$\Pi_1 = \text{Pr}_t^X (X(T) > K),$$
$$\Pi_2 = \text{Pr}_t^B (X(T) > K),$$

where the apex underlines that in the first case we are using as probability measure the one that makes $e^{rT}/X(T)$ a martingale, whereas in the second case the ratio $X(T)/e^{rT}$ must be a martingale.

The two expectations above can be evaluated using the characteristic function of $Z(T)$. Let us define:

$$\varphi_1(\gamma) = \int_{-\infty}^{+\infty} e^{i\gamma z} \mathbb{P}_t^X (Z(T) \in dz),$$
$$\varphi_2(\gamma) = \int_{-\infty}^{+\infty} e^{i\gamma z} \mathbb{P}_t^B (Z(T) \in dz),$$

where $\mathbb{P}_t^X$ and $\mathbb{P}_t^B$ denotes the density function of $Z(T)$ under the two different numéraires (stock and money market account) and where we omit the dependence of the characteristic function on the option expiry T. The functions $\varphi_1(\gamma)$ and $\varphi_2(\gamma)$ are the Fourier transforms of the probability density functions $\mathbb{P}_t^X (Z(T) \in dz)$ and $\mathbb{P}_t^B (Z(T) \in dz)$. It can be shown (e.g., Kendall (1994) or Duffie, Pan and Singleton (1998)) that the probabilities Π_1 and Π_2 can be recovered using the inversion formulas

$$\Pi_1 = \frac{1}{2} + \frac{1}{\pi} \int_0^{+\infty} \text{Re} \left(\frac{e^{-i\gamma k} \varphi_1(\gamma)}{i\gamma} \right) d\gamma, \qquad (6.33)$$

$$\Pi_2 = \frac{1}{2} + \frac{1}{\pi} \int_0^{+\infty} \text{Re} \left(\frac{e^{-i\gamma k} \varphi_2(\gamma)}{i\gamma} \right) d\gamma, \qquad (6.34)$$

and Re($\cdot$) stands for the real part of its argument.

Integrals (6.33) and (6.34) can be computed by quadrature. To this end, some care is necessary if we are to tackle instability issues linked to the oscillatory nature of the integrand function, due to the presence of the complex exponential function. Recalling our previous discussion, the trapezoidal rule is expected to perform much better than other Newton–Cotes rules and should be comparable to Gaussian quadrature. However a more efficient approach has been proposed in Geman and Eydeland (1995) and in Carr and Madan (1998). These authors, instead of computing Π_1 and Π_2 separately, calculate the Fourier transform of an adjusted call option price with respect to the logarithmic strike price k. These authors introduce a dumping parameter $\alpha > 0$ and define the following quantity:

$$c_\alpha(e^k, T) := e^{\alpha k} c(e^k, T), \qquad (6.35)$$

where α has to be chosen so that c_α (e^k, T) is square-integrable and therefore admits the Fourier transform $\mathcal{F}[c_\alpha](\gamma)$:

$$\mathcal{F}[c_\alpha](\gamma) = \int_{-\infty}^{+\infty} e^{i\gamma k} c_\alpha(e^k, T)\, dk.$$

It can be shown that this quantity can be expressed in terms of the characteristic function $\varphi_2(\gamma)$ by

$$\mathcal{F}[c_\alpha](\gamma) = \frac{e^{-rT} \varphi_2(\gamma - \alpha i - i)}{\alpha^2 + \alpha - \gamma^2 + i(2\alpha + 1)\gamma}. \qquad (6.36)$$

If the characteristic function of $z(T)$ is known in closed form, we also have an analytical expression for $\mathcal{F}[c_\alpha](\gamma)$ at our disposal. Similarly, for a put-option we define

$$p_\alpha(e^k, T) := e^{-\alpha k} p(e^k, T).$$

Its Fourier transform $\mathcal{F}[p_\alpha](\gamma) = \int_{-\infty}^{+\infty} e^{i\gamma k} e^{-\alpha k} p(e^k, T)\, dk$ can be written as:

$$\mathcal{F}[p_\alpha](\gamma) = \frac{e^{-rT} \varphi_2(\gamma + \alpha i - i)}{\alpha^2 - \alpha - \gamma^2 + i(-2\alpha + 1)\gamma}. \qquad (6.37)$$

As a last step, Fourier inversion yields the option prices

$$c(e^k, T) = \frac{e^{-\alpha k}}{\pi} \int_0^{+\infty} e^{-i\gamma k} \mathcal{F}[c_\alpha](\gamma)\, d\gamma, \qquad (6.38)$$

$$p(e^k, T) = \frac{e^{\alpha k}}{\pi} \int_0^{+\infty} e^{-i\gamma k} \mathcal{F}[p_\alpha](\gamma)\, d\gamma. \qquad (6.39)$$

It is possible to prove that this method is viable provided that the moment of order $1 + \alpha$ exists and is finite for some $\alpha > 0$:

$$\mathbb{E}_t^*\left(X(T)^{1+\alpha}\right) < \infty.$$

It turns out that any $\alpha \in [1.5, 2]$ works quite well for most cases.

Computing (6.32) and (6.38) (or (6.39)) can be done by using either the trapezoidal rule or the Gauss–Legendre quadrature. The trapezoidal rule can be implemented very efficiently by means of the Fast Fourier Transform (FFT) algorithm. We end this section by illustrating this procedure. Let us consider the problem of computing the $N \times 1$ vector $\mathbf{H} = \{H_0, \ldots, H_n, \ldots, H_{N-1}\}$ given the $N \times 1$ vector $\mathbf{h} = \{h_0, \ldots, h_k, \ldots, h_{N-1}\}$, such that:

$$H_n = \sum_{j=0}^{N-1} e^{+ijn\frac{2\pi}{N}} h_j, \quad n = 0, \ldots, N-1. \tag{6.40}$$

Let $\mathbf{H} = \{H_0, \ldots, H_n, \ldots, H_{N-1}\}$ be the discrete Fourier transform of the vector $\mathbf{h} = \{h_0, \ldots, h_k, \ldots, h_{N-1}\}$. Vice versa, using the discrete inverse Fourier transform we can also recover $\mathbf{h}$ from $\mathbf{H}$ as:

$$h_k = \frac{1}{N} \sum_{j=0}^{N-1} e^{-ijk\frac{2\pi}{N}} H_j, \quad k = 0, \ldots, N-1. \tag{6.41}$$

The only difference between (6.40) and (6.41) is represented by the change of sign in the exponential and the multiplicative constant $1/N$.

In general, if we try to compute $\mathbf{H}$ from $\mathbf{h}$, or vice versa, we need N^2 multiplications involving complex quantities, plus additional $N(N-1)$ complex sums. Exploiting the fact that these computations are not independent of each other, in the 1960s Cooley and Tukey discovered an algorithm requiring only $N \ln_2(N)/2$ operations. This algorithm is known as the Fast Fourier Transform and its discovery has greatly stimulated the use of the Fourier transform in several technical disciplines. An important aspect of FFT is that the algorithm is based on a recursive procedure that allows one to express the FFT of length N as the sum of two FFT (each one of length $N/2$). This fact implies that the best choice for N is a power of 2.

In order to exploit the FFT algorithm, we discretize the inversion integral (6.38) using the trapezoidal rule with step η:[11]

$$c(e^k, T) \simeq \frac{e^{-\alpha k}}{\pi} \sum_{j=0}^{N-1} e^{-i\eta jk} \mathcal{F}[c_\alpha](j\eta) w_j \eta. \tag{6.42}$$

Here $w_1 = 1/2, w_2 = 1, \ldots, w_{N-2} = 1, w_{N-1} = 1/2$. This quadrature introduces two types of error: first, a truncation error due to the finiteness of the upper limit in the numerical integration; second, a sampling error due to the evaluation of the integrand at grid points only. The FFT returns the option value over a grid of N evenly spaced logarithmic strike prices $k_0, \ldots, k_{N-1}$, with $k_n = k_0 + n\lambda$, $k_0 = -N\lambda/2$, and $k_{N-1} = N\lambda/2$. By setting $k_n = k_0 + n\lambda$ into expression (6.42), we obtain:

---

[11] The trapezoidal rule can be applied to (6.32) or (6.39) as well.

Table 6.30. Pseudo-code for option pricing via the FFT algorithm

(1)	Assign $N, \lambda, \alpha, \varphi(\gamma)$;
(2)	Construct the Fourier transform of dampened option price $F[c_\alpha](\gamma)$ in (6.38);
(3)	Construct vector h with components (6.44);
(4)	Apply the FFT algorithm and obtain a vector H with N components;
(5)	Multiply the nth component of H by $\exp(-\alpha k_n)N/\pi$, where $k_n = -N\lambda/2 + n\lambda$.

$$c\left(e^{k_n}, T\right) \simeq \frac{e^{-\alpha k_n}}{\pi} \sum_{j=0}^{N-1} e^{-i\eta j(k_0 + n\lambda)} F[c_\alpha](j\eta)\eta w_j$$

$$= \frac{e^{-\alpha k_n}}{\pi} \sum_{j=0}^{N-1} e^{-i\eta j n\lambda} e^{-i\eta j k_0} F[c_\alpha](j\eta)\eta w_j,$$

for $n = 0, \ldots, N-1$. Finally, if we set

$$\lambda\eta \equiv \frac{2\pi}{N},$$

we can apply the FFT algorithm (6.41). The choice $\lambda\eta \equiv 2\pi/N$ highlights a trade-off arising between the accuracy of the integral, which is determined by the sampling rate η of the Fourier transform, and the degree of space refinement as represented by λ. A finer discretization of the strike price space comes with a rougher discretization step η in the transform plane (and vice versa). Unfortunately, numerical experiments are required to determine the best compromise between the two quantities involved in this trade-off. If we choose $\lambda\eta = 2\pi/N$, we then have $\eta k_0 = -N\lambda\eta/2 = -\pi$ and thus:

$$c\left(e^{k_n}, T\right) \simeq \frac{e^{-\alpha k_n}}{\pi} \sum_{j=0}^{N-1} e^{-ijn\frac{2\pi}{N}} h_j, \qquad (6.43)$$

where

$$h_j = e^{ij\pi} F[c_\alpha](j\eta)\eta w_j. \qquad (6.44)$$

Given the FFT algorithm, option prices can be computed according to the procedure described in Table 6.30.

6.8.1 MATLAB® and VBA® Algorithms

Matlab® includes built-in routines `fft(x)` and `ifft(x)` which implement discrete Fourier and inverse transforms. The Matlab® FFT code is based on FFTW (The Fastest Fourier Transform in the West) developed at MIT and available from http://www.fftw.org. The `fft(x)` Matlab® function operates the following sum

$$X(k) = \sum_{k=1}^{N} x(j)e^{-i\frac{2\pi}{N}(k-1)(j-1)}$$

and therefore if we need to compute (6.43), we need to construct a vector **x** having as element at position j exactly the quantity h_j given in (6.44).

Here, we illustrate the Matlab® implementation in the Gaussian case. Implementation in a stochastic volatility model is illustrated in the case-study "Fixing Volatile Volatility" in the second part of this book. In the Gaussian model, the risk neutral characteristic function is

$$\varphi(\gamma) = \exp\left(i\gamma\left(r - \frac{\sigma^2}{2}\right)(T - t) - \frac{1}{2}\gamma^2\sigma^2(T - t)\right).$$

Below, we use subfunctions to place all the functions required by the numerical inversion in a single M-file that is named `FFT_Pricing_CallPut.m`. This file can be run from the Matlab® command window. The function

```
function [K, Y]
    = FFT_Pricing_CallPut(S,rf,sg,t,alpha,npower)
```

returns two vectors K and Y containing strikes and corresponding option prices. The arguments correspond to the spot price (S), the risk-free rate (rf), the volatility (sg), the time to maturity (t), the dumping parameter (alpha), the exponent for determining the power of 2 (npower).[12] In particular, if alpha is positive (negative) we are pricing a call (put) option. Standard values for npower are 12 or 13, whilst the absolute value of alpha can be 1.5.

```
function [K, Y]
    = FFT_Pricing_CallPut(S,rf,sg,t,alpha,npower)
%find the number of points
N = 2^npower;
%spacing in the strike
lambda = 0.01;
% parameters for the cf grid (0,N*eta)
eta = 2*pi/(N*lambda);
% create grid for the cf
g = (0:N-1) * eta;
%create grid for the strike
k0 = -N*lambda/2;
k = k0 + (0:N-1) * lambda;
%compute the ch. function
fc = ft_option(rf,sg,t,alpha,g);
w = [0.5 ones(1,N-2) 0.5]; % trapezoidal rule
h2 = exp(-i*k0*g) .* fc .* w*eta;
%Fourier inversion
g = fft(h2);
g2 = real( exp(-alpha.*k) / pi .* g);
```

---

[12] The FFT computes the sum using 2^n power points.

```
K = S * exp(k); K = K';
Y = exp(-rf*t) * S * g2; Y = Y';
plot(K,Y);
axis([S*0.4 S*2 0 S])
xlabel('strike')
ylabel('Option price')
% fourier transform of the modified option price
function y = ft_option(rf,sg,t,a,g)
y1 = ft_pdf(rf,sg,t, g - i*(a+1));
y2 = a^2 + a - g.^2 + i*(2*a+1)*g;
y = y1./ y2;
% characteristic function of the gaussian density
function res = ft_pdf(rf,sg,t,g)
mu = rf-sg*sg/2;
res = exp(i*t*mu.*g-0.5*t*sg*sg.*g.^2);
```

Example Writing in the Matlab® command window,

```
>>[K, callprice,putprice]
  = FFT_Pricing_CallPut(1,0.05,0.2,1,1.5,12);
```

we can generate Fig. 6.14 that shows that in the range of strikes (0.6 × S, 1.4 × S) the error (Black–Scholes price-FFT inversion) appears to be of order 10^{-16}.

The inversion of the Fourier transform of the option can be performed using the quadl integration function. The following code available in the Matlab® module quad_Pricing_CallPut.m illustrates the procedure. In the following code, the

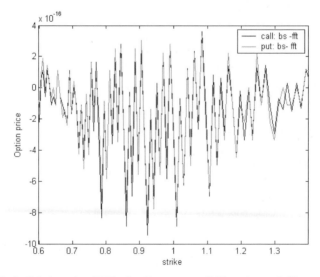

Fig. 6.14. Black–Scholes price–FFT price. Parameters: $X(0) = 1$, $r = 0.05$, $\sigma = 0.2$, $T = 1$ year, $\alpha = 1.5$, $N = 2^{12}$.

argument gmaxx is introduced in order to truncate the upper limit of integration.

```
function [callprice,putprice,errcall,errput]
  = quad_Pricing_CallPut(S,K,rf,sg,t,alpha,gmaxx)
fun = inline('real(exp(-i*x*logK).*ft_option(rf,sg,t,
             alpha,x))','x','rf','sg','t','logK',
             'alpha');
optprice = quad(fun,0,gmaxx,10^-6,'trace off',rf,sg,t,
                log(K),alpha)
                *exp(-rf*t-alpha*log(K))*S/pi;
  if alpha>0
    callprice = optprice;
    %put-call parity
    putprice = callprice-S+exp(-rf*t)*K;
  else
    putprice = optprice;
    callprice = putprice+S-exp(-rf*t)*K';
  end
  [bscall, bsput]=blsprice(S,K,rf,t,sg,0);
  errcall = bscall-callprice;
  errput = bsput-putprice;
```

Example Let us write in the Matlab® command window

```
>>[cp ,pp,errcall,errput]
  =quad_Pricing_CallPut(1,0.9, 0.05,0.2,1 ,1.5,50);
```

and we get

cp	0.16699453941906
pp	0.02310102146970
errcall	−0.00000005533490
errput	−0.00000005533490

The reader can verify how the error depends on the truncation.

We have also implemented the FFT inversion in VBA® translating the C code provided in Antia (2002). The VBA® function is

```
function fft(N As Integer,cg As Variant,iflg As Integer)
As Integer
```

where N is the number of points, which must be a power of 2; cg is an array of length $2N$ containing the data points (real and complex part), iflg is a flag to decide whether to calculate forward (iflg = 1) or inverse transform (iflg = -1). After execution, the FFT algorithm replaces the original data points in the cg vector by either the Fourier or inverse transform according to the selected value for iflg. The function returns 0 provided the execution has been successful; the number 611 if $N < 2$; and 631 if N is not a power of 2.

Numerical examples are presented in the next section along with a description of Matlab® and VBA® codes for specific models. Notice that VBA® does not support complex calculations, so that the implementation requires a preliminary step defining what is a complex number and then how to perform operations on it, such as division between two complex numbers. For this purpose, we have defined a complex number as a Variant array, with the first component being the real part and the second one the imaginary part. Then we have constructed the basic complex functions (available in the VBA® module mComplexFunctions). These are adapted from C routines available in Press et al. (1992), pp. 948–950. As an example, we report the code for computing the difference of two complex numbers A and B:

```
Function Csub(A As Variant, B As Variant) As Variant
    Dim c As Variant
    ReDim c(2)
    c(1) = A(1) - B(1) 'Compute the real part of the difference'
    c(2) = A(2) - B(2) 'Compute the imaginary part of the difference'
    Csub = c
End Function
```

Arguments A and B in CSub are complex numbers represented by vectors with two elements, containing (respectively) the real and the imaginary part of the two complex numbers. Csub returns a vector with two elements, containing the real and the imaginary part of the difference. Examples on the use of the main complex functions are given in the spreadsheet ComplexFunctions.

6.8.2 Options Pricing with Lévy Processes

A large class of models can be processed by using characteristic functions of the involved random quantities. An example is provided by Lévy processes, which display a number of interesting features. First, they are the most direct generalization of model based on Brownian motion (BM); second, they are analytically tractable; third, they are general enough to include a wide variety of patterns, so that they can account for smile and skew effects occurring in option prices; fourth, the i.i.d. structure of Lévy processes simplifies the estimation of the corresponding parameters under the historical probability measure. Any Lévy process is fully determined by the characteristic function of its increments

$$\mathbb{E}_t^*\left(e^{i\gamma z(T)}\right) = e^{m\Delta + \psi_\Delta(\gamma)}, \tag{6.45}$$

where $\Delta = T - t$ and m is the drift parameter. Equivalently, we can specify the price process in terms of the characteristic exponent $\psi_\Delta(\gamma)$ of the logarithmic increments, which is defined as the logarithm of the characteristic function. In Table 6.31, we list a few parametric Lévy processes and their associated characteristic exponent. The normal model is a benchmark assumption: we have the purely diffusive Brownian motion, which gives rise to the geometric Brownian motion (GBM) process for the

Table 6.31. Characteristic exponents of some parametric Lévy processes: G (Gaussian), NIG (Normal Inverse Gaussian), M (Meixner), VG (Variance Gamma), CGMY (Carr–Geman–Madan–Yor), DE (Double Exponential), JD (Jump Diffusion or Merton), S (Stable)

Model (parameters)	$\psi_\Delta(\gamma)$
$G(\sigma)$	$-\frac{\sigma^2}{2}\gamma^2\Delta$
$NIG(\alpha,\beta,\delta)$	$-\delta\Delta(\sqrt{\alpha^2-(\beta+i\gamma)^2}-\sqrt{\alpha^2-\beta^2})$
$M(\alpha,\beta,\delta)$	$2\delta\Delta\ln(\frac{\cos(\beta/2)}{\cosh((\alpha\gamma-i\beta)/2)})$
$VG(\sigma,v,\theta)$	$-\frac{\Delta}{v}\ln(1-i\theta v\gamma+(\sigma^2 v/2)\gamma^2)$
$CGMY(C,G,M,Y)$	$C\Delta\Gamma(-Y)((M-i\gamma)^Y-M^Y+(G+i\gamma)^Y-G^Y)$
$DE(\sigma,\lambda,p,\eta_1,\eta_2)$	$-\frac{1}{2}\sigma^2\gamma^2\Delta+\lambda\Delta(\frac{(1-p)\eta_2}{\eta_2+i\gamma}+\frac{p\eta_1}{\eta_1+i\gamma}-1)$
$JD(\sigma,\alpha,\lambda,\delta)$	$-\frac{1}{2}\sigma^2\gamma^2\Delta+\lambda\Delta(e^{i\gamma\alpha-\gamma^2\delta^2/2}-1)$
$S(\kappa,\alpha,\beta)$	$-\kappa^\alpha|\gamma|^\alpha\Delta(1-i\beta\,\mathrm{sign}(\gamma)\tan(\frac{\alpha\pi}{2}))$

Table 6.32. Parameter restrictions of some parametric Lévy processes

Model	Parameters restriction
G	$\sigma>0$
NIG	$\alpha>0,\ -\alpha<\beta<\alpha,\delta>0$
M	$\alpha>0,\ -\pi<\beta<\pi,\delta>0$
VG	$v>0,\ G>0,\ M>0$
CGMY	$C>0,\ G>0,\ M>0,\ Y<2$
DE	$\sigma>0,\ \lambda>0,\ p>0,\ \eta_1>0,\ \eta_2>0$
JD	$\sigma>0,\ \lambda>0,\ \delta>0$

price of the underlying. The model introduced by Merton (1976) and the double exponential model developed in Kou (2002) are jump-diffusion processes that account for the presence of fat tails in the empirical distribution of the underlying asset. The remaining models reported in Table 6.31, are pure jump processes with finite variation that can display both finite and infinite activity. They are subordinated Brownian motions: in other words, they can be interpreted as Brownian motions subject to a stochastic time change which is related to the level of activity in the market. In particular, stable processes display the additional feature that their distribution does not depend on the monitoring interval, modulo a scale factor. The parameters of the different models must satisfy some constraints, as given in Table 6.32. So far, the drift parameter m in (6.45) has been left unspecified. Moreover, due to the incompleteness of the market, we have to choose a martingale measure for the risk-neutral pricing of derivatives. In particular, a mathematical tractable choice consists in choosing the value of m such that the stock price discounted by the money-market account is a martingale, i.e. $\mathbb{E}_t^*[X(T)/B(T)]=X(t)/B(t),\ \forall T\geq 0$. A simple algebraic manipulation shows that m must be set equal to

$$m=r-\frac{\psi_\Delta(-i)}{\Delta}, \tag{6.46}$$

where r denotes the constant risk-free rate.

Tables 6.33 and 6.34 illustrate the main VBA® functions that allow one to price under Lévy dynamics. In these functions, dT stands for the time to maturity of the option, rf represents the risk-free interest rate, model is an integer number that allows the user to select the pricing model according to the order given in Table 6.31, g is the Fourier parameter γ in the definition of the Fourier transform, parameters is a row vector containing the parameters characterizing the chosen model. For example, if the chosen model is given by the Meixner process (see Table 6.31), then model = 3 and parameters is a 3 × 1 row vector containing the numerical

Table 6.33. VBA® functions for the inversion of the Fourier transform

```
Function cfLevy(model As Integer, dT As Double, g As Variant,
               parameters As Variant) as Variant
Function cfrn(model As Integer, rf As Double, dT As Double,
              g As Variant, parameters As Variant) as Variant
Function cfrncall(model As Integer, rf As Double,
                  dT As Double, g As Variant, aa As Double,
                  parameters As Variant) as Variant
Function cfrnput(model As Integer, rf As Double, dT As Double,
                 g As Variant, aa As Double,
                 parameters As Variant) as Variant
Function TableIFT(choice As Integer, model As Integer,
                  spot as double, rf As Double, dT As Double,
                  n As Integer, dx As Double, aa As Double,
                  parameters As Variant) as Variant
Sub macroIFT()
```

Table 6.34. Structure of VBA® modules containing the main VBA® functions for the inversion of the Fourier transform

mComplexFunctions	Operations between complex numbers		
Module Name	Function	Formula	Sub-routines
mcfLevy	cfrn	(6.45) and (6.46)	cfLevy
			cfNig
			cfMeixner
			cfVarianceGamma
			cfCgmy
			cfDe
			cfMerton
mcfcallput	cfrncall cfrnput	(6.36) and (6.37)	cfrn
mFFT	fft	(6.40) and (6.41)	
mFFTInversion	TableIFT	(6.43)	cfrn
			cfrncall
			cfrnput
			mFFT

Table 6.35. Parameters setting for Lévy models

Model	Parameters				
$G(\sigma)$	0.18850				
$NIG(\alpha, \beta, \delta)$	6.18820	−3.89410	0.16220		
$M(\alpha, \beta, \delta)$	0.39770	−1.49400	0.34620		
$VG(\sigma, v, \theta)$	0.01440	0.20000	−0.14000		
$CGMY(C, G, M, Y)$	0.02440	0.07650	7.55150	1.29450	
$DE(\sigma, \lambda, \pi, \eta_1, \eta_2)$	0.14163	0.04534	0.08982	0.24672	0.40000
$JD(\sigma, \alpha, \lambda, \delta)$	0.13358	−0.54976	0.11870	0.25651	
$S(\kappa, \alpha, \beta)$	0.13358	1.50000	0.11870		

values for the parameter set (α, β, δ). The order in which the parameters are read is the one given in the first column of Table 6.31. Quantity n represents the number of points necessary to invert the Fourier transform. It must be a power of 2, not greater than 2^{11}. Lag dx is the spacing at which the density of the option prices are returned. Parameter aa is the dumping value α given in formula (6.35). The argument choice in the function `TableIFT` is an integer number that must be set equal to 1 if the user wants to obtain the density function or equal to 2 (resp. 3) if the user wants to compute call (resp. put) option prices. Finally, spot is the current spot price of the underlying asset. Function `cfLevy` returns the complex number `Variant` representing the characteristic function in the selected model. Routine `cfrn` performs risk-neutralization on the characteristic function returned by `cfLevy` according to formula (6.46). Function `cfrncall` returns the Fourier transform of the dampened call option price. The output of function `TableIFT` is an $n \times 2$ array. This figure varies according to the value of parameter choice. If choice $= 2$, the array contains the possible future spot prices and the corresponding values of the density function. If choice $= 2$ (or 3) the array contains the possible strike prices and the corresponding call (or put) prices. This function works only if $n = 2^j$, where j is an integer no greater than 11. In order to perform the FFT inversion with j larger than 11, we need to run the macro `macroIFT`. This macro reads the relevant information in the spreadsheet IFFT and then prints the output starting from cell I36. As a numerical example, we consider the parameter set provided by Schoutens (2003, pp. 82) and reported in Table 6.35 for the reader's convenience. The example is implemented in the Excel file Levy.xls. Figure 6.15 illustrates the way we set parameters to be passed to the VBA® functions. Parameters stem from calibrating the considered models to market option prices for varying strikes and times to maturity. We consider a risk-free rate equal to 3.7% and a 1 year time to maturity. The spot price is 25.67. Figure 6.16 compares the risk-neutral densities of logarithmic returns with reference to NIG and Gaussian models. In the Gaussian case, the volatility has been selected so that the log-returns in the two models display the same variance, that is $\sigma = 22.0078\%$.[13] In particular, Figure 6.16 displays the two densities in a logarithmic scale. In the Gaussian model, this curve is a parabola, whereas the NIG model produces an asymmetric shape and a nearly linear decay in the left-hand tail.

---

[13] For both models, $\mathbb{E}(x(T)) = x(0)e^{rT}$.

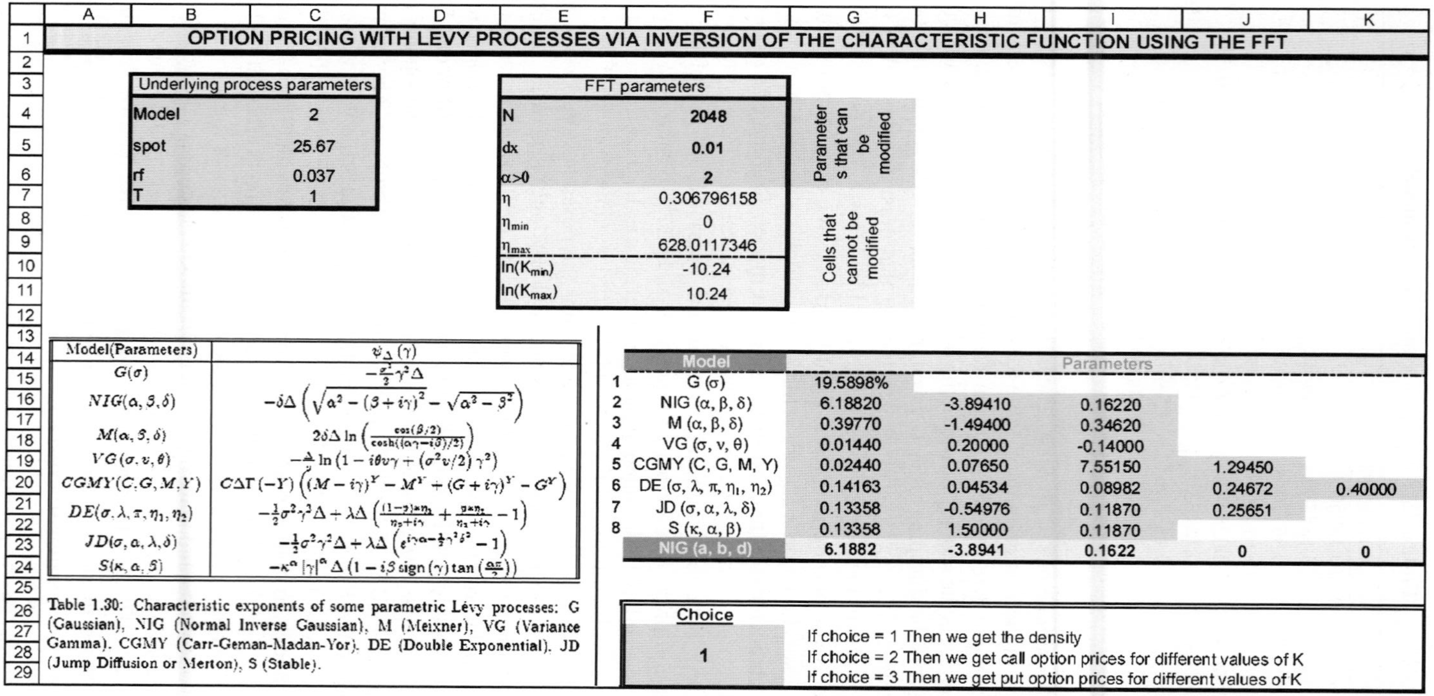

Fig. 6.15. Excel screen for setting the parameters in the Lévy models and in the FFT algorithm.

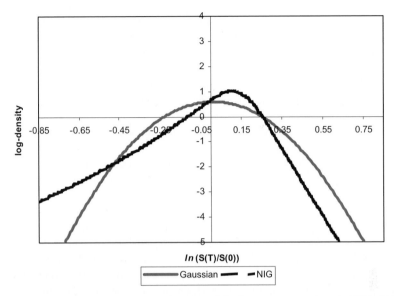

Fig. 6.16. Log-density: Gaussian and NIG models. Parameters are given in Table 6.35.

Table 6.36. Call and put option prices in the Black–Scholes and in the NIG model

K	$c_{NIG}(K, T)$	$p_{NIG}(K, T)$	$c_{BS}(K, T)$	$p_{BS}(K, T)$
21.4414	5.5710	0.5635	5.3114	0.3039
22.7672	4.5074	0.7776	4.2856	0.5559
24.1751	3.4588	1.0858	3.3255	0.9524
25.6700	2.4677	1.5352	2.4677	1.5352
27.2573	1.5964	2.1936	1.7415	2.3388
28.9428	0.9183	3.1398	1.1629	3.3845
30.7326	0.4741	4.4204	0.7313	4.6776

This behavior reflects into higher prices for out-of-the money put options compared to the Black–Scholes price. Indeed, OTM puts and ITM calls, as it can be argued from the put-call parity, depend on the thickness of the left-hand tail of the underlying asset distribution density. This phenomenon is illustrated in Table 6.36 where option prices are reported for varying strikes. There, the volatility parameter corresponding to the Gaussian model has been set in a way that at-the-money options in the Black–Scholes and in the NIG have the same price, that is $\sigma = 19.5898\%$. Similar results can be obtained with reference to the other models by selecting the appropriate routine in the spreadsheet `Levy.xls`.

6.9 Comments

Standard references for quadrature methods are Davis and Rabinowitz (1975), Stoer and Bulirsch (1980), Evans and Swartz (2000) and Antia (2002). A detailed discussion on adaptive methods can be found in Gander and Gautschi (2000). Details on the CEV model can be found in Cox and Ross (1976), Emanuel and MacBeth (1982), Goldenberg (1991) and Cox (1996). Schroder (1989) shows that the transition density of the CEV process can be written in terms of the non-central chi-square distribution function. A good approximation to it is given by Sankaran (1963) and discussed by Johnson and Kotz (1995). The CIR model has been introduced by Cox, Ingersoll and Ross (1985). See also Feller (1951), Lamberton and Lapeyre (1996) and Cairns (2004). Closed form expressions for European options on coupon bonds have been obtained by Jamshidian (1991) and Longstaff (1993). A listing of pricing formulae for different types of barrier options can be found in Rubinstein and Reiner (1991). A mathematically oriented discussion of the barrier option pricing problem is contained in Rich (1994). In a nutshell, there are several approaches to barrier option pricing: (a) the probabilistic method, see Kunitomo and Ikeda (1992); (b) the Laplace Transform technique, see Pelsser (2000), Sbuelz (1999, 2005), Jamshidian (1997), Geman and Yor (1996); (c) the Black–Scholes PDE, which can be solved using separation of variables, see Hui, Lo and Yuen (2000) or finite difference schemes, see Boyle and Tian (1998) and Zvan, Vetzal and Forsyth (2000), (d) binomial and trinomial trees, see Boyle and Lau (1994), Cheuk and Vorst (1996), Figlewski and Gao (1999), Ritchken (1995), Tian (1999); (e) Monte Carlo simulations with various enhancements, see Andersen and Brotherton-Ratcliffe (1996), Baldi, Caramellino and Iovino (1999), Beaglehole, Dybvig and Zhou (1997); (f) quadrature methods, see Aitsahlia and Lai (1997), Sullivan (2000), Andricopoulos et al. (2003), Duan et al. (2003), Fusai and Recchioni (2001); (g) approximated formulae, see Broadie, Glasserman, and Kou (1997, 1999), Hörfelt (2003); (g) Wiener-Hopf methods, see Fusai, Abrahams and Sgarra (2006). The Fourier transform approach to option pricing has been used in several papers, for example Heston (1993), Bates (1991), Duffie, Pan and Singleton (1998). Biffis and Millossovich (2006) price guaranteed annuity options using affine processes and Fourier inversion. Useful references are Carr et al. (2005), Cerny (2003), Lewis (2000) and Lipton (2001). The Fast Fourier algorithm has been introduced in finance by Carr and Madan (1999). For a thorough introduction to Lévy processes see Sato (2000) and Applebaum (2004). Applications to finance can be found in Merton (1976), Kou (2002), Carr et al. (2003), Schoutens (2003), Cont and Tankov (2004). In pricing using Lévy models, we have used the risk-neutral measure. Another possible choice is to choose a different martingale measures, such as the Esscher transform, as advocated at first in Gerber and Shiu (1994).

7

The Laplace Transform[*]

Key words: integral transform, numerical inversion, PDE, ODE

In this chapter, we illustrate the use of the Laplace transform in option pricing. Using the Laplace transform method we can transform a PDE into an ordinary differential equation (ODE) that in general is easier to solve. The solution of the PDE can be then obtained inverting the Laplace transform. Unfortunately when we consider interesting examples, such as pricing Asian options, usually it is difficult to find an analytical expression for the inverse Laplace transform. Then the necessity of the numerical inversion. For this reason, in this chapter we also discuss the problem of the numerical inversion, presenting the Fourier series algorithm that can be easily implemented in MATLAB® or VBA®. The numerical inversion is often disbelieved generically referring to its "intrinsic instability" or for "its inefficiency from a computational point of view". So the aim of this chapter is also to illustrate that the numerical inversion is feasible, is accurate and is not computational intensive. For these reasons, we believe that the Laplace transform instrument will gain greater importance in the Finance field, as already happened in engineering and physics.

In Sect. 7.1 we define the Laplace transform and we give its main properties. In Sect. 7.2, we illustrate the numerical inversion problem. Section 7.3 illustrates a simple application to finance.

7.1 Definition and Properties

In this section we give the basic definition and the properties of the Laplace transform. We say that a function F is of exponential order, if there exist some constants, M and k, for which $|F(\tau)| \leq Me^{k\tau}$ for all $\tau \geq 0$. The Laplace transform $\widehat{F}(\gamma)$ of a

---

[*] with Marina Marena.

function $F(\tau)$ is defined by the following integral:

$$\widehat{F}(\gamma) = \mathcal{L}(F(\tau)) = \int_0^{+\infty} e^{-\gamma\tau} F(\tau)\, d\tau \qquad (7.1)$$

where γ is a complex number and $F(\tau)$ is any function which, for some value of γ, makes the integral finite. The integral (7.1) then exists for a whole interval of values of γ, so that the function $\widehat{F}(\gamma)$ is defined. The integral converges in a right-plane $\mathrm{Re}(\gamma) > \gamma_0$ and diverges for $\mathrm{Re}(\gamma) < \gamma_0$. The number γ_0, which may be $+\infty$ or $-\infty$, is called the *abscissa of convergence*.

Not every function of τ has a Laplace transform, because the defining integral can fail to converge. For example, the functions $1/\tau$, $\exp(\tau^2)$, $\tan(\tau)$ do not possess Laplace transforms. A large class of functions that possess a Laplace transform are of exponential order. Then the Laplace transform of $F(\tau)$ surely exists if the real part of γ is greater than k. In this case, k coincides with the abscissa of convergence γ_0. Also there are certain functions that cannot be Laplace transforms, because they do not satisfy the property $\widehat{F}(+\infty) = 0$, e.g. $\widehat{F}(\gamma) = \gamma$. An important fact is the uniqueness of the representation (7.1), i.e. a function $\widehat{F}(\gamma)$ cannot be the transform of more than one continuous function $F(\tau)$. We have indeed:

Theorem 1 *Let $F(\tau)$ be a continuous function, $0 < \tau < \infty$ and $\widehat{F}(\gamma) \equiv 0$, for $\gamma_0 < \mathrm{Re}(\gamma) < \infty$. Then we have $F(\tau) \equiv 0$.*

In Table 7.1 we give the most important properties of the Laplace transform. In particular, we stress the linearity property

$$\mathcal{L}\big(aF_1(\tau) + bF_2(\tau)\big) = a\mathcal{L}(F_1(\tau)) + b\mathcal{L}(F_2(\tau)),$$

and the Laplace transform of a derivative

$$\mathcal{L}(\partial_\tau F(\tau)) = \gamma\mathcal{L}(F(\tau)) - F(0).$$

In Table 7.2 we give several examples of the Laplace transform $\widehat{F}(\gamma)$ and the corresponding function $F(\tau)$.

If the Laplace transform is known, the original function $F(\tau)$ can be recovered using the inversion formula (Bromwich inversion formula), that can be represented as an integral in the complex plane. We have the following result:

Theorem 2 *If the Laplace transform of $F(\tau)$ exists and has abscissa of convergence with real part γ_0, then for $\tau > 0$*

$$F(\tau) = \mathcal{L}^{-1}(\widehat{F}(\gamma)) = \lim_{R\to\infty} \frac{1}{2\pi i} \int_{a-iR}^{a+iR} \widehat{F}(\gamma) e^{\tau\gamma}\, d\gamma,$$

where a is another real number such that $a > \gamma_0$ and i is the imaginary unit, $i = \sqrt{-1}$.

Table 7.1. Basic properties of the Laplace transform

Property	Function	Laplace transform
Definition	$F(\tau)$	$\widehat{F}(\gamma) = \int_0^{+\infty} e^{-\gamma\tau} F(\tau)\, d\tau$
Linearity	$aF_1(\tau) + bF_2(\tau)$	$a\widehat{F}_1(\gamma) + b\widehat{F}_2(\gamma)$
Scale	$aF(a\tau)$	$\widehat{F}(\gamma/a)$
Shift	$e^{a\tau} F(\tau)$	$\widehat{F}(\gamma - a)$
Shift	$\begin{cases} F(\tau - a), & \tau > a \\ 0, & \tau < a \end{cases}$	$e^{-a\gamma}\widehat{F}(\gamma)$
Time derivative	$\frac{\partial F(\tau)}{\partial \tau}$	$\gamma\widehat{F}(\gamma) - F(\tau)\vert_{\tau=0}$
Differentiation	$\frac{\partial^n F(\tau)}{\partial \tau^n}$	$\gamma^n \widehat{F}(\gamma) - \gamma^{n-1} F(0) + \cdots$ $- \gamma^{n-2} F'(0) - \cdots - F^{(n-1)}(0)$
Integral	$\int_0^\tau F(s)\, ds$	$\frac{\widehat{F}(\gamma)}{\gamma}$
Multiplication by polynomials	$\tau^n F(\tau)$	$(-1)^n \widehat{F}^{(n)}(\gamma)$
Convolution	$\int_0^\tau F(s)G(\tau - s)\, ds$	$\widehat{F}(\gamma)\widehat{G}(\gamma)$
Ratio of polynomials	$\sum_{k=1}^n \frac{P(\alpha_k)}{Q'(\alpha_k)} e^{\alpha_k \tau}$	$\sum_{k=1}^n \frac{P(\gamma)}{Q(\gamma)}$
	$P(x)$ polynomial of degree $< n$; $Q(x) = (x - a_1)(x - a_2)\cdots(x - a_n)$ where $a_1 \neq a_1 \neq \cdots \neq a_n$	
Final value	$\lim_{\tau\to\infty} F(\tau)$	$\lim_{\gamma\to 0} \gamma\widehat{F}(\gamma)$
Initial value	$\lim_{\tau\to 0} F(\tau)$	$\lim_{\gamma\to\infty} \gamma\widehat{F}(\gamma)$
Inversion	$\lim_{k\to\infty} \frac{1}{2\pi i} \int_{a-ik}^{a+ik} f(\gamma)e^{\tau\gamma}\, d\gamma$	$\widehat{F}(\gamma)$
	where c is the real part of the rightmost singularity in the image function	

Table 7.2. Some Laplace transforms and their inverses. The function $\delta(t)$ is the delta-Dirac function, the function $J_n(x)$ is the Bessel function of the first kind and of order n, Erfc(x) is the complementary error function, i.e. Erfc$(x) = 2N(-\sqrt{2}x)$, where $N(x)$ is the cumulative normal distribution

	$\widehat{F}(\gamma)$	$F(\tau)$
1	1	$\delta(\tau)$
2	$e^{-a\gamma}$	$\delta(\tau - a)$
3	$\frac{1}{\gamma}$	1
4	$\frac{1}{\gamma^2}$	τ
5	$\frac{1}{\gamma^n},\ n > 0$	$\frac{\tau^{n-1}}{\Gamma(n)}$
6	$\frac{1}{(\gamma-a)^n},\ n > 0$	$\frac{\tau^{n-1} e^{a\tau}}{(n-1)!}$
7	$\frac{1}{\sqrt{\gamma - a} + b}$	$e^{a\tau}\left(\frac{1}{\sqrt{\pi\tau}} - be^{b^2\tau}\, \text{Erfc}(b\sqrt{\tau})\right)$
8	$\frac{e^{-\vert a\vert\sqrt{\gamma}}}{\sqrt{\gamma}}$	$\frac{e^{-a^2/4\tau}}{\sqrt{\pi\tau}}$
9	$e^{-\vert a\vert\sqrt{\gamma}}$	$\frac{a e^{-a^2/4\tau}}{2\sqrt{\pi\tau^3}}$
10	$\frac{e^{-a\sqrt{\gamma}}}{\sqrt{\gamma}(\sqrt{\gamma}+b)}$	$e^{b(b\tau+a)}\, \text{Erfc}(b\sqrt{\tau} + \frac{a}{2\sqrt{\tau}})$
11	$\frac{e^{-a/\gamma}}{\gamma^{n+1}}$	$(\frac{\tau}{a})^{n/2} J_n(2\sqrt{a\tau})$

The real number a must be selected so that all the singularities of the image function $\widehat{F}(\gamma)$ are to the left of the vertical line $\gamma = \gamma_0$. The integral in the complex plane can be sometimes evaluated analytically using the Cauchy's residue theorem. But this goes beyond an elementary treatment of the Laplace transform and we refer the reader to some textbooks on complex analysis, such as Churchill and Brown (1989). Moreover, this analytical technique often fails and the Bromwich's integral must be integrated numerically.

7.2 Numerical Inversion

Aim of this section is to illustrate how simple and accurate can be the numerical inversion of the Laplace transform. The general opinion that the inversion of the Laplace transform is an ill-conditioned problem[1,2] is due to one of the first tentatives of inversion that reduce the inversion problem to the solution of an ill-conditioned linear system. If we consider a quadrature formula for the integral defining the Laplace transform, we get

$$\widehat{F}(\gamma) = \sum_{i=1}^{n} w_i e^{-\gamma \tau_i} F(\tau_i). \tag{7.2}$$

Writing this equation for n different values of γ, where γ is supposed to be a *real* number, we are left with an $n \times n$ linear system to be solved wrt the n unknown values $F(\tau_i)$. Unfortunately, the solution of this linear system can change abruptly given little changes in $\widehat{F}(\gamma)$. The ill-conditioning of the above inversion method is common to all numerical routines that try the inversion computing the Laplace transform only for real values of the parameter γ. The exponential kernel that appears in the definition of the Laplace transform smooths out too much the original function. Therefore, to recover $F(\tau)$ given values of the Laplace transform on the real axis can be very difficult. This problem occurs when $\widehat{F}(\gamma)$ is the result of some physical experiment, so that it can be affected by measurement errors. Instead, this problem does not arise when the Laplace transform is known in closed form as a complex function. In this case instead of discretizing the integral defining the forward Laplace transform, we

---

[1] The concept of well-posedness was introduced by Hadamard and, simply stated, it means that a well-posed problem should have a solution, that this solution should be unique and that it should depend continuously on the problem's data. The first two requirements are minimal requirements for a reasonable problem, and the last ensures that perturbations, such errors in measurement, should not unduly affect the solution.

[2] "The inversion of the Laplace transform is well known to be an ill-conditioned problem. Numerical inversion is an unstable process and the difficulties often show up as being highly sensitive to round-off errors", Kwok and Barthez (1989). "The standard inversion formula is a contour integral, not a calculable expression .... These methods provide convergent sequences rather than formal algorithms; they are difficult to implement (many involve solving large, ill-conditioned systems of linear equations or analytically obtaining high-order derivatives of the transform) and none includes explicit, numerically computable bounds on error and computational effort", Platzman, Ammons and Bartholdi (1988).

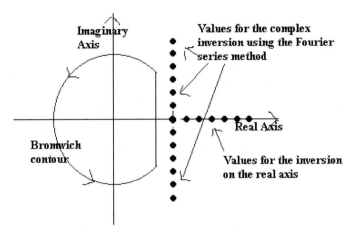

Fig. 7.1. Sample points for the inversion with the Fourier series method using the Bromwich contour and sample points for the inversion on the real axis using the definition of Laplace transform.

can operate the inversion using directly the Bromwich contour integral, and then using values of the transform in the complex plane. The different approach of inverting the Laplace transform on the real axis or on the complex plane is illustrated in Fig. 7.1. This section describe a very effective Laplace inversion algorithm that involves complex calculations.[3]

Letting the contour be any vertical line $\gamma = a$ such that $\widehat{F}(\gamma)$ has no singularities on or to the right of it, the original function $F(\tau)$ is given by the inversion formula:

$$F(\tau) = \frac{1}{2\pi i} \int_{a-i\infty}^{a+i\infty} e^{\gamma\tau} \widehat{F}(\gamma)\,d\gamma, \quad \tau > 0. \tag{7.3}$$

Alternatively, setting $a+iu = \gamma$ and using the identity from complex variable theory, $e^\gamma = e^a(\cos(u) + i\sin(u))$, $\mathrm{Re}(\widehat{F}(a+iu)) = \mathrm{Re}(\widehat{F}(a-iu))$, $\mathrm{Im}(\widehat{F}(a+iu)) = -\mathrm{Im}(\widehat{F}(a-iu))$, $\sin(u\tau) = -\sin(-u\tau)$, $\cos(u\tau) = \cos(-u\tau)$, and from the fact that the integral in (7.3) is 0 for $\tau < 0$, we get

$$F(\tau) = \frac{2e^{a\tau}}{\pi} \int_{0}^{+\infty} \mathrm{Re}\big(\widehat{F}(a+iu)\big) \cos(u\tau)\,du \tag{7.4}$$

and

$$F(\tau) = -\frac{2e^{a\tau}}{\pi} \int_{0}^{+\infty} \mathrm{Im}\big(\widehat{F}(a+iu)\big) \sin(u\tau)\,du.$$

$F(\tau)$ can be calculated from (7.4) by performing a numerical integration (quadrature). Since there are many numerical integration algorithms, the remaining goal is

---

[3] Certain computer languages such as Matlab®, Mathematica, Fortran and C++ have automatic provision for doing complex calculations. In VBA® or C we need instead to define a new type of variable and to say how operations between complex numbers must be performed.

to exploit the special structure of the integrand in (7.4) in order to calculate the integral accurately and efficiently.

The algorithm we describe is named Fourier series method and has received great attention recently in finance, for the simplicity of implementation and the accuracy in the numerical results. The underlying idea of the method is to discretize (7.4) using the trapezoidal rule. Then the inversion is given as a sum of infinite terms. The convergence of the series is accelerated using the Euler algorithm. This algorithm allows one to compute a series with great accuracy using a limited number of terms (in several examples founded in the literature no more than 30).

7.3 The Fourier Series Method

The Fourier series algorithm has been originally proposed by Dubner and Abate (1968) and then improved by Abate and Whitt (1992b). It is essentially a trapezoidal rule approximation to (7.4). An essential feature of this method is that an expression for the error in the computed inverse transform is available. Therefore, one can control the maximum error in the inversion technique. Since the trapezoidal rule is a quite simple integration procedure, its use can appear surprising. It turns out to be surprisingly effective in this context with periodic and oscillating integrands, because the errors tend to cancel. In particular, it turns out to be better than familiar alternatives such as Simpson's rule or Gaussian quadrature for inversion integrals.

If we apply the trapezoidal rule with step size Δ to the expression in (7.4), we get

$$F(\tau) \simeq F_\Delta^{DA}(\tau) = \frac{\Delta e^{a\tau}}{\pi} \operatorname{Re}(\widehat{F}(a)) + \frac{2\Delta e^{a\tau}}{\pi} \sum_{k=1}^{\infty} \operatorname{Re}\left(\widehat{F}(a + ik\Delta)\right) \cos(k\Delta\tau).$$

If we set $\Delta = \pi/(2\tau)$ and $a = A/(2\tau)$, we can eliminate the cosine terms and we obtain an alternating series

$$F_\Delta^{DA}(\tau) = \frac{e^{A/2}}{2\tau} \operatorname{Re}\left(\widehat{F}\left(\frac{A}{2\tau}\right)\right) + \frac{e^{A/2}}{\tau} \sum_{k=1}^{\infty} (-1)^k \operatorname{Re}\left(\widehat{F}\left(\frac{A + 2k\pi i}{2\tau}\right)\right). \quad (7.5)$$

The choice of A has to be made in such a way that a falls at the left of the real part of all the singularities of the function $\widehat{F}(\gamma)$ ($a = 0$ suffices when F is a bounded continuous probability density). Assuming that $|F(\tau)| < M$, Abate and Whitt (1992b) show that the discretization error can be bounded by

$$\left|F(\tau) - F_\Delta^{DA}(\tau)\right| < M \frac{e^{-A}}{1 - e^{-A}} \simeq M e^{-A}, \quad (7.6)$$

so that we should set A large in order to make the error small. In order to obtain a discretization error less than $10^{-\delta}$, we can set $A = \delta \ln 10$. However, increasing A can make the inversion (7.5) harder, due to roundoff errors. Thus A should not be chosen too large. In practice, Abate and Whitt (1992b) suggest to set A equal to 18.4.

The remaining problem consists in computing the infinite sum in (7.5). If the term $\mathrm{Re}(\widehat{F}((A + 2k\pi\mathrm{i})/(2\tau)))$ has a constant sign for all k, it can be convenient to consider an accelerating algorithm for alternating series. Abate and Whitt (1992b) propose the use of the Euler algorithm. This algorithm consists in summing explicitly the first n terms of the series and then in taking a weighted average of additional m terms. In practice, the Euler algorithm estimates the series using $E(\tau; n, m)$, where

$$F_{\Delta}^{\mathrm{DA}}(\tau) \approx E(\tau; n, m) = \sum_{k=0}^{m} \binom{m}{k} 2^{-m} s_{n+k}(\tau), \qquad (7.7)$$

and where $s_n(\tau)$ is the nth partial sum:

$$s_n(\tau) = \frac{\mathrm{e}^{A/2}}{2\tau} \mathrm{Re}\left(\widehat{F}\left(\frac{A}{2\tau} \right) \right) + \frac{\mathrm{e}^{A/2}}{\tau} \sum_{k=1}^{n} (-1)^k \mathrm{Re}\left(\widehat{F}\left(\frac{A + 2k\pi\mathrm{i}}{2\tau} \right) \right). \qquad (7.8)$$

As pointed out in Abate and Whitt (1992b, p. 46), in order for Euler summation to be effective, $a_k = \mathrm{Re}(\widehat{F}((A + 2k\pi\mathrm{i})/(2\tau)))$ must have three properties for sufficiently large k: (a) to be of constant sign, (b) to be monotone, (c) the higher-order differences $(-1)^m \Delta^m a_{n+k}$ are monotone. On a practical side, these properties are not checked, so that the algorithm is used in a heuristic way. Usually, $E(\tau; n, m)$ approximates the true sum with an error of the order of 10^{-13} or less with the choice $n = 38$ and $m = 11$, i.e. using just 50 terms. The direct computation of the series can require more than 10,000 terms. The Abate–Whitt algorithm gives excellent results for functions that are sufficiently smooth (say, twice continuously differentiable). However, the inversion algorithm performs less satisfactorily for points at which the function $f(t)$ or its derivative is not differentiable.

Example Let us test the algorithm with the series $\sum_{k=1}^{+\infty} (-1)^k / k$, that converges to $-\ln 2 = -0.6931471805599453$. Computing the sum using 100,000 terms, we get -0.6931421805849445, i.e. a five digits accuracy. Using the Euler algorithm with $n = 19$ and $n + m = 30$, we get -0.693147180559311, i.e. a ten digits accuracy! This is illustrated in Fig. 7.2.

The procedure for the numerical inversion is then resumed in Table 7.3.

7.4 Applications to Quantitative Finance

In this section we illustrate how the Laplace transform method can be useful in solving linear parabolic equations. We consider two examples: (a) pricing a call option in the standard Black–Scholes model; (b) pricing an Asian option in the square-root model.

7.4.1 Example

For this, let us consider the Black–Scholes PDE satisfied by the price $F(\tau, X)$ of a derivative contract having time to maturity $T - t$

Euler Algorithm for the series $\Sigma(1\text{-})^k/k$

m   10

n	$(-1)^k/k$	s_n
1	-1	-1
2	0.5	-0.5
3	-0.333333	-0.833333
4	0.25	-0.583333
5	-0.2	-0.783333
6	0.166667	-0.616667
7	-0.142857	-0.759524
8	0.125	-0.634524
9	-0.111111	-0.745635
10	0.1	-0.645635
11	-0.090909	-0.736544
12	0.083333	-0.653211
13	-0.076923	-0.730134
14	0.071429	-0.658705
15	-0.066667	-0.725372
16	0.0625	-0.662872
17	-0.058824	-0.721695
18	0.055556	-0.66614

n	$(-1)^k/k$	s_n	Extra terms	Bin. Coeff.	$\binom{m}{k}2^{-m}s_{n+k}(\tau)$
19	-0.052632	-0.718771	0	1	-0.000653097
20	0.05	-0.668771	1	10	-0.006996
21	-0.047619	-0.71639	2	45	-0.029484488
22	0.045455	-0.670936	3	120	-0.08372041
23	-0.043478	-0.714414	4	210	-0.137965796
24	0.041667	-0.672747	5	252	-0.175402705
25	-0.04	-0.712747	6	210	-0.138281301
26	0.038462	-0.674286	7	120	-0.083358164
27	-0.037037	-0.711323	8	45	-0.029689836
28	0.035714	-0.675609	9	10	-0.006934487
29	-0.034483	-0.710091	10	1	-0.000660897
30	0.033333	-0.676758			

Euler Sum	-0.69314718056

Fig. 7.2. Euler algorithm for computing $\sum_{k=1}^{\infty} \frac{(-1)^k}{k}$.

$$\partial_t F + rx\partial_x F + \frac{1}{2}\sigma^2 x^2 \partial_{xx} F = rF, \qquad (7.9)$$

$$F(T, x) = \phi(x),$$

and appropriate boundary conditions. Let us define

$$\tau = \frac{\sigma^2}{2}(T - t), \quad z = \ln x,$$

and let us introduce the new function

$$F(t, x) = f(\tau, z).$$

Table 7.3. Pseudo-code for implementing the numerical inversion of the Laplace transform

---

Define the Laplace Transform $\widehat{F}(\frac{A}{2\tau})$
Assign A, n, m
Compute s_j in (7.8), $j = 1, m + m$.
Using $s_n, \ldots, s_{n+m}$ compute $E(\tau; n, m)$

Then f solves the PDE

$$-\partial_\tau f(\tau, z) + \left(\frac{r}{\sigma^2/2} - 1\right)\partial_z f(\tau, z) + \partial_{zz} f(\tau, z) - \frac{r}{\sigma^2/2} f(\tau, z) = 0, \qquad (7.10)$$

with initial condition $f(0, z) = F(T, e^z)$. In the following, we consider as payoff function

$$f(0, z) = F(T, e^z) = (e^z - e^k)_+,$$

i.e. a plain vanilla option (and therefore $f(\tau, z) \to e^z - e^k$ as $z \to +\infty$ and $f(\tau, z) \to 0$ as $z \to -\infty$). If we Laplace transform the above partial differential equation with constant coefficients, the result will be an algebraic equation in the transform of the unknown variable. Indeed, from the properties illustrated in Table 7.1, we have

$$\mathcal{L}(f(\tau, z)) = \int_0^\infty e^{-\gamma\tau} f(\tau, z)\,d\tau = \hat{f}(\gamma, z),$$

$$\mathcal{L}(\partial_\tau f(\tau, z)) = \int_0^\infty e^{-\gamma\tau} \partial_\tau f(\tau, z)\,d\tau = \gamma\hat{f}(\gamma, z) - f(0, z),$$

$$\mathcal{L}(\partial_z f(\tau, z)) = \int_0^\infty e^{-\gamma\tau} \partial_z f(\tau, z)\,d\tau = \partial_z\hat{f}(\gamma, z),$$

$$\mathcal{L}(\partial_{zz} f(\tau, z)) = \int_0^\infty e^{-\gamma\tau} \partial_{zz} f(\tau, z)\,d\tau = \partial_{zz}\hat{f}(\gamma, z).$$

Therefore, we have the means of turning the PDE (7.9), for the linearity of the Laplace transform, into the second-order ordinary differential equation (ODE):

$$-(\gamma\hat{f}(\gamma, z) - (e^z - e^k)_+) + \left(\frac{r}{\sigma^2/2} - 1\right)\partial_z\hat{f}(\gamma, z)$$
$$+ \partial_{zz}\hat{f}(\gamma, z) - \frac{r}{\sigma^2/2}\hat{f}(\gamma, z) = 0.$$

Then setting $m = r/(\sigma^2/2)$ we get

$$\partial_{zz}\hat{f}(\gamma, z) + (m - 1)\partial_z\hat{f}(\gamma, z) - (m + \gamma)\hat{f}(\gamma, z) + (e^z - e^k)_+ = 0 \qquad (7.11)$$

with boundary conditions given by the Laplace transform of the boundary conditions of the original PDE:

Table 7.4. Laplace transform and PDE

Original space $\left\{ \begin{matrix} \text{PDE} \\ +\text{IC} \\ +\text{BC's} \end{matrix} \right\}$ Solution

analytical   ↗↖   numerical

$\mathcal{L}$-transform   ↖↗ $\mathcal{L}^{-1}$-transform

Image space $\left\{ \begin{matrix} \text{ODE} \\ +\text{BC's} \end{matrix} \right\} \rightarrow$ Solution

$$\hat{f}(\gamma, z) \rightarrow \mathcal{L}\left(e^z - e^{-m\tau}e^k\right) = \frac{e^z}{\gamma} - \frac{e^k}{\gamma + m} \quad \text{as } z \rightarrow +\infty, \qquad (7.12)$$

$$\hat{f}(\gamma, z) \rightarrow \mathcal{L}(0) = 0 \quad \text{as } z \rightarrow -\infty. \qquad (7.13)$$

The initial condition of the PDE has been included in the ODE, where now there is the appearance of the term $(e^z - e^k)_+$. Therefore, instead of solving the PDE (7.10) we are left with the second-order differential equation in (7.11), that actually is simpler to solve. Then, the problem will be to recover the solution of the PDE from the solution of the ODE, i.e. to find the inverse Laplace transform. The procedure is illustrated in Table 7.4.

In order to solve (7.11), let us define

$$\hat{f}(\gamma, z) = \exp(\alpha z)\hat{g}(\gamma, z),$$

where $\alpha = (1 - m)/2$. Then $\hat{g}(\gamma, z)$ solves

$$\partial_{zz}\hat{g}(\gamma, z) - (b + \gamma)\hat{g} + e^{-\alpha z}\left(e^z - e^k\right)_+ = 0,$$

with $b = \alpha^2 + m = (m - 1)^2/4 + m$. We can solve this ODE separately in the two regions $z > k$ and $z \le k$ to get

$$\hat{g}(\gamma, z) = \begin{cases} \frac{e^{-(\alpha-1)z}}{\gamma} - \frac{e^{-\alpha z + k}}{\gamma + m} + h_1(\gamma, z)A_1 + h_2(\gamma, z)A_2, & z > k, \\ h_1(\gamma, z)B_1 + h_2(\gamma, z)B_2, & z \le k, \end{cases}$$

where

$$h_1(\gamma, z) = e^{-\sqrt{b + \gamma}z}, \qquad h_2(\gamma, z) = e^{+\sqrt{b + \gamma}z}.$$

Here A_1, A_2, B_1 and B_2 are constants to be determined according to the boundary conditions (7.12) and (7.13) and requiring that $\hat{f}(\gamma, z)$ is continuous and differentiable at $z = k$ (*smooth pasting conditions*). We observe that the singularities of $\hat{g}(\gamma, z)$ are 0, $-m$ and $-b$. Therefore the abscissa of convergence of $\hat{g}(\gamma, z)$ is given by

$$\gamma_0 = \max(0, -m, -b).$$

Given that when $\gamma > \gamma_0$, $\lim_{z \to +\infty} e^{\alpha z}h_1(\gamma, z) = 0$ and $\lim_{z \to +\infty} e^{\alpha z}h_2(\gamma, z) = \infty$, we must set $A_2 = 0$. Similarly, when $z < k$ we need to set $B_1 = 0$. We are therefore left with

$$\hat{g}(\gamma, z) = \begin{cases} \dfrac{e^{-(\alpha-1)z}}{\gamma} - \dfrac{e^{-\alpha z + k}}{\gamma + m} + h_1(\gamma, z)A_1, & z > k, \\ h_2(\gamma, z)B_2, & z \le k, \end{cases}$$

and now we determine A_1 and B_2 with the additional conditions

$$\lim_{z \to k^+} \hat{f}(\gamma, z) = \lim_{z \to k^-} \hat{f}(\gamma, z),$$

$$\lim_{z \to k^+} \partial_z \hat{f}(\gamma, z) = \lim_{z \to k^-} \partial_z \hat{f}(\gamma, z).$$

With some tedious algebra, we get

$$A_1(\gamma) = \frac{e^{(1-a+\sqrt{b+\gamma})k}(\gamma - (a - 1 + \sqrt{b+\gamma})m)}{2\gamma\sqrt{b+\gamma}(\gamma + m)},$$

$$B_2(\gamma) = \frac{e^{(1-a-\sqrt{b+\gamma})k}(\gamma - (a - 1 - \sqrt{b+\gamma})m)}{2\gamma\sqrt{b+\gamma}(\gamma + m)},$$

and finally we obtain the following expression for the function $\hat{f}(\gamma, z)$

$$\hat{f}(\gamma, z) = e^{az}\left[\left(\frac{e^{-(\alpha-1)z}}{\gamma} - \frac{e^{-\alpha z + k}}{\gamma + m} \right) 1_{(z>k)} \right.$$
$$\left. + \frac{e^{-\sqrt{b+\gamma}|z-k|+(1-a)k}(\gamma - (a - 1 + \sqrt{b+\gamma}\,\mathrm{sgn}(z-k))m)}{2\gamma\sqrt{b+\gamma}(\gamma + m)} \right],$$

$$\tag{7.14}$$

where $\mathrm{sgn}(z) = 1_{(z\ge 0)} - 1_{(z<0)}$.

We can also easily obtain the Laplace transform of the Delta and the Gamma of the option differentiating with respect to $x = e^z$ the Laplace transform.

Numerical Inversion

The numerical inversion has been implemented in MATLAB® and in VBA®. In MATLAB®, we have built the following functions

```
function [lt] = ltbs(spot, strike, sg, rf, gamma)
function [euler] = AWBS(spot, strike, expiry, sg, rf,
aa, terms, extraterms)
```

The function `ltbsm` returns the Laplace transform in (7.14), taking as inputs the spot price (`spot`), the strike (`strike`), the volatility (`sg`), the risk-free rate (`rf`) and the Laplace parameter γ (`gamma`). The function `AWBS` performs the numerical inversion (Fourier series with Euler summation) returning the Black–Scholes price. The parameter `aa` is the constant A that determines the discretization error in (7.6), `terms` is the number of terms n we use to estimate s_n, and `extraterms` is the additional number of terms m needed to perform the Euler summation. Similar functions have been constructed in VBA® for Excel. Here below, we give the Matlab® code.

```
    function [optprice] = AWBS(spot, strike, expiry, sg,
rf, aa, terms, extraterms)
    tau = expiry * sg * sg / 2;
    sum = 0;
    %%compute the LT at gamma = aa / (2 * tau)
    lt = ltbs(spot, strike, sg, rf, aa / (2 * tau));
    sum = lt* exp(aa / 2) / (2 * tau);
    %apply the Euler algorithm
    k = [1:terms + extraterms];
    arg = aa / (2 * tau)+i*pi.*k / tau;
    term = ((-1) .^k) .* ltbs(spot, strike, sg, rf, arg)
* exp(aa / 2) / tau;
    csum = sum+cumsum(term);
    sumr = real(csum(terms:terms+extraterms));
    j=[0:extraterms];
    bincoeff = gamma(extraterms+1)./(gamma(j+1).
* gamma(extraterms-j+1));
    %extrapolated result
    optprice = (bincoeff*sumr')*(2) ^(-extraterms);

    function [lt] = ltbs(spot, strike, sg, rf, gamma)
    m = 2 * rf / (sg * sg); a = (1 - m) / 2; b = a ^2 + m;
    z = log(spot); k = log(strike);
    %%%FORMULA 14: NUMERATOR
    term0 = (b+gamma).^0.5;
    %'the numerator
    if spot >strike
     term1 = term0;
    else
     term1 = -term0;
    end
    term1 = a - 1+term1;
    term1 = m*term1;
    num = gamma-term1;
    %'the denominator
    den = 2.*gamma .* term0.*(m+gamma);
    %'the exponential term
    term2 = exp(k*(1-a)-term0*abs(z-k));
    result = term2.*num./ den;
    if spot > strike
    %'exp(-(a-1)*z)/gamma
     cterm1 = exp(-(a - 1) * z)./gamma;
    %'exp(-a*z+k)/(gamma+m)
     cterm2 = exp(-a * z + k)./(gamma +m);
    %'A1*h1
     result = cterm1-cterm2+result;
    end
    lt = exp(a * z).*result;
```

Table 7.5. Pricing of a call option: analytical Black–Scholes (3rd column) and numerical inversion of the Laplace transform (4th and 5th columns). Parameters: strike $= 100, r = 0.05$, $\sigma = 0.2$

Expiry	Spot	BS	$A = 18.4, n = 15, m = 10$	$A = 18.4, n = 50, m = 10$
0.001	90	0.00000	0.00000	0.00000
0.001	100	0.254814	0.254814	0.254814
0.001	110	10.00500	10.00500	10.00500
0.5	90	2.349428	2.349428	2.349428
0.5	100	6.888729	6.888729	6.888729
0.5	110	14.075384	14.075384	14.075384
1	90	5.091222	5.091222	5.091222
1	100	10.450584	10.450584	10.450584
1	110	17.662954	17.662954	17.662954
5	90	21.667727	21.667727	21.667727
5	100	29.13862	29.13862	29.13862
5	110	37.269127	37.269128	37.269128
20	90	57.235426	57.235426	57.235427
20	100	66.575748	66.575748	66.575749
20	110	76.048090	76.048090	76.048091
		m.s.e.	0.00000147	0.00000203

In Table 7.5 we report the exact Black–Scholes price and the one obtained by numerical inversion. The numbers in Table 7.5 can be obtained running the MATLAB® module `main`.

7.4.2 Example

As a second example, we consider the use of the Laplace transform with respect to the strike and not with respect to the time to maturity. This different approach is possible when the moment generating function (m.g.f.) of the underlying variable is known in closed form. The m.g.f. of a random variable Z is defined as $\mathbb{E}_0[e^{-\gamma Z}]$. In particular, if Z is a non-negative r.v. and admits density $f_Z(z)$, we have

$$\mathbb{E}_0[e^{-\gamma Z}] = \int_0^{+\infty} e^{-\gamma z} f_Z(z)\, dz,$$

and hence the interpretation of the m.g.f. as Laplace transform of the density function. Notice that the existence of the m.g.f. is not always guaranteed because it is required that the m.g.f. is defined in a complete neighborhood of the origin. For example, this is not the case when Z is lognormal.

If the m.g.f. of the random variable Z is known, we can also obtain the Laplace transform of a call option written on $Z(t)$. Let us consider a contingent claim with payoff given by $\alpha(Z(t) - Y)_+$, where α and Y are constants. By no-arbitrage arguments, the option price is:

$$C\big(Z(0), t, Y\big) = \alpha e^{-rt} \int_0^{+\infty} (z - Y)^+ f_Z(z)\, dz$$

$$= \alpha e^{-rt} \int_Y^{+\infty} (z - Y) f_Z(z)\, dz, \tag{7.15}$$

where $f_Z(z)$ is the risk-neutral density of $Z(t)$. Let us define the Laplace transform wrt Z of the above price

$$c\big(Z(0), t; \gamma\big) = \mathcal{L}\big[C\big(Z(0), t, Y\big)\big] = \int_0^{+\infty} e^{-\gamma Y} C\big(Z(0), t, Y\big)\, dY.$$

Replacing (7.15) in this formula and using a change of integration, we get

$$c\big(Z(0), t; \gamma\big) = \alpha e^{-rt} \int_0^{+\infty} e^{-\gamma Y} \int_x^{+\infty} (z - Y) f_Z(z)\, dz\, dY$$

$$= \alpha e^{-rt} \int_0^{+\infty} \left(\int_0^z e^{-\gamma Y} (z - Y)\, dY \right) f_Z(z)\, dz$$

$$= \alpha e^{-rt} \int_0^{+\infty} \left(\int_0^z \big(z e^{-\gamma Y} - Y e^{-\gamma Y}\big)\, dY \right) f_Z(z)\, dz$$

$$= \alpha e^{-rt} \int_0^{+\infty} \frac{e^{-\gamma z} + \gamma z - 1}{\gamma^2} f_Z(z)\, dz$$

$$= \alpha e^{-rt} \left(\frac{\mathbb{E}_0[e^{-\gamma Z(t)}]}{\gamma^2} + \frac{\mathbb{E}_0[Z(t)]}{\gamma} - \frac{1}{\gamma^2} \right).$$

Using the fact that the Laplace inverse of $1/\gamma$ is 1 and the Laplace inverse of $1/\gamma^2$ is Y, we can write the option price as follows

$$C\big(Z(0), t, Y\big) = \alpha e^{-rt} \left(\mathcal{L}^{-1}\left(\frac{\mathbb{E}_0[e^{-\gamma Z_t}]}{\gamma^2} \right) + \mathbb{E}_0[Z_t] - Y \right), \tag{7.16}$$

and the pricing problem is reduced to the numerical inversion of $\mathbb{E}_0[e^{-\gamma Z_t}]/\gamma^2$.

As a concrete example, let us consider the square root process

$$dX(t) = rX(t)\, dt + \sigma\sqrt{X(t)}\, dW(t),$$

and our aim is to price a fixed strike Asian call option, having payoff

$$\frac{1}{t}\left(\int_0^t X(u)\, du - Kt \right)_+.$$

In order to obtain the price of the Asian option, we compute the moment generating function of $\int_0^t X(u)\, du$:

$$v\big(X(0), t; \gamma\big) = \mathbb{E}_0\big[e^{-\gamma \int_0^t X(u)\, du}\big]. \tag{7.17}$$

By the Feynman–Kac theorem, $v(X(0), t; \gamma)$ is the solution of the PDE:

$$-\partial_t v + rx\,\partial_x v + \frac{1}{2}\sigma^2 x\,\partial_{xx} v = \gamma x v$$

with initial condition

$$v\big(X(0), 0; \gamma\big) = 1.$$

To solve this PDE, we exploit the linearity of the drift and variance coefficients and, following Ingersoll (1986), pp. 397–398, we consider a solution of the type:

$$v(X, t;\ \gamma) = e^{-A(t;\ \gamma)X - B(t;\gamma)}.$$

Replacing this function in the PDE, it is then easy to show that $B(t; \gamma) = 0$ and

$$A(t; \gamma) = \frac{2\gamma(\exp(t\lambda) - 1)}{\lambda + r + (\lambda - r)\exp(t\lambda)}, \tag{7.18}$$

where $\lambda = \sqrt{r^2 + 2\gamma\sigma^2}$. Therefore, using (7.16), we can write the price of the Asian option as

$$\alpha e^{-rt}\left(\mathcal{L}^{-1}\left(\frac{e^{-A(t;\gamma)X - B(t;\gamma)}}{\gamma^2}\right) + \mathbb{E}_0\left[\int_0^t X(u)\,du\right] - X\right),$$

where $\mathcal{L}^{-1}$ is the Laplace inverse. In particular, we have:

$$\begin{aligned}
\mathbb{E}_0\left[\int_0^t X(u)\,du\right] &= \int_0^t \mathbb{E}_0[X(u)]\,du \\
&= \int_0^t X(0)e^{ru}\,du \\
&= X(0)\frac{e^{rt} - 1}{r}.
\end{aligned}$$

Numerical inversion

Table 7.6 provides some numerical example. In the numerical inversion of the Laplace transform we have used $A = 18.4$, and the Euler algorithm has been applied using a total of $20 + 10$ terms.

Table 7.6. Prices of an Asian option in the square-root model

K	$\sigma = 0.1$	$\sigma = 0.3$	$\sigma = 0.5$
0.9	0.137345	0.15384	0.18691
0.95	0.09294	0.12001	0.15821
1	0.05258	0.09075	0.13253
1.05	0.02268	0.06640	0.10987
1.1	0.00687	0.04696	0.09016

These examples have been obtained writing the Matlab® functions

```
AWSR(spot, strike, expiry, sg, rf, aa, terms,
extraterms)
ltsr(spot, expiry, sg, rf, gamma)
```

The function AWSR performs the numerical inversion of the Laplace transform according to the Abate–Whitt algorithm. The function ltsr returns the quantity $v(X, t; \gamma)/\gamma^2$. The complete code is given here below.

```
function [optprice]=AWSR(spot, strike, expiry, sg, rf,
aa, terms, extraterms)
    X = strike*expiry;
    sum = 0;
    %%compute the LT at gamma = aa / (2 * strike)
    lt = ltsr(spot , expiry , sg , rf , aa / (2 * X) );
    sum = lt* exp(aa / 2) / (2 * strike);
    %apply the Euler algorithm
    k = [1:terms + extraterms];
    arg = aa / (2 * X)+i*pi.*k / X;
    term = ((-1) .^k) .* ltsr(spot, expiry, sg, rf, arg)
* exp(aa / 2) / X;
    csum = sum+cumsum(term);
    sumr = real(csum(terms:terms+extraterms));
    j = [0:extraterms];
    bincoeff = gamma(extraterms+1)./(gamma(j+1).
* gamma(extraterms-j+1));
    %extrapolated result
    euler = (bincoeff*sumr')*(2) ^(-extraterms);
    %apply the final formula
    optprice = exp(-rf*expiry)*(euler+spot*(exp(rf*expiry)
-1)/rf - X)/expiry;

    function [lt] = ltsr(spot, expiry, sg, rf, gamma)
    lambda = sqrt(rf^2+2*gamma*sg*sg);
    numerator = 2*gamma.*(exp(expiry.*lambda)-1);
    denominator = lambda+rf+(lambda-rf).*exp(expiry.
*lambda);
    lt = exp(-spot*numerator./denominator)./gamma.^2;
```

7.5 Comments

A good introduction to the Laplace transform topic can be found in Dyke (1999), whilst a classical but more advanced treatment is Doetsch (1970). Extensive tables for analytical inversion of the Laplace transform are available: see for example Abramowitz and Stegun (1965). Davies and Martin (1970) provide a review and a comparison of some numerical inversion available through 1979. More recently Duffy (1993) compares three popular methods to numerically invert the Laplace transform. The methods examined in Duffy are (a) the Crump inversion method, Crump (1970); (b) the Weeks method that integrates the Bromwich's integral by using Laguerre polynomials, Weeks (1966); (c) the Talbot method that deforms the Bromwich's contour so that it begins and ends in the third and second quadrant of the γ-plane, Talbot (1979). If the locations of the singularities are known, these schemes may provide accurate results at minimal computational expense. However, the user must provide a numerical value for some parameters and therefore an automatic inversion procedure is not possible. At this regard, a recent paper by Weideman (1999) seems to give more insights about the choice of the free parameters. Another simple algorithm to invert Laplace transforms is given in Den Iseger (2006). In general this algorithm outperforms the Abate–Whitt algorithm in stability and accuracy. The strength of the Den Iseger algorithm is the fact that in essence it boils down to an application of the discrete FFT algorithm. However, the Den Iseger algorithm may also perform unsatisfactorily when the function or its derivative has discontinuities. Other interesting numerical inversion algorithms can be found in Abate, Choudhury and Whitt (1996), Garbow et al. (1988a, 1988b). Finally, we mention the often quoted Gaver–Stehefest algorithm, Gaver, Jr. (1966) and Stehfest (1970), a relatively simple numerical inversion method using only values of the Laplace transform on the real axis but requiring high precision.[4]

The numerical inversion of multidimensional Laplace transforms is studied in Abate, Choudhury and Whitt (1998), Choudhury, Lucantoni and Whitt (1994), Singhal and Vlach (1975), Singhal, Vlach and Vlach (1975), Vlach and Singhal (1993), Chpt. 10, Moorthy (1995a, 1995b). Among the others, papers that discuss the instability of the numerical inversion are Bellman, Kalaba and Lockett (1966), Platzman, Ammons and Bartholdi (1988), Kwok and Barthez (1989), Craig and Thompson (1994). An useful source for the solution of ordinary differential equations is Ince (1964).

Selby (1983) and Buser (1986) have introduced the Laplace transform in finance. Useful references are Shimko (1991) and Fusai (2001), that have lots of examples on which to practice. Laplace transform has been used in finance for pricing (a) barrier options, Geman and Yor (1996), Pelsser (2000), and Sbuelz (1999, 2005), Davydov and Linetsky (2001a, 2001b); (b) interest rate derivatives, Leblanc and Scaillet (1998) and Cathcart (1998); (c) Asian options, Geman and Yor (1993), Geman and Eydeland (1995), Fu, Madan and Wang (1998), Lipton (1999), and Fusai (2004); (d) other exotic options (corridor, quantile, parisian and step options), Akahori (1995), Ballotta (2001), Ballotta and Kyprianou (2001), Chesney et al. (1995), Chesney et al. (1997), Dassios (1995), Hugonnier (1999), Linetsky (1999), Fusai (2000), Fusai and Tagliani (2001); (e) credit risk, Di Graziano and Rogers (2005); (f) options on hedge funds, Atlan, Geman and Yor (2005). A review can be found in Craddock, Heath and Platen (2000). Useful formulae related to the Laplace trans-

---

[4] A Matlab® implementation can be found at http://www.mathworks.com/matlabcentral/fileexchange/loadFile.do?objectId=9987&objectType=file

form of the hitting time distribution and to exponential functionals of the Brownian motion can be found in Yor (1991), Rogers (2000), Borodin and Salminen (2002), Salminen and Wallin (2005).

Finally, we mention the web page mantained by Valko,[5] a useful reference for finding the most important algorithms for the numerical inversion of the Laplace transform.

---

[5] http://pumpjack.tamu.edu/valko/public_html/Nil/index.html

8

Structuring Dependence using Copula Functions[*]

The Latin word "copula" denotes linking or connecting between parts. This word has been adopted in statistics to denote a class of functions allowing to build cross-dependent multivariate distributions. Although the terms "correlation" and "dependence" are often used interchangeably, the former is a rather particular kind of dependence measure between random variables. As such, it suffers from inconveniences due to its limitation in capturing other forms of dependence. For instance, it is not difficult to find examples of dependent variables displaying zero correlation. The problem of modeling dependence structures is that this feature does not always show out of the joint distribution function under consideration. It would be of some help to separate the statistical properties of each variable from their dependence structure. Copula functions provide us with a viable way to achieve this goal.

This chapter is organized as follows. Section 8.1 introduces the notion of copula and related definitions. Section 8.2 presents an overview of major concepts of dependence and examines their link to copulas. Sections 8.3 and 8.4 exhibit the most important families of copulas together with their properties. Section 8.5 is devoted to the statistical inference of copula functions. Section 8.6 discusses Monte Carlo simulation techniques. Section 8.7 concludes with a few remarks and comments.

8.1 Copula Functions

A copula is a mathematical function C representing a joint distribution F as a function of the corresponding marginal distributions $F_j, j = 1, \ldots, n$, i.e., $F(x_1, \ldots, x_n) = C(F_1(x_1), \ldots, F_n(x_n))$. We begin with a definition and then present a theoretical result stating the claimed property.

Definition (Copula function) An n-copula (function) is a real-valued function C from the unit cube $[0, 1]^n$ onto the unit interval $[0, 1]$, such that:

(1) *Groundness*: $C(\mathbf{u}) = 0$ if at least one coordinate u_j is zero;

---

[*] with Davide Meneguzzo.

(2) *Reflectiveness*: $C((1, \ldots, 1, u_j, 1, \ldots, 1)) = u_j$;
(3) *N-increasing property*: For all $\mathbf{u}^1 = (u_1^1, \ldots, u_n^1)$ and $\mathbf{u}^2 = (u_1^2, \ldots, u_n^2)$ in $[0, 1]^n$ with $\mathbf{u}^1 \leq \mathbf{u}^2$, i.e., $u_i^1 \leq u_i^2$ for all i, the C-volume of the hypercube with corners $\mathbf{u}^1$ and $\mathbf{u}^2$ is positive, i.e.,[1]

$$\sum_{i_1=1,2} \cdots \sum_{i_n=1,2} (-1)^{i_1+i_2+\cdots+i_n} C\left(u_1^{i_1}, \ldots, u_n^{i_n}\right) \geq 0.$$

It is easy to see that these properties follow from the fact that the composite function $C \circ (F_1, \ldots, F_n)$ is a cumulative distribution function (c.d.f.); moreover, any copula function is itself the joint c.d.f. of n standard uniform random variables (r.v.'s) $U_1, \ldots, U_n$, i.e., $C(u_1, \ldots, u_n) = \mathbb{P}(U_1 \leq u_1, \ldots, U_n \leq u_n)$.

The following theorem is a key result. It states a relation between distributions and copulas.

Theorem (Sklar, 1959) *Let F be an n-dimensional c.d.f. with marginal c.d.f.'s $F_1, \ldots, F_n$. Then, there exists a function $C : \mathbb{R}^n \to [0, 1]$ such that for all y in $\mathbb{R}^n$:*

$$F(y_1, \ldots, y_n) = C\left(F_1(y_1), \ldots, F_n(y_n)\right).$$

If all margins $F_1, \ldots, F_n$ are continuous, then C is unique. Conversely, if C is a n-copula and $F_1, \ldots, F_n$ are c.d.f.'s, then the function defined by $F(y_1, \ldots, y_n) = C(F_1(y_1), \ldots, F_n(y_n))$ is a joint n-dimensional c.d.f. with marginals c.d.f.'s $F_1, \ldots, F_n$.

This theorem elucidates the main idea of dependence modelling using copula functions: the statistical properties of margins and their association structure can be disentangled, so that one may model one independently of the other.

An immediate corollary of Sklar's theorem states that the copula corresponding to an n-dimensional distribution function F with continuous margins $F_1, \ldots, F_n$ can be computed from these functions as

$$C(u_1, \ldots, u_n) = F\left(F_1^{-1}(u_1), \ldots, F_n^{-1}(u_n)\right), \tag{8.1}$$

where F_i^{-1} is the quasi-inverse function of F_i as defined in subsection "Transformation Methods" in chapter "Static Monte Carlo".

---

[1] The case $n = 2$ is self-evident: the C-volume is the area

$$\begin{aligned}
V_C &= (-1)^{1+1} C(u_1^1, u_2^1) + (-1)^{1+2} C(u_1^1, u_2^2) + (-1)^{2+1} C(u_1^2, u_2^1) \\
&\quad + (-1)^{2+2} C(u_1^2, u_2^2) \\
&= C(u_1^1, u_2^1) - C(u_1^1, u_2^2) - C(u_1^2, u_2^1) + C(u_1^2, u_2^2).
\end{aligned}$$

This is exactly the probability that a bivariate random variable with distribution function C assumes a value in the rectangle with vertices (u_1^1, u_2^1), (u_1^1, u_2^2), (u_1^2, u_2^1) and (u_1^2, u_2^2).

8.2 Concordance and Dependence

This section discusses the relationship between copula functions and *association measures* for pairs of random variables. These notions allow one to model the statistical dependence between prices, indices and other financial quantities. It is worth pointing out that most notions involving statistical distributions can be formulated in terms of the corresponding copulas.

8.2.1 Fréchet–Hoeffding Bounds

It is clear that the copula function of mutually independent variables is given by the following function defined in $[0, 1]^n$:

$$C^{\perp}(u_1, \ldots, u_n) = \prod_{j=1}^{n} u_j.$$

This is known as the *product copula*.

What about the extreme cases of perfect positive and negative dependence? The answer to this question is linked to the Fréchet–Hoeffding theorem stating upper and lower bounds for copula functions.

Definition A copula C_1 is *smaller* than a copula C_2, and we write $C_1 \prec C_2$, if:
1. $C_1(\mathbf{u}) \leq C_2(\mathbf{u})$, and
2. $\overline{C}_1(\mathbf{u}) \leq \overline{C}_2(\mathbf{u})$,
   for all $\mathbf{u}$ in $[0, 1]^n$, where $\overline{C}$ denotes the *joint survival function*

$$\overline{C}(u_1, \ldots, u_n) = \mathbb{P}(U_1 > u_1, \ldots, U_n > u_n). \tag{8.2}$$

Theorem (Fréchet–Hoeffding, 1957) *If C is an n-copula, then*

$$C^- \prec C \prec C^+,$$

where

$$C^-(u_1, \ldots, u_n) = \max\left(\sum_{j=1}^{n} u_j - n + 1, 0\right),$$

$$C^+(u_1, \ldots, u_n) = \min(u_1, \ldots, u_n).$$

Functions $C^{\perp}$ and C^+ are n-copulas for all $n \geq 2$, whereas C^- is not a copula for $n \geq 3$. To see this, we compute the volume of the n-cube $[1/2, 1]^n$ using the lower copula bound:

$$\max(1 + \cdots + 1 - n + 1, 0) - n \max(1/2 + 1 + \cdots + 1 - n + 1, 0)$$
$$+ \binom{n}{2} \max(1/2 + 1/2 + 1 + \cdots + 1 - n + 1, 0)$$
$$+ \max(1/2 + \cdots + \cdots + 1/2 - n + 1, 0)$$
$$= 1 - n/2 + 0 + \cdots + 0.$$

This number is negative for $n \geq 3$, hence C^- is not a copula according to the definition above. However, C^- is the best possible lower bound in the sense that for any $n \geq 3$ and any u in $[0, 1]^n$, there exists an n-copula C such that $C(\mathbf{u}) = C^-(\mathbf{u})$.

The Fréchet–Hoeffding theorem has a clear interpretation in the bivariate case: two random variables X_1 and X_2 exhibit copula C^- (resp. C^+) if and only if they are discordant (resp. concordant) functions of a common r.v. Y, i.e., $X_1 = f_1(Y)$ and $X_2 = f_2(Y)$ with f_1 increasing and f_2 decreasing (resp. both f_1 and f_2 increasing).

8.2.2 Measures of Concordance

Roughly speaking, two random variables are said to be concordant if large values of one come with large values of the other and small values of one come with small values of the other. Two samples (x_i, y_i) and (x_j, y_j) of a continuous random vector (X, Y) are concordant if $(x_i - x_j)(y_i - y_j) > 0$. This means that either $x_i < x_j$ and $y_i < y_j$ or $x_i > x_j$ and $y_i > y_j$. Since dependence between variables X_1 and X_2 is driven by their copula C, we interchangeably use symbols $\kappa_{X_1 X_2}$ and κ_C to denote a measure of concordance as defined as follows:

Definition (Measure of concordance) A function κ defined on pairs of random variables is a *measure of concordance* if for every pair X_1 and X_2 of r.v.'s it satisfies the following properties:

(1) *Completeness*: κ_{X_1, X_2} is well defined;
(2) *Normality*: $-1 \leq \kappa_{X_1, X_2} \leq 1$;
(3) *Symmetry*: $\kappa_{X_1, X_2} = \kappa_{X_2, X_1}$;
(4) *Weak nullity*: $\kappa_{X_1, X_2} = \kappa_{C^\perp} = 0$ if X_1 and X_2 are statistically independent;
(5) *Specularity*: $\kappa_{-X_1, X_2} = \kappa_{X_1, -X_2} = -\kappa_{X_1, X_2}$;
(6) *Monotony*: for all pairs of copulas such that $C_1 \prec C_2$, then $\kappa_{C_1} \leq \kappa_{C_2}$;
(7) *Continuity*: if $(X_{1,n}, X_{2,n})$ has copula C_n and $(C_n)_{n \geq 1}$ converges pointwise to C, then $\lim_{n \to \infty} \kappa_{C_n} = \kappa_C$.

It can be shown that κ is invariant under any strictly increasing transformation T of the underlying r.v.'s:

$$\kappa(T(X_1), T(X_2)) = \kappa(X_1, X_2).$$

Kendall's tau and the Spearman's rho are measures of concordance playing an important role within nonparametric statistics. They provide us with viable alternatives to the linear correlation coefficient as a measure of association for nonelliptical distributions. (As will be shown below, in this case the linear correlation coefficient turns out to be incapable of properly representing the notion of dependence.) We define these measures through their copulas as follows:

1. *Kendall's tau*:

$$\tau = 4 \iint_{[0,1]^2} C(u_1, u_2) \, dC(u_1, u_2) - 1$$

$$= 1 - 4 \iint_{[0,1]^2} \frac{\partial C(u_1, u_2)}{\partial u_1} \frac{\partial C(u_1, u_2)}{\partial u_2} \, du_1 \, du_2.$$

It can be shown that Kendall's tau matches the difference between the probability of concordance and the probability of discordance for two independent random vectors (X_1, Y_1) and (X_2, Y_2) sharing a common joint distribution function F and copula C:

$$\tau = \mathbb{P}\big((X_1 - X_2)(Y_1 - Y_2) > 0\big) - \mathbb{P}\big((X_1 - X_2)(Y_1 - Y_2) < 0\big).$$

2. *Spearman's rho*:

$$\varrho = 12 \iint_{[0,1]^2} C(u_1, u_2) \, du_1 \, du_2 - 3$$

$$= 12 \iint_{[0,1]^2} u_1 u_2 \, dC(u_1, u_2) - 3.$$

Spearman's rho can also be expressed in terms of probabilities of concordance and discordance. More precisely, given three i.i.d. vectors (X_1, Y_1), (X_2, Y_2) and (X_3, Y_3) sharing a common copula C, we have

$$\varrho = 3\big[\mathbb{P}\big((X_1 - X_2)(Y_1 - Y_3) > 0\big) - \mathbb{P}\big((X_1 - X_2)(Y_1 - Y_3) < 0\big)\big].$$

Example Independent variables feature the product copula $C^{\perp}$. Since the partial derivative $\partial C^{\perp}/\partial u_1 \, \partial u_2 = 1$, we have

$$\tau_{C^{\perp}} = 1 - 4 \iint_{[0,1]^2} u_1 u_2 \, du_1 \, du_2 = 0,$$

$$\varrho_{C^{\perp}} = 12 \iint_{[0,1]^2} u_1 u_2 \, du_1 \, du_2 - 3 = 0.$$

Kendall's tau and Spearman's rho are linked through a well-established relationship. The attainable region for the two concordance measures is defined as the set of pairs (τ, ϱ) that are compatible with this relation. For positive values of τ, ϱ lies in the interval $0.5[3\tau - 1, 1 + 2\tau - \tau^2]$; for negative values of τ, ϱ lies in the interval $0.5[\tau^2 + 2\tau - 1, 1 + 3\tau]$. Figure 8.1 exhibits a graph of the resulting region in the $(\tau - \varrho)$-Cartesian system.

8.2.3 Measures of Dependence

In the previous section, we have introduced Kendall's tau and Spearman's rho as measures of concordance. A major drawback of this notion is that, according to Property (4) in the definition of concordance measure, statistical independence is not implied by a vanishing concordance measure. In this section, we introduce the notion of *measure of dependence* as a way to overcome this difficulty. A measure of dependence provides us with an indication about the extent to which two variables are mutually related. The key idea is to assess a distance between the copula associated to a given pair of random variables and the product copula $C^{\perp}$. The next definition gathers a minimal number of properties for a function of two random variables to be a measure of dependence.

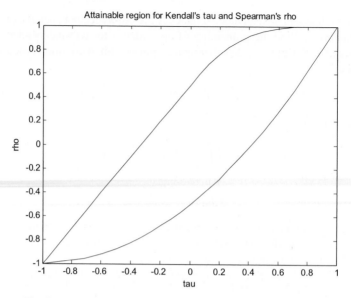

Fig. 8.1. Attainable region for Kendall's tau and Spearman's rho.

Definition (Measure of dependence) A function δ defined on pairs of random variables is a *measure of dependence* if it satisfies the following properties:

(1) *Completeness, symmetry, specularity, monotony* and *continuity*, as in the definition of concordance;

(2) *Normality* and *positiveness*: $0 = \delta_{C^\perp} \leq \delta_C \leq \delta_{C^+} = 1$;

(3) *Strong nullity*: $\kappa_{X_1,X_2} = \kappa_{C^\perp} = 0$ if, and only if, X_1 and X_2 are statistically independent.

The most popular dependence measures are defined as follows:

1. *Hoeffding's phi*:

$$\phi = 3\sqrt{10 \iint_{[0,1]^2} \left(C(u_1, u_2) - u_1 u_2\right)^2 du_1 \, du_2}.$$

2. *Schweitzer–Wolff's sigma*:

$$\sigma = 12 \iint_{[0,1]^2} |C(u_1, u_2) - u_1 u_2| \, du_1 \, du_2.$$

8.2.4 Comparison with the Linear Correlation

Pearson's linear correlation coefficient ρ between two random variables X_1 and X_2 is defined by:

$$\rho(X_1, X_2) = \frac{\text{Cov}(X_1, X_2)}{\sqrt{\text{Var}(X_1) \, \text{Var}(X_2)}}.$$

This is not a measure of dependence. In particular:

(1) A null ρ is compatible with some structure of dependence between variables.

Example The *cubic copula* is defined as

$$C(u_1, u_2) = u_1 u_2 + \alpha[u_1(u_1 - 1)(2u_1 - 1)][u_2(u_2 - 1)(2u_2 - 1)],$$

with $\alpha \in [-1, 2]$. This copula differs from the product copula. Consequently, the two variables exhibit a dependence structure. However, the corresponding linear correlation coefficient is zero:

$$\text{Cov}(X_1, X_2) = \iint_{[0,1]^2} \left(C(u_1, u_2) - u_1 u_2 \right) du_1\, du_2$$

$$= \alpha \int_{[0,1]} [u_1(u_1 - 1)(2u_1 - 1)]\, du_1 \int_{[0,1]} [u_2(u_2 - 1)(2u_2 - 1)]\, du_2$$

$$= \alpha \left(\frac{1}{2}u_1^4 - u_1^3 + \frac{1}{2}u_1^2 \right) \Big|_0^1 \times \left(\frac{1}{2}u_2^4 - u_2^3 + \frac{1}{2}u_2^2 \right) \Big|_0^1 = 0.$$

It is interesting to notice that the Hoeffding's phi detects the aforementioned dependence between X_1 and X_2:

$$\phi = 3\sqrt{10 \iint_{[0,1]^2} \left(C(u_1, u_2) - u_1 u_2 \right)^2 du_1\, du_2}$$

$$= 3\sqrt{10 \iint_{[0,1]^2} \left(2u_1^3 - 3u_1^2 + u_1 \right)^2 \left(2u_2^3 - 3u_2^2 + u_2 \right)^2 du_1\, du_2}$$

$$= 3\sqrt{10 \int_{[0,1]} \left[4u_1^6 - 12u_1^5 + 13u_1^4 - 6u_1^3 + u_1^2 \right] du_1}$$

$$\times \sqrt{\int_{[0,1]} \left[4u_2^6 - 12u_2^5 + 13u_2^4 - 6u_2^3 + u_2^2 \right] du_2} = 0.0451.$$

(2) ρ needs not span the whole interval $[-1, 1]$. We may show that for a suitable pair of margins, the linear correlation ranges over a strict subinterval of $[-1, 1]$ across all possible copulas.

Example Wang (1999) shows that the minimum and maximum values attained by the linear correlation of lognormal variables $X_1 \sim \mathcal{LN}(\mu_1, \sigma_1)$ and $X_2 \sim \mathcal{LN}(\mu_2, \sigma_2)$ are given by

$$\rho_- = \frac{e^{-\sigma_1 \sigma_2} - 1}{(e^{\sigma_1^2} - 1)^{1/2}(e^{\sigma_2^2} - 1)^{1/2}} \leq 0,$$

$$\rho_+ = \frac{e^{\sigma_1 \sigma_2} - 1}{(e^{\sigma_1^2} - 1)^{1/2}(e^{\sigma_2^2} - 1)^{1/2}} \geq 0.$$

If we set $\sigma_1 = 1$ and $\sigma_2 = 3$, the correlation coefficient ranges over the interval $[-0.008, 0.16]$.

(3) A null ρ is compatible with perfect dependence between variables.

Example Consider the following copula:

$$
C = \begin{cases}
u_1 & \text{if } 0 \le u_1 \le \frac{1}{2}u_2 \le \frac{1}{2}, \\
\frac{1}{2}u_2 & \text{if } 0 \le \frac{1}{2}u_2 < u_1 \le 1 - \frac{1}{2}u_2, \\
u_1 + u_2 - 1 & \text{if } \frac{1}{2} \le 1 - \frac{1}{2}u_2 \le u_1 \le 1.
\end{cases}
$$

We have $\text{Cov}(U_1, U_2) = 0$, but $\mathbb{P}\{U_2 = 1 - |2U_1 - 1|\} = 1$, meaning that one variable can be perfectly forecasted from the knowledge of the other.

8.2.5 Other Notions of Dependence

Copula functions allow one to control tail dependence. This quantity measures the extent of dependence between r.v.'s arising from extreme observations. From a geometric perspective, tail dependence represents the concentration on the lower and upper quadrant tail of the joint distribution function of two r.v.'s X_1 and X_2. More formally:

Definition (Upper and lower tail dependence) Given r.v.'s X_1 and X_2 with marginal distributions F_1 and F_2, the upper and lower tail dependence numbers are defined as

$$
\lambda_U = \lim_{u \uparrow 1} \mathbb{P}\big[X_2 > F_2^{-1}(u) \big| X_1 > F_1^{-1}(u)\big],
$$

$$
\lambda_L = \lim_{u \downarrow 0} \mathbb{P}\big[X_2 \le F_2^{-1}(u) \big| X_1 \le F_1^{-1}(u)\big].
$$

If λ_U (resp. λ_L) is positive, then the two random variables are said to be asymptotically dependent in the upper (resp. lower) tail; if, instead, λ_U (resp. λ_L) vanishes, then they are said to be asymptotically independent. This definition can be recast in terms of the copula function associated with the two variables:

$$
\lambda_U = \lim_{u \uparrow 1} \frac{1 - 2u + C(u, u)}{1 - u},
$$

$$
\lambda_L = \lim_{u \downarrow 0} \frac{C(u, u)}{u}.
$$

Example Consider the bivariate Clayton copula $C(u, u)$ with $\alpha > 0$ as defined in Sect. 8.7. The lower tail is given by:

$$
\begin{aligned}
\lambda_L^{\text{Clayton}} &= \lim_{u \downarrow 0} \frac{C(u, u)}{u} \\
&= \lim_{u \downarrow 0} \frac{(2u^{-\alpha} - 1)^{-1/\alpha}}{u} \\
&= \lim_{u \downarrow 0} -\frac{1}{\alpha}(2u^{-\alpha} - 1)^{-(1+\alpha)/\alpha}\left(-2\alpha u^{-(1+\alpha)}\right)
\end{aligned}
$$

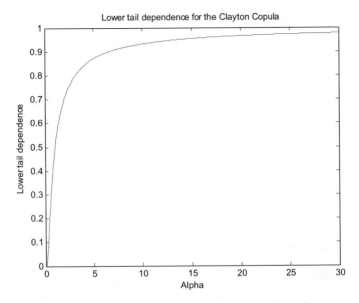

Fig. 8.2. Lower tail dependence of the Clayton copula varying α.

$$= \lim_{u \downarrow 0} 2\left(2u^{-\alpha} - 1\right)^{-(1+\alpha)/\alpha} \left(u^{-\frac{(1+\alpha)}{\alpha}\alpha}\right)$$

$$= \lim_{u \downarrow 0} 2\left(u^{\alpha}\left(2u^{-\alpha} - 1\right)\right)^{-(1+\alpha)/\alpha}$$

$$= \lim_{u \downarrow 0} 2\left(2 - u^{\alpha}\right)^{-(1+\alpha)/\alpha} = 2^{-1/\alpha} > 0.$$

Figure 8.2 shows the behavior of the lower tail dependence λ_L across different values for the parameter α.

The ability of copula functions to describe the way two random variables jointly behave under tail events is useful for the purpose of examining the joint behavior of excessive negative returns of financial time series. An important notion in this respect is the one of *positive quadrant dependence* proposed by Lehmann (1966). Roughly speaking, two r.v.'s are positive quadrant dependent if the probability of being simultaneously small is an upper bound for the probability of the same event under the assumption that the two r.v.'s are statistically independent. More precisely:

Definition (Positive quadrant dependence) Two random variables X_1 and X_2 are said to be positive quadrant dependent (PQD) if

$$\mathbb{P}(X_1 \leq x_1, X_2 \leq x_2) \geq \mathbb{P}(X_1 \leq x_1) \times \mathbb{P}(X_2 \leq x_2),$$

for all $(x_1, x_2) \in \mathbb{R}^2$.

The PQD inequality may be easily rewritten in terms of copulas: X_1 and X_2 are PQD provided that

$$C(u_1, u_2) \geq u_1 u_2,$$

for all $(u_1, u_2) \in [0, 1]^2$. By applying Bayes' rule, the PQD inequality may be rewritten as

$$\mathbb{P}(X_1 \leq x_1 | X_2 \leq x_2) \geq \mathbb{P}(X_1 \leq x_1).$$

A typical restriction arising in financial applications requires that this conditional probability be a non-increasing function of x_2. This can be interpreted as saying that the probability a price or index return X_1 assumes a small value does not rise with the increase of another price or index return X_2.

Definition (Left tail decreasing) We say that a random variable X_1 is left tail decreasing (LTD) in X_2, if the conditional distribution function $\mathbb{P}(X_1 \leq x_1 | X_2 \leq x_2)$ is a non-decreasing function of x_2, for all $x_1 \in \mathbb{R}$. This is equivalent to requiring that the ratio $C(u_1, u_2)/u_2$ be non-decreasing in u_2 for all $u_2 \in [0, 1]$.

The cumulative distribution function of both the minimum and the maximum of n random variables $X_1, \ldots, X_n$ with c.d.f.'s $F_1, \ldots, F_n$ can be expressed in terms of their copula C. Let $X_1, \ldots, X_n$ be r.v.'s with a common range J and set $m = \min(X_1, X_2, \ldots, X_n)$ and $M = \max(X_1, X_2, \ldots, X_n)$. Since $F_M(a) = \mathbb{P}(M \leq a) = \mathbb{P}(X_1 \leq a, X_2 \leq a, \ldots, X_n \leq a) = F(a, a, \ldots, a)$, we have:

$$F_{\max(X_1, X_2, \ldots, X_n)}(a) = C\big(F_1(a), F_2(a), \ldots, F_n(a)\big).$$

Moreover, $F_m(a) = \mathbb{P}(m \leq a) = 1 - \mathbb{P}(m > a) = 1 - \mathbb{P}(X_1 > a, X_2 > a, \ldots, X_n > a)$, so that

$$F_{\min(X_1, X_2, \ldots, X_n)}(a) = 1 - \overline{C}\big(F_1(a), F_2(a), \ldots, F_n(a)\big),$$

where $a \in J$ and $\overline{C}$ is the *survival copula* defined in formula (8.2).

8.3 Elliptical Copula Functions

Elliptical distributions share several properties with the multivariate normal distribution. Elliptical copulas are the corresponding copula functions.

Definition (Elliptical distributions) The n-dimensional random vector $\mathbf{Y}$ has an elliptical distribution with parameters μ, Σ and ϕ if there exists a vector $\mu \in \mathbb{R}^n$ and an $n \times n$ positive definite symmetric matrix Σ such that the characteristic function of $\mathbf{Y} - \mu$ is written as a function of the quadratic form $\mathbf{t}^{\top} \Sigma \mathbf{t}$, i.e.,

$$\varphi_{\mathbf{Y}-\mu}(t) = \phi\big(\mathbf{t}^{\top} \Sigma \mathbf{t}\big),$$

for some regular $\phi : \mathbb{R}_+ \to \mathbb{R}$. We denote this by $\mathbf{Y} \sim E_n(\mu, \Sigma, \phi)$.

Example Consider an n-dimensional multivariate Gaussian distribution $\mathbf{Y} \sim N_n(0, \mathbf{I}_n)$, with $\mathbf{I}_n$ representing the $n \times n$ identity matrix. Since vector components are independent univariate normals, each one with characteristic function given by $\exp(-t^2/2)$, the characteristic function of $\mathbf{Y}$ reads as

$$\exp\left(-\frac{1}{2}(t_1^2 + \cdots + t_n^2)\right) = \exp\left(-\frac{1}{2}\mathbf{t}^\top\mathbf{t}\right).$$

Consequently, the multivariate normal is elliptical.

Definition (Multivariate Gaussian copula) Let R be a symmetric, positive definite matrix with unit diagonal entries. The *multivariate Gaussian copula* (MGC) is defined as

$$C(u_1, \ldots, u_n; R) = \Phi_R\big(\Phi^{-1}(u_1), \ldots, \Phi^{-1}(u_n)\big), \tag{8.3}$$

where Φ_R denotes the standardized multivariate normal distribution with correlation matrix R.

In terms of density, we have the canonical representation

$$f(x_1, \ldots, x_n) = c\big(F_1(x_1), \ldots, F_n(x_n)\big) \times \prod_{j=1}^{n} f_j(x_j),$$

where c is the nth mixed derivative of copula C in expression (8.3), namely:

$$c\big(F_1(x_1), \ldots, F_n(x_n)\big) = \partial^n_{u_1, \ldots, u_n} C(u_1, \ldots, u_n)\big|_{u_1 = F_1(x_1), \ldots, u_n = F_1(x_n)}, \tag{8.4}$$

and f_j is a standard normal density $f_j(x_j) = \frac{\mathrm{d}}{\mathrm{d}x} F_j(x)|_{x=x_j}$. The function c is called *copula density*.

The copula density is the ratio between the joint density f and the product of all marginals f_j. We prove this result in the bivariate case. Let $X_1 \sim F_1$ and $X_2 \sim F_2$ be continuous random variables in $\mathbb{R}$ with joint c.d.f. F. We define probability integral transforms as $U_1 = F_1(X_1)$ and $U_2 = F_2(X_2)$. Consequently, $X_1 = F_1^{-1}(U_1)$ and $X_2 = F_2^{-1}(U_2)$, where we adopt generalized inverse functions (see chapter "Static Monte Carlo"). If J denotes the Jacobian of the multivalued function $(F_1^{-1}(U_1), F_2^{-1}(U_2))$, the change of variable formula leads to

$$\begin{aligned}
c(u_1, u_2) &= f\big(F_1^{-1}(u_1), F_2^{-1}(u_2)\big) \det J \\
&= f\big(F_1^{-1}(u_1), F_2^{-1}(u_2)\big) \\
&\quad \times \big(\partial_{U_1} F_1^{-1}(u_1)\partial_{U_2} F_2^{-1}(u_2) - \partial_{U_2} F_1^{-1}(u_1)\partial_{U_1} F_2^{-1}(u_2)\big) \\
&= f\big(F_1^{-1}(u_1), F_2^{-1}(u_2)\big)\big(\partial_{U_1} F_1^{-1}(u_1)\partial_{U_2} F_2^{-1}(u_2)\big) \\
&= f\big(F_1^{-1}(u_1), F_2^{-1}(u_2)\big)\big(\partial_{X_1} F_1(x_1)\partial_{X_2} F_2(x_2)\big)^{-1} \\
&= \frac{f(F_1^{-1}(u_1), F_2^{-1}(u_2))}{f_1(F_1^{-1}(u_1))f_2(F_2^{-1}(u_2))}.
\end{aligned}$$

If the two r.v.'s are statistically independent, the numerator in the last expression matches the denominator and the copula degenerates into the unit constant. From the definition of MGC we can easily determine the corresponding density.

Example In the case of a multivariate normal distribution, we have:

$$\frac{1}{(2\pi)^{n/2}|R|^{1/2}} \exp\left(-\frac{1}{2}\mathbf{x}^\top R^{-1}\mathbf{x}\right)$$

$$= c\big(\Phi(x_1), \ldots, \Phi(x_n)\big) \prod_{j=1}^{n}\left(\frac{1}{\sqrt{2\pi}}\exp\left(-\frac{1}{2}x_j^2\right)\right),$$

and thus

$$c\big(\Phi(x_1), \ldots, \Phi(x_n)\big) = \frac{\frac{1}{(2\pi)^{n/2}|R|^{1/2}}\exp(-\frac{1}{2}\mathbf{x}^\top R^{-1}\mathbf{x})}{\prod_{j=1}^{n}(\frac{1}{\sqrt{2\pi}}\exp(-\frac{1}{2}x_j^2))}. \tag{8.5}$$

By letting $u_j = \Phi(x_j)$, we can write:

$$c(u_1, \ldots, u_n) = \frac{1}{|R|^{1/2}}\exp\left(-\frac{1}{2}\omega^\top(R^{-1}-I)\omega\right),$$

with $\omega = (\Phi^{-1}(u_1), \ldots, \Phi^{-1}(u_n))^\top$.

In the case of distributions with a Gaussian copula, there is a relationship between Kendall's tau and Pearson's linear correlation rho:

$$\tau = \frac{2}{\pi}\arcsin\rho.$$

A similar result can be obtained for the Spearman's rho:

$$\varrho = \frac{6}{\pi}\arcsin\rho.$$

Figure 8.3 shows the surface of the Gaussian copula density as depicted in equation (8.5) for the bivariate case, with $\rho = 0.5$. Figure 8.4 exhibits the contour plot of the MGC density for different values of correlation. These figures can be obtained by running the MATLAB® code BivGCDens.m; see Table 8.1.

Definition (Student T-copula) For a symmetric and positive definite matrix R with unit diagonal entries, let $T_{R,v}$ denote the standardized multivariate Student's t distribution with correlation matrix R and $v \geq 1$ degrees of freedom:

$$T_{R,v}(y_1, \ldots, y_n) = \int_{-\infty}^{y_1} \cdots \int_{-\infty}^{y_n} \frac{\Gamma(\frac{v+n}{2})|R|^{-1/2}}{\Gamma(\frac{v}{2})(v\pi)^{n/2}}$$

$$\times \left(1 + \frac{1}{v}\mathbf{x}^\top R^{-1}\mathbf{x}\right)^{-(v+n)/2} dx_1 \cdots dx_n.$$

The *multivariate Student T-copula* (MTC) is defined as:

$$C(u_1, \ldots, u_n; R, v) = T_{R,v}\big(T_v^{-1}(u_1), \ldots, T_v^{-1}(u_n)\big),$$

where T_v^{-1} is the inverse of the univariate Student's t cumulative distribution function with v degrees of freedom.

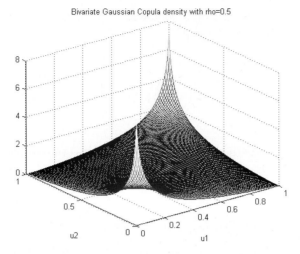

Fig. 8.3. Density of a two-dimensional Gaussian copula with $\rho = 0.5$.

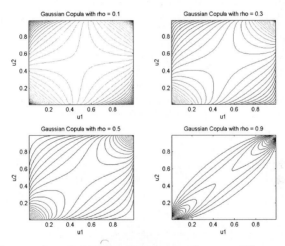

Fig. 8.4. Contour plots of a bivariate Gaussian copula across different values of ρ.

The corresponding copula density can be computed using formula (8.4) as

$$c(u_1, \ldots, u_n; R) = |R|^{-1/2} \frac{\Gamma(\frac{v+n}{2})}{\Gamma(\frac{v}{2})} \left(\frac{\Gamma(\frac{v}{2})}{\Gamma(\frac{v+1}{2})} \right)^n \frac{(1 + \frac{1}{v}\boldsymbol{\omega}^\top R^{-1} \boldsymbol{\omega})^{-(v+n)/2}}{\prod_{j=1}^n (1 + (\varsigma_j^2)/v)^{-(v+1)/2}},$$
(8.6)

where $\omega_j = T_v^{-1}(u_j)$. It can be proved that the Student t copula exhibits identical upper and lower tail dependence coefficients, namely:

$$\lambda_{\mathrm{U}} = \lambda_{\mathrm{L}} = 2\overline{T}_{v+1}\left(\frac{\sqrt{v+1}\sqrt{1-\rho}}{\sqrt{1+\rho}} \right),$$

Table 8.1.

```
function c = BivGCDens(rho)

ul = 0.01:0.01:0.99;
u2 = ul;

for i=1:length(ul)
    for j=1:length(u2)

        c(i,j) = (1/(sqrt(1-rho*2))) * exp(-((norminv(ul(i))^2 +
norminv(u2(j))^2 -
...
            2*rho*norminv(ul(i))*norminv(u2(j)))/(2*(1 - rho^2)))
+ ...
            ((norminv(ul(i))^2 + norminv(u2(j))*2)/2 ));
    end
end

[X,Y] = meshgrid(ul);
surf(Y,X,c);

xlabel('ul');
ylabel('u2');
title('Bivariate Gaussian Copula density');
```

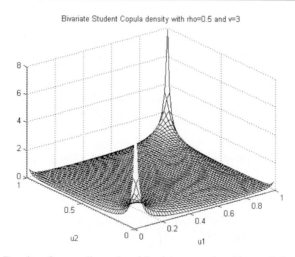

Bivariate Student Copula density with rho=0.5 and v=3

Fig. 8.5. Density of a two-dimensional Student t-copula with $\rho = 0.5$ and $v = 3$.

where $\overline{T}_{v+1}$ denotes the survival probability for a univariate Student t with $v + 1$ degrees of freedom. Clearly, this coefficient is increasing with ρ, decreasing with v, and vanishes as the number of degrees of freedom diverges to infinity, provided that $\rho < 1$. Figure 8.5 displays the surface of a T-Student copula density in two

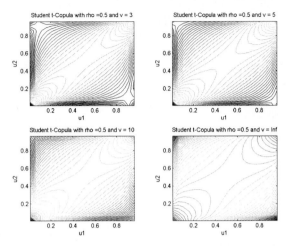

Fig. 8.6. Contour plots of a two-dimensional Student t-copula with $\rho = 0.5$ across different values of ν.

dimensions under the assumption that $\rho = 0.5$ and $\nu = 3$. Figure 8.6 shows the contour plot of this density across different values of ν.

These figures can be obtained by running the code TCDens.m; see Table 8.2.

8.4 Archimedean Copulas

Archimedean copulas constitute an important class of copula functions due to their analytical tractability and ability to reproduce a large spectrum of dependence structures.

An Archimedean copula *generator* is a convex and strictly decreasing function φ from the unit interval $[0, 1]$ onto $\mathbb{R}_+$ with $\varphi(1) = 0$. The pseudo-inverse of φ is defined as a function $\varphi^{[-1]}$ from $[0, \infty]$ onto $[0, 1]$ such that:

$$\varphi^{[-1]}(z) = \begin{cases} \varphi^{-1}(z) & \text{if } 0 < z \le \varphi(0), \\ 0 & \text{if } \varphi(0) \le z < \infty. \end{cases} \tag{8.7}$$

If $\varphi(0) = \infty$, then the pseudo-inverse collapses into an ordinary inverse function, that is $\varphi^{[-1]} = \varphi^{-1}$, and the generator is said to be *strict*. In this instance, an explicit expression for the copula function can be obtained.

Definition (Archimedean copula) Let φ be an Archimedean copula generator. The function

$$C(u_1, u_2) = \varphi^{[-1]}\big(\varphi(u_1) + \varphi(u_2)\big)$$

is an Archimedean copula.

Archimedean copulas are *symmetric*:

Table 8.2.

```
function c = TCDens(v,rho)

u1 = 0.01:0.01:0.99;
u2 = u1;

const = (1/sqrt(1-
rho^2))*(gamma((2+v)/2)/gamma(v/2))*(((gamma(v/2))/
gamma((v+1)/2))*2);

for i = 1:length(u1)
     for j = 1:length(u2)

        varsigma1_sqr = tinv(u1(i),v)^2;
        varsigma2_sqr = tinv(u2(j),v)^2;
        qF = varsigma1_sqr + varsigma2_sqr -
2*rho*tinv(u1(i),v)*tinv(u2(j),v);

        c(i,j) = 0.5*(1+qF/(1-rho*2)/v)^M-(v+2)/2)/...
              ((1+varsigma1_sqr/v)*(1+varsigma2_sqr/v))^(-
(v+1)/2)*const;

     end
end

[X,Y] = meshgrid(u1);
surf (Y,X,c) ;

xlabel('u1');
ylabel('u2');
title('Bivariate Student Copula density');
```

$$C(u_1, u_2) = C(u_2, u_1),$$

and *associative*

$$C\big(u_1, C(u_2, u_3)\big) = C\big(C(u_1, u_2), u_3\big).$$

Moreover, their Kendall's tau can be computed in terms of the copula generator as:

$$\tau = 1 + 4 \int_0^1 \frac{\varphi(u)}{\varphi'(u)} \, du.$$

Recall that a function $f(t)$ is *completely monotone* on a given interval if it is continuous and has derivatives of all orders with alternating signs, namely if:

$$(-1)^k \frac{d^k}{dt^k} f(t) \geq 0,$$

for all t in the interior of the interval and all positive integers k. The next result allows us to generalize Archimedean copulas to the multivariate case.

Theorem (Kimberling, 1974) *Let φ be an Archimedean copula generator. The function $C : [0, 1]^n \to [0, 1]$ defined by:*

$$C(u_1, \ldots, u_n) = \varphi^{[-1]}\big(\varphi(u_1) + \cdots + \varphi(u_n)\big),$$

is a copula if and only if $\varphi^{[-1]}$ is completely monotone on $[0, \infty]$.

We now list a few parametrized families of Archimedean copulas.

Example (Gumbel copula) The Gumbel n-copula is defined by the strict generator $\varphi(u) = (-\ln(u))^\alpha$, with $\alpha \in [1, \infty)$. For all $\alpha > 1$, the copula reads as:

$$C(u_1, \ldots, u_n) = \exp\left\{ -\left[\sum_{i=1}^{n} (-\ln u_i)^\alpha \right]^{1/\alpha} \right\}.$$

The Kendall's tau can be computed as $\tau = 1 - \alpha^{-1}$. It can be shown that Gumbel copulas have upper tail dependence $\lambda_U = 2 - 2^{1/\alpha}$ and lower tail dependence vanishing as α diverges to infinity. Figure 8.7 shows the surface of a Gumbel copula density for the bivariate case with $\alpha = 10$. Figure 8.8 exhibits contour plots of Gumbel copula densities for varying assignments of parameter α. These figures can be obtained by running code `BivGDens.m`; see Table 8.3.

Example (Clayton copula) The Clayton n-copula is defined by formula (8.7) with $\varphi(u) = u^{-\alpha} - 1$, for any $\alpha > 0$. It can be written in analytic terms as:

$$C(u_1, \ldots, u_n) = \left[\sum_{i=1}^{n} u_i^{-\alpha} - n + 1 \right]^{-1/\alpha}.$$

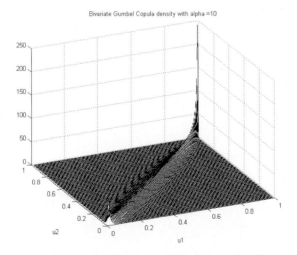

Fig. 8.7. Density of a two-dimensional Gumbel copula with $\alpha = 10$.

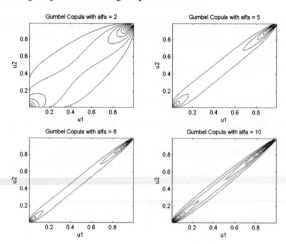

Fig. 8.8. Contours of a two-dimensional Gumbel copula for different values of α.

Table 8.3.

```
function c = BivGDens(alfa)

u1 = 0.01:0.01:0.99;
u2 = u1;

for i=1:length(u1)
    for j=1:length(u2)
        u1tilde = - log(u1(i));
        u2tilde = - Iog(u2(j));
        w = (u1tilde^alfa) + (u2tilde^alfa);
        pdf = ((u1tilde*u2tilde)^(alfa-1))*((w^(1/alfa))
+ alfa - 1)/(w^(2-(1/alfa)))/(u1(i)*u2(j));
        cdf = exp(-((-log(u1(i)))^alfa + (-
log(u2(j)))^alfa)*(1/alfa));
        c(i, j) = pdf*cdf;
    end
end

[X,Y] = meshgrid(u1);
surf (Y,X,c);

xlabel('u1');
ylabel('u2');
title('Bivariate Gumbel Copula density');
```

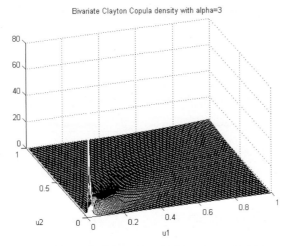

Fig. 8.9. Density of a two-dimensional Clayton copula with $\alpha = 3$.

The Kendall's tau can be computed as $\tau = \alpha/(\alpha + 2)$. Unfortunately, the Spearman's rho does not admit an analytical expression. The lower tail dependence is $\lambda_{\mathrm{L}} = 2^{-1/\alpha}$, showing that Clayton copulas exhibit asymptotic lower tail dependence provided that $\alpha > 0$. We can also show that this copula exhibits an asymptotically independent upper tail:

$$
\begin{aligned}
\lambda_{\mathrm{U}} &= \lim_{u \uparrow 1} \frac{1 - 2u(2u^\alpha - 1)^{1/\alpha}}{1 - u} \\
&= \lim_{u \uparrow 1} \frac{-2 + 2(2u^{-\alpha} - 1)^{-1/\alpha - 1} u^{-\alpha - 1}}{-1} \\
&= \lim_{u \uparrow 1} \frac{-2 + 2(2 - u^a)^{-1/\alpha - 1}}{-1} \\
&= 0.
\end{aligned}
$$

Figure 8.9 shows the surface of a Clayton copula density in the bivariate case with $\alpha = 3$. Figure 8.10 exhibits contour plots of Clayton copula densities across varying assignments of parameter α. We see that increasing values for α are accompanied by a stronger tail dependence. These figures can be obtained by running code BivCDens.m; see Table 8.4.

Example (Frank copula) The Frank n-copula is defined by generator $\varphi(u) = -\ln(\frac{\exp(-\alpha u)-1}{\exp(-\alpha)-1})$ with $\alpha \neq 0$. As a result,

$$
C(u_1, \ldots, u_n) = -\frac{1}{\alpha} \ln\left\{ 1 + \frac{\prod_{i=1}^{n}(e^{-\alpha u_i} - 1)}{(e^{-\alpha} - 1)^{n-1}} \right\}.
$$

For $n \geq 3$, the generator is strict provided that $\alpha > 0$. Kendall's tau is given by $\tau = 1 - 4\alpha^{-1}[1 - D_1(\alpha)]$ and the Spearman rho is computed as $\varrho = 1 - 12\alpha^{-1}[D_1(\alpha) -$

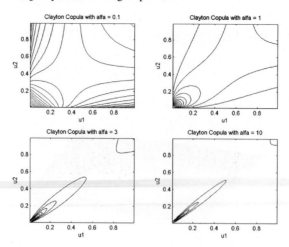

Fig. 8.10. Contour plots of a two-dimensional Clayton copula across different values of α.

Table 8.4.

```
function c = BivCDens.m(alfa)

u1 = 0.01:0.01:0.99;
u2 = u1;

for i=1:length(u1)
    for j =1:length(u2)
        c(i,j) = (1 + alfa) * (u1(i)^(-alfa-1))
                * (u2(j)^(-alfa-1)) *...
                (((u1(i)^(-alfa)) + (u2(j)^(-alfa)) -
1)^((-1/alfa)-2));
    end
end

[X,Y] = meshgrid(u1);
surf(Y,X,c);

xlabel('u1');
ylabel('u2');
title('Bivariate Clayton Copula density');
```

$D_2(\alpha)$], where D denotes Debye's function

$$D_j(x) = \frac{j}{x^j} \int_0^x t^j \left(\exp(t) - 1\right)^{-1} dt.$$

This copula is neither lower nor upper tail dependent, i.e., $\lambda_L = \lambda_U = 0$. Figure 8.11 shows the surface of a Frank copula density in the bivariate case for $\alpha = 3$. Figure 8.12 exhibits contour plots of Frank copula densities for varying assignments

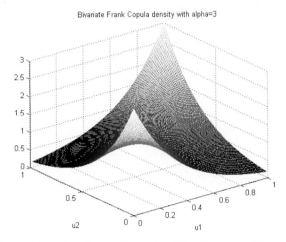

Fig. 8.11. Density of a two-dimensional Frank copula with $\alpha = 3$.

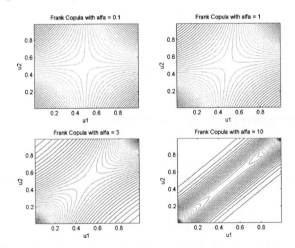

Fig. 8.12. Contour plots of a two-dimensional Frank copula for different values of α.

of parameter α. These figures can be obtained by running code `BivFDens.m`; see Table 8.5.

8.5 Statistical Inference for Copulas

We present three methods for estimating parameters identifying a copula within a given family:

A. Exact maximum likelihood (EML);
B. Inference functions for margins (IFM);
C. Nonparametric kernel (NK).

Table 8.5.

```
function c = BivFDens(alfa)

ul = 0.01:0.01:0.99;
u2 = ul;

for i=1:length(u1)
    for j =1:length(u2)

    A = -alfa * (exp(-alfa)-1) * exp(-alfa*ul(i)) *
exp(-alfa*u2(j));
    B = ((exp(-alfa)-1) + (exp(-alfa*ul(i)) -1) *
(exp(-alfa*u2(j)) - 1))^2;

    c(i,j) = (A/B);

    end
end

[X,Y] = meshgrid(u1);
surf (Y,X,c);

xlabel('u1');
ylabel('u2');
title('Bivariate Frank Copula density');
```

We consider a strictly stationary n-dimensional stochastic process $\mathbf{X}$. Input data consists of a time series $(X_{1,t}, X_{2,t}, \ldots, X_{n,t})_{t=1,\ldots,T}$ of sample vectors recorded at times $t = 1, \ldots, T$. These data may represent observed returns of n financial asset prices observed at consecutive dates. In what follows, we consider continuous distributions only. In particular, both copula and marginal distributions are assumed to be continuous. By inspecting the canonical representation (8.4), estimating a multivariate statistical model requires three steps:

(1) Select marginal distributions and estimate them on univariate data;
(2) Select a family of copula functions representing the dependence structure of multivariate data;
(3) Estimate the copula parameters.

The first step uses classical estimation methods from univariate statistics. The second step is the most delicate one: though little work has been published on this issue, we briefly examine it later on in this chapter. Here, we are more concerned with the last step. This can be performed by maximizing the likelihood function resulting from the previous steps. The EML method solves (1) and (2) in one fell swoop. However, this comes at the price of having a burdensome computational algorithm. IMF splits univariate from dependence structure estimations and therefore is more reasonable in the context of a copula-based model description. NK assumes no particular form for the distribution functions involved in the estimation process, but it requires large data sets.

8.5.1 Exact Maximum Likelihood

Let $\{x_{1t}, x_{2t}, \ldots, x_{nt}\}_{t=1,\ldots,T}$ be the sample data matrix. The logarithmic likelihood function is given by

$$\mathcal{L}(\theta) = \sum_{t=1}^{T} \ln c\big(F_1(x_{1t}), F_2(x_{2t}), \ldots, F_n(x_{nt})\big) + \sum_{t=1}^{T} \sum_{j=1}^{n} \ln f_j(x_{jt}), \qquad (8.8)$$

where θ gathers all parameters defining bo the the marginals and the copula.

The maximum likelihood estimator is obtained by maximizing the function $\mathcal{L}$ above on a compact domain Θ for the parameter choice:

$$\widehat{\theta}_{\mathrm{MLE}} = \arg\max_{\theta \in \Theta} \mathcal{L}(\theta).$$

Under suitable regularity conditions, this estimator is consistent, asymptotically efficient, and asymptotically normal, that is:

$$\sqrt{T}(\widehat{\theta}_{\mathrm{MLE}} - \theta_0) \to \mathcal{N}\big(0, \Im^{-1}(\theta_0)\big).$$

Here, θ_0 is the exact value for the unknown parameter and $\Im(\theta_0)$ denotes Fisher's information matrix defined entry by entry as:

$$\Im(\theta_0)_{i,j} = \mathbb{E}[\partial_{x_i}\mathcal{L}(\theta_0)\partial_{x_j}\mathcal{L}(\theta_0)].$$

Example (Multivariate Gaussian copula) Let C be a multivariate Gaussian copula as defined in formula (8.5). The corresponding log-likelihood function is defined as follows:

$$\mathcal{L}(\theta) = -\frac{T}{2}\ln|R| - \frac{1}{2}\sum_{t=1}^{T}\omega_t^{\top}\big(R^{-1} - I\big)\omega_t,$$

where $\omega_t = (\Phi^{-1}(x_{1t}), \ldots, \Phi^{-1}(x_{nt}))^{\top}$, Φ denotes the standard normal c.d.f., R is a symmetric and positive definite matrix, and the parametric space is $\Theta = \mathbb{R}^{n \times n}$. Since $[\partial/\partial R^{-1}]\mathcal{L}(\theta) = 2^{-1}[TR - \sum_{t=1}^{T}\omega_t^{\top}\omega_t]$, then the maximum likelihood estimator is computed as:

$$\widehat{\theta}_{\mathrm{MLE}} = \widehat{R}_{\mathrm{MLE}} = \frac{1}{T}\sum_{t=1}^{T}\omega_t^{\top}\omega_t.$$

8.5.2 Inference Functions for Margins

ML may require an excessive computational effort as it requires to jointly estimate both parameters of the marginal distributions and the ones that identify the copula. If we take a closer look at the log-likelihood function (8.8), we see that estimating a copula requires to specify parametric univariate marginal distributions. These parameters can be estimated on univariate time series and then plugged into the full likelihood. This latter finally depends on the copula parameters. This method exploits the fundamental role played by copula functions: that is to separately specify univariate margins and a dependence structure. This observation leads to the following estimation procedure developed by Joe and Xu (1996):

Algorithm (IFM)

1. Define log-likelihood functions L for the joint distribution, L_j for the jth marginal, and L_c for the copula function c:

$$\mathcal{L}(\boldsymbol{\omega}, \boldsymbol{\theta}) = \sum_{t=1}^{T} \ln c\Big(F_1(x_{1t}; \theta_1), F_2(x_{2t}; \theta_2), \ldots, F_n(x_{nt}; \theta_n); \boldsymbol{\omega}\Big)$$

$$+ \sum_{t=1}^{T} \sum_{j=1}^{n} \ln f_j(x_{jt}; \theta_j),$$

$$\mathcal{L}_j(\theta_j) = \sum_{t=1}^{T} \ln f_j(x_{jt}; \theta_j),$$

$$\mathcal{L}_c(\boldsymbol{\omega}) = \sum_{t=1}^{T} \ln c\Big(F_1(x_{1t}; \widehat{\theta}_1), \ldots, F_n(x_{nt}; \widehat{\theta}_n); \boldsymbol{\omega}\Big).$$

(Here vector $\boldsymbol{\theta} = (\theta_1, \ldots, \theta_n)$ collects all parameters of the marginal distributions and vector $\boldsymbol{\omega}$ gathers all copula parameters.)

2. For each univariate marginal, estimate parameters θ_j over a compact domain Θ_j by maximum likelihood (ML):

$$\widehat{\theta}_j = \arg \max_{\theta_j \in \Theta_j} \mathcal{L}_j(\theta_j).$$

3. Estimate the copula parameters $\boldsymbol{\omega}$ over a compact domain Ω by ML:

$$\widehat{\boldsymbol{\omega}} = \arg \max_{\boldsymbol{\omega} \in \Omega} \mathcal{L}_c(\boldsymbol{\omega}).$$

4. Calculate $\widehat{\boldsymbol{\beta}}_{\text{IFM}} = (\widehat{\boldsymbol{\omega}}, \widehat{\boldsymbol{\theta}})$.

The IFM estimator solves the set of equations

$$\left(\frac{\partial \mathcal{L}_1}{\partial \theta_1}, \ldots, \frac{\partial \mathcal{L}_n}{\partial \theta_n}, \frac{\partial \mathcal{L}_c}{\partial \boldsymbol{\omega}} \right) = 0,$$

whereas the MLE results from solving

$$\left(\frac{\partial \mathcal{L}}{\partial \theta_1}, \ldots, \frac{\partial \mathcal{L}}{\partial \theta_n}, \frac{\partial \mathcal{L}}{\partial \omega}\right) = 0.$$

Therefore, in general the two estimators do not agree. IFM estimation can also be adopted as a starting point to speed up the optimization for a classical ELM procedure. To address the issue of the asymptotic efficiency of the IFM estimator compared to the MLE, we need to consider the corresponding asymptotic covariance matrices. Under regularity conditions, the IFM estimator turns out to be asymptotically normal, namely

$$\sqrt{T}\left(\widehat{\boldsymbol{\beta}}_{\mathrm{IFM}} - \boldsymbol{\beta}_0\right) \to \mathcal{N}\left(0, \hbar^{-1}(\boldsymbol{\beta}_0)\right).$$

Here $\hbar(\boldsymbol{\beta}_0)$ represents Godambe's information matrix

$$\hbar(\boldsymbol{\beta}_0) = D^{-1} V\left(D^{-1}\right)^{\top},$$

$D = E[\nabla_{\boldsymbol{\beta}} s(\boldsymbol{\beta})]$, $V = E[s(\boldsymbol{\beta}) s(\boldsymbol{\beta})^{\top}]$, and $s(\beta) = (\partial \mathcal{L}_1 \partial \theta_1, \ldots, \partial \mathcal{L}_n \partial \theta_n, \partial \mathcal{L}_c \partial \omega)^{\top}$ is a score vector.

Estimating the covariance matrix h requires a computation of several partial derivatives. Joe (1997) suggests using the Jacknife method or other bootstrap methods to achieve this goal. In a time series context, it may be useful to adopt a block-bootstrap, especially when the time series in hand shows a low autocorrelation.

8.5.3 Kernel-based Nonparametric Estimation

A potential problem arising while adopting the IFM method is that the number of possible arrangements can be very high and one can easily get lost looking for the best combination of marginals and the copula. This problem can be overcome by adopting a nonparametric approach to model margins and copulas. This framework allows for data to deliver distributions and a copula without calling for a subjective choice about their functional form.

We consider a nonparametric method to estimate copula functions in the context of multivariate stationary processes satisfying strong mixing conditions. In particular, we present a kernel-based approach providing differentiable estimates.

The starting point of this method is the usual representation $C(u_1, \ldots, u_n) = F(F_1^{-1}(u_1), \ldots, F_n^{-1}(u_n))$ as in formula (8.1). Here, $F_1^{-1}, \ldots, F_n^{-1}$ are the pseudo-inverse functions of the univariate c.d.f.'s $F_1, \ldots, F_n$. (For a clear definition of pseudo-inverse function, see chapter "Static Monte Carlo".) Consequently, estimating a copula function amounts to providing estimates of F and the marginals F_j. Each of these estimations is performed by means of a kernel-based method.

Let $\xi_j(u_j)$ denote the unique solution of the equation $F_j(y) = u_j$ for any arbitrary $u_j \in (0, 1)$. A kernel is any bounded, symmetric, and normalized (i.e., $\int k(x)\, dx = 1$) real-valued function $k(x)$ defined on $\mathbb{R}$. An n-dimensional kernel is defined in terms of kernels by:

$$k(\mathbf{x}) = \prod_{j=1}^{n} k_j(x_j),$$

where $\mathbf{x} = (x_1, \ldots, x_n)$.

Consider a bandwidth matrix H made by collecting positive functions $h_j(T)$ on the main diagonal in a diagonal matrix, that is:

$$H = \begin{pmatrix} h_1(T) & 0 & 0 \\ 0 & \ldots & 0 \\ 0 & 0 & h_n(T) \end{pmatrix}.$$

We assume that $|h| + (T|h|)^{-1} \to 0$ as T diverges to infinity. We consider an n-dimensional kernel of the form:

$$k(\mathbf{x}; h) = \prod_{j=1}^{n} k_j \left(\frac{x_j}{h_j} \right).$$

A single probability density function (p.d.f.) of Y_{jt} computed at y_j can be estimated as:

$$\widehat{f}_j(y_j) = \frac{1}{Th_j} \sum_{t=1}^{T} k_j \left(\frac{y_j - Y_{jt}}{h_j} \right),$$

while the joint p.d.f. of a random vector $\mathbf{Y}_t$ computed at $\mathbf{y} = (y_1, \ldots, y_n)^\top$ can be estimated as:

$$\widehat{f}(\mathbf{y}) = \frac{1}{T|h|} \sum_{t=1}^{T} k(\mathbf{y} - \mathbf{Y}_t; h).$$

By integrating on the appropriate domains, we obtain estimates for c.d.f.'s of both Y_{jt} and $\mathbf{Y}_t$:

$$\widehat{F}_j(y_j) = \int_{-\infty}^{y_j} \widehat{f}_j(x) \, dx,$$

$$\widehat{F}(\mathbf{y}) = \int_{-\infty}^{y_1} \int_{-\infty}^{y_2} \cdots \int_{-\infty}^{y_n} \widehat{f}(\mathbf{x}) \, d\mathbf{x}.$$

The copula is obtained through

$$\widehat{C}(\mathbf{u}) = \widehat{F}(\widehat{\boldsymbol{\xi}}),$$

where $\mathbf{u} = (u_1, \ldots, u_n)$, $\widehat{\boldsymbol{\xi}} = (\widehat{\xi}_1, \ldots, \widehat{\xi}_n)^\top$, and $\widehat{\xi}_j = \inf_{y \in \mathbb{R}} \{y: \widehat{F}_j(y) \geq u_j\}$.

Notice that $\widehat{\xi}_j$ is a kernel estimate of the quantile of the distribution of Y_{jt} with a probability level u_j.

Example (Gaussian kernel) For a Gaussian kernel $k_j(x) = \varphi(x)$, we obtain:

$$\widehat{F}_j(y_j) = T^{-1} \sum_{t=1}^{T} \Phi\left(\frac{y_j - Y_{jt}}{h_j}\right),$$

$$\widehat{F}(\mathbf{y}) = T^{-1} \sum_{t=1}^{T} \prod_{j=1}^{n} \Phi\left(\frac{y_j - Y_{jt}}{h_j}\right),$$

where Φ is the standard Gaussian c.d.f. in n dimensions.

8.6 Monte Carlo Simulation

The standard simulation problem can be cast as follows. We wish to simulate a sample of a random vector $\mathbf{X} = (X_1, \ldots, X_n)$ with distribution function $F_\mathbf{X}$ defined by assigning marginals $F_{X_1}, \ldots, F_{X_n}$ and a copula function C, that is:

$$F_\mathbf{X}(\mathbf{x}) = C\big(F_{X_1}(x_1), \ldots, F_{X_n}(x_n)\big),$$

where $\mathbf{x} = (x_1, \ldots, x_n)$. Since $(F_{X_1}(X_1), \ldots, F_{X_n}(X_n))$ is an n-dimensional uniform vector with distribution function C, then a sample of $\mathbf{X}$ can be obtained through the following:

Algorithm (Monte Carlo simulation)
1. Sample an n-dimensional uniform vector $\mathbf{U} = (U_1, \ldots, U_n) \sim C$;
2. Return $\mathbf{X} = (F_{X_1}^{-1}(U_1), \ldots, F_{X_n}^{-1}(U_n))$.

Therefore, all we need is to (1) simulate a random sample of a $[0, 1]^n$-valued random vector $\mathbf{U}$ with distribution function C, i.e., $F_\mathbf{U}(\mathbf{u}) = C(u_1, \ldots, u_n)$, and (2) compute the (pseudo) inverse functions $F_{X_1}^{-1}, \ldots, F_{X_n}^{-1}$. The latter problem has been solved in section "Inverse Function Method" within chapter "Static Monte Carlo" (Chapt. 1). We now address the problem of sampling from a distribution function on the hypercube $[0, 1]^n$ with uniform margins.

8.6.1 Distributional Method

We start with relation (8.1), which we reproduce here for the reader's convenience:

$$C(u_1, \ldots, u_n) = F\big(F_1^{-1}(u_1), \ldots, F_n^{-1}(u_n)\big). \tag{8.9}$$

Here F is *any* multivariate distribution function with continuous margins $F_1, \ldots, F_n$ and copula function C (Sklar theorem). In particular, we consider a distribution function F whose samples are easy to obtain. Of course, it would be pointless to set $F = F_\mathbf{X}$ as long as our original issue is just to simulate a sample of $F_\mathbf{X}$ using the algorithm above stated. From this relation, we may recast the problem as follows: look for random variables $U_1, \ldots, U_n$ such that

$$\mathbb{P}(U_1 \leq u_1, \ldots, U_n \leq u_n) = F\big(F_1^{-1}(u_1), \ldots, F_n^{-1}(u_n)\big). \tag{8.10}$$

Let $\mathbf{Y}$ be a random vector with multivariate distribution F. Then, the right-hand side in the last expression equals

$$\mathbb{P}\big(Y_1 \le F_1^{-1}(u_1), \dots, Y_n \le F_n^{-1}(u_n)\big),$$

which can be written as

$$\mathbb{P}\big(F_1(Y_1) \le u_1, \dots, F_n(Y_n) \le u_n\big).$$

By equating this expression to the left-hand side in formula (8.10), we see that $(U_1, \dots, U_n) \overset{d}{=} (F_1(Y_1), \dots, F_n(Y_n))$. This leads to the following algorithm.

Algorithm (Distributional method)

1. Given a copula function C, find a distribution F such that (8.9) holds true.
2. Simulate a sample $\mathbf{Y} = (Y_1, \dots, Y_n)$ from distribution F.
3. Return $(F_1(Y_1), \dots, F_n(Y_n))$, where F_i is the ith marginal of F.

Example (Gaussian copula simulation) To simulate a random vector $(u_1, \dots, u_n)$ distributed according to a Gaussian copula $C_{\mathcal{N}}$ with correlation matrix R, we start with the canonical representation:

$$C_{\mathcal{N}}(u_1, \dots, u_n) = \Phi_n\big(\Phi^{-1}(u_1), \dots, \Phi^{-1}(u_n)\big),$$

where Φ_n is the c.d.f. of an n-variate normal distribution function with linear correlation matrix R and Φ is the univariate standard normal c.d.f.; next, we apply the procedure described above, with $F_i = \Phi$ for all i. To generate a sample $\mathbf{Y}$ from Φ_n, it suffices to:

(1) Find the Cholesky decomposition of R, so that $AA^\top = R$, with A lower triangular;
(2) Generate a sample of n independent random variates $Z_1, Z_2, \dots, Z_n$ from $N(0, 1)$;
(3) Set $\mathbf{Y} = A\mathbf{Z}$ (with $\mathbf{Z} = (Z_1, Z_2, \dots, Z_n)^\top$ and $\mathbf{Y} = (Y_1, Y_2, \dots, Y_n)^\top$); and finally,
(4) Deliver $(\Phi(Y_1), \dots, \Phi(Y_n))$.

Figure 8.13 shows 1,000 random samples of a Gaussian copula with correlation matrix $R = \mathrm{diag}(0.5, \dots, 0.5)$. Function GCSimul.m (see Table 8.6) implements this sampling method.

Example (T-Student copula simulation) In order to simulate a random vector $(u_1, \dots, u_n)$ distributed according to a T-Student copula with correlation matrix R and v degrees of freedom, we start with the canonical representation:

$$C_{t\text{-Student}}(u_1, \dots, u_n) = T_{n,v,R}\big(T_v^{-1}(u_1), \dots, T_v^{-1}(u_n)\big),$$

where $T_{n,v,R}$ is the c.d.f. of an n-variate t-Student distribution function with v degrees of freedom and covariance matrix $\frac{v}{v-1}R$, and T_v denotes a univariate t-Student

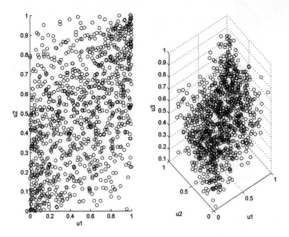

Fig. 8.13. 1,000 samples from a Gaussian copula with $\rho = 0.5$.

Table 8.6.

```
function u = GCSimul (Corr, simul)

y = randn(size(Corr,1), simul); % indipendent gaussian random
variables
A = chol(Corr); % Cholesky factorization
x = (A'*y); % Correlated gaussian random variables
u = normcdf(x); % gaussian Copula simulation
```

c.d.f. Then, we apply the usual procedure with $F_i = T_\nu$, for all i. To generate a sample $\mathbf{Y}$ from $T_{n,\nu,R}$, it suffices to set

$$\mathbf{Y} = \frac{\sqrt{\nu}}{\sqrt{\chi_\nu^2}} \mathbf{Z},$$

where $\mathbf{Z} \sim \mathcal{N}(\mathbf{0}, \Sigma)$, with Σ_{hk}: $R_{hk} = \Sigma_{hk}/\sqrt{\Sigma_{hh}\Sigma_{kk}}$, and χ_ν^2 is a chi-squared sample with ν degrees of freedom (see chapter "Static Monte Carlo" for the simulation of these variables), under the assumption that χ_ν^2 and $\mathbf{Z}$ are independent. Finally, we deliver $(T_\nu(Y_1), \ldots, T_\nu(Y_n))$. Figure 8.14 shows 1,000 random simulations from a t-Student copula, with $\nu = 3$ and $R = 0.5$ for each element. Function TCSimul.m (see Table 8.7) requires as inputs the correlation matrix R, the degree of freedom ν, and the number n of simulations. It delivers a sample of an n-vector with distribution $C_{t\text{-Student}}$.

8.6.2 Conditional Sampling

We want to generate a sample of uniformly distributed r.v.'s whose joint distribution is a copula function $C = C(u_1, u_2, \ldots, u_n)$. Let $C_k(u_1, u_2, \ldots, u_k) =$

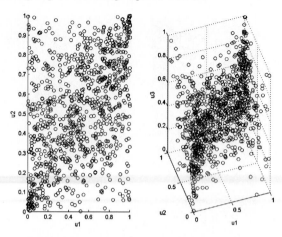

Fig. 8.14. 1,000 samples from a Student t-copula with $\rho = 0.5$ and $\nu = 3$.

Table 8.7.

```
function u = TCSimul(Corr, V, simul)

z = randn(size(Corr, 1), simul); % indipendent gaussian random
variables
A = chol(Corr); % Cholesky factorization
y = (A*z); % Correlated gaussian random variables

s = chi2rnd(V,[1 simul]); % Random numbers from the chi-square
distribution

x = (sqrt(V./s)'*ones(1,size(Corr,1)))'.*y; % Multivariate
Student-t distribution simulation

u = tcdf(x,V); % Student Copula simulation
```

$C(u_1, u_2, \ldots, u_k, 1, \ldots, 1)$ denote the k-dimensional margin of C, for all $k = 1, \ldots, n$ (with obvious conventions for $k = 1, k = n$). The conditional distribution of U_k given $U_1, \ldots, U_{k-1}$ is given by

$$F_{U_k|U_1,\ldots,U_{k-1}}(u_k|u_1,\ldots,u_{k-1}) = \mathbb{P}(U_k \leq u_k|U_1 = u_1,\ldots,U_{k-1} = u_{k-1})$$
$$= \frac{\partial^{k-1}_{u_1\ldots u_{k-1}} C_k(u_1,\ldots,u_k)}{\partial^{k-1}_{u_1\ldots u_{k-1}} C_{k-1}(u_1,\ldots,u_{k-1})}.$$

Naturally, we assume that both the numerator and the denominator exist and the latter does not vanish.

Algorithm (Conditional sampling)

1. Simulate a random sample u_1 from $\mathcal{U}(0, 1)$;

2. Simulate a random sample u_2 from $F_{.U_2|U_1}(\cdot|u_1)$;

$\vdots$

n. Simulate a random variate u_n from $F_{U_n|U_1,...,U_{n-1}}(\cdot|u_1,\ldots,u_{n-1})$.

To simulate a sample u_k from $F_{U_k|U_1,...,U_{k-1}}(u_k|u_1,\ldots,u_{k-1})$, we draw v from $\mathcal{U}[0, 1]$ and set $u_k = F^{-1}_{U_k|U_1,...,U_{k-1}}(v|u_1,\ldots,u_{k-1})$. This often requires us to numerically solve the equation $v = F_{U_k|U_1,...,U_{k-1}}(u_k|u_1,\ldots,u_{k-1})$ in the unknown variable u_k. Clearly, this task may be computationally intensive.

In the case of Archimedean copulas the conditional sampling method may be rewritten by using the following results:

Theorem *Let* $C(u_1, u_2, \ldots, u_n) = \varphi^{-1}(\varphi(u_1) + \varphi(u_2) + \cdots + \varphi(u_n))$ *be an Archimedean n-variate copula with generator* $\varphi(\cdot)$*. Then, for* $k = 2, \ldots, n$*,*

$$F_{U_k|U_1,...,U_{k-1}}(u_k|u_1,\ldots,u_{k-1})$$
$$= \frac{\varphi^{-1(k-1)}(\varphi(u_1) + \varphi(u_2) + \cdots + \varphi(u_k))}{\varphi^{-1(k-1)}(\varphi(u_1) + \varphi(u_2) + \cdots + \varphi(u_{k-1}))}, \tag{8.11}$$

where $\varphi^{-1(k)}$ *denotes the kth ordinary derivative of the inverse function* φ^{-1}*.*

We now apply this result to the Clayton, Gumbel, and Frank copulas.

Example (Clayton n-copula simulation) For a Clayton copula, the main ingredients involved in the sampling procedure are:

- Copula generator: $\varphi(u) = u^{-\alpha} - 1$.
- Inverse generator: $\varphi^{-1}(t) = (t + 1)^{-1/\alpha}$.
- Ordinary derivatives: $\varphi^{-1(k)}(t) = (-1)^k \frac{(\alpha+1)(\alpha+2)\cdots(\alpha+k-1)}{\alpha^k}(t + 1)^{-1/\alpha-k}$.

For instance, $\varphi^{-1(1)}(t) = -\frac{1}{\alpha}(t + 1)^{-1/\alpha-1}$, $\varphi^{-1(2)} = \frac{1}{\alpha}\frac{\alpha+1}{\alpha}(t + 1)^{-1/\alpha-2}$. This provides us with the following algorithm.

Algorithm

1. Simulate n independent uniformly distributed random variables $v_1, v_2, \ldots, v_n$;
2. Set $u_1 = v_1$;
3. Compute $F_{U_2|U_1}(u_2|u_1) = \frac{\varphi^{-1(1)}(\varphi(u_1)+\varphi(u_2))}{\varphi^{-1(1)}(\varphi(u_1))}$ according to expression (8.11), where $\varphi(u_1) = u_1^{-\alpha} - 1$ and $\varphi(u_1) + \varphi(u_2) = u_1^{-\alpha} + u_2^{-\alpha} - 2$;
4. Set $u_2 = F^{-1}_{U_2|U_1}(v|u_1) = (v_1^{-\alpha}(v_2^{-\alpha/(\alpha+1)} - 1) + 1)^{-1/\alpha}$;

$\vdots$

$2n$. Set $u_n = \{(u_1^{-\alpha} + u_2^{-\alpha} + \cdots + u_{n-1}^{-\alpha} - n + 2)(v_n^{\alpha/(\alpha(1-n)-1)} - 1) + 1\}^{-1/\alpha}$;
$2n + 1$. Return $(u_1, \ldots, u_n)$.

Figure 8.15 shows 1,000 random samples from a Clayton copula using the algorithm described above. These results can be obtained by running the `ClayCSim.m` (see Table 8.8) routine. Input figures are the number N of simulations, dimension M of the copula and parameter α. The program returns an $N \times M$ vector of random numbers drawn from a Clayton copula.

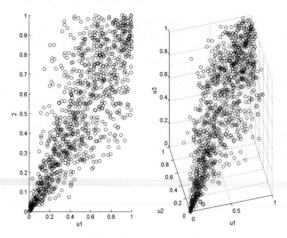

Fig. 8.15. 1,000 samples from a Clayton copula with $\alpha = 3$.

Table 8.8.

```
function v = ClayCSim(N,M,alfa)

for i=1:N

    u = rand(1,M);

    for j=2:M

        v(i,1) = u(1);
        u_vec = [v(i,1: j-1) u(j)];
        k = length(u_vec);
        v(i,j) = ((sum(u_vec(1:k-1).^(-alfa),2) - k +2) *...
                  ((u_vec(k)^(alfa/((alfa* (1-k))-1)))) - 1) +
1)^(-1/alfa);
    end
end
```

Example (Gumbel n-copula simulation) For a Gumbel copula, the main ingredients involved in the sampling procedure are:

- Copula generator: $\varphi(u) = (-\ln(u))^{\alpha}$.
- Inverse generator: $\varphi^{-1}(t) = \exp(-t^{1/\alpha})$.
- Ordinary derivatives: $\varphi^{-1(1)}(t) = -e^{-t^{1/\alpha}} \frac{1}{\alpha} t^{1/\alpha(1-\alpha)}$, $\varphi^{-1(2)}(t) = \frac{1}{\alpha^2} e^{-t^{1/\alpha}} \times t^{(1-2\alpha)/\alpha}(w+1-\alpha), \dots$.

This provides us with the following algorithm.

Algorithm

1. Simulate n independent uniformly distributed random variables $v_1, v_2, \dots, v_n$;

2. Set $u_1 = v_1$;
3. Compute $F_{U_2|U_1}(u_2|u_1) = \frac{\varphi^{-1(1)}(\varphi(u_1)+\varphi(u_2))}{\varphi^{-1(1)}(\varphi(u_1))}$ according to expression (8.11), where $\varphi(u_1) = (-\ln(u_1))^\alpha$ and $\varphi(u_1) + \varphi(u_2) = (-\ln(u_1))^\alpha + (-\ln(u_2))^\alpha$;
4. Set $u_2 = F_{U_2|U_1}^{-1}(v_2|u_1)$, which needs to be computed numerically;

$\vdots$

$2n$. Set u_n by numerically solving a nonlinear equation.
$2n+1$. Return $(u_1, \ldots, u_n)$.

Example (Frank n-copula simulation) For a Franck copula, the main ingredients involved in the sampling procedure are:

- Copula generator: $\varphi(u) = \ln(\frac{\exp(-\alpha u)-1}{\exp(-\alpha)-1})$.
- Inverse generator: $\varphi^{-1}(t) = -\frac{1}{\alpha}\ln(1 + e^t(e^{-\alpha}-1))$.
- Ordinary derivatives: $\varphi^{-1(1)}(t) = -\frac{1}{\alpha}w, \ldots, \varphi^{-1(k)}(t) = (-1)^k\frac{1}{\alpha}g_k(w)$, where $w = \frac{e^t(e^{-\alpha}-1)}{1+e^t(e^{-\alpha}-1)}$, and $g_k(w) = w(w-1)\,\partial_w g_{k-1}(w)$.

This provides us with the following algorithm.

Algorithm
1. Simulate n independent uniformly distributed random variables $v_1, v_2, \ldots, v_n$;
2. Set $u_1 = v_1$;
3. Compute $F_{U_2|U_1}(u_2|u_1) = \frac{\varphi^{-1(1)}(\varphi(u_1)+\varphi(u_2))}{\varphi^{-1(1)}(\varphi(u_1))}$ according to expression (8.11), where $\varphi(u_1) = \ln(\frac{\exp(-\alpha u_1)-1}{\exp(-\alpha)-1})$ and $\varphi(u_1) + \varphi(u_2) = \ln(\frac{(\exp(-\alpha u_1)-1)(\exp(-\alpha u_2)-1)}{(\exp(-\alpha)-1)^2})$.
4. Set $u_2 = F_{U_2|U_1}^{-1}(v|u_1) = -\frac{1}{\alpha}\ln\{1 + \frac{v_2(1-e^{-\alpha})}{v_2(e^{-\alpha u_1}-1)-e^{-\alpha u_1}}\}$;
5. Compute $F_{U_3|U_1,U_2}(u_3|u_1,u_2) = \frac{\varphi^{-1(1)}(\varphi(u_1)+\varphi(u_2)+\varphi(u_3))}{\varphi^{-1(1)}(\varphi(u_1)+\varphi(u_2))}$ according to expression (8.11), where $\varphi(u_1) + \varphi(u_2) = \ln(\frac{(\exp(-\alpha u_1)-1)(\exp(-\alpha u_2)-1)}{(\exp(-\alpha)-1)^2})$ and $\varphi(u_1) + \varphi(u_2) + \varphi(u_3) = \ln(\frac{(\exp(-\alpha u_1)-1)(\exp(-\alpha u_2)-1)(\exp(-\alpha u_3)-1)}{(\exp(-\alpha)-1)^3})$;
6. Set $u_3 = F_{U_3|U_1,U_2}^{-1}(v_3|u_1,u_2)$, which need be computed numerically by solving a polynomial equation of the second order in the variable $x = e^{-\alpha u_3} - 1$;

$\vdots$

$2n$. Set u_n by solving a polynomial equation of degree $k-1$.
$2n+1$. Return $(u_1, \ldots, u_n)$.

Figure 8.16 shows 1,000 random samples distributed according to a Frank copula as obtained through the algorithm just described.

8.6.3 Compound Copula Simulation

This is a simulation method for Archimedean copulas involving the Laplace transform and its inverse function.

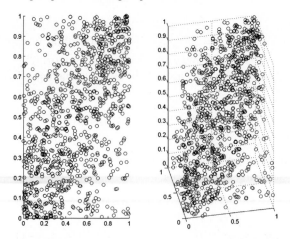

Fig. 8.16. 1,000 samples from a Frank copula with $\alpha = 3$.

Definition (Laplace transform) The Laplace transform of a non-negative random variable X with distribution function $F(x)$ and density function $f(x)$ is defined by:

$$\mathcal{L}_X(t) = \mathbb{E}\big[e^{-tX}\big] = \int_0^\infty e^{-tx}\, dF(x) = \int_0^\infty e^{-tx} f(x)\, dx = \mathcal{L}_f(t),$$

for $t \geq 0$.

Definition (Inverse Laplace transform) The inverse Laplace transform of a function $\gamma : \mathbb{R}_+ \to [0, 1]$ is defined as the function $\psi : \mathbb{R}_+ \to [0, 1]$ which solves:

$$\mathcal{L}_\psi(t) = \int_0^\infty e^{-tx} \psi(x)\, dx = \gamma(t),$$

for $t \geq 0$.

Let φ be an Archimedean copula generator and $\varphi^{[-1]}$ its pseudo inverse as defined in (8.7). Then, compound copula simulation algorithm can be written as follows.

Algorithm (Compound copula simulation)
1. Generate $X_1, \ldots, X_n$ i.i.d. uniformly distributed r.v.'s on $(0, 1)$;
2. Generate a random variable Y that is independent on $X_1, \ldots, X_n$ and whose Laplace transform is $\varphi^{[-1]}$;
3. The random samples from the Archimedean copula with generator φ are defined as:

$$U_i = \varphi^{[-1]}\left(-\frac{1}{Y} \ln X_i\right), \quad 1 \leq i \leq n.$$

In the following table, we exhibit generator, inverse generator and Laplace transform for some Archimedean copulas. This is all we need in order to apply the Compound copula simulation algorithm.

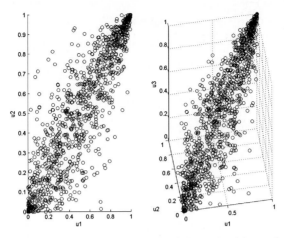

Fig. 8.17. 1,000 samples from a Gumbel copula with $\alpha = 3$.

	$\varphi(t)$	$\varphi^{[-1]}(s)$	Y-distribution
Clayton	$(t^{-\alpha} - 1)$	$(1 + s)^{-1/\alpha}$	Gamma$(1/\alpha)$
Gumbel	$(-\ln t)^{\alpha}$	$\exp(-s^{1/\alpha})$	Alpha-stable$(1/\alpha)$

We see that, in order to generate a Clayton and a Gumbel copula, we need to draw samples from a Gamma (see chapter "Static Monte Carlo") and an alpha-stable distribution. A common procedure to get a draw from a *Stable*$(1, 0, 0)$ with parameter β is based on the following result. Let υ be a uniform sample on $(-\frac{\pi}{2}, \frac{\pi}{2})$ and let ξ be an independently drawn exponential sample with mean 1. Then

$$\gamma = \frac{\sin(\beta\upsilon)}{(\cos \upsilon)^{1/\beta}} \left[\frac{\cos((1 - \beta)\upsilon)}{\xi} \right]^{(1-\beta)\beta},$$

is a random sample from a stable distribution.

Figure 8.17 shows 1,000 random samples from a Gumbel copula using the algorithm described above. The results can be obtained running the function GumbC-Sim.m; see Table 8.9.

The inputs of the code are the number N of simulations, the dimension M of the copula and α. The program returns an N random vectors with a Gumbel copula using the compound copula algorithm.

8.7 Comments

The term "copula function" was first adopted in Sklar (1959), although related results can be traced back to Hoeffding (1940). The body of research origin is Fréchet (1951), who discovered the lower and the upper bounds for bivariate copulas. For a complete discussion on the early years and a detailed presentation of the mathematical theory of copula functions, the reader can consult Nelsen (1999) and Joe (1997).

Table 8.9.

```
function u = GumbCSim(N,M,alfa)

for 1=1:M

    % Uniform [-pi/2,pi/]

    V = pi*rand(1,1) - (pi/2);

    % Indipendent exponential with mean 1

    xi = exprnd(1);

    % 1/alfa stable

    beta = 1/alfa;
    Z = (sin(beta*V)/(cos(V)^(1/beta)))*(((1/xi)*(cos((1-
beta)*V)))^((1-beta)/beta));

    R = rand(1,N);

    for j=1:N
        u(i,j) = exp((-abs((-(1/Z)*log(R(j)))/(beta)))));
    end

end
```

Cherubini, Luciano and Vecchiato (2004) provide an excellent monograph on the use of copulas in quantitative finance. Other interesting applications can be found in Jouanin, Riboulet and Roncalli (2003). Our treatment follows several sources. The notions of concordance and dependence measures are detailed in Embrechts, Lindskog and McNeil (2001). The concept of positive quadrant dependence is from Lehmann (1966). Mikusinski, Sherwood and Taylor (1992) deal with properties of comonotonic and countermonotonic variables. Nelsen (1999) contains a proof of the statement that C^- is the pointwise greatest lower bound within the class of copula functions. Hu, Müller and Scarsini (2003) provide interesting counterexamples to show the subtlety of the notion of positive dependence. Schweizer and Wolff (1981) detail the homonymous index. The Kimberling (1974) theorem is the main result for Archimedean copulas. The IFM method has been introduced by Joe and Xu (1996). The nonparametric kernel method is presented in Scaillet (2000), who also derives corresponding asymptotic distributions. Extending the static notion of copula functions to a dynamic framework is not a trivial task. Seminal papers in this respect are van den Goorbergh, Genest and Werker (2003), Fermanian and Wegkamp (2004), Kallsen and Tankov (2004), and Tankov (2005). Cherubini and Luciano (2001) provide applications to the risk assessment of financial portfolios, while Fermanian and Scaillet (2004) illustrate some pitfalls arising upon modelling with copula functions.

An application to credit risk and basket option pricing is developed through a case-study in Part II of the present book (see also Meneguzzo and Vecchiato (2004)). Rebonato (1999) details several techniques for tracking volatile market conditions within implied option pricing models. Extreme events have been extensively studied in Embrechts, Klüppelberg and Mikosch (1997), Longin (1996, 2000), Longin et al. (2001), Longin and Solnik (2001), Poncet and Gesser (1997), among others.

Part II

Problems

Portfolio Management and Trading

9

Portfolio Selection: "Optimizing" an Error[*]

Key words: estimation risk, simulation, optimization, asset allocation

Markowitz mean-variance theory (MV) provides a classic solution to the portfolio selection problem. The risk of a portfolio (as measured by the variance of its return) can be reduced by combining assets whose returns are imperfectly correlated. Diversification, however, is not boundless.

Implementation of this approach nonetheless requires knowledge of both the expected returns on all assets comprised in a portfolio and their covariances; an information set which by definition is not available. A common way to circumvent this problem is thus to use sample estimates of such measures in the optimization procedure (the so-called plug-in approach). It follows that a prototypical investor is not only exposed to market risk, but also to estimation risk. The latter can, therefore, be defined as the loss of utility which arises from forming portfolios on the basis of sample estimates rather than true values. Clearly, even if the true moments of the asset return distributions of a portfolio were known with certainty, MV optimized portfolios would not beat other portfolios in every future investment period, since return realizations usually differ from their expected values. However, over an appropriately large investment period, MV would provide on average the optimal portfolio composition.

The aim of this case is to discuss a few relevant problems caused by the exposure to estimation risk when investors deal with portfolio selection and illustrate the resampling technique as originally proposed by Michaud (1998).[1] By relying on a statistical view of MV optimization, this method leads to a better understanding of estimation risk exposure. Sample measures provide the initial moment estimate to a

---

[*] with Giovanna Boi, Riccardo Grassi and Alessandra Palmieri.

[1] Resampled Efficiency optimization was co-invented by Richard Michaud and Robert Michaud, U.S. patent 6,003,018, worldwide patents pending. New Frontier Advisors, LLC (NFA) is exclusive worldwide licensee.

multivariate normal distribution which is used to generate asset returns. A number of independent draws are then sorted out of this multivariate population of returns in order to simulate new return series. For each resampled series, sample moments are calculated and corresponding efficient frontier portfolios are computed. The dispersion in the asset allocation of simulated portfolios comes from the estimation risk that can affect the MV frontier constructed using the plug-in approach. For any given level of expected return, an average of the portfolio weights over the simulated portfolios yields the resampled efficient portfolios. Following the procedure in Herold and Maurer (2002), the resampling approach is empirically tested and in-sample and out-of-sample performances are compared with plug-in approach results. A detailed discussion of the pros and cons of the resampling procedure can be found in Scherer (2002), whilst alternative procedures for coping with estimation risk are presented in great detail in Meucci (2005) and in Brandt (2006). Other important references are Best and Grauer (1991), Brandt (2006), Britten-Jones (1999), Chopra and Ziemba (1993) and Jobson and Korkie (1980). Extensions to dynamic portfolio strategies have been proposed by Lacoste, El Karoui and Jeanblanc (2005), Karatzas et al. (1986), Karatzas, Lehoczky and Shreve (1987), Merton (1971), Portait, Bajeux-Besnainou and Jordan (2001, 2003), and Portait and Nguyen (2002).

The structure of this case is as follows. In Sect. 9.1, classic MV portfolio optimization is presented together with major issues – such as lack of diversification across assets and instability of the optimal portfolio both in time and along the efficient frontier – which may arise from exposure to estimation risk. Section 9.2 details the resampling technique. Section 9.3 describes the MATLAB$^{\circledR}$ functions we used to perform resampling. Section 9.4 presents the results of the in-sample/out-of-sample analysis.

9.1 Problem Statement

In modern finance, Markowitz's MV portfolio selection technique provides the paradigmatic solution to the problem of optimally allocating capital among risky assets. According to this approach, in each period an investor chooses a portfolio $\omega = [w_1, w_2, \ldots, w_K]$, such that portfolio variance is minimized given a predetermined level m of expected return. Therefore the investor's problem, assuming no short-selling, may be summarized as follows:

$$\min_{\omega} \omega \Sigma \omega'$$
$$\text{sub:} \quad \omega\mu = m,$$
$$\omega\mathbf{1} = 1, \tag{9.1}$$
$$\omega \geq \mathbf{0},$$

where μ is the $(K \times 1)$ vector of expected returns, Σ the $(K \times K)$ variance–covariance matrix of returns, $\mathbf{1}$ the $(K \times 1)$ vector with all elements equal to 1. Thus, in each period the investor trades off portfolio expected return with portfolio variance. The minimum variance frontier represents the combination variance–expected returns, constructed considering portfolios that have minimum variance for a given level of expected return. The efficient frontier is the upward sloping portion of the minimum variance frontier and every investor will choose a portfolio on the efficient frontier on the basis of his personal attitude towards risk. A strongly risk averse agent will prefer low variance and low expected return efficient portfolios. A less risk averse agent will choose higher expected return and therefore riskier portfolios.

The inputs to the classical portfolio selection model are the expected return vector μ and the variance–covariance matrix Σ. However, since these parameters are not known with certainty, they need to be estimated from sample data, thereby exposing the asset allocation choice to an estimation risk. The optimization problem then becomes

$$\min_{\omega} \omega \hat{\Sigma} \omega'$$
$$\text{sub:} \quad \omega \hat{\mu} = m,$$
$$\omega \mathbf{1} = 1, \tag{9.2}$$
$$\omega \geq \mathbf{0},$$

where $\hat{\mu}$ and $\hat{\Sigma}$ represent the estimates of μ and Σ. Therefore, estimation risk refers to the difference between the optimal solution to (9.1) and the optimal solution to (9.2). As an example, suppose we build a portfolio using assets with equal expected returns and variances–covariances. If, because of estimation errors, sample estimates differ across assets, MV optimization will favour some assets over others, giving to the former a higher weight vis-à-vis the latter. As a result, over-weighted assets will be those with large estimated expected returns, low variances and low correlations. Alas, these assets are likely to be the ones most affected by estimation errors!

To summarize, the consequences of estimation risk on MV optimized portfolios are of three kinds:

(a) *Low degree of diversification.* MV portfolios often involve very extreme positions. In particular, as the number of assets grows, the weight on each single asset does not tend to zero as suggested by a naive notion of diversification;

(b) *Sudden shifts* in the allocation of the optimal weights along the efficient frontier, i.e., the composition of the optimal portfolio is very different for individuals that differ slightly in their attitude towards risk;

(c) *High sensitivity* of portfolio weights to small variations in expected returns. Since little changes in expected returns can completely alter the composition of MV optimal portfolios, while modifications in the variance–covariance matrix have smaller impact, it follows that errors in the sample estimates of expected return have great bearing on allocation choices. For instance, Chopra and Ziemba (1993) find that errors in mean estimates are about ten times as important as errors in variance estimates. Errors in variance estimates are in turn about twice as important as errors in covariance estimates.

As consequence, MV optimized efficient portfolios constructed using sampled means and sampled variance–covariance matrices from a given population generally score badly once their performance is verified using out of sample data. In other words, their *ex-post* reward/risk ratio is lower than expected.

9.2 Model and Solution Methodology

In order to improve MV optimization and address estimation risk, Michaud (1998) proposes a statistical procedure based on a well-known statistical procedure named bootstrapping, Maddala and Li (1996). More specifically, a *resampling* procedure aimed at the construction of a *region* of statistically equivalent efficient portfolios is introduced, and the concept of *resampled efficiency* defined. In this way, the traditional curve representing the efficient frontier becomes a region the area of which may be viewed as a measure of the uncertainty affecting the construction of the efficient frontier.

More precisely, Michaud's resampling procedure requires:

1. Collecting T historical returns on a set of K asset classes, i.e. on investments such as stocks, bonds, real estate, or cash.
2. Computing sample means $\hat{\mu}$ and the sample variance–covariance matrix $\widehat{\Sigma}$.
3. Finding optimal weights for a set of Z mean-variance efficient portfolios. Target expected returns to solve problem (9.1) are fixed in the following way. Let $\mathbb{E}r_{\mathrm{M}}$ be the higher, among asset classes, expected return and let $\mathbb{E}r_{\mathrm{GMV}}$ be the expected return on the global minimum variance portfolio.[2] Then divide the range $(\mathbb{E}r_{\mathrm{GMV}}, \mathbb{E}r_{\mathrm{M}})$ in $Z - 1$ sub-intervals

$$\left\{ \mathbb{E}r_{\mathrm{GMV}}, \mathbb{E}r_{\mathrm{GMV}} + \delta, \ldots, \mathbb{E}r_{\mathrm{GMV}} + (Z - 1) \times \delta = \mathbb{E}r_{\mathrm{M}} \right\},$$

where

$$\delta = \frac{\mathbb{E}r_{\mathrm{M}} - \mathbb{E}r_{\mathrm{GMV}}}{(Z - 1)}.$$

The portfolio whose expected return is $\mathbb{E}r_{\mathrm{GMV}}$ is called portfolio with rank 1, while the one whose expected return is $\mathbb{E}r_{\mathrm{M}}$ is called portfolio with rank Z. Portfolios with intermediate expected return are then named accordingly.

4. Assuming that asset returns are from a multivariate normal distribution, with mean and variance–covariance matrix equal to the sample ones and estimated in Step 2.
5. Simulating N independent draws (i.e., resampling) for each asset class from the multivariate normal distribution as defined in Step 4, with each draw consisting in T random numbers for each asset class.
6. For each simulation $i = 1, \ldots, N$, re-estimating a new set of optimization inputs, μ^i and Σ^i, and finding Z simulated efficient portfolios (i.e., repeating N times Step 3).

---

[2] This means that $\mathbb{E}r_{\mathrm{M}} = \max_{i=1,\ldots,n} \hat{\mu}_i$. Instead, $\mathbb{E}r_{\mathrm{GMV}}$ is the expected return of the portfolio that solves the selection problem $\min_{\omega} \omega \widehat{\Sigma} \omega'$, under the constraints $\omega \mathbf{1} = 1$ and $\omega \geq \mathbf{0}$. The constraint on the target expected return is not included.

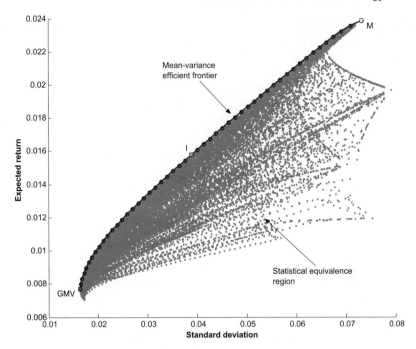

Fig. 9.1. Mean-variance efficient frontier and statistical equivalence region.

7. For each rank j, $j = 1, \ldots, Z$, computing the average composition across the N simulations. The Z portfolios with this average composition are called resampled portfolios of rank j.

Figure 9.1 provides a graphical representation of the statistical equivalence region and gives an immediate idea on how estimation errors can affect the determination of an efficient frontier. Note that Fig. 9.1 is constructed with the following additional steps.

(a) Calculate the expected return and the standard deviation of each of the Z mean-variance efficient portfolios using the sample mean and the sample variance–covariance matrix obtained in Step 2. Plot the MV efficient frontier.

(b) Calculate the expected return and the standard deviation of each of the $(N \times Z)$ resampled portfolios using the original sample mean and variance–covariance matrix. Plot these $(N \times Z)$ combinations variance–expected return on the same graph as the true MV efficient frontier.

Clearly, *resampled* portfolios will lie below the MV efficient frontier, as they are sub-optimal with respect to optimized portfolios based on sample estimates of the mean and the variance–covariance matrix. Nonetheless, all of these portfolios may still be considered statistically equivalent to portfolios plotted along the MV efficient frontier.

This procedure can be very useful for two main reasons. First, it provides a means of testing whether portfolios can be considered to be statistically equivalent. This may be useful in asset allocation as it may increase the stability of the optimal portfolios in time (issue (c) mentioned in the previous section) and avoid costly re-balancing. Second, the adoption of resampled portfolios may be a way of increasing portfolio diversification and stability along the frontier (problems (a) and (b) mentioned in the previous section), while maintaining coherence with the postulations of the Markowitz theory.

9.3 Implementation and Algorithm

In this section we present the MATLAB® functions used to perform the analysis and simulations. Functions used are:

- `effront.m`
- `resampfront.m`
- `simul.m`
- `stateqregion.m`
- `confregion.m`
- `resampstats.m`

Function `effront.m` computes the efficient frontier solving a standard quadratic programming problem. Optimal portfolios satisfy the expected return ranking constraint, i.e. they are equally distant in terms of expected return. A short-selling constraint is also imposed. Note that we are presenting a very simple code just to highlight basic computational steps. The reader is invited to extend the code. For instance, by adding a procedure to check if the variance–covariance matrix is semi-definite positive, or by inserting new constraints so as to limit the exposure of some asset classes.[3]

`Resampfront.m` is a loop function. It repeatedly applies `effront.m` by relying on time series generated by `mvnrnd.m`.[4] Note that the `resampfront.m` code strictly follows on a step-by-step basis the procedure described in Sect. 9.2. Using codes presented here, both our in-sample and out-of-sample analyses can be viably conducted. In-sample analysis is performed using the m-files `stateqregion.m`, `confregion.m` and `resampstats.m`. The function `stateqregion.m` jointly plots the mean-variance set and the statistical equivalence region.

---

[3] Note that if the global minimum variance portfolio and the maximum expected return portfolio coincide, the efficient frontier reduces to a single point. This event is unlikely but definitely possible. As the resampling procedure is based on averaging the simulated portfolios, we require that the algorithm always generates the same number of portfolios. Therefore, when only one efficient portfolio exists `effront.m` builds a $(K \times Z)$ matrix of weights which replicates the existing portfolio weights Z times. A standard mean-variance algorithm would output a single weight vector $(K \times 1)$, resulting in a dimensionality error for the resampling code.

[4] `mvnrnd.m` is the multivariate random number generator built in MATLAB®.

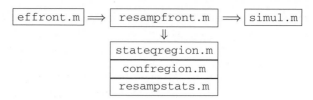

Fig. 9.2. Flow-chart.

Command `confregion.m` finally plots confidence regions for a given set of resampled portfolios. Confidence regions are computed following the procedure described in Michaud (1998), which may be summarized as follows:

1. For each portfolio on the efficient frontier, 500 statistically equivalent portfolios are generated by simulation;
2. Resampled portfolios are computed as a mean across simulated portfolios;
3. The variance of a generic simulated portfolio, ω_S, relative to the correspondent resampled portfolio, ω_R, is defined as

$$rv = (\omega_S - \omega_R)\,\Sigma\,(\omega_S - \omega_R)';$$

4. Portfolios belonging to the $\alpha\%$ confidence region are those for which $rv \leq rv^*$, where rv^* is the α percentile of the distribution of relative variances.

Function `resampstats.m` generates a structure array[5] which collects sample statistics on the distribution of portfolio weights. Function `simul.m` is a loop function used to simultaneously construct the MV frontier and perform the resampling procedures (results presented in next section are based on `simul.m`). The following flow-chart (Fig. 9.2) presents the logic relationships between functions described in this section.

To perform in-sample analysis:

1. Load a return time series, `RetSeries`, into the workspace. `RetSeries` needs to be a $(T \times K)$ matrix, where T indicates the number of returns and K the number of assets. Returns in `RetSeries` start from more recent observations going back into the past. Fix the number of portfolios `NumPortf` to be generated along the efficient frontier and the number `N` of simulations to be run.
2. Call function `resampfront.m` to generate the mean-variance and the resampled frontiers.

```
>> [Wrsp,ERrsp,SDrsp,Wmv,ERmv,SDmv,Wmv_S]
   = resampfront(RetSeries,NumPortf,N)
```

3. Call function `stateqregion.m` to generate a plot of the statistical equivalence region.

```
>> stateqregion(RetSeries, ERmv, SDmv, Wmv_S)
```

---

[5] Structure arrays are particular arrays having fields. Each field can store data of different type and dimension.

4. Call function confregion.m. This function generates a plot of the confidence region for a set of resampled portfolios – specify the portfolio set by mean of the input vector (PortfSet), specify the confidence level (ConfLevel).

```
>> confregion(Wrsp,ERrsp,SDrsp,ERmv,SDmv,Wmv_S,
                RetSeries,PortfSet,ConfLevel)
```

5. Call function resampstats.m to obtain sample statistics about the distribution of portfolio weights.

```
>> [Stats] = resampstats(Wmv, Wmv_S, Wrsp, PortfSet,
                ConfLevel)
```

Note that Steps 3–5 can be performed independently. For instance, one can decide to obtain simply confidence regions performing Steps 1, 2 and 4.

To perform out-of-sample simulation:

1. Load a return time series, RetSeriesTotal – postscript Total has been chosen to indicate a time series which spans the entire simulation period.
2. Call function simul.m – this function will repeatedly call mean variance and resampling procedure comparing the performance of the two methods and storing results period by period. Input data T indicates the time length of the in-sample period; N and *NumPortf* respectively indicate the number of simulations to perform in resampling procedure and the number of portfolios defining mean variance and resampled set.

```
>> [ASRmv,ASRrsp,ATOmv,ATOrsp]
  = simul(RetSeriesTotal,T,N,NumPortf)
```

9.4 Results and Comments

In this section, we provide a description of data used to perform simulations and discuss our results. In Sect. 9.4.1, we analyze the relationship between mean-variance and resampled portfolios. Section 9.4.2 presents results of the out-of-sample simulation.

Table 9.1 shows the input asset classes used together with benchmarks chosen to represent them. We consider the MSCI equity indices of six countries (Canada, US, Germany, Japan, France, UK),[6] plus Merril Lynch bond indices for the Euro and US area. All returns are expressed in US dollars. Stock Benchmark indices are downloaded from Datastream, and bond benchmarks from Bloomberg.

For each asset class we compute monthly logarithm returns for the period 1/1/85 to 8/31/02, for a total of $T = 200$ observations. We start our analysis by generating

---

[6] MSCI provides global equity indices, which, over the last 30 years, have been the most widely used international equity benchmark indices by institutional investors. Additional information on the construction of these indices can be found at www.msci.com.

Table 9.1. Asset classes for empirical analysis

Asset class	Benchmark
Canada	MSCI Canada
US	MSCI US
Germany	MSCI Germany
Japan	MSCI Japan
France	MSCI France
UK	MSCI UK
Euro Bonds	ML Euro aggregate
US Bonds	ML US aggregate

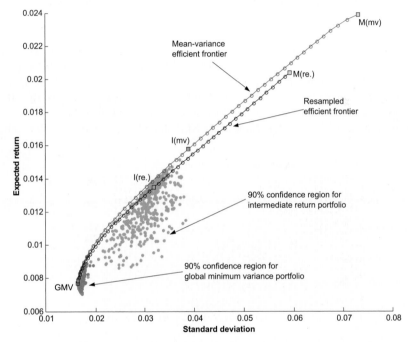

Fig. 9.3. 90% confidence regions for global minimum variance and intermediate return portfolios.

mean-variance and resampled efficient frontiers (see Figs. 9.1 and 9.3) with respect to the first 50 observations, i.e., $T = 50$. We then perform a rolling out-of-sample simulation by comparing the performance of mean-variance and resampled efficient portfolios. The in-sample period is respectively set equal to 30, 60 and 90 periods. The out-of-sample period is kept fixed and equal to $T = 100$. Further details of these simulations will be given in Sect. 9.4.2 where we describe the results obtained.

9.4.1 In-sample Analysis

The results of the in-sample analysis are presented in Figs. 9.1, 9.3 and 9.4. Table 9.2 supports the results obtained.

Figure 9.1 shows the mean-variance efficient frontier together with the statistical equivalence region (SER). The first is computed using a standard quadratic programming algorithm; the second is obtained applying the procedure described in Sect. 9.3. Sample data is of length $T = 50$ and asset classes are those described in Table 9.1. As previously mentioned, 25500 statistically equivalent portfolios are "evaluated" using the true mean vector and variance–covariance matrix. As a consequence they all plot under the efficient frontier. This is a rather obvious result. Less obvious is the high level of dispersion that characterizes the statistically equivalent portfolios with respect to the mean-variance set. This result brings us to at the heart of the problem: even small changes in the sample data can cause a drastic change of the mean-variance efficient curve. In the words of Michaud, one can say that Fig. 9.1 *"dramatically illustrates the enormous, even startling, variability implicit in efficient frontier portfolio estimation"*. The size of the SER suggests that much of the effort put in the optimization procedure can be worthless or even dangerous. Note that not only the dimension but also the shape of the SER should be considered. SER exhibits

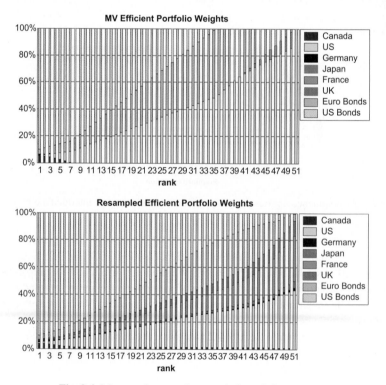

Fig. 9.4. Mean-variance and resampled portfolio weights.

a characteristic "comet" shape. Dispersion increases moving from the lower to the upper part of the efficient frontier (i.e., moving from lower expected return portfolios to the higher expected return portfolios). In order to understand this, examine Fig. 9.3 displaying the mean-variance efficient frontier together with the resampled efficient frontier (REF). Let us focus on three specific portfolios on the mean-variance efficient frontier: GMV (the Global Minimum Variance portfolio; which ranks as the 1st portfolio), I (the Intermediate Return portfolio; which ranks as the 26th portfolio) and M (the Maximum Expected Return portfolio; which ranks as the 51st portfolio). Each of these three portfolios has a corresponding matching portfolio on the resampled frontier. These matching portfolios all plot along a curve, the REF, which is located below the efficient frontier (a fact given the construction[7]). The distance between portfolios on the mean-variance frontier and their match on the resampled efficient frontier is then increasing as one moves towards higher return variance (that is, to the right) in the standard deviation–expected return (SD, ER) space. Moreover, the resampled frontier is "shorter" than the mean-variance frontier, since resampling rules out extreme and poorly diversified portfolios. As a result, the method is implicitly providing investors with a prudential and safer set of portfolios, and this implicit protection is increasing in the horizontal axis. When low levels of risk are considered, mean-variance and resampled portfolios are very similar in terms of reward/risk ratio (for instance, look at the case of GMV portfolios). When intermediate or high levels of risk are considered, differences in terms of reward/risk ratio between mean-variance and resampled portfolios are instead remarkable (as for the I and M portfolios).

In Fig. 9.3, we also identify the 90% confidence interval regions for the Global Minimum Variance and the Intermediate portfolios. These confidence regions differ from each other. The region surrounding the GMV portfolio (the dark nebula) appears very dense and compact. On the contrary, the confidence region around the Intermediate portfolio (the green nebula) circumscribes a larger area. Admittedly, the confidence region around portfolio M (Maximum Expected Return portfolio), which is not represented here, would cover an even wider portion of the (SD, ER) space. These results are consistent with the higher level of estimation error that characterizes intermediate and extreme portfolios vis-à-vis the GMV portfolio. It is thus easy to understand why the statistical equivalence region in Fig. 9.1 exhibits a 'comet' shape: the simulation procedure highlights the growing level of instability to which one is exposed when moving upward and to the right on the efficient frontier.

The analysis just illustrated is further supported by data provided in Table 9.2, where a comparison between resampled and mean-variance weights, respectively for portfolios GMV, I and M is presented. At the GMV portfolio level, differences in the portfolio weights between resampling and mean-variance are practically negligible. Moving to the intermediate return level, differences in portfolio weights become re-

---

[7] Each resampled portfolio is determined by averaging the weights of the rank associated simulated portfolios. As these portfolios are sub-optimal by definition (as they are evaluated using the original sample mean vector and var/cov matrix), the resampled portfolios will be sub-optimal too.

Table 9.2. Distribution statistics

	Resampled weights	St. dev.	5% perc.	Median	95% perc.	MV weights
Global minimum variance portfolio						
Canada	3.96	2.15	0.36	4.02	7.44	5.13
US	0.79	1.57	0	0	4.21	0
Germany	1.17	1.44	0	0.60	4.36	1.11
Japan	0.49	0.74	0	0	2.04	0.17
France	0.06	0.26	0	0	0.4	0
UK	0.33	0.78	0	0	1.96	0
Euro Bonds	3.23	2.65	0	2.95	8.06	3.53
US Bonds	89.96	3.24	84.19	90.39	94.57	90.05
Intermediate (expected) return portfolio						
Canada	1.14	4.46	0	0	6.44	0
US	19.18	19.33	0	13.64	49.54	35.65
Germany	0.31	1.56	0	0	0.41	0
Japan	1.09	4.5	0	0	7.35	0
France	8.86	14.88	0	0	43.67	0
UK	4.71	12.03	0	0	39.78	0
Euro Bonds	22.92	20.33	0	23.04	50.63	36.37
US Bonds	41.78	19.55	17.09	40.81	64.62	27.99
Maximum expected return portfolio						
Canada	0.10	5.45	0	0	0	0
US	43.20	44.62	0	0	100	100
Germany	2.00	14.07	0	0	0	0
Japan	5.00	21.90	0	0	0	0
France	33.00	47.26	0	0	100	0
UK	11.00	31.45	0	0	100	0
Euro Bonds	4.90	4.26	0	0	0	0
US Bonds	0.80	3.41	0	0	0	0

markable. Resampling tends to preserve diversified portfolios, whilst mean-variance optimization tends to concentrate allocation (only 3 asset classes are invested in). The striking case is the M portfolio case. The resampling portfolio shows a tendency towards the 'star' asset class (US stocks = 43% of the portfolio), but diversification is still preserved. In the mean-variance setting, diversification is instead completely abolished and an extreme 100% US stock portfolio is built. Alas, a portfolio with such characteristics is likely to maximize estimation errors and will get poor out-of-sample performances.

Table 9.2 also collects information on the statistical properties of the portfolio weights. First, consider standard deviations. These increase moving from GMV to M. Now consider the percentile intervals. Again their magnitude increases moving from GMV to M. This is consistent with the affirmation that estimation risk affects extreme portfolios more than low-risk ones.

Figure 9.4 exhibits efficient portfolio weights generated both in mean-variance and resampling settings. Lack of diversification and tendency towards extreme positions are evident. Two main observations can be drawn. First, mean-variance and resampled portfolios are quite similar for low levels of expected return, while they tend to diverge moving to higher expected return levels. This justifies previous comments on Fig. 9.3. Second, instability of portfolio weights grows moving upward along the frontier. Note, for instance, the abnormal difference in weights between rank 50 and rank 51 in the mean-variance setting: at least 15% of portfolio composition is changed.

9.4.2 Out-of-sample Simulation

In the previous section, we have illustrated problems that may arise in asset allocation when investors are exposed to estimation risk. The analysis we have conducted was based on an in-sample approach. We now move to an out-of-sample approach. Using data on the eight asset classes, we implement a rolling out-of-sample analysis. Our aim is to test which asset allocation strategy performs better between the mean-variance approach and the resampled procedure. Three simulations are performed. They differ for the size of the estimation period. Three different sampling lengths are considered: $T = 30$, $T = 60$ and $T = 90$. In each case, the out of sample period length remains the same: $T = 100$. Simulations are performed considering three portfolios: GMV, Intermediate return and Maximum (expected) return portfolios. For each period optimal portfolios are computed, then the sample period is moved forward one month and optimization is repeated. Using the optimal weights computed for the previous period together with the actual returns of the asset classes, the algorithm computes realized returns generated by the optimal portfolios. Realized returns are then averaged across the out of sample period and normalized for the average risk of the portfolio. In this way, an (average) realized Sharpe ratio is computed for all three portfolios in both the cases of the mean-variance and resampling strategies. Note that resampling is carried out by generating return series of a length equal to that of the sample period considered.[8] The number of simulations for resampling the frontier is set equal to 500. Thus, a resampled portfolio with rank j is obtained as an average of 500 simulated portfolios, always with same rank j. Tables 9.3 and 9.4 present the results. A comparison of the mean-variance and resampling techniques can be done by considering both the out-of-sample performance and the turnovers. The out-of-sample performance is summarized by the average Sharpe ratios reported in Table 9.3. For $T = 30$, resampled portfolios perform better than mean-variance portfolios in all cases. This makes sense. The estimation period in this case is quite short and it is therefore reasonable to expect that the effect of estimation risk will be more significant, penalizing mean-variance and favoring resampling methods. Note that the over-performance of the resampling procedure with respect to mean-variance increases when moving from GMV to M. We can justify this result by recalling that

---

[8] For instance, when the estimation period is of length $T = 30$, the simulated series have length $T = 30$.

Table 9.3. Sharpe ratios for different horizons

	Mean-variance			Resampling		
	Return	Risk	Sharpe ratio	Return	Risk	Sharpe ratio
$T = 30$						
GMV	0.5587	1.0828	0.5160	0.5443	1.0941	0.5249
I	0.4822	2.1829	0.2209	0.5628	1.9390	0.2903
M	0.1376	4.5968	0.0299	0.4688	3.7186	0.1261
$T = 60$						
GMV	0.5536	1.1574	0.4783	0.5597	1.1631	0.4812
I	0.6561	1.9438	0.3376	0.5682	1.9009	0.2989
M	0.7969	3.9009	0.2043	0.5352	3.4432	0.1554
$T = 90$						
GMV	0.5454	1.2255	0.4450	0.5470	1.2294	0.4449
I	0.5684	2.0567	0.2764	0.5144	1.9775	0.2601
M	0.7668	3.9053	0.1963	0.4294	3.5797	0.1200

Table 9.4. Portfolio turnovers

	Mean-variance turnover	Resampling turnover	Differential turnover
$T = 30$			
GMV	3.3453	3.0274	0.3179
I	13.2883	8.2368	5.0516
M	17.1717	13.3697	3.8020
$T = 60$			
GMV	1.5135	1.5050	0.0084
I	6.7243	5.3784	1.3459
M	16.1616	10.5455	5.6162
$T = 90$			
GMV	1.1279	1.1205	0.0074
I	4.9018	4.5062	0.3956
M	5.0505	9.1131	−4.0626

estimation risk affects more extreme portfolios (refer to Fig. 9.2). If we move to the second and third simulations, $T = 60$ and $T = 90$ respectively, results go in the opposite direction: mean-variance optimization over-performs resampling in all cases (except the case of the GMV portfolio in simulation 2 where resampling obtains a small over-performance). Thus, increasing the sample size leads to a better performance of mean-variance portfolio selection with respect to the resampling technique. This result could depend on the following:

1. By increasing sample size, estimation risk decreases and the quadratic programming procedure is more likely to produce portfolios that are real winners.
2. The improvement that resampling can give to the performance of a portfolio "*critically depends on the relevance of the inputs for the forecast horizon*" as stressed by Michaud (1998).

3. Sample period considered: in strong trending markets, like the one considered, to assume extreme positions can be highly rewarding.

Therefore, we are unable to conclude that resampled portfolios out-perform mean-variance portfolios in terms of average out-of-sample Sharpe ratios. This lack of exhaustive indications should be nonetheless read in conjunction with the considerations that follow on the marginal cost of portfolio selection in the two cases. In particular, consider the (average) turnover of the portfolios. This quantity measures the rate of trading activity across portfolio assets and, hence, it represents the percentage of portfolio that is bought and sold in exchange for other assets. There are several ways to calculate this quantity. In the present study turnover has been computed as the sum of the absolute values of purchases and sales during a pre-set time period, divided by 2. Let $\omega(t-1)$ and $\omega(t)$ be K-dimensional column vectors representing portfolio weights at time $t-1$ and t respectively. Then, the portfolio turnover from time $t-1$ to time t, $TO(t-1,t)$, is defined as

$$TO(t-1,t) = \frac{\sum_{i=1}^{K} |\omega_i(t-1) - \omega_i(t)|}{2},$$ (9.3)

where $\omega_i(t)$ and $\omega_i(t+1)$ indicate the weight of the generic asset class i at time $t-1$ and t respectively.

As an example, let us consider the case in which

$$K = 2, \qquad \omega'(t-1) = \begin{bmatrix} 1 \\ 0 \end{bmatrix}, \qquad \omega'(t) = \begin{bmatrix} 0 \\ 1 \end{bmatrix}.$$

Therefore,

$$TO(t-1,t) = \frac{|1-0| + |0-1|}{2} = 1$$

i.e., a one hundred percent turnover moving from time $t-1$ to time t. Formula (9.3) can be easily converted in MATLAB® code. Using the functions sum and abs, we can write

```
TO = sum(abs(W_minus1-W))/2.
```

This syntax has been used in the simul.m code.

In all cases (except for portfolio M when $T = 90$) the differential turnover is positive (i.e., mean-variance portfolios exhibit higher turnovers compared to resampled portfolios). Therefore, the over-performance obtained in simulations 2 and 3 through mean-variance quadratic optimization was obtained at the cost of a higher turnover. As trading is costly, this result can modify the judgement on which of the two methods is really optimal.

10

Alpha, Beta and Beyond[*]

Key words: beta estimation, OLS, robust estimate, Bayesian method, Kalman filter, shrinkage, backtesting

Although academics and practitioners continue to debate its relevance,[1] the beta, which is a measure of stock sensitivity to market movements, has become the best known and most widely employed measure for market risk. Similarly, the use of the beta to estimate expected returns, finds in the Capital Asset Pricing Model (CAPM) the theoretical foundation for justifying the current practice in the investment industry (for a presentation of the CAPM we refer to Barucci (2003), Cochrane (2001) and Sharpe, Alexander and Bailey (1999)). Indeed, as early as in 1982, Gitman and Mercurio (1982) were already reporting that slightly more than 50 percent of managers were familiar with the beta and were using it in their financial activities. Sixteen years later, Bruner et al. (1998) confirmed that betas had, by that time, become the dominant technique to estimate the firm cost of equity. Similar evidence was then provided by Block (1999) during the following year. According to his survey, at the turn of the century, more than 30 percent of respondents considered the beta an important tool in the valuation process and were using it in their business.

Besides providing evidence on the affirmation of the beta, these surveys show that managers, traders and analysts in the investment management industry generally purchase beta estimates from commercial providers. And since multifactor return models have become increasingly popular among practitioners, they have extended this habit to the purchase of other return factor estimates. In this manner, commercially prepared estimations have acquired unparalleled importance in the financial industry.

A review of a variety of commercial providers (such as Bloomberg and Reuters) reveals that, in estimating betas, ordinary least squares (OLS) (or some adjusted versions of it) is the preferred technique. The use of OLS is justified by the fact that

---

[*] with Samuele Marafin, Francesco Martinelli and Carlo Pozzi.
[1] See the seminal work by Fama and French (1992).

square error minimization is the best way to estimate parameters in linear models, provided some stringent assumptions hold. Unfortunately, these assumptions often fail to verify, as many stylized facts complicate empirical estimations. As a result, many techniques have been devised to circumvent different problems and should today be considered when using simple or multiple regression models as an aid to decision-making activities.

In this chapter, we provide a review of some relatively simple and more advanced estimation methodologies. In Sect. 10.1, we begin with OLS and then we describe how stylized facts can be dealt with an appropriate estimation methodology. First, we tackle the problem of the occurrence of exceptionally-large or exceptionally-small return observations (outliers), which confer non-Gaussian features to the distribution of data samples (normality assumption is fundamental in OLS estimation). So in Sect. 10.2.2 robust regression is reviewed as a possible solution for this problem. Then, we consider complications given by small data samples and in Sects. 10.2.3 and 10.2.4 we discuss shrinkage and Bayesian estimates as a means of fortifying estimation by using information in the cross-section of multiple return time series. Next, changes in the correlation structure across return factors are taken into account. Indeed, the way return factors influence observed returns evolves in time and so do factor sensitivities (i.e., the regression parameters that linearly link observed returns to return factors). Simple methods, such as rolling regressions, which update beta estimates and keep track of factor changes can be applied; but they tend to average out past factor sensitivities rather than predicting where these are going to be in the future. Therefore, we propose an adaptive procedure, like the exponential smoothing in Sect. 10.2.5 and the Kalman filter in Sect. 10.2.6, so as to consider the tendency of the beta to vary over time. Finally in Sect. 10.4, all considered estimation approaches are comparatively examined. OLS, shrinkage, Bayesian and robust estimation, exponential smoothing and Kalman filter are juxtaposed in order to draw inference on their estimation performance. The comparison between them is provided by an out-of-sample analysis to gauge their predicting power in value at risk modelling. This analysis concludes the chapter.

10.1 Problem Statement

We now introduce the market model, wherein the return on equity assets is related on a market index in a linear way

$$r_{i,t} = \alpha_i + \beta_i r_{I,t} + \varepsilon_{i,t}, \tag{10.1}$$

where $r_{i,t}$ denotes the return on the ith stock, $r_{I,t}$ the return on a market index (such as the S&P 500) and $\varepsilon_{i,t}$ the part of stock return which is not explained by the market index, all referred at time t. This model is also known as the one factor model and is different from the CAPM for at least two reasons. Unlike the CAPM, it is not an equilibrium model that aims to explain how asset prices are set. Second, in (10.1) we use a market index such as the S&P 500 as factor, whilst the CAPM involves the market portfolio, i.e. the portfolio composed of all securities (not only the financial ones) in the market place.

The standard estimation assumptions underlying equation (10.1) are:

$$\mathbb{E}(\varepsilon_{i,t}\varepsilon_{j,t}) = \begin{cases} \sigma_{\varepsilon_i}^2, & i = j, \\ 0, & i \neq j, \end{cases} \forall t,$$

$$\mathbb{E}(\varepsilon_{i,t}\varepsilon_{i,s}) = 0, \quad t \neq s, \forall i,$$

$$\mathbb{E}(\varepsilon_{i,t}r_{I,t}) = 0, \quad \forall i, t.$$

The coefficient β_i measures how stock returns respond to changes in the market index and therefore constitutes an important element in risk-management analysis. Beta is mathematically defined as

$$\beta_i = \frac{\sigma_{i,I}}{\sigma_I^2},$$

where $\sigma_{i,I}$ is the unconditional covariance of the ith stock returns with returns on the market index I, and σ_I^2 is the unconditional variance of the market I. The formula for β_i arises from the following remark. If we consider the equation $r_{i,t} = \alpha_i + \beta_i r_{I,t} + \varepsilon_{i,t}$, we observe that for given α_i and β_i we get a value for the residual $\varepsilon_{i,t}$. Let us define a measure of goodness of the linear relationship as the percentage of the variance that is explained by the regression, i.e. $R^2 = 1 - \text{Var}(\varepsilon_{i,t})/\text{Var}(r_{i,t})$. The term in the numerator is a measure of the amount of randomness in the residual, which is zero if and only if the approximation (10.1) is exact. The term in the denominator is a measure of the amount of randomness in the original invariants, as it is proportional to the average of the variances of all the invariants. The model (10.1) is viable if the R^2 approaches one. A value close to zero indicates that the factor model performs poorly. Therefore, if we look for the value of $\{\alpha_i, \beta_i\}$ that maximizes R^2, then we obtain $\beta_i = \sigma_{i,I}/\sigma_I^2$ and $\alpha_i = \mathbb{E}(r_{i,t}) - \beta_i\mathbb{E}(r_{I,t})$. Greater details can be found in Meucci (2005).

From the financial point of view, $\text{Var}(r_{i,t})$ measures the total riskness of the asset i, $\text{Var}(\varepsilon_{i,t})$ measures the risk that can be diversified away and $\beta_i^2 \text{Var}(r_{I,t})$ tracks the source of asset risk which is systematic in the market (indeed is related to the variance of the market index) and therefore cannot be diversified away. Given the market index variance, investors thus expect a remuneration which is explained by beta, because such a term measures the non-diversifiable risk of an equity. In particular, the CAPM states that a higher beta, i.e. a larger exposure to market risk, requires a higher expected return. Residuals ($\varepsilon_{i,t}$) tend instead to offset each other when more stocks are pooled into an equity portfolio. As such, the risk they bear (observe that, given our assumptions, the covariance of residuals with the market index is zero) is unimportant because it is easy and inexpensive to diversify it away through portfolio management. In the literature on the CAPM, this type of risk is also often called *idiosyncratic* risk, since it is the specific risk of a single equity.

10.2 Solution Methodology

Once applied to empirical data, the single-index model may pose challenges to analysts confronting the task of finding the best estimates for betas. In particular, the difficulty is knowing which way is preferable to estimate β, considering that different techniques are available to solve different problems. As discussed in the introduction, non-normality in the data sample, small samples, variations in risk factors and different possible estimation periods represent common complications. We discuss these challenges and their solutions. But first we need to clarify how OLS estimation determines β in order to understand why the complications just cited limit its validity as an estimation technique. In the following our attention will be concentrated on the single index model, although multifactor models are becoming popular in the financial industry because they allow a better understanding of the multivariate facets of risk, see Meucci (2005), Sects. 3.4, 4.2.2 and 7.3.

10.2.1 Constant Beta: OLS Estimation

Linear regression estimation requires minimizing the sum of squared residuals between predicted and observed values of the dependent variable. Let us suppose to have collected T historical observations on the stock and on the market. The OLS estimators $\hat{a}_{i,OLS}$ and $\hat{\beta}_{i,OLS}$ of the unknown parameters a_i and β_i in the simple linear model admit the following matrix representation

$$[\hat{a}_{i,OLS}, \hat{\beta}_{i,OLS}]' = (\mathbf{X'X})^{-1}\mathbf{X'r},$$

where $\mathbf{X}$ is a $T \times 2$ matrix, having in the first column all elements equal to 1 and in the second column the returns on the market index. $\mathbf{r}$ is a $T \times 1$ vector containing the stock returns.

In econometric investigation, it is often preferable to choose an estimation method that produces unbiased, efficient and consistent estimators. These three criteria denote the statistical properties that an estimator of $\boldsymbol{\beta}$ should have. Admittedly, OLS estimators are widely employed because – besides being fairly easy to determine – they enjoy the above quoted properties under certain general conditions. They are unbiased, since their expected value is equal to the parameter they estimate, i.e. $\mathbb{E}(\hat{\beta}_{i,OLS}) = \beta_i$ and $\mathbb{E}(\hat{a}_{i,OLS}) = a_i$. They are efficient, as the volatility of their value, measured through variance, attains a certain minimum threshold. Indeed, according to the Gauss–Markov Theorem, OLS are the best linear unbiased estimator (BLUE): they have the smallest variance among all linear unbiased estimates of β. In particular, the variance–covariance matrix of these estimators is

$$\mathrm{Var}([\hat{a}_{i,OLS}, \hat{\beta}_{i,OLS}]') = \hat{\sigma}_{\varepsilon_i}^2 (\mathbf{X'X})^{-1},$$

where $\hat{\sigma}_{\varepsilon_i}^2$ is the mean squared error of the regression model,

$$\hat{\sigma}_{\varepsilon_i}^2 = \frac{(\mathbf{r} - \hat{\mathbf{r}})'(\mathbf{r} - \hat{\mathbf{r}})}{T - 2}.$$

Moreover, $\hat{a}_{i,OLS}$ and $\hat{\beta}_{i,OLS}$ are also consistent because they converge in probability to the real value they are meant to estimate as the size of the sample increases to infinity.

Once the normality assumption is made about the random error in (10.1), the maximum likelihood estimator for β_i and a_i is the same as the OLS estimator. Also, since $[\hat{a}_{i,OLS}, \hat{\beta}_{i,OLS}]$ is a linear function of the normally distributed random vector $\mathbf{r}$, we have

$$[\hat{a}_{i,OLS}, \hat{\beta}_{i,OLS}]' \sim \mathcal{N}\left([a, \beta]', \hat{\sigma}_{\varepsilon_i}^2 (\mathbf{X}'\mathbf{X})^{-1}\right).$$

Given these properties, OLS should produce better estimates when long periods of data are used, although there will be a bigger possibility of introducing a bias due to old data. Instead, when short-term risk exposures need to be modelled, since sensitivities to different sources of risk (i.e., the parameters to be estimated) may change over time, determining them as an average across the entire time span of the data sample cancels out every temporal variation and darkens the impact of recent market dynamics on risk factors. Averaging observations across the entire dataset may then have an opposite effect vis-à-vis the one just cited. Moreover, with OLS estimations, a single large movement in the variables occurring during the period analyzed tends to have a persistent effect on betas and does not decrease in time until it drops out of averages (the so-called "ghost feature"). Finally, when exceptional observations (outliers) in a sample become numerous (hence, the sample under investigation is significantly non-normal), OLS estimators lose their statistical qualities and expose beta estimates to serious errors.

10.2.2 Constant Beta: Robust Estimation

Let us begin reviewing solutions to overcome OLS drawbacks by focusing on the presence of outliers that can have a substantial influence on the values of least squares estimates.[2] In financial markets, outliers are observed both among equity and market returns. However, because of portfolio effects, they are found with considerably greater frequency in the return time series of individual equities. Outliers can occur because an individual stock or the market (or both) makes an unusual move, because a split or reverse split occurs and is not removed from the data, or because of a data error. Outliers occur more frequently for companies with small capitalizations than for companies with large market caps. In many cases, the deletion of a small number of outliers, sometimes even a single outlier, changes the value of the OLS beta. Such sensitivity of the OLS beta to outliers can result in quite misleading interpretations of the risk and return characteristics of a company. We argue here that practitioners should not rely solely on OLS betas but should also compute *resistant* betas that are not significantly influenced by a small fraction of outliers. By now a number of alternatives exist to least squares that are robust according to various statistical criteria. Here we consider some of them. Two remedies are possible in this case. Under a first approach, outliers are just ignored and data samples are simply cleaned from undesirable observations defined with reference to an adequate benchmark (e.g., observations which are too distant from their mean or median). Alternatively, outliers

---

[2] See for example Judge et al. (1988), Huber (1981), Rousseeuw and Leroy (1987), and Hampel (1986).

can be *winsorized,* that is altered and transformed into more suitable values for further quantitative treatment. In this chapter, we privilege a winsorizing procedure as suggested in Martin and Simin (1999).

Consider the linear model (10.1). A robust regression estimate $[\mathbf{a}_{i,M}, \boldsymbol{\beta}_{i,M}]$ is defined as[3]

$$[\mathbf{a}_{i,M}, \boldsymbol{\beta}_{i,M}] = \arg\min_{a_i, \beta_i} \sum_{t=1}^{T} \rho\left(\frac{r_{i,t} - a_i - \beta_i r_{I,t}}{\mathbf{s}}\right), \tag{10.2}$$

where $\mathbf{s}$ is a robust scale estimate for the residuals and ρ is a symmetric *loss* function. Dividing the residual by the scale estimate $\mathbf{s}$ makes the estimator invariant with respect to the scale of the error ϵ_i. We recall two measures resistant (robust) to outliers: the median and the trimmed mean. The median is the 50th percentile of the sample, which only changes slightly if you add a large perturbation to any value. The idea behind the trimmed mean is to ignore a small percentage of the higher and lower values of a sample and then to determine the center of the sample. Another robust regression methodology uses an iteratively reweighted least squares algorithm, with the weights at each iteration calculated by applying an appropriate function to the residuals from the previous iteration. This algorithm gives lower weight to points that do not fit well. The results are less sensitive to outliers in the data as compared with ordinary least squares regression.

Least Median of Squares

One of the earliest types of robust regression is called median regression, which has the advantage of reducing the influence of the outliers. This algorithm minimizes the median of ordered squares of residuals to obtain the regression coefficients:

$$[\mathbf{a}_{i,\mathrm{med}}, \boldsymbol{\beta}_{i,\mathrm{med}}] = \arg\min_{a_i, \beta_i} \operatorname*{median}_{t}\left(\frac{r_{i,t} - a_i - \beta_i r_{I,t}}{\mathbf{s}}\right)^2.$$

Least Trimmed Squares

One way to eliminate possible outliers is to run the analysis on trimmed or winsorized distributions. This means setting the values smaller than the α quantile to that of the α quantile on one tail and setting those values larger extreme than the $1 - \alpha$ quantile to those of that quantile. The winsorized distribution is then estimated by minimizing the sum of squared absolute residuals. Let us define q_α and $q_{1-\alpha}$ to be the relevant quantiles. The least trimmed squares estimator is given by

$$[\mathbf{a}_{i,\mathrm{lts}}, \boldsymbol{\beta}_{i,\mathrm{lts}}] = \arg\min_{a_i, \beta_i} \sum_{t=1}^{T} \rho(r_{i,t} - a_i - \beta_i r_{I,t}),$$

---

[3] For a complete description of the overall computational methodology, see Yohai, Stahel and Zamar (1991).

where

$$\rho(\varepsilon) = \begin{cases} q_\alpha^2 & \text{if } \varepsilon < q_\alpha, \\ \varepsilon^2 & \text{if } q_\alpha < \varepsilon < q_{1-\alpha}, \\ q_{1-\alpha}^2 & \text{if } \varepsilon > q_{1-\alpha}. \end{cases}$$

Iteratively Reweighted Least Squares

Another robust regression analysis is performed with Iteratively Reweighted Least Squares, where we minimize a weighted sum of squared errors. If the residual is small the weight is equal to one, but if it is larger the weight is equal to a tuning constant divided by the absolute value of the scale. These are the so-called "Huber weights" referred to in Huber (1981). Since the weights are recalculated in each step we speak of an iteratively reweighted least squares algorithm. The iterations will be stopped when the parameter estimates and/or the residuals do not change significantly anymore. Other different weighting procedures can be used. For example, with biweights, all cases with small residuals are downweighted and cases with large residuals are assigned zero weights, thereby eliminating their influence altogether. Because Huber weights sometimes have problems with extreme outliers and because biweights sometimes have trouble converging on one solution, Huber weights are first used in the early iterations and then biweights are used during later iterations of the regression until final convergence is reached. Other functions to weight the residual are the 'Andrews' function, the 'Cauchy', the 'Fair', the 'Huber', the 'Logistic', the 'Talwar' or the 'Welsch' functions.

10.2.3 Constant Beta: Shrinkage Estimation

When sample observations are relatively few but regard multiple assets, *shrinkage* estimation can improve the determination of single-asset betas. A mathematical presentation can be found in Judge et al. (1988), whilst the importance of such an approach in finance is detailed in Meucci (2005). An empirical examination is performed in Lavely, Wakefield and Barrett (1980). Shrinkage exploits the information in the cross-section of betas, thereby providing less extreme estimations for a single asset. Using OLS, basically we construct a beta vector just grouping the betas individually estimated. This procedure produces a beta vector that is too dispersed, i.e. too sensible to sampling errors. Let us consider n assets. For each asset we run an OLS regression separately. The resulting n OLS estimates are grouped in a vector β whose expected length can be proved to be

$$\mathbb{E}(\hat{\beta}'\hat{\beta}) = \beta'\beta + \hat{\sigma}_{\varepsilon_i}^2 \sum_{j=1}^{n} \frac{1}{\lambda_j},$$

where λ_j are the eigenvalues of the matrix $\mathbf{X}'\mathbf{X}$. If this matrix has small eigenvalues ($\mathbf{X}'\mathbf{X}$ is said to be ill-conditioned), then the estimated betas can be very dispersed in relation to the true ones. The ill-conditioning problem can become serious as we increase the number of assets with respect to the number of observations. In order to avoid this problem, shrinkage adjusts single-asset OLS estimates and constrains their single values near a mean target value. In general, this procedure works well if the number of assets under examination is greater than three and observations are relatively limited. The simplest shrinkage technique is given by the so-called James–Stein estimator

$$\beta_{i,JS} = \bar{\beta} + \alpha_i (\hat{\beta}_{i,OLS} - \bar{\beta}). \tag{10.3}$$

In this equation, $\hat{\beta}_{i,OLS}$ is a single-stock OLS beta estimate, $\bar{\beta}$ is a mean determined across all stocks, and α_i is the shrinkage parameter. The shrinkage parameter produces a biased estimator, but with a lower variance with respect to the OLS estimator: what we pay for introducing the bias we get back reducing the variance. Admittedly, the value of α_i makes the most of this technique and analysts should pay particular attention to its determination. For example, Sharpe, Alexander and Bailey (1999, pp. 503–505) report some ad-hoc procedure to adjust historical common stock betas that have been adopted by investment firms. For example, a common practice is to adjust the historical beta toward a target value of 1 giving a 2/3 weight on the OLS estimate and a 1/3 weight on the value of 1. Here instead it is considered the case of the shrinkage parameter equals

$$\alpha_i = 1 - \frac{(k-3)v}{(\hat{\beta}_{i,OLS} - \bar{\beta})'(\hat{\beta}_{i,OLS} - \bar{\beta})},$$

where k is the number of assets ($k > 3$), v is the pooled variance of betas obtained through k individual regressions (with assets in excess of at least three stocks), with

$$v = \frac{1}{k}(\hat{\beta}_{i,OLS} - \bar{\beta})'(\hat{\beta}_{i,OLS} - \bar{\beta}).$$

This adjustment procedure has solid theoretical foundations in the work by James and Stein (1961), and can be also justified using a Bayesian approach to estimation, as described in the next subsection. For details see Judge et al. (1988).

10.2.4 Constant Beta: Bayesian Estimation

The Bayesian procedure extends the shrinkage estimation technique by integrating prior information about model parameters into the model fitting process. The Bayesian procedure extends the classical viewpoint that is based only on the likelihood of the sample data given assumed parameter values, so that prior information about model parameters becomes part of the model fitting process. Bayesian estimates are then a combination of prior beliefs and sample information. The idea is to express uncertainty about the true value of a model parameter with a prior density that describes one's beliefs about this true value. Then sample information is added and is summarized in a likelihood function that is used to update the prior density to a posterior density, using the Bayes' rule,[4] see Fig. 10.1.

---

[4] For an introduction to Bayesian estimation and application in finance we refer to Meucci (2005).

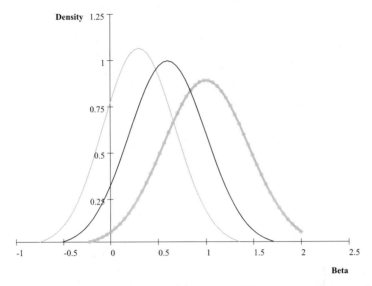

Fig. 10.1. Combining prior (density in the right) and likelihood (density to the left) to obtain the posterior (density in the middle).

The basic formula used in Bayesian analysis reads

$$f(\boldsymbol{\theta}|\mathbf{r}) \propto h(\boldsymbol{\theta})L(\mathbf{r}|\boldsymbol{\theta}), \quad \boldsymbol{\theta} \in \boldsymbol{\Omega} \subseteq \mathbf{R}^{m}.$$

Here, a posterior distribution f of the parameter set $\boldsymbol{\theta}$, given the observations $\mathbf{r}$, is related to the prior distribution $h(\boldsymbol{\theta})$ and to the likelihood function $L(\mathbf{r}|\boldsymbol{\theta})$, i.e. the distribution of $\mathbf{r}$ given $\boldsymbol{\theta}$. The posterior distribution is therefore a synthesis of the overall information in the problem: model specification, observations and prior distribution. If many parameters are involved, it is possible to integrate all from their joint posterior distribution. However, the objective is to obtain the marginal distribution of each parameter, since marginal distributions can be used to infer about single parameters.

Therefore, once the posterior distribution is computed, in order to obtain an estimate for an unknown parameter $\boldsymbol{\theta}$, it is necessary to define a function $l(\boldsymbol{\theta}, \hat{\boldsymbol{\theta}})$ that associates a possible loss in the estimate $\hat{\boldsymbol{\theta}}$ given that the true parameter is $\boldsymbol{\theta}$. The Bayesian estimate will try to minimize the expected loss, using the posterior distribution:

$$\min_{\hat{\boldsymbol{\theta}}} \mathbb{E}[l(\boldsymbol{\theta}, \hat{\boldsymbol{\theta}})] = \min_{\hat{\boldsymbol{\theta}}} \int_{\Omega} l(\boldsymbol{\theta}, \hat{\boldsymbol{\theta}}) f(\boldsymbol{\theta}|\mathbf{r}) \, d\boldsymbol{\theta}.$$

For instance, with a quadratic loss function, $l(\boldsymbol{\theta}, \hat{\boldsymbol{\theta}}) = (\boldsymbol{\theta} - \hat{\boldsymbol{\theta}})'\mathbf{A}(\boldsymbol{\theta} - \hat{\boldsymbol{\theta}})$ with $\mathbf{A}$ a semidefinite positive matrix, the parameter $\boldsymbol{\theta}$ is estimated by the mean of the posterior distribution, i.e. $\hat{\boldsymbol{\theta}} = \int_{\Omega} f(\boldsymbol{\theta}|\mathbf{r}) \, d\boldsymbol{\theta}$.

In finance, Bayesian estimators for betas were first advocated by Vasicek (1973). After their academic introduction, they have become relatively well known and the financial industry has begun to use them. In this project, we follow Vasicek (1973) and we consider the single index model,

$$r_{i,t} = a_i + \beta_i r_{I,t} + \epsilon_{i,t}$$
$$= \eta + \beta_i (r_{I,t} - \bar{r}_I) + \epsilon_{i,t},$$

where $\eta = a_i + \beta_i \bar{r}_I$ and $\epsilon_{i,t} \sim \mathcal{N}(0, \sigma_{\epsilon_i}^2)$. Vasicek (1973) assumes that the prior distribution of betas is normal with parameters b_i and s_i^2 and that the prior density of η and $\sigma_{\epsilon_i}^2$ is uninformative and independent of the prior distributions of the betas:

$$h\left(\eta, \sigma_{\epsilon_i}^2\right) \propto \frac{1}{\sigma_{\epsilon_i}}, h(\beta_i) \propto \exp\left(-\frac{(\beta_i - b_i)^2}{2s_i^2}\right)$$

$$\implies \quad h\left(\beta_i, \eta, \sigma_{\epsilon_i}^2\right) \propto \frac{1}{\sigma_{\epsilon_i}} \exp\left(-\frac{(\beta_i - b_i)^2}{2s_i^2}\right).$$

The posterior density of β_i, η, $\sigma_{\epsilon_i}^2$ can be obtained by exploiting the fact that residuals are normal. The marginal posterior density of β_i is obtained as:

$$f\left(\beta_i | \bar{r}_i, v, b; \sigma_{\epsilon_i}^2, s_i^2\right) \propto \exp\left(-\frac{(\beta_i - b_i)^2}{2s_i^2}\right)\left(T - 2 + \frac{v(\beta_i - b_i)^2}{s_i^2}\right)^{-(T-1)/2},$$

where $\bar{r}_i = \frac{1}{T}\sum_{t=1}^{T} r_{i,t}$ and $v = \sum_{t=1}^{T}(r_{I,t} - \bar{r}_I)^2$. For T larger than 20, the posterior distribution of β_i is approximately normal with mean and variance,

$$\mathbb{E}[\beta_i] = (1 - w_i)\beta_{i,OLS} + w_i b_i,$$

$$\text{Var}[\beta_i] = \frac{1}{1/s_i^2 + 1/\sigma_i^2}$$

with

$$w_i = \frac{1}{s_i^2} \frac{1}{1/s_i^2 + 1/\sigma_i^2}, \quad \sigma_i^2 = \frac{\sigma_{i,OLS}^2}{v},$$

$$\sigma_{i,OLS}^2 = \frac{1}{T-2}\sum_{t=1}^{T}(r_{i,t} - \hat{a}_{i,OLS} - \hat{\beta}_{i,OLS} r_{I,t})^2.$$

With a Bayesian approach, the estimation of betas closely depends on the variance of the distribution of prior betas. Figure 10.2 shows that when s_i^2 is low, i.e. there is a strong confidence in the prior, the weight w_i is high and $\mathbb{E}[\beta_i]$ as an estimator converges to the prior estimate b_i. On the other hand, with larger values of s_i^2, w_i tends to be lower and $\mathbb{E}[\beta_i]$ tends to the OLS estimate, i.e. only sample information is exploited. In substance, the greater the variance of the prior, the lower the informational value of the view, the greater the reliance of the model on classic OLS estimates. Vasicek (1973) suggests using the cross-sectional distribution of betas as a source to extract plausible values of the parameters b_i and s_i^2 in the prior density.

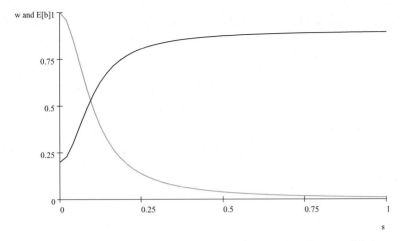

Fig. 10.2. Behavior of $\mathbb{E}[\beta_i]$ (thick size) and w_i varying s_i when $\beta_{OLS} = 0.9, b_i = 0.2,$ $\sigma_i^2 = 0.1^2$.

10.2.5 Time-Varying Beta: Exponential Smoothing

So far we have not considered the possibility that true betas can change over time. Since the appearance of the market model, the question of the beta stability has been indeed of great concern in the financial community. The first work in this sense is by Blume (1975) and since then considerable general evidence exists that the beta stability assumption is invalid. For example, during a recession financial leverage of firms in relatively poor shape may increase sharply relative to other firms, causing their stock betas to rise. At the same time, the decrease in the uncertainty about the growth prospects of the firms can cause their betas to decrease. Other reasons like technology or taste shocks, can change the relative share of different sectors in the economy, inducing fluctuations in the betas of firms in these sectors. Blume indeed presented theoretical and empirical arguments documenting a tendency of the beta coefficient towards a steady-state value, as extreme values are likely to be moderated over time.

With the exponential smoothing technique we introduce this aspect in the estimation. Observe that, by construction, OLS estimations assign equal weights to all observations in a time series. Hence, if betas follow a temporal path, OLS betas are not a good choice, since it is preferable to assign different weights to return observations across time. In order to do this, a common approach among practitioners is to rely on the exponentially-weighted moving average (EWMA) procedure. With this type of estimation, recent observations are privileged vis-à-vis older ones, thereby taking into account the dynamic ordering in returns. When EWMA beta estimations are fitted on squared return time series, unusually large returns have an immediate impact on estimated betas. The effect of return innovations, however, gradually diminishes over time, since a smoothing parameter λ allows for weighting past observations in the estimation according to their age. More precisely, in an EWMA model, the larger the value of λ, the greater the weight given to old observations and the smoother the impact of innovations on estimation results. To illustrate this procedure in detail, let us compute an EWMA estimate for the volatilities of the index and asset returns, as well as for the correlation between them. Assuming we use the same smoothing parameter λ, we have the following conditional estimations for their respective variances ($i = I$) and covariance ($i \neq I$)

$$\sigma_{iI,t}^2 = (1 - \lambda) \sum_{j=1}^{\infty} \lambda^{j-1} r_{i,t-j} r_{I,t-j}$$

$$= (1 - \lambda) r_{i,t-1} r_{I,t-1} + \lambda \sigma_{iI,t-1}^2.$$

With these conditional values it is then possible to determine a time varying beta for the time series through

$$\beta_{i,t} = \frac{\sigma_{iI,t}^2}{\sigma_{I,t}^2}.$$

EWMA beta estimates depend on two separate components. A first component tracks the reaction to market innovations and is controlled by $(1-\lambda) r_{i,t-1}^2$. The impact of an unusually large observation on this component is thus smaller the greater the smoothing parameter. A second component tracks instead the persistence in volatility and is modeled by $\lambda \sigma_{i,t-1}^2$. Through this term, the larger the λ, the greater the persistence in return volatility. As a result, with λ values closer to 1, little role is assigned to market events while much of the variance $\sigma_{i,t}^2$ is driven by the observed behavior of return volatility in the past. Conversely, with λ values closer to 0, market events acquire importance while past return volatility has less of an explanatory role.

The choice of λ, however, does not need to be discretional. Its selection can in fact be assisted by some assumptions on the distributional properties of stock returns. If these are conditionally normally distributed, an optimal value for λ can be found by constructing and maximizing the likelihood function of the sample with respect to it. We have opted for the suggestion in the Technical Document of RiskMetrics (1996) of setting λ equal to 0.97 for daily data and 0.94 for monthly data. The starting values in the above recursion when $t = 0$ can be setted equal to the sample variance, using a pre-sample dataset.

10.2.6 Time-Varying Beta: The Kalman Filter

In this paragraph we describe the Kalman filter (KF) algorithm and its use in estimating the beta at time t, based on the information available at time t. In the Kalman filter formulation we model a dynamical system using a state or observation equation and a transition equation. The state and transition equations represent the so-called state space formulation of the model. The state space form is a powerful tool which opens the way to handling a wide range of time series models. Once the model has been put in its state space form, the Kalman filter may be applied, and this leads to algorithms for prediction and smoothing. The Kalman filter also opens the way to the maximum likelihood estimation of the unknown parameters in the model. This can be done by the prediction error decomposition as shown in this section. More in detail, the KF algorithm is a recursive procedure that allows to compute the optimal estimator of the state variable at time t (the beta), using the information available until time t. Note that beta is an unobservable quantity, and then we try to use all information in the data up to time t to estimate the unobservable state variable in t.

The observation equation specifies how the return process depends on the parameter β_t, given the return on the index. Basically, this is the classical linear equation (10.1):

$$r_t = a + \beta_t r_{I,t} + \epsilon_t, \tag{10.4}$$

where we now allow for a time-variation in beta and where, for notational simplicity, we have omitted the dependence on the index i. As in (10.1), ϵ_t are the measurement errors that we assume to be normally distributed, $\epsilon_t \sim \mathcal{N}(0, \sigma_\epsilon^2)$. The transition equation then specifies how the beta evolves over time. We assume:

$$\beta_t = \bar{\beta} + \phi(\beta_{t-1} - \bar{\beta}) + \eta_t, \tag{10.5}$$

where $\eta_t \sim \mathcal{N}(0, \sigma_\eta^2)$ and $\mathbb{E}(\epsilon_t \eta_t) = 0$. The above transition equation encopasses several specifications: (a) the OLS specification is obtained with $\sigma_\eta^2 = 0$ and $\phi = 0$; (b) the random coefficient model assuming that beta fluctuates randomly around a mean value, $\beta_t = \bar{\beta} + \eta_t$, is obtained setting $\phi = 0$; (c) the random walk model, $\beta_t = \beta_{t-1} + \eta_t$, is obtained setting $\bar{\beta} = 0$ and $\phi = 1$; (d) the mean reverting model, $\beta_t = (1 - \phi)\bar{\beta} + \phi\beta_{t-1} + \eta_t$, finally assumes that next period beta will be a weighted average of this period's coefficient and its long term mean value $\bar{\beta}$.

Let us denote $\bar{\beta}_{t|t-1}$ as an estimator of the unknown state variable β_t, based on the available information (i.e. the observed return on the asset and on the market) up to time $t - 1$. The available information in t includes r_t and $r_{I,t}$ and their past values but excludes β_t. We say that $\bar{\beta}_{t|t-1}$ is the optimal estimator in the sense of mean square error if it minimizes $\Sigma_{t|t-1} = \mathbb{E}_{t-1}[(\beta_t - \bar{\beta}_{t|t-1})^2]$. Similarly, $\bar{\beta}_t \equiv \bar{\beta}_{t|t}$ is the optimal mean square estimator if it minimizes $\Sigma_t \equiv \Sigma_{t|t} = \mathbb{E}_t[(\beta_t - \bar{\beta}_t)^2]$ and now the conditioning takes into account all available information up to time t. It is well known that, minimizing the mean square forecast error provides the conditional mean of β_t as optimal estimators:

$$\bar{\beta}_{t|t-1} = \mathbb{E}_{t-1}(\beta_t) \quad \text{and} \quad \bar{\beta}_t = \mathbb{E}_t(\beta_t). \tag{10.6}$$

The Kalman filter algorithm provides a procedure for computing the estimators $\bar{\beta}_{t|t-1}$ and $\bar{\beta}_t$ and their mean square errors $\Sigma_{t|t-1}$ and $\Sigma_{t|t}$. To this aim the KF algorithm is based on two steps: (1) prediction equations that allow us to estimate $\bar{\beta}_{t|t-1}$ and $\Sigma_{t|t-1}$; (2) updating equations that, given the new information, allow for the estimation of $\bar{\beta}_t$ and $\Sigma_{t|t}$. In particular, for the univariate linear model, the prediction equations are

$$\bar{\beta}_{t|t-1} = (1 - \phi)\bar{\beta} + \phi\bar{\beta}_{t-1|t-1}, \tag{10.7}$$

$$\Sigma_{t|t-1} = \phi^2 \Sigma_{t-1|t-1} + \sigma_\eta^2, \tag{10.8}$$

whilst the updating equations are

$$\bar{\beta}_t = \bar{\beta}_{t|t-1} + \frac{\Sigma_{t|t-1} r_{I,t}}{f_t} v_t,$$ (10.9)

$$\Sigma_t = \Sigma_{t|t-1}\left(1 - \frac{\Sigma_{t|t-1} r_{I,t}^2}{f_t}\right),$$ (10.10)

$$v_t = r_t - \mathbb{E}_{t-1}(r_t) = r_t - (\alpha + \bar{\beta}_{t|t-1} r_{I,t}),$$ (10.11)

$$f_t = r_{I,t}^2 \Sigma_{t|t-1} + \sigma_\epsilon^2.$$ (10.12)

Notice that the above equations say that the optimal estimates are linear in the observation. However, this is true only when the disturbances in the state space model are normally distributed. Otherwise (10.6) is still true, i.e. the optimal mean square estimator is equal to the conditional mean of the state vector, but now the Kalman filter algorithm provides linear estimators that are suboptimal, i.e. they minimize the mean square error only among all linear estimators. But the truly optimal estimator can be nonlinear.

Once we have obtained the optimal estimate of the state vector and of its variance, we need to estimate the parameters of the model, i.e. α, ϕ, σ_η^2, σ_ϵ^2, $\bar{\beta}$ and the initial state of the system i.e. β_0 and Σ_0. If the disturbances and the initial state are normally distributed, these parameters can be obtained by maximizing the maximum likelihood function through the so-called "prediction error decomposition". When the observations are independent, their joint density can be constructed using a series of conditional probability density functions:

$$p(\mathbf{r}) \equiv p(r_t, r_{t-1}, \dots, r_1, r_0)$$
$$= p(r_t | r_{t-1}, \dots, r_1, r_0) p(r_{t-1} | r_{t-2}, \dots, r_1, r_0) \cdots p(r_1 | r_0).$$

Then the likelihood function for T observations can be obtained from the conditional probability density function:

$$\mathcal{L}\left(\mathbf{r} | r_0; \alpha, \phi, \sigma_\eta^2, \sigma_\epsilon^2, \bar{\beta}, \beta_0, \Sigma_0\right) = \prod_{t=1}^{T} p(r_t | r_{t-1}, \dots, r_1, r_0),$$

and $p(r_1 | r_0) = p(r_1)$ is the unconditional density of r_1. Given the assumption of normality of ϵ_t, we obtain that $p(r_t | r_{t-1}, \dots, r_1, r_0)$ is normal with mean $\mathbb{E}_{t-1}(r_t)$ and variance $\text{Var}_{t-1}(r_t)$. In particular,

$$\mathbb{E}_{t-1}(r_t) = \alpha + \bar{\beta}_{t|t-1} r_{I,t},$$
$$\text{Var}_{t-1}(r_t) = \mathbb{E}_{t-1}\left(r_t - \mathbb{E}_{t-1}(r_t)\right)^2$$
$$= \mathbb{E}_{t-1}\left((\beta_t - \bar{\beta}_{t|t-1}) r_{I,t} + \epsilon_t\right)^2$$
$$= r_{I,t}^2 \Sigma_{t|t-1} + \sigma_\epsilon^2.$$

The prediction error decomposition form of the log-likelihood function is then:

$$\ln \mathcal{L}\big(\mathbf{r}|r_0; \alpha, \phi, \sigma_\eta^2, \sigma_\epsilon^2, \bar{\beta}, \beta_0, \Sigma_0\big) = \ln \prod_{t=1}^{T} p(r_t|r_{t-1}, \ldots, r_1, r_0)$$

$$= -\frac{T}{2} \ln 2\pi - \frac{1}{2} \sum_{t=1}^{T} \ln |f_t| - \frac{1}{2} \sum_{t=1}^{T} \frac{v_t^2}{f_t},$$

where v_t and f_t are defined in (10.11) and in (10.12). The maximum likelihood estimate of the model parameters can be done performing the maximization of $\ln \mathcal{L}$ with respect to $\theta = (\alpha, \phi, \sigma_\eta^2, \sigma_\epsilon^2, \bar{\beta}, \beta_0, \Sigma_0)$. The Kalman filter algorithm thus consists of the following steps:

```
Step 1 (initialization step):
assign a starting value to β₀ and Σ₀.
Step 2 (prediction step):
using equations (10.7) and (10.8) compute β̄₁|₀ and Σ₁|₀.
Step 3 (updating step):
using equation (10.9) and (10.9) compute β̄₁ and Σ₁.
Step 4:
repeat step 2 and compute β̄₂|₁ and Σ₂|₁.
Step 5:
repeat step 3 and compute β̄₂ and Σ₂.
Repeat the procedure up to last observation available.
```

The state space formulation allows us to model the beta dynamics in different ways (mean reverting, random walk and random coefficient). The best specification can be then chosen using an *Information Criterion*, i.e. a measure of the distance between the true model and the Kalman estimates. Here we recall the *Akaike Information Criterion* (AIC) and the *Bayes Information Criterion* (BIC), defined as

$$\text{AIC} = -\frac{2}{T} \ln \widehat{\mathcal{L}} + 2\frac{k}{T}, \qquad \text{BIC} = -\frac{2}{T} \ln \widehat{\mathcal{L}} + k\frac{\ln T}{T},$$

where $\widehat{\mathcal{L}}$ is the maximized likelihood function, k the number of estimated parameters, and T the number of observations. The AIC and the BIC try to strike a balance between goodness of fit and parsimonious specification of the model. The model that minimizes the AIC or the BIC is the preferred one; although there is a tendency of AIC to choose overparameterized models. The Information Criterion is useful if we compare non-nested models like the random coefficient model versus the random walk model. When we compare nested models, such as random coefficient versus mean reverting, other tests can be used, like the likelihood ratio test or the Wald test. A detailed description can be found in Harvey (1994).

10.2.7 Comparing the models

In this section, we describe how to compare the different models examining their ability in predicting future betas and in estimating value at risk (VaR). As suggested in Alexander (2001), we consider statistical and operational criteria for evaluating the success of a forecast. Statistical evaluation methods compare model predictions with observable quantities such as return. An operational evaluation method is the backtesting of a value at risk model by counting losses greater than those predicted by the VaR measure. In both cases, the most important test of the forecasting power is in out-sample with a certain amount of the historic data that should be withheld from the period used to estimate the model, so given the one-period ahead forecasts, we can compare the accuracy of the different forecasting models.

Statistical Criteria

We measure the predictive ability of the alternative beta estimators, computing at time t for each estimator a return forecast for time $t + 1$ and comparing it to the realized return in the successive period. Several statistical measures of accuracy of forecasts can be computed, like mean square error, mean error, mean absolute error and mean absolute percentage error. Another statistical procedure is to perform a regression of the realized return on the return forecast. If the beta model is correctly specified, the constant from this regression should be zero and the slope coefficient should one. The R^2 from this regression will assess the amount of variation in squared returns that is explained by the successive forecast of β.

Operational criteria

Operational evaluation methods are mainly based on the profit and loss (P&L) generated by a trading strategy or on backtesting a VaR model by counting the VaR violations, i.e. the times the actual return is lower than the VaR trigger. We recall that the $\text{VaR}_{\alpha,\tau}$ is that number such that the probability of losing $\text{VaR}_{\alpha,\tau}$ or more over the next τ days is

$$\Pr_t\left(\frac{\pi_{t+\tau} - \pi_t}{\pi_t} < -\text{VaR}_{\alpha,\tau}\right) = \alpha,$$

where π_t denotes the value of the portfolio at time t. Any percentage loss over a period of length τ will be less than the value at risk with $(1-\alpha)100\%$ confidence level. If we add to the linear structure of the model a normality assumption on the distribution of the residuals and of the factors, we can use the model for VaR calculations under the normality assumption. In particular we have

$$\text{VaR}_{\alpha,\tau} = -z_\alpha\sqrt{\tau\sigma^2}, \tag{10.13}$$

where z_α is the quantile of order α of the $\mathcal{N}(0, 1)$ distribution, and σ^2 is the variance of return in the unit of time. The linear structure of factor models allows for a simple computation of the portfolio variance and for its decomposition in risk due to fundamental and specific factors.[5] Indeed, for the single asset, we have

$$\text{Var}(r_{i,t}) = \beta_{i,t}^2 \, \text{Var}(r_{I,t}) + \text{Var}(\varepsilon_{i,t}),$$

i.e. we can decompose the risk of the asset, measured by $\text{Var}(r_{i,t})$, in its two main components: market risk, $\beta_{i,t}^2 \, \text{Var}(r_{I,t})$, and specific risk, $\text{Var}(\varepsilon_{i,t})$. If in addition we assume that factors and residuals have a normal distribution and are serially uncorrelated, then:

$$r_{i,t} \sim \mathcal{N}\left(0, \beta_{i,t}^2 \, \text{Var}(r_{I,t}) + \text{Var}(\varepsilon_{i,t})\right).$$

This decomposition is valid at aggregate level as well. If we denote with $\mathbf{w}$ the $N \times 1$ vector of weights of the portfolio, then the portfolio return r_{wt}, is

$$r_{w,t} = \sum_{i=1}^{N} w_i r_{it} = a_i + \left(\sum_{i=1}^{N} w_i \beta_{i,t}\right) r_{I,t} + \sum_{i=1}^{N} w_i \varepsilon_{it},$$

and then the variance of the portfolio return to be used in the VaR formula (10.13) is

$$\text{Var}(r_{wt}) = \mathbf{w}' \boldsymbol{\beta}_t \boldsymbol{\beta}_t \mathbf{w} \, \text{Var}(r_{It}) + \mathbf{w}' \mathbf{D} \mathbf{w}.$$

The total number of exceptional losses, actual loss greater than predicted by the VaR measure, may be regarded as a random variable that has a binomial distribution. Then the expected number of exceptional losses is np, where the probability of "success" of an exceptional loss is p and the variance of the exceptional losses is $np(p-1)$. Using the fact that a binomial distribution is approximately normal when n is large and p is small, the $(1-\alpha)$ confidence interval for the number of exceptional losses is approximately:

$$\left(np - z_{1-\frac{\alpha}{2}}\sqrt{np(1-p)}, \; np + z_{1-\frac{\alpha}{2}}\sqrt{np(1-p)}\right).$$

This backtesting is named Kupiec Test (1995) and it provides an unconditional evaluation, because the forecast performance is examined over the sample period without reference to the information available at each point in time. A conditional backtesting has been proposed by Christoffersen (1998). He observes that violations should occur not only 1% or 5% of time, but they should also be independent and identically distributed (i.i.d.) over time. The corresponding test is referred as conditional coverage, with respect to the unconditional test that just compares the nominal coverage, for which the ordering of zeros and ones in the indicator sequence does not matter and only the total number of ones plays a role. However, in our empirical application we have only considered the Kupiec Test.

---

[5] The decomposition is possible for the assumption that factors and residuals are not correlated, $\text{Cov}(r_{I,t}, \varepsilon_t) = 0$.

10.3 Implementation and Algorithm

Functions and market data are stored in the folder BetaEstimation. Before running the code, you must include this folder and its subfolders in the MATLAB® directory list. To do so select Set Path from the File Menu and follow the instructions. The Beta Estimation folder contains the following MATLAB® scripts to estimate the beta according to the different procedures:

```
Par1OLS, Par1EWMa, Par1BetaBAYES, Par1BetaTrimming,

Par1BetaMedRobust, Par1BetaRobust, Par01BetaKALMAN,

KalmanRunOpt, Kalmanottimilsqnlfmin, Par02BetaKALMAN,

ForecastingBeta, LMSregor, ForecastingtrimBeta,

Backesting.
```

The scripts for plotting the estimated beta are: Graph1Ols, Graph1EWMA, Graph1bayes, Graph1robusta, Graph1kalman, Graphforec, GraphVaR.

To run the code from the Excel spreadsheet you need to have the Excel link. Estimating the Kalman filter requires the MATLAB® Optimization Toolbox.

Par1OLS.m and Graph1Ols.m – These return the beta estimation using the linear regression model (ordinary least squares (OLS)) from a rolling-window of 260 days. The input prices are those quoted by Bloomberg. The output structure also contains some regression diagnostics, like the R^2 statistic, the adjusted R^2 statistic, the Student's t statistics, p-values for each t-statistic. The Graph1Ols.m script plots the beta estimate using a rolling-window of 260 day periods.

Par1EWMa.m and Graph1EWMA.m – These return the beta estimation using the exponential smoothing model EWMA from a rolling-window of 260 day period. The input prices are market prices and the λ factor that controls the smoothing effect. In the sheet Engine.xls of the worksheet Nasdaqcomp there is a cell named "lambda" where we can set the value of this parameter, see Fig. 10.3. The default value is 0.94. The script Graph1EWMA.m plots the beta estimate using a rolling-window of 260 day period.

Par1BetaBAYES.m and Graph1bayes.m – These return the beta estimation according to the Bayesian procedure using a rolling-window of 260 days. The input prices are the market data and the parameters of the prior distribution, i.e. the mean b_i and the variance s_i^2, as shown in Fig. 10.4. Graph1bayes.m plots the estimated beta.

Fig. 10.3. Engine to estimate the beta according to EWMA.

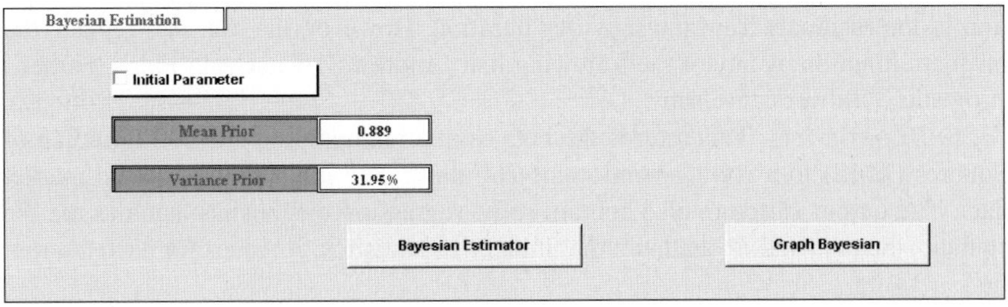

Fig. 10.4. Engine to estimate the beta according to the Bayesian procedure (Vasicek model).

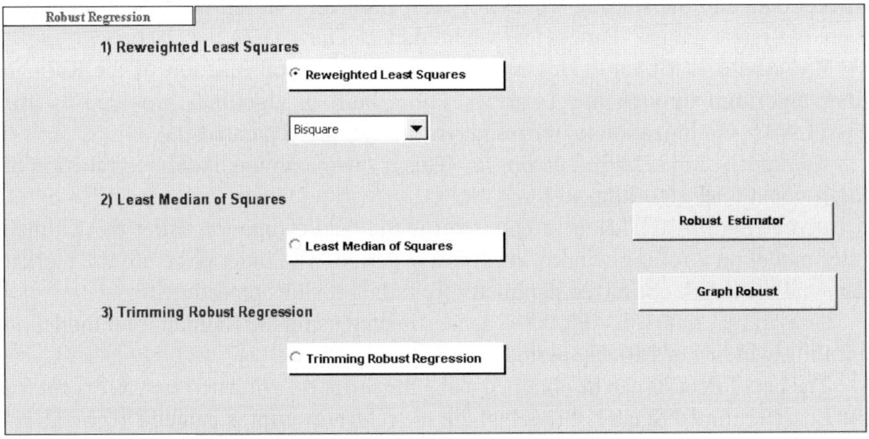

Fig. 10.5. Engine to estimate the beta according to the robust procedure (trimmed mean, reweighted least squares, LMS).

For the robust procedure, the estimation can be run according to three different procedures: trimming, reweighted least squares and least median of squares. The corresponding scripts are given here below. The choice of the robust methodology can be done choosing the desired specification as illustrated in Fig. 10.5. The script Graph1robust.m generates the plot of the estimated beta.

Par1BetaTrimming.m – This returns the beta estimation using the Least Trimmed Squares from a rolling-window 260 day period. The input prices are the market data. The output structure contains also some regression diagnostics, such as the R^2 statistic, the adjusted R^2 statistic, the Student's t statistics, p-values for each t statistic.

Par1BetaMedRobust.m – This returns the beta estimation applying the robust regression model to a rolling-window of 260 days. The input prices are the market data. The output structure also contains some regression diagnostics: such as the R^2 statistic, the adjusted R^2 statistic, the Student's t statistics, p-values for each t statistic. The robust fit function uses an iteratively reweighted least squares algorithm, with the weights at each iteration calculated by applying the bisquare function to the residuals from the previous iteration. However, the user can replace the bisquare function by any of the following list: 'Andrews', 'Cauchy', 'Fair', 'Huber', 'Logistic', 'Talwar', 'Welsch'.

LMSregor.m – This returns the beta estimation applying the Least Median of Squares (LMS) to a rolling-window of 260 days. The input prices are the market data. The output structure also contain some regression diagnostics: such as the R^2 statistic, the adjusted R^2 statistic, the Student's t statistics, p-values for each t statistic.

Par01BetaKALMAN.m – This returns the value of the likelihood function in the Kalman filter algorithm using a rolling-window of 260 days. The inputs are the market data and the starting values for the parameters. In the worksheet "Engine" there is a range named "Parameter" where they can be set, see Fig. 10.6.

KalmanRunOpt.m – This maximizes the likelihood function of the Kalman filter algorithm through the lsqnonlin.m built in algorithm provided by the MATLAB® Optimization toolbox and returns optimized parameters;

kalmanottimilsqnlfmin.m – This is the maximum likelihood function of the Kalman filter algorithm.

Par02BetaKALMAN.m – This returns the beta estimation using the Kalman filter model on a rolling-window of 260 day period. The input price are the market data and the return optimized parameter from the previous procedure.

Graph1kalman.m – Plots the beta estimated using the Kalman filter model on a rolling-window of 260 day period.

The beta forecasts can be obtained using the script Forecasting.m, except for the Least Trimmed Squares Model where the relevant script is named Forecast-ingtrim.m. The different forecasts can be visualized using the script Graph-forec.m.

Backtesting.m and GraphVaR.m – This executes a backtesting on the different models and then computes the Kupiec Test (1995) described in the previous section. The VAR forecasts are plotted using the script GraphVaR.m.

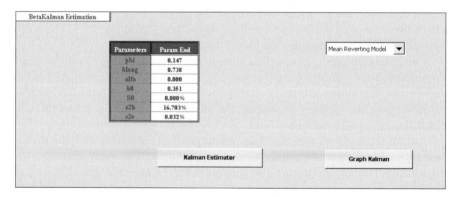

Fig. 10.6. Engine to estimate the beta according to the Kalman filter (mean reverting, random walk, random coefficient).

10.4 Results and Comments

The data employed in this study have been downloaded from Bloomberg and refer to daily closing prices of the 100 stocks belonging to the S&P100 index, that we consider to represent the market index. The period under examination runs between January 4th, 1999 and December 4th, 2004, that represents the sample period. The out of sample period is taken to be December 4th, 2004 to April 2005. Market data are stored in the sheet Nasdaqcomp. In the same Excel file, in the sheet Engine you can select the optional parameters for the different estimation procedures.

We report here only the results relative to ALLSTATE CORP. The user can conduct the analysis on the remaining stocks. The setup for the different estimation procedures is as follows. In the Bayesian estimation we fix the mean and the variance of the prior equal to the mean and to the standard deviation of the betas computed across all securities. The parameter λ in the EWMA procedure has been set equal to 0.94. In order to choose among the different possible specifications in the Kalman filter, we have computed the AIC and the BIC statistics, see Table 10.1. These statistics do not appear to be significantly different, although the RW specification seems to be the preferred one. Taking into account the economic interpretation underlying the different specifications, we have decided to use the mean-reverting process in the testing period.

In the first out of sample test, we use the estimated beta and the current market return to make forecasts on the future return

$$\hat{r}_{it} = \hat{\alpha}_{i,t-1} + \hat{\beta}_{i,t-1} r_{I,t},$$

and then we regress this forecast on the realized return

$$r_{i,t} = \delta + \gamma \hat{r}_{it} + \eta_t.$$

The results of this regression are reported in Table 10.2, together with the mean error (ME), the mean absolute error (MAD), the mean square error (MSE), the R^2

Table 10.1. Choosing the Kalman filter specification

Model parameter	RW	RC	MR
ϕ	1.000	0.000	0.147
$\bar{\beta}$	0.000	0.737	0.738
α	0.025%	0.021%	0.022%
β_0	0.849	0.613	0.351
Σ_0	0.000%	0.000%	0.000
σ_η^2	0.006%	17.093%	16.703%
σ_ϵ^2	0.036%	0.032%	0.032%
$\hat{L}$	3836.482	3844.710	3845.057
Number of observation, T	1246	1246	1246
Number of parameters, k	5	6	7
AIC	−6.150	−6.162	−6.161
BIC	−6.129	−6.137	−6.132

Table 10.2. Results of the return forecasting exercise

	ME	MAD	MSE	R^2	$\hat{\gamma}$	$\hat{\delta}$
OLS 1 yr	0.110%	0.886%	0.013%	35.842%	0.347	0.001
EWMA 1 yr	0.082%	0.888%	0.013%	34.820%	0.324	0.001
Bayes 1 yr	0.032%	0.878%	0.013%	35.837%	0.346	0.001
Robust regression	0.104%	0.886%	0.013%	35.883%	0.343	0.001
Kalman filter	0.049%	0.902%	0.013%	36.098%	0.285	0.001

Table 10.3. Results of the VAR forecasting exercise

	Obs.	Sign.	Expected violations	Conf. interval	Num. viol.
OLS 1 yr	138	5.000%	7	(2.689, 11.111)	8
EWMA 1 yr	138	5.000%	7	(2.689, 11.111)	9
Bayes 1 yr	138	5.000%	7	(2.689, 11.111)	8
Robust regression	138	5.000%	7	(2.689, 11.111)	8
Kalman filter	138	5.000%	7	(2.689, 11.111)	8

of the regression and the estimated values for δ and γ. An unbiased forecast should return $\hat{\delta} = 0$ and $\hat{\gamma} = 1$. The different models appear to provide similar forecasting ability, but all of them appear biased ($\hat{\gamma}$ is on average approximately equal to 0.30).

The second forecasting exercise consists in counting the VAR violations for the different models, as discussed in Sect. 10.2.7. The results are given in Table 10.3. Due to the low number of observations, we are (unfortunately) unable to distinguish between the performance of the different models: all of them show a number of violations (8 or 9) in line with the expectations (between 3 and 11).

Automatic Trading: Winning or Losing in a kBit[*]

Key words: random walk, AR(1), GARCH, bootstrapping, technical analysis, Monte Carlo simulation

Technical analysis focuses on historical information concerning price movements in order to forecast future price trends. In this manner, technical analysts argue that changes in the psychology of the market can be used to profit from trading. With technical analysis profits can be obtained in two different ways. Under a first approach, the analyst purely relies on charting and "reads" historical price trajectories to find clues on trend reversals. Under a second approach, the role of the analyst is reduced. An automatic trading system uses historical information to implement a trading process governed by a set of well-defined rules, while traders just need to choose the strategy to implement.

Despite its long history (one of the first attempts to forecast stock prices is attributed to Charles H. Dow in the late 1800s), academics have traditionally regarded technical analysis with a mixture of suspicion and contempt (see, for instance, Fama and Blume (1966), Fama (1969) and Jensen and Benington (1970)). In fact the efficient market paradigm holds that in capital markets all relevant information is reflected in current prices. Therefore, for traditional researchers, trading rules based on past prices are meaningless. But with the accumulation of evidence on the inefficiency of many marketplaces,[1] the idea that some technical rules can be used to systematically gain profits from trading has acquired momentum. An important contribution in this sense has come from Brock et al. (1992) (BLL henceforth). Using a set of simple technical rules, these authors first investigated sixty years of the Dow Jones Index and achieved statistically significant profits. Since then, several authors have tried to replicate BLL's results on different data sets and periods. For instance, Hudson, Dempsey and Keasey (1996) have used data from the United Kingdom.

---

[*] with Carlo Pozzi and Federico Roveda.

[1] A list of references can be found in Lo, MacKinlay and Zhang (1997), pp. 43–44.

Bessembinder and Chan (1995) have tested emerging markets in the Asian area. More recently, Isakov and Hollistein (1999) have studied the Swiss market, while Detry and Grégoire (2001) have concentrated on several European indexes.

In this chapter, we provide a review of a few well-known techniques. Our treatise is organized as follows. Section 11.2 illustrates the chosen technical strategies and the statistical tests conducted to assess the statistical significance of their actual profits. In particular, we focus on the MACD-H indicator and a development of the moving average crossover. According to the procedure suggested by BLL, we test our results using a bootstrap methodology. Section 11.3 presents the VBA® code used to run our strategy. In Sect. 11.4, we discuss our results testing the proposed technical rule on the five largest capitalized stocks in the US equity market.

11.1 Problem Statement

The technical strategy we analyze in this chapter is named *moving average convergence divergence* (MACD). In our treatise, we depart from BLL, who consider two of the most simple and known classes of technical rules based on the crossover of two moving averages,[2] and present the MACD technique which is a trend-following momentum indicator originally developed by Appel and Hitschler (1980). Illustrations of this indicator can also be found in Murphy (1999) and – in a more detailed fashion – in Thorp (2000). Given its relatively old age, the MACD rule has long been known to technical analysts. For this reason, we deem that it may be regarded as the methodology free of data snooping biases.

The MACD indicator combines two exponential moving averages of past prices into two lines: the MACD line and the signal line. The MACD line is constructed as the difference between two exponential moving averages computed using last m and n closing prices, where m and n are integers such that $n > m$.[3] It follows that the MACD line crosses the zero line each time there is a crossover between the two moving averages. To see this, let P_t be the market closing price at day t, and let us define a long-period and a short-period exponential moving average, $EMAL_t$ and $EMAS_t$ respectively, as

$$EMAL_t = \frac{1}{n}P_t + \left(1 - \frac{1}{n}\right)EMAL_{t-1}, \qquad EMAL_0 = P_0, \quad t = 1, 2, \ldots,$$

$$EMAS_t = \frac{1}{m}P_t + \left(1 - \frac{1}{m}\right)EMAS_{t-1}, \qquad EMAS_0 = P_0, \quad t = 1, 2, \ldots,$$

and let the MACD line, $MACD_t$, be defined as the difference between $EMAL_t$ and $EMAS_t$,

---

[2] The rule is: "initiates buy (sell) signals when the short run moving average is above (below) the long run moving average".

[3] Exponential averages are a standard tool in technical analysis. They are preferable to arithmetic averages because they respond more quickly to changes in price. Indeed more weight is placed on the most recent prices, whilst arithmetic averages give the same weight to all recorded prices.

$$MACD_t = EMAS_t - EMAL_t, \qquad MACD_0 = 0, \quad t = 1, 2, \ldots .$$

It is easy to observe that, since the fast moving average ($EMAS_t$ on m-periods) reflects the short-period trend in the market while the slow moving average ($EMAL_t$ on n-periods) reflects the long-period tendency, when the short-period moving average is above the long-period moving average, $MACD_t$ is positive. This may be interpreted as a bullish market phase. The opposite is true when the fast moving average is dominated by the long one. The MACD is thus a trend-follower algorithm. It performs at its best during strong trending periods, but tends to lose money during periods of choppy trading (i.e., when prices move sideways for several weeks). A trigger or signal line, named SL_t, is also used. SL_t is a k-period exponential moving average of the MACD line

$$SL_t = \frac{1}{k} MACD_t + \left(1 - \frac{1}{k}\right) SL_{t-1}, \qquad SL_0 = 0, \quad t = 1, 2, \ldots .$$

Taking the difference between the MACD and the signal line we obtain the MACD-Histogram (MACDH) indicator, the importance of which consists in highlighting variations in the spread between the fast and the slow lines,

$$MACDH_t = MACD_t - SL_t.$$

When the histogram lies above (below) the zero line but starts to fall (increase) towards the zero line, then the up trend (down trend) is losing momentum (the distance between fast and slow moving averages is decreasing. According to the MACD rules, a buy (sell) signal is generated each time the MACD line moves above (below) the signal line or alternatively when the MACDH crosses the zero line from below (above). In this way, one is always active in the market, since on the closing of a position, a new one can be opened with the opposite sign. For this reason this trading system is classified into the category *Stop* and *Reverse*.

Histogram zero crossovers can also be used to detect early exit signals from existing positions. Indeed, MACD tracks whether the market is gaining or losing momentum, measuring the acceleration or deceleration between moving averages and signaling if a stock or index has been overbought or oversold. Therefore, MACD is not only a lagging indicator, but also a leading indicator which can be used to detect excesses of demand or supply in the market. Prudence is however recommended, since using MACD turns in order to initiate new positions can be quite dangerous as it implies opening positions against prevailing market trends (Murphy (1999)).

Unlike other oscillators, like the *relative strength index* or the *stochastic index*, MACD it is not constrained between upper and lower bounds, but it can hit new highs or lows as long as price trends are gaining momentum. In the finance practice, the most popular parameters in the MACD and MACDH computations are $m = 12$, $n = 26$ and $k = 9$ periods. They are also set as default values in many software packages. In this chapter, the VBA® code takes into account the possibility of having one set of numbers for buy signals and another for sell signals. We follow here the original suggestion of Appel and Hitschler (1980) who recommended using an asymmetric combination of 8-17-9 for buy signals and 12-26-9 for sell signals.

11.2 Model and Solution Methodology

11.2.1 Measuring Trading System Performance

The possibility of discerning between different trading strategies (TS) is provided by the measurement of their performance. A detailed analysis of this topic is given by Schwager (1996), Chapter 21. In these pages, as standard measure of performance, we use the cumulative sum of all profits and losses generated by a TS. We define this net measure as the difference between all the profits (total gross profit) and losses (total gross loss) obtained over a given period of time. With these two quantities, we also calculate a risk measure for our TS by dividing the total gross loss by the total gross profit. This ratio is called *profit factor* – namely, the expected reward on one unit of capital invested in the TS. Profit factors are then employed to rank alternative TS. Once this is done, a common rule among practitioners is to discard strategies with profit factors below two. The total number of trades has also a bearing on trading performance. Another rule of thumb in this case is to consider ten to twenty trades per year.

Other useful quantities may supplement comparisons between TS: (1) the *average trade*, (2) the *average winning trade*, (3) the *average losing trade*, (4) the *average winning/losing trade ratio*, (5 and 6) the *largest winning trade* and the *largest losing trade*. Let us examine them separately. (1) The average trade is the average net profit across all trades. (2 and 3) The average winning (losing) trade is the average profit (loss) across all winning (losing) trades. Low values for the average winning trade imply high vulnerability to mistakes. (4) The average winning/losing trade ratio is an additional measure of TS risk and so are the following two. (5 and 6) In fact, the largest winning (losing) trade should be read in conjunction with the gross profit (loss) of a TS: if a large part of it comes from a single trade, a strategy may be considered as risky. Therefore, a single trade should not exceed $1/4$ of gross profits (losses). However, in the opposite case, tracking the largest loss is not recommended as a unique measure of risk because it may bias performance judgements by magnifying the importance of a single negative event. This is particularly significant in the evaluation of managers with long track records, since large losses occur with more frequency over longer periods of time.

A significant historical performance measure is then *maximum drawdown*. This quantity measures the largest decline (*peak-to-valley*) from the highest equity value, tracked over s days prior to the sample period considered. In other words, it represents the top cumulative loss an investor would have incurred, had they invested at the worst possible prior time (i.e., the prior equity peak) and stayed long through the sample period considered. Maximum drawdown can thus be seen as the largest margin necessary to cover the highest possible loss produced by a TS. This measure is computed as

$$MD_T = -\max_{0 \leq t \leq T}\left\{\max_{0 \leq s \leq t}\pi_s - \pi_t\right\},$$

where π_t represents cumulative profits up to time t. As a descriptive statistic, maximum drawdown has one key positive aspect and some important limitations. Vis-à-vis volatility, it is an observable quantity measuring a capital requirement. This can be expected to be smaller either when the upward price drift is steeper or when the variability of the process is lower. But being derived from a single string of data, maximum drawdown is exposed to large errors. For this reason, it may be biased with respect to means determined across multiple periods. Moreover, the longer the time series considered, the greater a drawdown is likely to be. A fact which suggests that it would be wise to compare drawdowns only across TS with track records of the same length.

The maximum number of losing trades is yet another measure of risk. This number counts how many consecutive losing trades an investor should accept before abandoning a TS. The psychological importance of this number is evident. It is thus preferable to establish it before beginning a trading scheme.

Finally, an important benchmark is the *one way break-even trading* (BET) measure. This quantity, suggested by Bessembinder and Chan (1998), represents the level of transaction costs that eliminates all profits achieved through a TS in excess of a simple buying and holding strategy. It is computed as

$$bet = \frac{2\bar{\pi}}{N},$$ (11.1)

where $\bar{\pi}$ is the average profit of the strategy and N is the total number of signals (purchases and sales).

11.2.2 Statistical Testing

As suggested in BLL, it is possible to conduct statistical tests on TS at two levels of depth. At a first stage, one can compare profits coming from buy and sell trades. If technical analysis does not have any power in predicting market trends, then profits on buy days should not differ from profits on sell days. Therefore we can use standard statistical tests and compare means and variances on buy versus sell days. Similar tests can be conducted to compare profits generated by the technical strategy and by a passive one, for instance the buy-and-hold position. For example, by using the t-statistic we can see if the mean buy and mean sell profits are significantly different. The t-statistic is

$$t^{(b,s)} = \frac{\mu_b - \mu_s}{(\sigma_b^2/N_b + \sigma_s^2/N_s)^{1/2}},$$

where μ_b (μ_s), σ_b^2 (σ_s^2) and N_b (N_s) are respectively the mean profit, the number of signals and the variance of profits for buys (sells). The test which checks if the global mean daily profit is significantly different from the daily unconditional profit is given by the t-statistic

$$t^{(str,b\&h)} = \frac{\mu_{str} - \mu_{b\&h}}{(\sigma_{str}^2/N_{str} + \sigma_{b\&h}^2/N_{b\&h})^{1/2}},$$

where "str" and "b&h" refer to the strategy and to the buy-and-hold position.

These tests assume normal, stationary and time-independent distributions. These assumptions are however too strong compared to the stylized facts which affect stock return distributions and time series (such as their high kurtosis and their conditional heteroskedasticity). Standard statistical tests can thus give misleading results. As a result, at a second stage, a possible solution is to bootstrap data, following the analyses of Efron (1979), Freedman and Peters (1984), and Efron and Tibshirani (1986). A review of bootstrap applications in finance can be found in Maddala and Li (1996). In the following pages we present an approach which proceeds as follows:

1. Assume a return generating model (e.g., random walk, AR(1), GARCH(1, 1), etc.) and estimate it on a data set.
2. Standardize residuals from the estimated model. Under the null hypothesis that the return model is the true generating process, standardized residuals should be realizations of i.i.d. innovations.
3. Re-sample with replacement estimated standardized residuals and use them as innovations to simulate a new price history of the same length as the original.
4. Apply the trading strategy to the bootstrapped series.
5. Repeat the procedure n times (BLL set $n = 500$) and obtain the empirical distributions of the mean profit of the TS under the chosen return generating model.
6. Compute the fraction of simulated strategies which have greater mean or standard deviation than the strategy applied to the original series. These fractions can be interpreted as p-values. For example, if only 1% of the simulations generated by the chosen model gives a mean return higher than the TS applied on the original series, then the p-value is equal to 1%. In other words, the probability that the chosen model generates the mean return obtained by the trading strategy on the original series is only 1%. Therefore, the mean returns of the trading rule cannot be explained by the assumed model.

By using the bootstrapping procedure, standardized residuals are not restricted to a particular distribution and we can test the technical strategy against different models. Moreover, when we reject the null hypothesis for a given model specification, we can test another one with few assumptions. Here we start testing the random walk (RW) model. RW simply assumes independence and identical distribution for returns, but does not assume their normal distribution,

$$r_t = \varepsilon_t, \quad \varepsilon_t \text{ are i.i.d.}$$

With the RW model, the bootstrapping procedure can be performed directly on the original return series. If this model is rejected, in order to investigate if serial correlation in returns can explain the profits from the trading rule, we can then consider an AR(1) model,

$$r_t = a + br_{t-1} + \varepsilon_t, \quad t = 1, \ldots, T,$$
$$\varepsilon_t \text{ are i.i.d.:} \quad \mathbb{E}(\varepsilon_t) = 0, \quad \text{Var}(\varepsilon_t) = \sigma_\varepsilon^2.$$

This model can be estimated with OLS. If $\hat{a}$, $\hat{b}$ and $\hat{\sigma}_\varepsilon$ are the estimated parameters, we can then compute standardized residuals from

$$\hat{\varepsilon}_t = \frac{r_t - \hat{a} - \hat{b}r_{t-1}}{\hat{\sigma}_\varepsilon}.$$

Subsequently, we can re-sample from $\hat{\varepsilon}_t$, $\eta_t = \hat{\varepsilon}_{k(t)}$, $t = 1, \ldots, T$, where $k(t)$ is a uniform random number over the integers $1, \ldots, T$, and we construct new return and price histories,

$$\tilde{r}_t = \hat{a} + \hat{b}\tilde{r}_{t-1} + \eta_t,$$
$$\tilde{P}_t = \tilde{P}_{t-1} \exp(\tilde{r}_t).$$

We apply the TS on each simulation and we compute the fraction of simulated returns which have higher mean or standard deviation than the ones obtained by applying the same TS on the original series. These fractions can be interpreted as p-values. If this model is also rejected, we can ask if conditional changes in variances can be a possible explanation for trading rule profits. Therefore we can re-specify the return model as a GARCH(1, 1):[4]

$$r_t = \mu + \varepsilon_t, \quad \varepsilon_t \text{ are i.i.d.} \sim N(0, h_t),$$
$$h_t = \omega + \alpha \varepsilon_{t-1}^2 + \beta h_{t-1}.$$

Once the GARCH(1, 1) has been estimated, we standardize estimated residuals

$$\hat{z}_t = \frac{r_t - \hat{\mu}}{\sqrt{\hat{h}_t}},$$

and then re-sample them with replacement

$$\eta_t = \hat{z}_{k(t)}$$

to form a new simulated series

$$\tilde{r}_t = \hat{\mu} + \sqrt{\hat{h}_t}\,\eta_t,$$
$$\hat{h}_t = \hat{\omega} + \hat{\alpha}\varepsilon_{t-1}^2 + \hat{\beta}h_{t-1},$$
$$\tilde{P}_{t+1} = \tilde{P}_t \exp(\tilde{r}_t).$$

Each simulation is based on n replications of the chosen return model. For each simulation, the same trading rule is applied. Then we compute the p-values and we decide if any of the specified models can explain the trading rule profits.

11.3 Code

The trading system and the bootstrap methodology presented above have been implemented in VBA®. We first run a historical analysis of our strategy. This can be done with the VBA® macro `StartTradingSystem()`. Once this macro is called, the user form "Trading System" is shown. See Fig. 11.1.

In the form, the user has to indicate:

---

[4] Brooks et al. (1998) have considered a more general GARCH-M model, in order to take into account changes in the conditional mean.

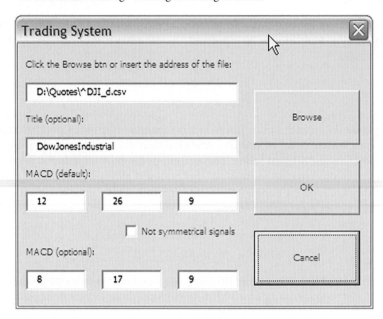

Fig. 11.1. Excel user form for starting the Trading System.

1. The path name of the file containing the historical data (the Browse button helps in finding the correct folder path). Data must be stored as comma separated files (.csv) and with the same field structure as downloads from Yahoo! Finance. Therefore, the .csv file must have seven columns that respectively refer to: date, opening price, maximum price, minimum price, closing price, volume and adjusted closing price. The macro only imports dates and adjusted closing prices, which will serve as a basis for the strategy to be implemented. The macro sorts data in the time series starting from the oldest one.
2. The name of the worksheet where the results of the strategy will be printed out. If no name is specified, the macro assigns by default the name "MACD". This worksheet will be created anew by the VBA® macro, so no other existing worksheet should have the same name as the one assigned here.
3. Parameters m, n and k to be used for the computation of the MACD. By default, the program assigns 12-26-9. However, the user can chose different periods and even adopt an asymmetric strategy, specifying different parameters to be considered for sale and purchase signals, as suggested by Appel and Hitschler (1980).
4. By clicking on OK, the form runs the following macros:
   a. `UploadingData()` to import data into a temporary worksheet.
   b. `TradingRules()` to apply the strategy. The VBA® code is very simple:

```
If MACDH(t) > 0 And MACDH(t-1) < 0 Then
Close short and Go long
ElseIf MACDH(t) < 0 And MACDH(t-1) > 0 Then
Close long and Go short
End If
```

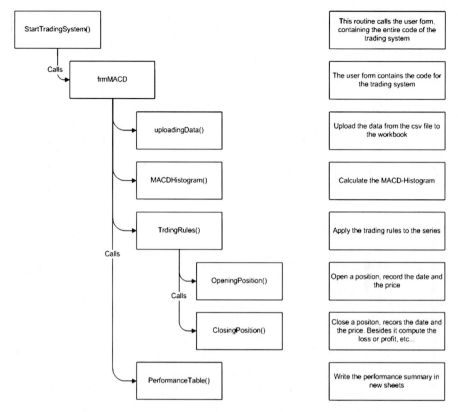

Fig. 11.2. Flow chart describing the macro StartTradingSystem().

c. PerformanceTable() to print the performance summary in the new worksheet.

In particular, MACDH(t) in the macro is computed by the VBA® function MACD-Histogram(vMACDHistogram() As Double, intSMA As Integer, intLMA As Integer, intSignal As Integer, Price() As Double). The entire procedure is described in the flowchart which appears in Fig. 11.2. The worksheet containing the performance summary, described in Sect. 11.2.1, is shown in Fig. 11.3. Finally, the value of the account, once the position has been closed, is reported in columns I (entire strategy), J (long position) and K (short positions).

The VBA® macro StartBootstrapSimulation() performs the bootstrap testing. Once this macro is run, the user form "Bootstrap Simulation" is shown. This user form appears in Fig. 11.4. In particular, in this one the user has to indicate:

1. The path name of the file containing historical data.
2. The name of the worksheet where the results of the bootstrap simulation will be shown.
3. The parameters to be used for the computation of the MACD.

Data:		
Time series dates	23/01/2006 17.19	
MACD: 12 * 26 * 9	1/3/2000 - 12/31/2004	
Drive \Path\Filename	C:\Gianluca\Presano\Roncoroni\TechnicalAnalysis\C01_01_2000_31_12_04.csv	

////////// All Trades \\\\\\\\\\

Total Net Profit	-7.?		
Gross profit	56.92	Gross loss	-64.73
Total # of trades	54	Percent profitable	0.37037
Number winning trades	20	Number losing trade	34
Largest winning trade	7.88	Largest losing trade	-4.12
Average winning trade	2.846	Average losing trade	-1.90382
Ratio avg win/avg los	1.494886451	Avg trade (win & loss)	-0.14463
Max Consecutive Winners	3	Max Consecutive Losers	4
Avg # bars in winners	55.15	Avg # bars in losers	21.17647
Max account drawdown	-16.64	Account size required	16.64
Profit factor	0.879344971	Return on account	-0.46935

////////// Long Trades \\\\\\\\\\

Total Net Profit	4.5		
Gross profit	30.99	Gross loss	-26.49
Total # of trades	27	Percent profitable	0.444444
Number winning trades	12	Number losing trade	15
Largest winning trade	5.55	Largest losing trade	-4
Average winning trade	2.5825	Average losing trade	-1.766
Ratio avg win/avg los	1.462344281	Avg trade (win & loss)	0.166667
Max Consecutive Winners	3	Max Consecutive Losers	5
Avg # bars in winners	51.75	Avg # bars in losers	17.53333
Max account drawdown	-10.28	Account size required	10.28
Profit factor	1.169875425	Return on account	0.43743

Equity Curve Date	Equity Curve	Equity Curve Long		Equity Curve Short	
10/01/2000	-2.56	09/02/2000	0.37	10/01/2000	-2.56
09/02/2000	-2.19	03/05/2000	2.49	17/03/2000	-4.33
17/03/2000	-3.96	21/06/2000	-0.13	02/06/2000	-8.45
03/05/2000	-1.84	07/09/2000	3.92	12/07/2000	-11.23
02/06/2000	-5.96	15/11/2000	-0.08	06/11/2000	-11.47
21/06/2000	-8.58	14/02/2001	0.7	11/12/2000	-13.76
12/07/2000	-11.36	13/06/2001	3.03	12/04/2001	-8.33
07/09/2000	-7.31	09/07/2001	0.8	22/06/2001	-11.34
06/11/2000	-7.55	10/12/2001	5.09	05/10/2001	-4.86
15/11/2000	-11.55	11/01/2002	4.1	26/12/2001	-6.84
11/12/2000	-13.84	12/04/2002	3.58	05/03/2002	-5.8
14/02/2001	-13.06	30/05/2002	0.59	17/05/2002	-5.69
12/04/2001	-7.63	16/09/2002	-1.19	12/08/2002	2.19
13/06/2001	-5.3	07/10/2002	-5.19	01/10/2002	1.14
22/06/2001	-8.31	09/12/2002	0.36	11/10/2002	-2.29
09/07/2001	-10.54	16/05/2003	5.66	14/03/2003	-0.22
05/10/2001	-4.06	27/06/2003	7.55	30/05/2003	-2
10/12/2001	0.23	21/07/2003	6.56	09/07/2003	-4.83
26/12/2001	-1.75	27/10/2003	6.61	18/09/2003	-6.8
11/01/2002	-2.74	03/02/2004	5.51	08/10/2003	-9.24
05/03/2002	-1.7	22/03/2004	6.49	19/12/2003	-9.74
12/04/2002	-2.22	16/04/2004	5.78	19/03/2004	-11.16
17/05/2002	-2.11	07/07/2004	4.59	29/03/2004	-12.83
30/05/2002	-5.1	06/08/2004	2.72	07/06/2004	-10.71
12/08/2002	2.78	23/09/2004	1.52	04/08/2004	-9.91
16/09/2002	1	31/12/2004	1.22	10/08/2004	-10.89
01/10/2002	-0.05		4.5	02/11/2004	-12.31
07/10/2002	-4.05				
11/10/2002	-7.48				
09/12/2002	-1.93				
14/03/2003	0.14				
16/05/2003	5.44				
30/05/2003	3.66				
27/06/2003	5.55				
09/07/2003	2.72				
21/07/2003	1.73				

Fig. 11.3. Excel worksheet illustrating the performance measures of the trading rule.

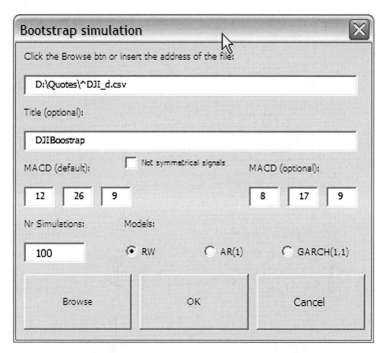

Fig. 11.4. Excel user form for starting the Boostrap Simulation.

4. The number of simulations.
5. The model assumed as return generator process. The user can choose between RW, AR(1) and GARCH(1, 1).
6. By clicking the OK button, the user form runs the following macros:
   a. `UploadingData()` to import data into a temporary worksheet;
   b. It estimates the return model and runs the bootstrap simulation. In particular, the function `RandomWalk(vSimulated() As Double, CurrentPrice As Double, vReturn() As Double)` estimates the random walk model; the AR(1) model is estimated using OLS by the function `AutoRegressive(vSimulated() As Double, CurrentPrice As Double, vReturn() As Double`; the GARCH(1, 1) model is estimated by the function `GARCH(vSimulated() As Double, CurrentPrice As Double, vReturn() As Double, dblMhu As Double, dblVar0 As Double, dblOmega As Double, dblAlpha As Double, dblBeta As Double)`. The estimation is performed by maximizing the likelihood function using the Excel Solver. The above functions return the vector `vSimulated()` containing a new bootstrapped series;
   c. It prints the performance measures of the trading rule applied to the bootstrap simulation in the new worksheet.

   The entire procedure is described in the flowchart appearing in Fig. 11.5. The worksheet with the results of bootstrap simulations is shown in Fig. 11.6. Results are

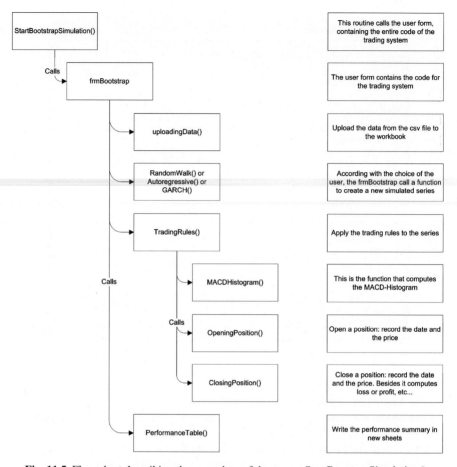

Fig. 11.5. Flow chart describing the procedure of the macro StartBoostrapSimulation().

presented as follows: in the first column, the number of the simulation is reported. Then we have the number of buy days, the number of short days, the conditional one day return for the strategy, for purchases, for sales, the conditional daily standard deviation for the strategy, for purchases, for sales, then the ratio between the conditional return and standard deviation, for the strategy, for longs, for sales, and finally the one-way break-even trading.

11.4 Results and Comments

In this section we report the results of the trading system previously described. In particular, we consider either the symmetric case with default parameters 12-26-9, or the asymmetric case with parameters 12-26-9 for opening long positions and 8-17-9 for opening short positions. Time series data have been downloaded from Yahoo!

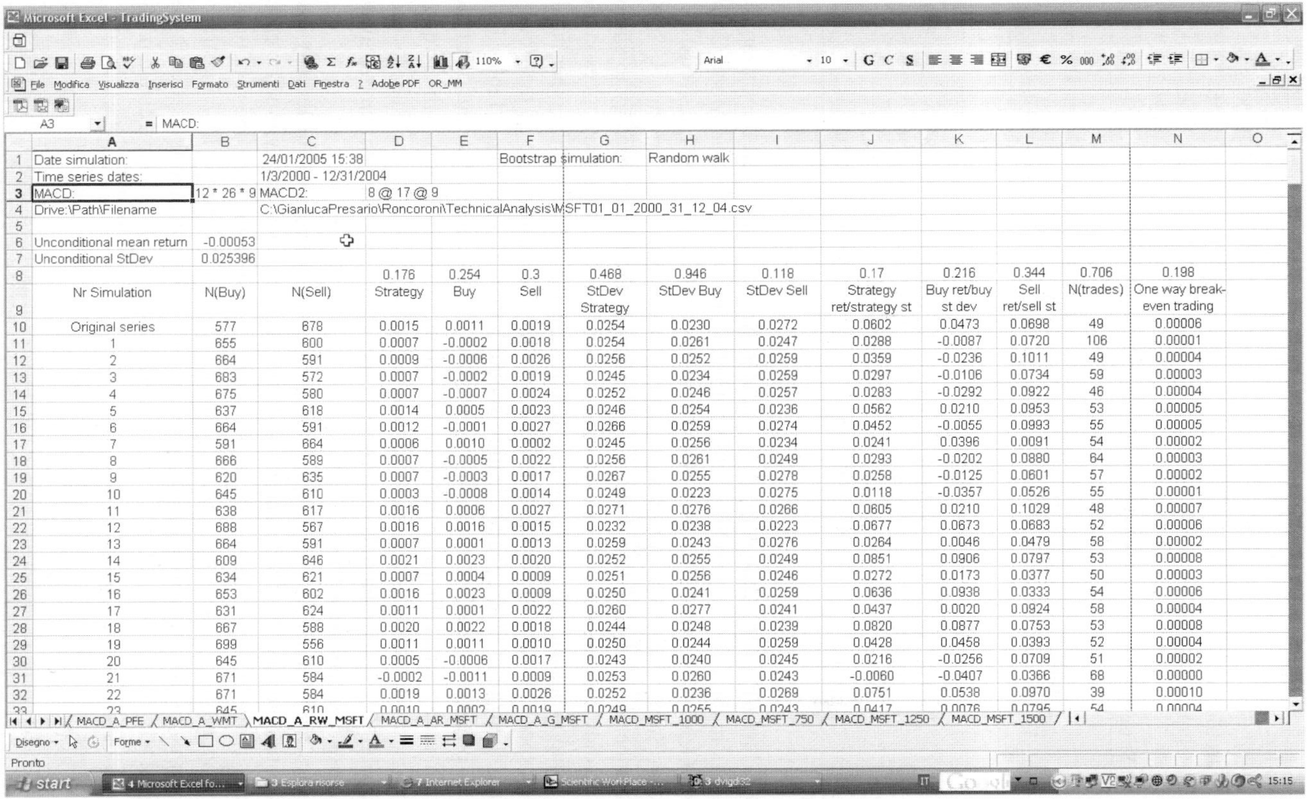

	A	B	C	D	E	F	G	H	I	J	K	L	M	N
1	Date simulation:		24/01/2005 15:38			Bootstrap simulation:		Random walk						
2	Time series dates:		1/3/2000 - 12/31/2004											
3	MACD:	12 * 26 * 9	MACD2:	8 @ 17 @ 9										
4	Drive:\Path\Filename		C:\GianlucaPresario\Roncoroni\TechnicalAnalysis\MSFT01_01_2000_31_12_04.csv											
5														
6	Unconditional mean return	-0.00053												
7	Unconditional StDev	0.025396												
8				0.176	0.254	0.3	0.468	0.946	0.118	0.17	0.216	0.344	0.706	0.198
9	Nr Simulation	N(Buy)	N(Sell)	Strategy	Buy	Sell	StDev Strategy	StDev Buy	StDev Sell	Strategy ret/strategy st	Buy ret/buy st dev	Sell ret/sell st	N(trades)	One way break-even trading
10	Original series	577	678	0.0015	0.0011	0.0019	0.0254	0.0230	0.0272	0.0602	0.0473	0.0698	49	0.00006
11	1	655	600	0.0007	-0.0002	0.0018	0.0254	0.0261	0.0247	0.0288	-0.0087	0.0720	106	0.00001
12	2	664	591	0.0009	-0.0006	0.0026	0.0256	0.0252	0.0259	0.0359	-0.0236	0.1011	49	0.00004
13	3	683	572	0.0007	-0.0002	0.0019	0.0245	0.0234	0.0259	0.0297	-0.0106	0.0734	59	0.00003
14	4	675	580	0.0007	-0.0007	0.0024	0.0252	0.0246	0.0257	0.0283	-0.0292	0.0922	46	0.00004
15	5	637	618	0.0014	0.0005	0.0023	0.0246	0.0254	0.0236	0.0562	0.0210	0.0953	53	0.00005
16	6	664	591	0.0012	-0.0001	0.0027	0.0266	0.0259	0.0274	0.0452	-0.0055	0.0993	55	0.00005
17	7	591	664	0.0006	0.0010	0.0002	0.0245	0.0256	0.0234	0.0241	0.0396	0.0091	54	0.00002
18	8	666	589	0.0007	-0.0005	0.0022	0.0256	0.0261	0.0249	0.0293	-0.0202	0.0880	64	0.00003
19	9	620	635	0.0007	-0.0003	0.0017	0.0267	0.0255	0.0278	0.0258	-0.0125	0.0601	57	0.00002
20	10	645	610	0.0003	-0.0008	0.0014	0.0249	0.0223	0.0275	0.0118	-0.0357	0.0526	55	0.00001
21	11	638	617	0.0016	0.0006	0.0027	0.0271	0.0276	0.0266	0.0605	0.0210	0.1029	48	0.00007
22	12	688	567	0.0016	0.0016	0.0015	0.0232	0.0238	0.0223	0.0677	0.0673	0.0683	52	0.00006
23	13	664	591	0.0007	0.0001	0.0013	0.0259	0.0243	0.0276	0.0264	0.0046	0.0479	58	0.00002
24	14	609	646	0.0021	0.0023	0.0020	0.0252	0.0255	0.0249	0.0851	0.0906	0.0797	53	0.00008
25	15	634	621	0.0007	0.0004	0.0009	0.0251	0.0256	0.0246	0.0272	0.0173	0.0377	50	0.00003
26	16	653	602	0.0016	0.0023	0.0009	0.0250	0.0241	0.0259	0.0636	0.0938	0.0333	54	0.00006
27	17	631	624	0.0011	0.0001	0.0022	0.0260	0.0277	0.0241	0.0437	0.0020	0.0924	58	0.00004
28	18	667	588	0.0020	0.0022	0.0018	0.0244	0.0248	0.0239	0.0820	0.0877	0.0753	53	0.00008
29	19	699	556	0.0011	0.0011	0.0010	0.0250	0.0244	0.0259	0.0428	0.0458	0.0393	52	0.00004
30	20	645	610	0.0005	-0.0006	0.0017	0.0243	0.0240	0.0245	0.0216	-0.0256	0.0709	51	0.00002
31	21	671	584	-0.0002	-0.0011	0.0009	0.0253	0.0260	0.0243	-0.0060	-0.0407	0.0366	68	0.00000
32	22	671	584	0.0019	0.0013	0.0026	0.0252	0.0236	0.0269	0.0751	0.0538	0.0970	39	0.00010
33	23	645	610	0.0010	0.0002	0.0019	0.0249	0.0255	0.0243	0.0417	0.0076	0.0795	54	0.00004

Fig. 11.6. Excel worksheet showing the results of the bootstrap simulation.

Table 11.1. Summary description of the data set

Stock	Ticker	Period	Number of obs.
Citigroup Inc.	C	1/3/00–12/31/04	1256
Exxon Mobil Corp.	XOM	1/3/00–12/31/04	1256
General Electric Co.	GE	1/3/00–12/31/04	1256
Microsoft Corp.	MSFT	1/3/00–12/31/04	1256
Pfizer Inc.	PFE	1/3/00–12/31/04	1254
Wal-Mart Stores Inc.	WMT	1/3/00–12/31/04	1256

Table 11.2. Descriptive statistics of daily logarithmic returns

Stock	Mean	Std. dev.	Skew.	Kurt.	JB
C	0.00029	0.02228	−0.28379	8.01079	1330
XOM	0.00034	0.01586	0.02424	6.48061	634
GE	−0.00016	0.02146	0.07536	6.06037	491
MSFT	−0.00053	0.02541	−0.20537	9.21079	2026
PFE	−0.00009	0.02003	−0.29621	5.57845	365
WMT	−0.00016	0.02099	0.13127	5.51836	335

Finance (http://finance.yahoo.com). In particular, we have examined the six largest cap stocks in the US market (Citigroup Inc. (Yahoo Ticker: C), Exxon Mobil Corp. (XOM), General Electric Co. (GE), Microsoft Corp. (MSFT), Pfizer Inc. (PFE) and Wal-Mart Stores Inc. (WMT)), over a period of five years from January 3, 2000 to December 31, 2004. Notice that the application of the trading strategy to single stocks avoids measurement errors due to non-synchronous reporting of prices in the index components and the problems related to trading the index.[5] On the other hand, we are exposed to data snooping bias related to the arbitrary choice of the stocks, although their large market capitalization should reduce this problem. Tables 11.1 and 11.2 provide summary information on the data set and on the distribution of the logarithm return series. In particular, the last column in Table 11.2 gives the Jarque–Bera statistic that measures deviations from normality. This is the case for all stocks.[6] Indeed, all of them have fat tails (a kurtosis higher than 3) and an asymmetric shape (positive or negative skewness).

Tables 11.3, 11.4 and 11.5 report some performance measures for the buy and hold position and for the TS. In particular Table 11.4 considers the symmetric case, whilst in Table 11.5 we consider the asymmetric case. In both cases, the results are very disappointing in respect to the buy-and-hold strategy. Figure 11.7 illustrates the dynamic of the cumulative profits of the trading rule applied to the six different listings. Only for Microsoft Corp. the strategy does generate an appreciable profit factor of 1.7488 in the symmetric case and a remarkable 2.0017 in the asymmetric

---

[5] This is issue is discussed in detail in Sweeney (1988).

[6] In particular, under the null hypothesis of a normal distribution, the Jarque–Bera statistic is distributed as a chi-square with 2 degrees of freedom. The critical value at 1% level for the Jarque–Bera statistic is 9.2103. Therefore, we reject the hypothesis of normal distribution at the 1% significance level.

Table 11.3. Summary performance measures of the buy-and-hold position

YT	TNP ($)	MAD ($)	PF	AWT ($)	AWL ($)
C	14.76	2.74	1.04	0.63	0.63
XOM	−8.32	17.90	0.98	0.57	0.55
GE	17.60	0.71	1.07	0.43	0.44
MSFT	−24.96	24.41	0.92	0.47	0.50
PFE	−3.40	19.53	0.99	0.52	0.53
WMT	−11.90	13.53	0.98	0.82	0.79

YT = Yahoo ticker; TNP = Total net profit; MAD = Maximum account drawdown;
PF = Profit factor; AWT = Average winning trade; AWL = Average losing trade.

Table 11.4. Summary performance measures of the symmetric trading rule 12-26-9

YT	TNP ($)	MAD ($)	PF	AWT ($)	AWL ($)
C	−7.81	16.64	0.88	2.85	1.90
XOM	−13.17	13.34	0.72	1.52	1.04
GE	−5.13	27.80	0.92	2.45	2.05
MSFT	23.79	8.13	1.75	2.42	1.22
PFE	−22.18	25.91	0.61	2.08	1.55
WMT	−49.78	63.74	0.50	2.53	2.57

YT = Yahoo ticker; TNP = Total net profit; MAD = Maximum account drawdown;
PF = Profit factor; AWT = Average winning trade; AWL = Average losing trade.

Table 11.5. Summary performance measures of the asymmetric trading rule 12-26-9; 8-17-9

YT	TNP ($)	MAD ($)	PF	AWT ($)	AWL ($)
C	−5.01	22.20	0.93	2.70	1.96
XOM	−10.47	15.23	0.78	1.25	1.09
GE	−28.25	29.49	0.64	2.34	1.84
MSFT	29.89	7.44	2.00	2.13	1.10
PFE	−39.62	39.62	0.50	1.73	1.89
WMT	−48.14	73.88	0.56	2.84	2.46

YT = Yahoo ticker; TNP = Total net profit; MAD = Maximum account drawdown;
PF = Profit factor; AWT = Average winning trade; AWL = Average losing trade.

case, whilst in the buy-and-hold case the profit factor is only 0.92. These results seem to indicate that for Microsoft Corp. the proposed technical strategy can identify profitable opportunities. In order to investigate their statistical significance, we shall conduct some tests.

At first, we conduct a standard t-test to compare the daily mean profit of the active (TS) and passive (buy-and-hold) strategies. In practice, the test is conducted on the series obtained as difference between daily profits under the two alternatives. Therefore, we test if the mean difference is equal to zero. The results of this test are given in Table 11.6. No statistically significant difference is obtained. Similar results are obtained in the asymmetric case and for the t-test that compares long

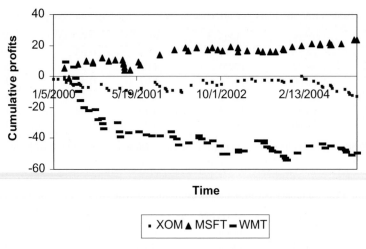

Fig. 11.7. Equity lines of the trading rule for Exxon Mobil Corp. (XOM), Microsoft Corp. (MSFT) and Wal Mart Stores Inc. (WMT).

Table 11.6. Summary performance measures of the asymmetric trading rule 12-26-9; 8-17-9

	$\mu_{\text{ts}-\text{bh}}$	$\sigma_{\text{ts}-\text{bh}}$	N	t-test ($\mu_{\text{ts}-\text{bh}} \neq 0$)
C	−0.0054	1.1359	1254	−0.1690
XOM	−0.0033	0.7991	1254	−0.1484
GE	0.0106	1.1228	1254	0.3358
MSFT	0.0009	1.0384	1254	0.0307
WMT	0.0504	1.5450	1254	1.1555

versus short positions (these results are not reported here). Our results collide with the ones presented by BLL, who rejected the null hypothesis that technical rules do not have significant positive performance. However (as discussed in Sect. 11.2.2), since our results can be affected by the violation of the normal, stationary and time-independent assumptions, we conduct a bootstrap simulation along the lines of the previous section. Figure 11.8 illustrates the density function of simulated mean returns (1,500 simulations) under the GARCH(1, 1) and of average profit obtained on the original series (black triangle on the horizontal axis).

In Tables 11.7, 11.8, 11.9, we report the *p*-values of the bootstrap simulation for the three models (RW, AR(1), GARCH(1, 1)), i.e. the fraction of simulations generating a mean daily return or standard deviation or Sharpe ratio higher than the same statistics obtained from the actual series.[7] In particular, we distinguish results for the strategy and for the long and short only positions. For example, the number 15% in Table 11.7 under the headings Strategy and Mean indicates that only 15% of the simulated random walks generated a mean return as large as that from the original

---

[7] We compute the Sharpe ratio as ratio between mean return and its standard deviation. Therefore we assume a zero risk-free rate.

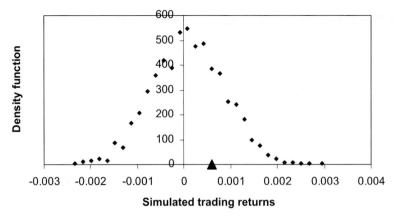

Fig. 11.8. Distribution of simulated mean returns (1500 simulations) under the GARCH(1, 1) and on the original series (black triangle on the horizontal axis).

Table 11.7. Summary performance measures of the asymmetric trading rule 12-26-9; 8-17-9

	Strategy	Buy	Sell
Mean	18%	25%	30%
Std. dev.	47%	95%	12%
Sharpe ratio	17%	22%	34%

Bootstrap results for MSFT: Random walk model: $r_t = \varepsilon_t$.

Table 11.8. Summary performance measures of the asymmetric trading rule 12-26-9; 8-17-9

	Strategy	Buy	Sell
Mean	16%	20%	30%
Std. dev.	48%	97%	12%
Sharpe ratio	15%	15%	34%

Bootstrap results for MSFT: AR(1) model: $r_t = a + br_{t-1} + \eta_t$. AR(1) parameters: $\hat{a} = -0.0005376$, $\hat{b} = -0.0257838$.

series. Considering different return models, we remark that we never obtain a p-value greater than 36% for the mean returns. BLL obtained instead much stronger results. Indeed, their p-values, computed across a much wider set of trading rules applied to the Dow Jones time series, were only occasionally greater than 11%. Instead, similarly to their results, the three models are unable to explain the volatility during buy and sell periods. This can be realized looking at the p-values for the TS and for the buy and sell positions. For example, 98% (7%) under the headings Std. Dev. and Buy (Sell) in Table 11.7 means that in 98% (7%) of the simulations, the buy (sell) volatility was higher (higher) than in the original series. If we consider the TS (buy + sell) the return models generate a volatility comparable to the original series (p-values of 47%, 50% for RW and AR(1)) or much higher (p-value of 84% for GARCH(1, 1)). These results are reflected in Sharpe ratios that are in general higher in the origi-

Table 11.9. Summary performance measures of the asymmetric trading rule 12-26-9; 8-17-9 for MSFT

	Strategy	Buy	Sell
Mean	20%	26%	36%
Std. Dev.	85%	98%	38%
Sharpe Ratio	17%	22%	37%

Bootstrap results for MSFT: GARCH(1, 1) model: $r_t = \mu + \varepsilon_t$, $h_t = \omega + \alpha \varepsilon_{t-1}^2 + \beta h_{t-1}$. GARCH(1, 1) parameters: $\hat{\mu} = 0.0002341$, $\hat{h}_0 = 0.0009691$, $\hat{\omega} = 0.0000014$, $\hat{\alpha} = 0.0863395$, $\hat{\beta} = 0.9186362$.

Table 11.10. Convergence of simulated p-values for the GARCH model

Number of simulations	Mean	St. dev.	Sharpe ratio
500	20%	84%	19%
750	20%	85%	18%
1000	21%	86%	19%
1250	19%	85%	18%
1500	20%	83%	19%

nal series vis-à-vis those generated by the return models (p-values of 15%, 13% and 19%). Therefore, the higher returns generated by the trading rule are not explained by a higher risk (measured by the standard deviation). In conclusion, our results on the performance of the TS are not so strong as in BLL, but they seem to suggest that the three chosen models fail to replicate the returns generated by the trading rule. In other words, the rule generates signals that are not likely to be explained by the return models. Nonetheless, these results are not robust to the presence of transaction costs. Indeed, if we compute the break-even transaction cost (Eq. (11.1)), only if we trade with transaction costs lower than 0.00266%, we can obtain a positive profit. Moreover, for the remaining stocks, whose results we do not report here, the return models always generate Sharpe ratios higher than in the original series.

Finally, we investigate the sensitivity of the bootstrap results to the number of simulations on our results. As remarked in BLL, this test is important since the asymptotic properties of the bootstrap applied to GARCH models are not known. In Table 11.10 we report the same figures as in Table 11.9 for the GARCH model, but varying the number of simulations. We can observe that the p-values obtained with 500 simulations are reliable. Considering additional simulations produces little improvement in the estimation of the p-values. Similar conclusions were also presented in BLL.

Vanilla Options

12

Estimating the Risk-Neutral Density[*]

Key words: log-normal mixture, implied volatility, calibration

Investors, risk-managers, monetary authorities and other financial operators are confronted with the need to assess market expectations concerning a number of fundamental macroeconomic variables, such as exchange and short-interest rates, stocks, commodities, stock indices. In simple terms, they need to estimate the probability distributions of future events. Information embedded in market prices of derivative assets provides central banks and operators with timely forward-looking information on market expectations regarding the underlying fundamental factors.

We tackle the issue of recovering the risk-neutral density function (henceforth PDF) as implied by market quotations and employ the resulting assessment about market expectations for trading purposes.

Section 1 describes the problem of recovering an implied risk-neutral density from option market prices and briefly examines basic procedures for extracting the PDF from option prices. In particular, we focus on a methodology based on a mixture of log-normals, as developed in Bahra (1997). We highlight the flexibility of this method to cope with the anomalies of the Black–Scholes theory, such as smile and smirk effects in the implied volatility curve. Sections 12.2 and 12.3 detail the solution methodology and provide two applications of the method. First, we recover the risk-neutral density compatible to a set of option prices written on the Standard & Poor's index S&P 500. Second, we propose and test a trading strategy exploiting differences between market and model prices as obtained from the estimated density function. We finally comment on the market's ability to forecast movements and disclose investor feeling as it can be inferred from our analysis.

---

[*] with Paolo Ghini.

12.1 Problem Statement

This section presents an overview of the most common methods used to extract risk-neutral density functions from option prices and focuses on a technique developed in Bahra (1997). Market participants' expectations are too heterogeneous and complex to be captured using simple descriptive statistics, such as means and other point estimates. However, this information can be extracted from prices of traded options. Indeed, due to a nonlinearity in their payoffs, option prices across different strikes for a common maturity allow the assigning of probabilities to a wide range of possible values taken by the underlying asset at that maturity. These probabilities represent a synthesis of market expectations about future trends as perceived by operators at a given point in time. Moreover, option prices can provide additional information compared to the one stemming from a time series analysis of the underlying price process.

Consider a European call option stricken at K and maturing in τ years. The fair value of this option is given by the risk-neutral expected value of its discounted payoff. This value reads as an integral over the exercise region:

$$c_t(K, \tau) = e^{-r\tau} \int_K^{\infty} (x - K)q_{t+\tau}(x)\, dx. \tag{12.1}$$

Here $q_{t+\tau}$ denotes the risk-neutral density of the underlying asset price at the expiration time $t + \tau$. Similarly, the fair price of a put option with equal features is

$$p_t(K, \tau) = e^{-r\tau} \int_0^K (K - x)q_{t+\tau}(x)\, dx. \tag{12.2}$$

The inverse problem consisting of the identification of a risk-neutral distribution $q_{t+\tau}$ implied by option prices was first addressed in a seminal paper by Breeden and Litzenberger (1978). These authors show that the risk-neutral density $q_{t+\tau}$ is recovered from option prices as

$$q_{t+\tau}(x) = e^{r\tau} \frac{\partial^2 c_t(K, \tau)}{\partial K^2}\bigg|_{K=x}. \tag{12.3}$$

Implementing this formula requires the knowledge of option prices for a continuum of strikes. Of course this is not possible in practice and infinitely many density functions are compatible to any given set of option prices over a finite range of strikes. However, some basic constraints have to be satisfied when constructing a risk-neutral density. For example, a well-defined risk neutral density is nonnegative, integrates to one, and prices exactly all calls and puts.

12.2 Solution Methodology

A possible way out is to derive a function $c(\cdot, \tau)$ by interpolating observed option prices. As observed by McCauley and Melick (1996a, 1996b) and Campa, Chang and Reider (1997, 1998) the most binding limitations of this approach occur on the over-the-counter (OTC) currency market where prices for very few strikes, say three to five, are usually quoted.

Several approaches to extract an implied density from option prices have been proposed in the literature. A comparison of different approaches can be found in Aparicio and Hodges (1998). We briefly recall:

1. Local volatility or implied tree models (Derman and Kani (1994), Dupire (1994), Rubinstein (1994), Jackwerth and Rubinstein (1996) and Jackwerth (1999)), where the future local volatility function of the underlying asset return is constructed in order to precisely recover option prices.
2. Interpolation of the implied volatility smile curve (Shimko (1993) and Dumas, Fleming and Whaley (1998)) where the volatility surface is assumed to be a parametric function. Parameters are fitted to calibrate option prices and the implied risk neutral density is then obtained by applying (12.3). Unfortunately, although very simple, this approach may deliver a density exhibiting kinks.
3. Stochastic volatility and jumps (Hull and White (1987), Melino and Turnbull (1990), Bates (1991), Heston (1993)), where the standard GBM process is extended to include a stochastic volatility component and jumps. The main problem with this approach consists of a lack of a closed form formula for option prices, so that calibration of the model can be highly time consuming.
4. Non-parametric approach (Aït-Sahalia and Lo (1998)) where the density is estimated by a kernel regression approach. However, this technique is based on time series price data so it cannot recover a risk-neutral density. Moreover, it is very demanding from a computational point of view.
5. Combination of parametric and non-parametric approach (Bedendo et al. (2005)). In particular, they focus the attention on the problem of fitting the tails.

An interesting way to reduce the degree of freedom allowed by relation (12.3), is to fit an implied density within a parametrized family of probability distributions. Bahra (1997), Melick and Thomas (1997) and Söderlind and Svensson (1997) look for a density obtained as a mixture of log-normal distributions. More precisely, the unknown density function is assumed to be a weighted average of log-normal densities with appropriate means and variances. The parameters are determined by minimizing the sum of the squared deviations of theoretical prices, computed by formulae (12.1) and (12.2), from market quotations. Bahra (1997) and Söderlind and Svensson (1997) investigate the ability of calibrated mixtures to provide useful indications about how monetary policies are conducted. Melick and Thomas (1997) retrieve an implied density function from American-style option prices on crude oil futures. After calibrating the model to market prices, these authors back out three curves describing possible future scenarios. For instance, during the crises preceding the Gulf war, the resulting density displays scenarios corresponding to alternative political evolutions of the situation: (a) a return to a pre-crises situation with a peaceful withdraw of Iraq from Kuwait; (b) a strong and relevant interruption of the Persian Gulf oil supplies due to the war; (c) uncertainty, with a continuation of unsettled conditions over relevant horizons.

Assuming a fixed structure for the terminal PDF, the question about the existence of a stochastic process for the underlying asset with the assigned marginal distribution arises. This problem has been investigated by Dupire (1994). A review of different approaches can be found in Rebonato (1999). Recently, Brigo, Mercurio and Rapisarda (2004) have shown the existence of such a process. This identification is relevant for pricing exotic and American options.

We now describe the method in greater detail. The mixture procedure consists of assuming a density function $q_{t+\tau}$ as a weighted sum of log-normal densities

$$q_{t+\tau}(x) = \sum_{i=1}^{k} w_i L(x; \alpha_i, \beta_i), \qquad (12.4)$$

where the weights w_i are normalized and positive

$$\sum_{i=1}^{k} w_i = 1 \quad \text{and} \quad w_i > 0, \quad \forall i.$$

Here $L(x; \alpha, \beta)$ represents a log-normal density with parameters α and β

$$L(x; \alpha, \beta) = \frac{1}{x\sqrt{2\pi\beta^2}} \exp\left(-\frac{1}{2\beta^2}(\ln x - \alpha)^2\right), \quad x > 0.$$

Each density in the mixture may be interpreted as a possible regime prevailing in the future. The most parsimonious parametrization is obtained for $k = 2$. In this case only five parameters need to be estimated. It is possible to make explicit dependence of density parameters α_i and β_i from the initial stock price, the drift coefficients μ_i and the volatility functions σ_i

$$\alpha_i = \ln S_t + \left(\mu_i - \frac{1}{2}\sigma_i^2\right)\tau,$$

$$\beta_i = \sigma_i \sqrt{\tau}.$$

The proposed parametric form turns out to be sufficiently flexible to capture the main statistical features in the market, such as skewness and fat tails in the price return distribution. The corresponding option pricing formulae read as

$$c_t(K, \tau) = e^{-r\tau} \int_K^\infty [wL(x; \alpha_1, \beta_1) + (1 - w)L(x; \alpha_2, \beta_2)](x - K)\, dx, \qquad (12.5)$$

$$p_t(K, \tau) = e^{-r\tau} \int_0^K [wL(x; \alpha_1, \beta_1) + (1 - w)L(x; \alpha_2, \beta_2)](K - x)\, dx. \qquad (12.6)$$

It is easy to show that these expressions can be represented as weighted averages of Black–Scholes formulae

$$c_t(K, \tau)$$
$$= e^{-r\tau} \left\{ w \left[e^{\alpha_1 + \beta_1^2/2} \mathcal{N}(d_1) - K\mathcal{N}(d_2) \right] \right.$$
$$\left. + (1-w) \left[e^{\alpha_2 + \beta_2^2/2} \mathcal{N}(d_3) - K\mathcal{N}(d_4) \right] \right\},$$

$$p_t(K, \tau)$$
$$= e^{-r\tau} \left\{ w \left[K\mathcal{N}(-d_2) - e^{\alpha_1 + \beta_1^2/2} \mathcal{N}(-d_1) \right] \right.$$
$$\left. + (1-w) \left[K\mathcal{N}(-d_4) - e^{\alpha_2 + \beta_2/2} \mathcal{N}(-d_3) \right] \right\},$$

where $\mathcal{N}$ denotes the cumulative normal distribution and

$$d_1 = \frac{-\ln K + \alpha_1 + \beta_1^2}{\beta_1}; \qquad d_2 = d_1 - \beta_1;$$

$$d_3 = \frac{-\ln K + \alpha_2 + \beta_2^2}{\beta_2}; \qquad d_4 = d_3 - \beta_2.$$

In the special case of a call option with a zero strike, we obtain the present value of the forward price $f_t(\tau)$:

$$c_t(0, \tau) = e^{-r\tau} \left\{ w e^{\alpha_1 + \beta_1^2/2} + (1-w) e^{\alpha_2 + \beta_2^2/2} \right\} = e^{-r\tau} f_t(\tau).$$

Absence of arbitrage opportunities requires that $f_t(\tau)$ must equal the mean of the PDF. These formulae provide model prices for any pair of density parameters. The model is calibrated by choosing a parameter set $\theta = \{\alpha_1, \alpha_2, \beta_1, \beta_2, w\}$ which minimizes the sum of squared differences between model and market prices for a set of calls $c^1, \ldots, c^n$ and puts $p^1, \ldots, p^m$ with a common maturity τ. The non-linear least squares problem reads as:

$$\min_{\theta} \left\{ \sum_{i=1}^{n} \left[c^i - c_t(K_i, \tau) \right]^2 \right.$$
$$+ \sum_{i=1}^{m} \left[p^i - p_t(H_i, \tau) \right]^2$$
$$\left. + \left[w e^{\alpha_1 + \beta_1^2/2} + (1-w) e^{\alpha_2 + \beta_2^2/2} - e^{r\tau} S_t \right]^2 \right\}, \qquad (12.7)$$

$$\text{sub} \quad \beta_1, \beta_2 > 0 \quad \text{and} \quad 0 \le w \le 1,$$

where $K_1, \ldots, K_n$ (resp. $H_1, \ldots, H_m$) are call (resp. put) strikes and S_t is the current spot price. Note that the third line of the target function includes the squared difference between theoretical forward price and expected price ($e^{r\tau} S_t$).

12.3 Implementation and Algorithm

In this section, we detail our implementation of the calibration procedure. Table 12.1 reports module names containing VBA® macros and functions.

The calibration has been conducted through the following steps.

Table 12.1. List of main VBA$^{®}$ macros and functions

Module name	Macro or function	Formula
mDensity	normalmixture	(12.4)
mOptionPrices	BlackScholes()	
mOptionPrices	Mixturecall()	(12.5)
mOptionPrices	Mixtureput()	(12.6)
mCalibrationMixture	Sub MixtureEstimation()	(12.7)

1. Collection of implied volatility, LIBOR rates, dividends and S&P 500 Index quotations. All of them are available in the spreadsheet Data of the Excel file RiskNeutralDensity.xls.

2. Transformation of implied volatility quotations into option prices. This is done in spreadsheet Mixture Calibration 3m and Mixture Calibration 6m. For each week, we obtain option prices using the Black–Scholes formula coded in the VBA$^{®}$ function BlackScholes(). Market option prices are computed in columns D–H.

3. Given the parameter set, for each week we compute theoretical call and put prices using the VBA$^{®}$ functions Mixturecall() and Mixtureput(). Theoretical option prices are computed in columns J–N.

4. Parameter calibration is performed using the macro MixtureEstimation() which can be run by pressing the button "Run Mixture Estimation" in the Excel spreadsheets Mixture Calibration 3m and Mixture Calibration 6m, see Fig. 12.1. The macro, for each week in the sample, runs the Excel Solver[1] and minimizes the sum of squared differences between market and model prices. These differences appear in column AF in the Excel spreadsheet. The user can also restrict the calibration period by specifying in cells AF3 and AF4 respectively the rows corresponding to the first and the last week. Remark that in order to ensure the respect of constraints such as positive variances, we have defined some dummy parameters, appearing in columns Y–AC, for which no restrictions are required. Original parameters are then functions of these parameters, chosen in a way to ensure that the constraints are fulfilled. The Excel Solver keeps on changing the content of cells Y–AC until the target function in column AF is minimized.

   In the calibration, for some weeks we have obtained unrealistic parameter values. In such cases, we have performed the optimization under a different starting value. Sometimes, a long time is required to obtain estimates comparable to the ones obtained in the nearby weeks. This happens for example on June 25, 1999.

---

[1] The Excel Solver is not always promptly available in the VBA$^{®}$ code. Therefore, in the VBA$^{®}$ window you must choose Tools, Riferimenti and then activate Solver.xla. If this file does not appear in the window, usually, it can be found in the directory where Microsoft Excel has been installed, and then in the sub-directory Library. In any case, you can locate the file by doing a file search.

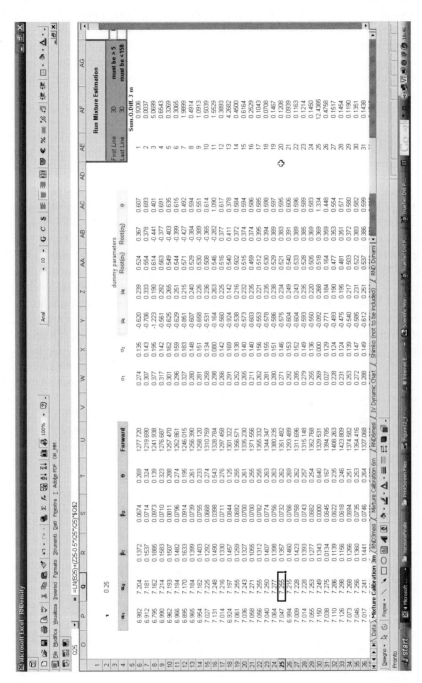

Fig. 12.1. Screenplot of the spreadsheet for calibrating the risk-neutral density.

5. Spreadsheets IV Dynamic Chart and RND Dynamic Chart include a dynamic chart of the implied volatility curve and the corresponding risk neutral density for each week in the sample.
6. Spreadsheets Strategy 3 m and Strategy 6 m build a trading strategy consisting of entering a long (resp. short) position in undervalued (resp. overvalued) options. The construction of the strategy has been done directly in the Excel spreadsheet without using VBA® code.

12.4 Results and Comments

Our data set covers the period ranging from January 1, 1999 to December 7, 2001. Data consist of 150 weekly observations on European-style options written on the S&P 500 Index and quoted Over-The-Counter (OTC) by a market maker.[2] We recall that the S&P 500 Index represents about the 80 percent of the entire US economy capitalization.

Option prices are determined each Friday morning depending on the closing values of the Index and interest rates of the previous day. For each week, quotations are available for five strike prices corresponding to At-The-Money (ATM) options and Out-of-The-Money (OTM) options with strikes 5 and 10 percent away from the current level of the index. Options are traded with "fixed time to maturity": each week, new contracts with three or six months to maturity are issued and quoted. Therefore, the data consists of the most liquid and deeply traded range of contracts and maturities. However, contracts cover only five strike-prices and observations for far from at-the-money options are not available. This problem implies that we cannot estimate with accuracy the tails of the risk-neutral density.

Quotations are expressed as percentage points of implied volatility (i.e., the volatility parameter entering the Black and Scholes formula), with respect to five different moneyness, according to standard rules in the OTC market. Table 12.2 provides a numerical example of the quotation mechanism. Quotations refer to OTM and ATM options.

The second line indicates the moneyness as defined as a percentage over the spot value of the index. For example, on Friday January 22, 1999, the implied volatility

Table 12.2. Example of implied volatilities on SPX 3 months options on January 3, 1999

22 Jan. 1999	OTM put	OTM put	ATM	OTM call	OTM call
Moneyness	0.9	0.95	1	1.05	1.1
Strike	1111.64	1173.40	1235.16	1296.92	1358.68
Implied vol.	35.40%	31.96%	28.97%	26.41%	23.93%
BS prices	32.10	45.42	65.36	44.12	21.18

S&P 500 Index: 1235.16; 3-months dividend yield: 1.33%;
3-months LIBOR rate: 4.97%.

---

[2] The dataset was kindly supplied by UBS-Warburg, London.

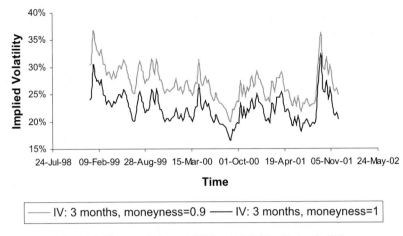

Fig. 12.2. Time evolution of ATM and OTM implied volatilities.

of a put contract issued on the SPX with a maturity of three months and a strike price equal to 0.90% of the S&P 500 index, is 35.40%.

Strike prices are computed in the third line by using the S&P 500 index value (1235.16 closing value on Thursday January 21, 1999). Therefore, given the implied volatility quotation, we can recover the corresponding option price in the fifth line, by applying the Black and Scholes formula.[3]

For each day in our data set, we solve the optimization problem in (12.7). Then, we test the quality of a trading strategy based on the calibrated density. Our analysis delivers several interesting findings.

Figure 12.2 shows that, over the period under investigation, implied volatilities from OTM options are higher than those corresponding to ATM options. Moreover, the movements in the two implied volatilities appear highly correlated. Since deep OTM put options provide valuation of the left-hand tail of the distribution, they quantify the risk of large adverse movements. This effect proves that the option market assigns to extreme events a higher probability of occurrence than the one predicted by the log-normal assumption. Consequently, the implied probability density cannot be log-normal. In this respect, Bates (1991) demonstrates that the 1987 crash modified the stock-index option pricing procedures employed by the operators. Indeed, the crash had the effect of increasing prices of OTM put options well beyond the value resulting from the Black–Scholes formula computed at a volatility level implied by ATM quotations. This phenomenon reflects the practice of hedging stock portfolios against the risk of a crash (static portfolio insurance), which leads to a strong demand for OTM put options. The writer of these options, while performing dynamic hedging, will soon find himself in a position to sell the underlying stock

---

[3] In the Black–Scholes formula the LIBOR rate has been converted to a continuously compounded interest rate and the SPX index has been adjusted by the continuous dividend yield. The LIBOR rate is assumed to be a risk-free interest rate and, in the example, it has been collected on Thursday January 21, 1999.

Table 12.3. Statistics of implied volatilities of 3-month options on SPX (from January 1, 1999, to December 31, 2001)

Moneyness	0.9	0.95	1	1.05	1.1
Max	36.59%	34.30%	32.47%	30.83%	29.35%
Mean	26.90%	24.55%	22.39%	20.44%	18.83%
Min	19.88%	18.14%	16.60%	15.30%	14.42%
St. dev.	3.06%	2.81%	2.61%	2.48%	2.28%

Table 12.4. Statistics of implied volatilities of 6-month options on SPX (from January 1, 1999 to December 31, 2001)

Moneyness	0.9	0.95	1	1.05	1.1
Max	34.18%	32.14%	29.75%	28.21%	27.10%
Mean	26.27%	24.50%	22.88%	21.37%	20.03%
Min	20.63%	19.18%	17.85%	16.64%	15.67%
St. dev.	2.86%	2.63%	2.42%	2.26%	2.11%

index with supply higher than the demand when the market is falling. Moreover, several stock exchanges provide a mechanism to limit running operations in case of large drops in quoted prices, thus preventing operators from performing the required hedging strategies whenever their need is particularly compelling. Hence, the relatively strong demand for OTM put options to cover the risk related to this kind of instances induces higher premia than the fair prices stemming from arbitrage pricing models.

A second empirical evidence is the asymmetry of the option smile: implied volatilities for ATM options are larger than implied volatilities for OTM call options. This behavior is confirmed by the average values reported in Tables 12.3 and 12.4. Figure 12.3 reproduces an implied PDF function as derived as above. The five parameters of the mixture density vary as option prices change to reflect changes in market expectations about the future. Mixtures can embed a great variety of shapes describing different scenarios, like the situation in which the market has a bimodal vision about the future value of an asset. This happens, for example, when market participants foresee the possibility of a shock affecting the indices. Tables 12.5 and 12.6 report some basic descriptive statistics concerning the series of the calibrated parameters. From the estimation of the mixture we can separate the contributions of the different densities to the mixture. Indeed the two components present very different mean levels, as illustrated in Fig. 12.4 for a typical day in our sample (1st June 2001). Moreover, the contribution to the mixture of the density with lower mean appears to be much less important (the parameter w has an average value of 0.3), nevertheless, it has a much larger standard deviation. This affects the kurtosis of the mixture, as we can see in Fig. 12.4 comparing the calibrated implied density with a normal with same mean and variance.

Figure 12.5 exhibits the time evolution of the expectations for a given maturity. We compare the implied distribution recovered from six-month options with the ones derived three months later from options maturing in three months. This comparison

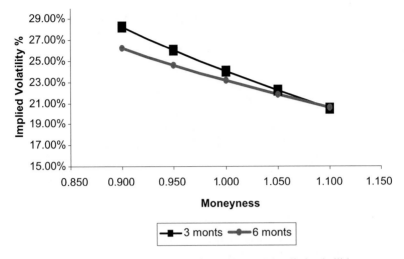

Fig. 12.3. Skew curve for 3-month and 6-month implied volatilities.

Table 12.5. Statistics of calibrated parameters of 3-month options on SPX (from January 1, 1999 to December 31, 2001)

Parameters	α_1	α_2	β_1	β_2	w
Max	7.248	7.384	0.189	0.097	0.640
Mean	7.065	7.236	0.124	0.068	0.302
Min	6.795	6.978	0.013	0.000	0.125
St. dev.	0.102	0.088	0.018	0.009	0.069

Table 12.6. Statistics of calibrated parameters of 6-month options on SPX (from January 1, 1999 to December 31, 2001)

Parameters	α_1	α_2	β_1	β_2	w
Max	7.416	7.280	0.202	0.296	0.857
Mean	7.265	7.064	0.089	0.195	0.602
Min	6.994	6.604	0.021	0.043	0.318
St. dev.	0.092	0.124	0.016	0.030	0.070

shows the way information improves over time.[4] Therefore hypothetical scenarios are clearer and the range of values that the index will be likely to assume at the maturity of the option narrows.

Another remarkable result concerns the time variation of parameters defining the mixture. This behavior might be due to the limited (five) number of traded options for a fixed maturity. Therefore modest changes in market's option prices cause large

---

[4] However, one should remember that recovered implied densities are risk-neutral. Nevertheless, assuming that market's risk-aversion is relatively stable over time, daily or weekly changes in the PDF should reflect only the change in investor expectations and not in their risk-adversion as well.

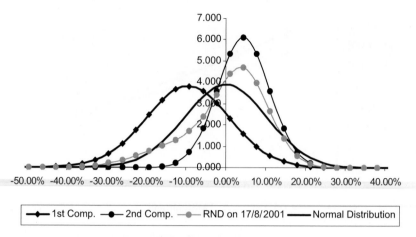

Fig. 12.4. Components of the 6 months risk-neutral density on the 1st of June 2001, and normal density with same first two moments as the mixture.

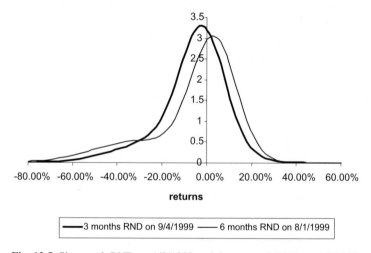

Fig. 12.5. Six-month RND on 1/9/1999 and three-month RND on 4/9/1999.

oscillations in the parameters of the distributions. Campa, Chang and Reider (1997, 1998) are confronted with the same issue while studying the OTC currency market. The time variation also seems to reflect the economic situation. Indeed in Figure 12.6 we see that the means of the two distributions move together over the entire sample. They increase up to October 2000 reflecting strong belief in an up-trending market. During the drop period affecting the New Economy in October 2000, market expectations reversed and average values began to decrease.

In order to test the empirical performance of the mixture, we propose a trading strategy that tries to exploit the differences between market and model prices. The strategy consists of entering a long (short) position in undervalued (overvalued) op-

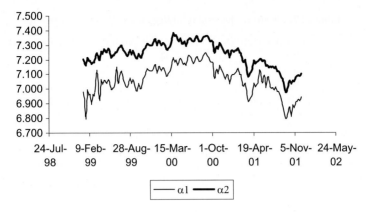

Fig. 12.6. Time variations of the means of the two components the log-normal mixture.

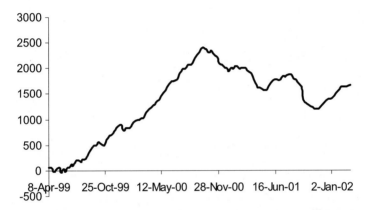

Fig. 12.7. Cumulative profits for options maturing in three months.

tions. Mispricing is measured as the absolute difference of market option price and model prices larger than a predetermined size. We test the strategy for different filter sizes, ranging from 0.01 to 0.2.

Since the calibrated lognormal mixture density tends to undervalue OTM put options,[5] we are often in a position to sell deep OTM puts (moneyness = 0.9) and to buy OTM options (moneyness = 0.95) at the same time. In other words, one becomes an insurer against large price falls in the market. This fact has determined very large losses for the options expiring in September 2001, the month of the terrorist attack on the Twin Towers. However, as shown in the profit and loss diagram reported in Fig. 12.7 for a filter size equal to 0.05, and in a greater detail in Tables 12.7 and 12.8, the strategy appears profitable if we do not consider filter sizes that are too large (i.e., approximately less than 0.15). However, higher average profits are associated to higher variability, as measured by the standard deviation.

---

[5] Bedendo et al. (2005) stress out that this problem occurs with parametric methods especially if the availability of observations is limited.

Table 12.7. Results of the trading strategy on SPX 3-month options

Filter size ($)	Cumulative profit ($)	Avg. weekly profit ($)	Std. dev. weekly profit
0.01	1785	11.74	48.27
0.05	1666	10.96	47.50
0.1	231	1.52	46.01
0.15	−610	−4.02	39.04
0.2	−1639	−10.78	38.02

Table 12.8. Results of the trading strategy on SPX 6-month options

Filter size ($)	Cumulative profit ($)	Avg. weekly profit ($)	Std. dev. weekly profit
0.01	2385	15.69	65.82
0.05	2265	14.90	56.44
0.1	494	3.25	49.30
0.15	−747	−4.91	48.92
0.2	−982	−6.46	49.86

13

An "American" Monte Carlo[*]

Key words: American options, Snell envelope, Monte Carlo simulation

We present a method to price American-style options using Monte Carlo simulation as proposed by Rogers (2002). This method is an alternative method to the famous simulation-based technique introduced by Longstaff and Schwartz (2001) (L&S). Roger's proposal shares the starting point of the analysis with L&S. They both start with the traditional dynamic programming equation, which conveys the idea that pricing an American-style option is a problem of knowing, at each point of time, whether it is worth to exercise the option immediately or to continue holding the option. As opposed to L&S who attempt to determine an optimal exercise policy, the method we adopt here transforms the dynamic programming equation in a dual problem, proposes a financial interpretation of this latter, and finally tries to solve it numerically.

Several pricing methods for American-style options have been proposed in the specialized literature (see, e.g., Lamberton and Lapeyre (1996) for a quick overview). In the case of options written on a large basket of underlying assets, these techniques tend to become quite inefficient and computationally slow. In order to overcome these difficulties, Carrière (1996) proposed a simulation-based method, which suffers, however, from numerical instability problems. Bouchard, Ekeland and Touzi (2004) proposed an alternative based on the notion of Malliavin derivative. However, it suffers from a computational drawback in terms of speed for large baskets. Broadie and Glasserman (1997) suggest a stochastic mesh method for overcoming these issues and their proposal has gained popularity among practitioners.

The present chapter is organized as follows. Section 1 quickly summarizes the American-style option pricing theory (see Carr, Jarrow and Myneni (1992), Jamshidian (1992), and Lamberton and Lapeyre (1996), among others) and singles out a computational issue involving conditional expectations. Section 13.2 introduces the dual

---

[*] with Igor Toder.

formulation of the result illustrated in the previous section. Section 13.3 details the general algorithm to implement the proposed methodology. Section 13.4 illustrates experimental results and concludes with a few comments.

13.1 Problem Statement

We assume a single-asset Black–Scholes framework where risk neutral price dynamics are given by

$$dS_t = r S_t \, dt + \sigma S_t \, dW_t,$$

starting at $S_0 = x$, and a deterministic money market account reads as $B_t = \exp(rt)$. Here, r denotes the continuously compounded risk-free rate of interest, σ is the constant volatility of instantaneous stock returns, and W indicates a standard Brownian motion.

We consider an American put option exercisable on time horizon $[0, T]$. Its *intrinsic value* is defined as the process $Z = (Z_t)_{0 \le t \le T}$ describing the payoff over the entire exercise period, that is $Z_t = \max(K - S_t, 0)$. We denote by $V = (V_t)_{0 \le t \le T}$ the arbitrage-free price process for this option. Our goal is to compute this figure under the assumption that the exercise time must belong to a finite set of dates $P = \{t_0, \ldots, t_N\}$, with $t_i = i\Delta$ and $iN = T$. We assume that at time $t_0 = 0$ all information is available to the option pricer. Though this particular contingent claim is usually referred to as a Bermuda option, we continue using the broader term "American option" in the sequel.

The time t value of the option under consideration can be expressed as a deterministic function $V(t, W)$ of time t and state W assumed by the Brownian noise W_t. It is a standard result that corresponding to any noise path $W_{t_0}, \ldots, W_{t_N}$, the function V solves a dynamic programming recursive equation

$$\begin{cases} V(T, W_T) = Z_T, \\ V(t_i, W_{t_i}) = \max\big(Z_{t_i}, \, e^{-r\Delta} \mathbb{E}\big(V(t_{i+1}, W_{t_{i+1}})| W_{t_i}\big)\big), & i = 0, \ldots, N-1. \end{cases}$$

The deterministic function $V : [0, T] \times \mathbb{R} \to \mathbb{R}_+$ is called the "*Snell envelope*" of the discrete-time process $(Z_{t_i})_{t_i \in P}$. The underlying idea justifying this algorithm is simple: at any time t_i, the option holder must decide whether he should exercise the option right and then collect the reward Z_{t_i}, or keep on holding the option. In this case, the standing position is worth the expected present value of what it will be worth one time period later, namely $e^{-r\Delta} \mathbb{E}(V(t_{i+1}, W_{t_{i+1}})| W_{t_i})$. Since the holder is assumed to behave in a way which would maximize the value of his position, he will choose the greatest between these two numbers. On the final day T, the option is worth its payoff Z_T. Once the recursion above is solved, the arbitrage-free price of the option is given by $V(t_0, W_{t_0})$. The key issue in this procedure is the computation of conditional expectation $\mathbb{E}(V(t_{i+1}, W_{t_{i+1}})| W_{t_i} = x)$, for all $x \ge 0$.

The method proposed by L&S consists of building an expected optimal exercise policy for the American-style option. They compute estimates for $\mathbb{E}(V(t_{i+1}, W_{t_{i+1}})| W_{t_i} = x)$ and use them to find, for each given path of the underlying, whether the anticipated exercise of the option is more rewarding than waiting until the next time step. This method comes up with a lower bound of the theoretical price.

In our setting we adopt the point of view of a trader on the selling side. It is common knowledge that the price of an option equals the price of its hedge. When a trader writes an option, his major concern consists of finding a hedging strategy for the option. The option price corresponds to the initial cost of the hedging strategy. If this cost is too high, then the trader will not make the deal since it is too expensive; if the price is too low, then the trader might not be able to hedge his option and he will not enter the deal. In contrast to L&S, we determine the option price by computing a hedging strategy instead of calculating an optimal exercise policy. This formulation enables us to get rid of the conditional expectation estimates mentioned above.

13.2 Model and Solution Methodology

We begin by introducing an alternative formulation to the Snell's envelope problem and interpret it in terms of a hedging strategy.

The discounted price process $(e^{-rt_i}\mathbf{V}(t_i, W_{t_i}))_{0 \leq i \leq N}$ solving the dynamic programming equation above can be proved to be a supermartingale. By a famous theorem due to Doob and Meyer, it can be then decomposed as follows:

$$\left(e^{-rt_i}\mathbf{V}(t_i, W_{t_i})\right) = \mathbf{V}(t_0, W_{t_0}) + M_{t_i}^* - A_{t_i}^*,$$

where $(M_{t_i}^*)_{0 \leq i \leq N}$ is a martingale starting at zero $(M_0^* = 0)$ and $(A_{t_i}^*)_{0 \leq i \leq N}$ is a predictable increasing process vanishing at $t = 0$. Predictability means that the value of the process A^* at time t_{i+1} is perfectly known one time step before, i.e., at time t_i.

The present case-study is grounded on a result stating that the price of an American-style option is given by

$$\mathbf{V}(t_0, W_{t_0}) = \inf_{M \in H_0^1} \mathbb{E}\left[\sup_{0 \leq i \leq N} \left(e^{-rt_i}Z_{t_i} - M_{t_i}\right)\right],$$

where H_0^1 denotes the set of martingales $M = (M_{t_i})_{0 \leq i \leq N}$ such that (1) $\mathbb{E}[\sup_{0 \leq i \leq N}|M_{t_i}|] < +\infty$ and (2) $M_{t_0} = 0$. We let M^* be the arg inf in the previous optimization problem. It can be proved that a conditional version of the previous result holds true:

$$\mathbf{V}(t_i, W_{t_i}) = \inf_{M \in H_0^1} \mathbb{E}\left[\sup_{i \leq j \leq N} \left(e^{-r(t_j - t_i)}Z_{t_j} - M_{t_j}\right) \mid W_{t_i}\right]$$

for all $0 \leq i \leq j \leq N$. Let us now explain how this formula can be used in practice for pricing purposes. We select a suitable martingale and then evaluate the expected value $\mathbb{E}[\sup_{0 \leq i \leq N} (e^{-rt_i}Z_{t_i} - M_{t_i})]$ through a Monte Carlo simulation.

To date, no theoretical rule exists to help the user in finding an acceptable martingale to begin with. In most cases, this selection is merely a "learning-by-doing" process. However, we will show below that even simple martingales usually deliver an algorithm producing accurate results. The problem of computing an optimal exercise strategy is now replaced by the problem of finding a good minimizing martingale M. We therefore see that this formulation provides us with an upper bound of the option price, whereas the method developed by L&S gives a lower bound.

We now interpret the preceding formula in terms of hedging; this interpretation in turn gives a good hint for finding an acceptable minimizing martingale. Once a martingale M has been chosen, we have that $\mathbf{V}(t_0, W_{t_0}) \leq \mathbb{E}[\sup_{0 \leq i \leq N} (e^{-rt_i} Z_{t_i} - M_{t_i})]$. This inequality implies that for any $t_k \in \mathcal{P}$, $e^{-rt_k} Z_{t_k} \leq \sup_{0 \leq i \leq N} (e^{-rt_i} Z_{t_i} - M_{t_i}) + M_{t_k}$. If a trader sells the option for a price $\eta_{t_0} = \mathbb{E}[\sup_{0 \leq i \leq N} (e^{-rt_i} Z_{t_i} - M_{t_i})]$ (which is greater than the theoretical price $\mathbf{V}(t_0, W_{t_0})$), then the intrinsic value of the option is bounded from above as follows:

$$e^{-rt_k} Z_{t_k} \leq \mathbb{E}(\eta_T - \eta_{t_0} | W_{t_k}) + (M_{t_k} + \eta_{t_0}),$$

where $\eta_T = \sup_{0 \leq i \leq N} (e^{-rt_i} Z_{t_i} - M_{t_i})$. We can interpret M as the discounted profit/loss (P&L) of a hedging strategy investing an initial amount η_{t_0}. Thus, the time t_k value of the hedging portfolio is $\eta_{t_0} + M_{t_k}$.

If $\mathbb{E}(\eta_T - \eta_0 | W_{t_k})$ is too high (in absolute deviation), then the (discounted) value of the hedging portfolio is not large enough to cover the (discounted) value $e^{-rt_k} Z_{t_k}$ claimed by the option buyer. The smaller the quantity $\mathbb{E}|\eta_T - \eta_0|$, the better the strategy. Interpreting M as the discounted value of a hedging portfolio gives a clue for guessing a suitable martingale.

13.3 Implementation and Algorithm

We now propose an algorithm to compute the price of an American-style option through the method previously described. We assume we have already decided upon the minimizing martingale M satisfying the equation given above. However, to date there is no theoretical rule which helps us select an optimal minimizing martingale of this kind and the final choice largely depends on the particular instance under consideration. For the sake of clarity, we first summarize the major quantities involved in the procedure:

- $N =$ Number of possible exercise dates
- $i =$ Date index
- $n =$ Number of sample paths
- $m =$ Sample index
- $K =$ Strike price
- $T =$ Time horizon
- $r =$ One-period risk-free rate of interest

Algorithm   1. Fix a strike price K, the risk-less rate r and the set of dates $t_0 < t_1 \leq t_2 \leq \cdots \leq t_N = T$ at which the option may be exercised;
2. Simulate n sample paths for the intrinsic value process Z and the minimizing martingale M. Here, $(Z_{t_i}^m)_{0 \leq i \leq N}$ and $(M_{t_i}^m)_{0 \leq i \leq N}$ denote the mth path for the process Z and M, respectively.

3. Let $\mathbf{X}$ be the n-dimensional vector recording the maximum value attained by the process $(e^{-rt_i}(Z_{t_i}) - M_{t_i})_{0 \le i \le N}$ on each sample path:

$$\mathbf{X}(m) = \max_{0 \le i \le N} \left[e^{-rt_i}\left(Z_{t_i}^m\right) - M_{t_i}^m \right].$$

4. At time t_i, for each simulated path m, compare the quantities $(e^{-rt_i}(Z_{t_i}^m) - M_{t_i}^m)$ and $\mathbf{X}(m)$. If the former is greater than the latter, store the value $(e^{-rt_i}(Z_{t_i}^m) - M_{t_i}^m)$ into the entry $\mathbf{X}(m)$.

5. If $t_{i+1} < T$, increase i by one unit and go to Step 4. Otherwise, return $\mathbf{V}(t_0, W_{t_0}) := \frac{1}{n} \sum_{m=1}^{n} \mathbf{X}(m)$ as an empirical estimate of the expected value $\mathbb{E}[\sup_{0 \le i \le N} (e^{-rt_i} Z_{t_i} - M_{t_i})]$.

13.4 Results and Comments

We consider an American put option written on a single asset paying no dividend. The option is struck at an exercise price K over the entire time horizon $[0, T]$. To find a minimizing martingale M, we proceed heuristically. Let $(BS_{t_i})_{0 \le i \le N}$ be the price process of a European put with the same features as the American put under analysis. Each price BS_{t_i} is given by the celebrated Black–Scholes formula. Also let $(\widetilde{BS_{t_i}})_{i \in [0,N]}$ denote the corresponding discounted price. One can prove that this process is a martingale under the risk neutral probability. However, we do not compute the price of the American put by using $\widetilde{BS_{t_i}}$ as a minimizing martingale M. Indeed, this choice turns out to be excessively crude for hedging purposes.

Notice that a trader may envisage a strategy hedging the American put since it becomes in-the-money. This could be performed by buying an equivalent European put as soon as the American put becomes in-the-money. Consequently the martingale M hedging the American put could be

$$M_{t_i} = [\widetilde{BS_{t_i}} - \widetilde{BS_{\tau_m}}] \mathbf{1}_{t_i \ge \tau_m},$$

where

$$\tau_m = \min \{t_k \in [0, T]: Z_{\tau_m} \ge 0\}$$

indicates the first time the intrinsic value becomes positive and $\mathbf{1}_{t_i \ge \tau_m}$ is the indicator function of the set $\{t_i: t_i \ge \tau_m\}$.

Results reported in Table 13.1 can be obtained by setting the following parameters in the routine `Rogers_full.m`:

- Strike price $K = 100$.
- Interest rate $r = 0.06$.
- Maturity (expressed in fraction of the year) $T = 0.5$.
- Volatility of the underlying process $\sigma = 0.4$.
- $b = 20$ exercise dates.
- $n = 500$ simulated paths.

Table 13.1. Binomial vs. simulated prices of standard American puts. $K = 100, r = 0.06,$ $T = 0.5, \sigma = 0.4, b = 20 , n = 500$

S_0	Bin. price	Mean price	Std. error	95% conf. interval
80	21.6059	21.7503	0.0263	[21.7340; 21.7666]
85	18.0374	18.1381	0.0263	[18.1218; 18.1544]
90	14.9187	14.9814	0.0267	[14.9649; 14.9979]
95	12.2314	12.2655	0.0220	[12.2519; 12.2791]
100	9.9458	9.9580	0.0200	[9.9457; 9.9704]
105	8.0281	8.0314	0.2027	[7.9057; 8.1570]
110	6.4352	6.4476	0.2302	[6.3049; 6.5902]
115	5.1283	5.1048	0.2447	[4.9531; 5.2565]
120	4.0611	4.0650	0.2547	[3.9072; 4.2228]

Table 13.2. Simulated prices of standard American puts. $S_0 = 120, K = 100, r = 0.06,$ $T = 0.5, \sigma = 0.4, b = 40$

Number of simulated paths	Mean price	Standard error
100	3.9001	0.3062
500	4.0429	0.2417
1000	4.0954	0.1336
5000	4.0912	0.0534
True price	4.0611	N/A

"Bin Prices" quoted on the first column correspond to numbers obtained by implementing a binomial-tree technique with a 1000-point time grid. These figures have been taken from Rogers (2002). We choose to simulate a relatively small number of trajectories to demonstrate the power of the algorithm in terms of accuracy under short computational time. Confidence intervals have been determined by launching the procedure 10 times.

Table 13.1 can be obtained by opening the file `Rogers_full.m`, inserting the parameters above, saving the file and launching the function `Rogers_full.m` with input variable `nb_simulations = 10`. As expected, the routine returns a mean price which is higher than the true price (except for $S_0 = 115$). We see that the value of the variance (and hence the accuracy of the confidence interval) deteriorates with the option getting more and more out-of-the-money. Part of this phenomenon can be explained by the fact that the number of simulations is very low and that the hedging strategy is far from being efficient in this case. This is particularly clear as the option becomes out-of-the-money.

A last experiment indicates that the convergence speed is low for options starting out-of-the-money. Figures reported in Table 13.2 can be obtained with the following parameters: $S_0 = 120 \ K = 100, r = 0.06, T = 0.5, \sigma = 0.4$ and $b = 40$. Again, standard errors have been determined by launching the routine 10 times and computing these descriptive statistics over the resulting sample of simulated prices. In summary, the pricing method presented here is innovative in that it avoids considering the issue of finding an optimal exercise policy for the option. However, it suffers from the complete lack of methodology in selecting a suitable minimizing martingale, whose choice might eventually require even more time than needed to compute the optimal exercise policy. The computer implementation is easy and the method delivers an upper bound for the option price. This feature may be coupled with the method developed by Longstaff and Schwartz for the purpose of determining a reasonable interval for the exact unknown price of American-style options.

14

Fixing Volatile Volatility[*]

Key words: stochastic volatility, fast Fourier transform, model calibration

Despite its simplicity and popularity among practitioners, it is commonly recognized that the Black–Scholes (BS) model only partly captures the complexity of financial markets and produces a persistent bias in pricing derivatives. A clear misspecification of this model is the assumption that stock returns are normally distributed. Indeed, nowadays there is abundant empirical evidence that return distributions exhibit sizable negative skewness and large kurtosis. Moreover, the assumption of constant volatility is particularly limiting and contrasts with volatility smiles and smirks implicit in market option prices.

In the last two decades, much effort has been put into the research on more realistic approaches that relax the most restrictive assumptions of the BS model. A large variety of sophisticated models have been proposed by assuming different processes for stock prices, interest rates and market prices of risks. For a review see Gatheral (2006). The drawback of these models is that closed-form formulae are rarely available and the advantage of a more realistic description can be offset by the costs of implementation and calibration. Among such alternative models, stochastic volatility (SV) models have attracted a great deal of interest. In an SV framework, volatility changes over time according to a random process usually assigned through a suitable stochastic differential equation (s.d.e.). In this manner, SV models manage to reproduce empirical regularities displayed by the risk-neutral density implied in option quotations.

In this case, we focus on the Heston (1993) model and an extension allowing for the inclusion of jumps in the stock return dynamics as proposed by Bates (1996) and by Bakshi, Cao and Chen (1997). We aim at testing the ability of the aforementioned models to fit market prices in comparison to the standard BS model. Our analysis focuses on option prices on the FTSE 100 index. However, model calibration requires

* with Matteo Bissiri, Andrea Bosio and Giacomo Le Pera.

analytical pricing formulas that are not explicitly available in the considered models. Fortunately, a closed-form formula for the characteristic function of the stock return distribution has been obtained by Heston (1993). Given this function, an expression for the Fourier transform of the option price with respect to the logarithm of the strike can be easily computed as in Carr and Madan (1999) and as illustrated in Chapter 6 "Quadrature Methods". Finally, option prices can be obtained by numerical inversion of the Fourier transform. This can be performed by the Fast Fourier Transform (FFT) technique, which reduces the computational burden quite significantly in comparison with standard algorithms for numerical integration (see Sect. 6.8 in Chapter 6). Following suggestions coming from several practitioners, particular attention has been placed on the choice of the loss function in the calibration stage. The performance of models under investigation has been also examined. We assessed the level of parameter consistency, measuring the extent to which they vary upon repeating the calibration process on a daily basis. We interpret large variations in the parameter figures as a clear indicator of model misspecification.

The chapter is organized in four sections. In Sect. 14.1, we briefly introduce SV models and their properties. Section 14.2.2 is dedicated to the description and calibration of the Heston model (HES) and its version with jumps (HESJ). Section 14.3 discusses the use of the FFT to compute option prices and briefly describes the Matlab® code. Finally, Sect. 14.4 exhibits the empirical performance of HES and HESJ models compared to the standard BS model.

14.1 Problem Statement

In the stochastic volatility model proposed by Heston (1993), stock price dynamics under real world measure are described by

$$dS(t) = \mu S(t)\,dt + \sigma(t)S(t)\,dW_s(t), \tag{14.1}$$

where the instantaneous variance $V(t) = \sigma^2(t)$ follows an s.d.e.

$$dV(t) = \alpha(\overline{V} - V(t))\,dt + \beta\sqrt{V(t)}\,dW_v(t). \tag{14.2}$$

Here the two Wiener innovations dW_s and dW_v are assumed to be correlated with

$$\text{corr}(dW_S(t), dW_V(t)) = \rho\,dt. \tag{14.3}$$

Equation (14.2) states that the instantaneous variance is described by a square-root diffusion with mean-reversion force α, dispersion coefficient β (the so-called *vol–vol* parameter), and long-run variance $\overline{V}$. In this square-root model, $V(t)$ is never negative provided that $V(0)$, α, and $\overline{V}$ are all positive. It can also be demonstrated that $V(t)$ remains strictly positive if $2\alpha\overline{V}/\beta^2 \geq 1$: in this case, the zero level represents a reflecting barrier.

As largely discussed in the specialized literature, e.g., Das and Sundaram (1999), the effect of stochastic volatility (as measured by the parameter β) is to increase the kurtosis of the stock return distribution, whereas a negative correlation parameter ρ generates negative skewness. However, it has not yet been established whether kurtosis and skewness introduced by stochastic volatility are suitable for the purpose of matching market prices or instead additional refinements of the model are required to achieve this goal. For instance, one can add flexibility to the model by assuming time-varying parameters. In this way, at-the-money implied volatilities are fitted exactly and a wider range of volatility surface shapes can be reproduced. As an alternative, the incorporation of a jump process into the stock dynamics has been invoked to account for the discrepancies between market prices/volatilities and those returned by stochastic volatility models. In particular, a purely diffusive process seems not adequate to explain the market quotes of deeply out-the-money options with a few days to expiry. The market fears large movements of the underlying and requires compensation for taking that risk. In this study we therefore consider a combination of the Heston model and a pure jump process (HESJ) as introduced by Bates (1996) and by Bakshi, Cao and Chen (1997). These authors assume normally distributed jumps occurring with a constant frequency λ. The resulting model is described by a pair of stochastic differential equations

$$dS(t) = \mu S(t)\, dt + \sigma S(t)\, dW_s(t) + J(t)\, dq(t), \tag{14.4}$$
$$dV(t) = \alpha\big(\overline{V} - V(t)\big)\, dt + \beta\sqrt{V(t)}\, dW_v(t),$$

where $dq(t)$ represents a Poisson random variable equal to zero (i.e., no jump) with probability $1 - \lambda\, dt$ or one (i.e., jump) with probability $\lambda\, dt$. The process $J(t)$ denotes a jump size which we suppose to be drawn from a given probability distribution. In particular, we assume that

$$\ln\big(1 + J(t)\big) \sim \mathcal{N}\left(\ln(1 + \mu_\lambda) - \frac{1}{2}\sigma_\lambda^2, \sigma_\lambda^2\right). \tag{14.5}$$

This model can also be seen as a stochastic volatility extension of the jump diffusion model introduced by Merton (1976). For the sake of simplicity, the jump component in the HESJ model is uncorrelated with the underlying and volatility processes even if this assumption is questionable. A further generalization may incorporate stochastic interest rates. However, these latter typically do not vary much over the lifetime of an option (say, 3–6 months). We use a term structure of discount factors $P(t, T)$ as built by linearly interpolating LIBOR rates. Note that the use of a term structure of discounting factors is consistent with the assumption of time-varying but deterministic interest rates. The stock price process can be simulated by means of Monte Carlo simulation in order to price exotic derivatives. However, the discretization of the square-root process for the variance is not straightforward and the Euler scheme converges to the true process only in the limit of very small time steps. A bias-free algorithm has been devised by Broadie and Kaya (2006) for the Heston model. Unfortunately, due to its complexity and lack of computational speed, it can be rarely used in practice. A less time consuming approach has been proposed by Kahl and Jäckel (2006) who adopt an implicit Milstein scheme for the variance. Finally, ad-hoc adjustments of the Euler scheme have been claimed to improve accuracy, while preserving computational efficiency, see Lord et al. (2006) and Andersen (2007). Moreover, for the purpose of model calibration, Monte Carlo simulation is quite cumbersome. For this reason we will use analytical (or semi-analytical) formulas for the vanilla instruments, as explained in the next section.

14.2 Model and Solution Methodology

14.2.1 Analytical Transforms

In the SV framework, the investor has to be rewarded not only for market risk but for volatility risk as well. In particular, Heston assumes that the dynamics of volatility under the risk-neutral measure can be obtained by taking a risk premium proportional to the instantaneous variance $\lambda(S, V, t) = \lambda_0 \sqrt{V(t)}$, so that the risk-adjusted processes in (14.2) become

$$
\begin{aligned}
dS(t) &= r(t)S(t)\,dt + \sigma(t)S(t)\,d\widetilde{W}_s(t), \\
dV(t) &= \left(\alpha\overline{V} - (\alpha + \lambda_0)V(t)\right)dt + \beta\sqrt{V(t)}\,d\widetilde{W}_v(t),
\end{aligned}
\tag{14.6}
$$

where $r(t)$ is the instantaneous risk-free rate of interest. A similar risk-adjustment can be done for the jump-diffusion model (14.4), once we assume, as in Merton (1976), that further randomness due to jumps represents nonsystematic risk and can be diversified away or, equivalently, that the corresponding market price of risk vanishes. Note that in equations (14.6), the identifiable parameters are $\alpha\overline{V}$ and $\alpha + \lambda_0$; therefore, it is not possible to univocally identify the parameter λ_0. To circumvent this issue, we assume $\lambda_0 = 0$ during the calibration process.

The time t price $c_T(k)$ of a call option expiring in $T - t$ years and stricken at a log-price $k = \ln K$ can be expressed as the discounted expected payoff under the risk-neutral measure

$$
c_T(k) = P(t, T) \int_k^{+\infty} \left(e^{s_T} - e^k\right) p(s_T)\,ds_T.
\tag{14.7}
$$

Here, $p(s_T)$ denotes the risk-neutral probability density of the log-price $s_T = \ln S_T$ at maturity. In the Heston model, the analytical representation for the characteristic function of $p(s_T)$, i.e., its Fourier transform, $f_T^{\text{HES}}(u)$ is known in closed form as

$$
f_T^{\text{HES}}(u) = \int_{-\infty}^{+\infty} e^{ius_T} p(s_T)\,ds_T
\tag{14.8}
$$

$$
= e^{(S+R+C+D+E)},
\tag{14.9}
$$

where:

$$
\begin{aligned}
S &= iu \ln(S(t)), \\
R &= -iu\left(\ln P(t, T) + \delta(T - t)\right),
\end{aligned}
$$

$$C = \frac{\alpha \overline{V}}{\beta^2}\left((\alpha - k + d)(T - t) - 2\ln\left(\frac{1 - g e^{d(T-t)}}{1 - g}\right)\right),$$

$$D = \frac{V(t)}{\beta^2}(\alpha - k + d)\left(\frac{1 - e^{d(T-t)}}{1 - g e^{d(T-t)}}\right),$$

$$k = iu\beta\rho,$$

$$d = \sqrt{(k - \alpha)^2 + \beta^2\left(iu + u^2\right)},$$

$$g = \frac{(\alpha - k + d)}{(\alpha - k - d)}.$$

Particular care must be taken when evaluating the multivalued logarithm with a complex argument. Typically, one can keep track of the branch switching in order to avoid discontinuities, which occur for sufficiently large maturities. A detailed discussion of this issue, as well as an improved algorithm for calculating the Heston characteristic function, can be found in Kahl and Jäckel (2006).[1] As an alternative, one can make the following replacements: $d \rightarrow -d$ and $g \rightarrow 1/g$. The resulting expression is exactly equivalent to the original Heston formula, but it has been claimed stable under a wider range of parameters, see Albrecher et al. (2007). When jumps are included into the stock return dynamics, the characteristic function becomes

$$f_T^{\text{HESJ}}(u) = f_T^{\text{HES}}(u)e^F = e^{(S+R+C+D+E+F)}, \tag{14.10}$$

where

$$F = \lambda(T - t)\left((i + \mu_\lambda)^{iu} e^{\frac{iu}{2}(iu-1)\sigma_\lambda} - 1\right) - \lambda iu\mu_\lambda(T - t).$$

Carr and Madan (1999) show that, if the characteristic function $f_T(u)$ is known, a simple expression for the Fourier transform with respect to the logarithm of the strike of a *modified* call price function is promptly available.

Since $c_T(k)$ converges to $\ln S_t$ for $k \rightarrow -\infty$, then it can be shown that it does not admit a Fourier transform. To solve this problem, Carr and Madan (1999) introduce a *dumping parameter* $Z > 0$ to obtain a modified square integrable function

$$\hat{c}_T(k) = e^{Zk} c_T(k), \quad Z > 0, \tag{14.11}$$

that therefore admits the Fourier transform. In particular, the Fourier transform of $\hat{c}_T(k)$ with respect to k can be expressed in terms of the characteristic function $f_T(u)$

$$\psi_T(v) = \int_{-\infty}^{+\infty} e^{ivk} \hat{c}_T(k)\, dk$$

$$= \frac{e^{-rT} f_T(v - (Z + 1)i)}{Z^2 + Z - v^2 + i(2Z + 1)v}. \tag{14.12}$$

The call option price can be obtained by the Fourier inversion integral

$$c_T(k) = \frac{e^{-Zk}}{\pi} \int_0^{+\infty} e^{-ivk} \psi_T(v)\, dv. \tag{14.13}$$

---

[1] In the Matlab® implementation of the Heston model, we have taken into account this issue.

Carr and Madan (1999) have also pointed out that for short maturities the integrand becomes highly oscillatory because the option price approaches its nonanalytic intrinsic value. They suggest an alternative normalization procedure, which focuses on the option time value. By differentiating the above expression with respect to the current spot price S_t, we can obtain an expression for the option delta Δ:

$$\Delta_T(k) = \frac{\partial c_T(k)}{\partial S(t)} = \frac{e^{-Zk}}{\pi} \int_0^{+\infty} e^{-ivk} \hat{\psi}_T(v) \, dv, \qquad (14.14)$$

where

$$\hat{\psi}_T(v) = \frac{1}{S(t)}(1 + Z + iv)\psi_T(v). \qquad (14.15)$$

Similar computations can be performed to obtain other Greeks, such as gamma, theta, kappa, and rho. The volatility sensitivity of the option (vega) can also be computed by taking the derivative of the characteristic function with respect to $\sqrt{V(t)}$ and then inverting it.

14.2.2 Model Calibration

Model calibration consists of finding a parameter vector (which we generally denote by θ) specifying unique dynamics for the BS, HES, and HESJ models. Ideally, we search for the best choice of model parameters which reproduce the market prices of the vanilla instruments used for hedging more exotic products. At the same time, we want to guarantee a realistic future dynamics of financial variables. In practice, we set up an optimization algorithm to select those numbers which minimize a given loss function (LF) measuring the distance between market and model figures (prices or implied volatilities).

We must stress that the price to pay for a more complex and realistic model is a nontrivial calibration. Moreover, the selection of relevant market data, the choice of a suitable loss function, and the optimization algorithm can be considered as part of the model specification, see the discussion in Christoffersen and Jacobs (2004).

Different metrics are used in the specialized literature, each one exhibiting its own pros and cons. For instance, the mean squared error (MSE) assigns much weight to high-valued options, namely those referring to in-the-money and long-maturity contracts; consequently, a limited subset of option prices effectively contributes to the LF.[2] More often, a proportional weight is assigned by using the percentage mean squared error (PMSE) metric. However, if the price of an option becomes close to zero (as is the case of out-of-the-money and short-term contracts), the associated weight can be extremely large and this may lead to numerical instability.

We recall that it is customary for traders to think in terms of implied volatilities rather than option prices. Compared to absolute prices, implied volatilities have the advantage of representing a homogeneous set of data across the entire range

---

[2] Price differences can be partly reduced by exclusively considering out-of-the-money calls and puts.

of moneyness. A metric based on market/model implied volatilities may then lead to a calibration procedure which is more stable in that it is less sensitive to small perturbations affecting market data and to varying initial guesses for the unknown parameters. Most common used loss functions belong to the family

$$LF(\theta, X) = \frac{1}{n} \sum_{i=1}^{n} w_i \big(\widehat{X}_i(\theta) - X_i \big)^p + \Omega(\theta, \theta_0),$$

where θ is the vector of model parameters, X_i ($i = 1, \ldots, n$) are the market quantities to be fitted (prices or volatilities) as collected in a single vector $\mathbf{X}$, and $\widehat{X}_i(\theta)$ ($i = 1, \ldots, n$) denote their model counterparts. The weights, w_i, are usually choosen in order to assign more relevance to at-the-money options, whose bid-ask spreads are typically smaller.[3] Finally, some authors suggest that a more stable calibration (although potentially inaccurate) is achieved by adding a function $\Omega(\theta, \theta_0)$ which penalizes large deviations of θ from the initial guess θ_0.

These considerations lead one to define the following three LFs:

$$MSE(\theta; \mathbf{X}) \equiv \frac{1}{n} \sum_{i=1}^{n} \big(\widehat{X}_i(\theta) - X_i \big)^2, \tag{14.16}$$

$$PMSE(\theta; \mathbf{X}) \equiv \frac{1}{n} \sum_{i=1}^{n} \left(\frac{\widehat{X}_i(\theta) - X_i}{\widehat{X}_i} \right)^2, \tag{14.17}$$

$$MAE(\theta; \mathbf{X}) \equiv \frac{1}{n} \sum_{i=1}^{n} |\widehat{X}_i(\theta) - X_i|. \tag{14.18}$$

If we fit volatilities, we may proceed as follows:

(1) From option market prices, using the Black–Scholes formula, we compute the corresponding market implied volatilities;
(2) From model option prices, as obtained by the HES or HESJ models, we compute model implied volatilities by numerically inverting the Black–Scholes formula with respect to the unknown volatility parameter.

Model calibration consists of finding a parameter set that minimizes a distance between market and model implied volatilities. A set of model parameters should be accepted only if price discrepancies do not exceed bid-ask spreads in order to avoid possible arbitrages when pricing exotic products.

The role of the optimization algorithm is also crucial. Finding a global minimum is rather difficult. Gradient-based optimizers are likely to return local minima depending on the initial guess. On the other hand, the better performance of global

---

[3] Possible choices consist in choosing as weigths the reciprocal of the bid-ask spread or even the Black–Scholes vega.

algorithms, such as simulated annealing, is counterbalanced by much longer calibration times. In this study, calibration has been carried out by minimizing the loss function through the nonlinear least squares algorithm lsqnonlin provided in the Matlab® Optimization Toolbox, which belongs to the class of local optimizers.

14.3 Implementation and Algorithm

Model calibration requires the numerical computation of theoretical option prices. A powerful technique, which lately has gained considerable attention in finance, is the Fast Fourier Transform (FFT) algorithm. This method allows for an accurate and efficient computation of the inversion integral (14.13). The procedure works as follows:

- Approximate the integral in (14.13) using the trapezoidal rule, with discretization step η and upper integration limit given by $a = N\eta$:

$$c_T(k) \approx \frac{e^{-zk}}{\pi} \sum_{j=1}^{N} e^{-i\eta(j-1)k} \psi_T\big(\eta(j-1)\big)\eta; \qquad (14.19)$$

- Divide the log-strike range $[-b, b]$ using a regular grid spacing centered at $k_u = 0$ (corresponding to an ATM option):

$$k_u = -b + \frac{2b}{N}(u-1), \qquad (14.20)$$

where $u = 1, \ldots, N$;
- Replace k in (14.19) with k_u defined in (14.20):

$$c(k_u) = \frac{e^{-zk_u}}{\pi} \sum_{j=1}^{N} e^{-i\frac{2b}{N}\eta(j-1)(u-1)} e^{ib\eta(j-1)} \psi_T\big(\eta(j-1)\big)\eta; \qquad (14.21)$$

- In order to improve the accuracy for large values of the step size η, we can incorporate Simpson's weights in (14.21):

$$c(k_u) = \frac{e^{-zk_u}}{\pi} \sum_{j=1}^{N} e^{-i\frac{2\pi}{N}(j-1)(u-1)} e^{ib\eta(j-1)} \psi[\eta(j-1)]\frac{\eta}{3}[3 + (-1)^j - \delta_{j,1}],$$

$$(14.22)$$

where $\delta_{j,1}$ is the Kronecker delta function;
- The sum (14.22) is efficiently obtained by the FFT algorithm, which reduces the number of operations from N^2 to $N\ln_2 N$; in particular the FFT algorithm returns option prices on the assigned log-strike grid, $k_1, \ldots, k_n$;
- The call option price for a given strike can be obtained by linear interpolation of grid option prices computed in the previous step.

14.3.1 Code Description

Functions and market data are stored in the folder HestonModels. Before running the code, you must include this folder and its subfolders in the Matlab® directory list. To do so, select Set Path from the File Menu and follow the instructions.

The HestonModel folder contains the following subfolders:

1. Common: It contains the main routines for calibration and all functions which are common to the different models;
2. BlackScholes: All BS-related functions;
3. Heston: All HES-related functions;
4. HestonJumps: All HESJ-related functions;
5. OptionData: Files with option quotes;
6. RatesData: Files with LIBOR rates and dividend;
7. Analysis: Results of the analysis.

Function `day.m` and function `historical.m`

These are the main routines for model calibration given the market data included in ".txt" files. `day.m` calibrates the model on a single day.

1. It reads two ".txt" files containing option quotes and LIBOR rates respectively. (The file names can be changed in the code.)
2. It calls the function `calibrate.m` which returns the estimated parameters, the value of the loss function, model prices and deltas, model volatilities, CPU time.
3. (Optional) It plots prices and volatilities as a function of strike and maturity.

`historical.m` performs the same task as `day.m`, but it scans all ".txt" file in a given folder and collects the results of calibration. The directory name can be modified in the code.

Function `compute.m`

It computes prices, deltas, and BS implied volatilities for a given model over a set of different strikes and maturities. Parameters of the model are passed as an input argument.

1. It organizes data by grouping options with the same maturity;
2. It interpolates the LIBOR curve to calculate the spot rate for a given maturity;
3. It computes prices, deltas and BS implied volatility by calling the corresponding function for each model.

Function `calibrate.m`

It calibrates model parameters on a set of market data, corresponding to options quotes with different strikes and maturities.

1. It organizes market prices by grouping options with similar maturity;
2. It interpolates the LIBOR curve to calculate the rate for each maturity;
3. It calculates market BS implied volatilities;
4. It minimizes a given loss function through the `lsqnonlin.m` built-in algorithm provided by the Matlab® Optimization toolbox and returns calibrated parameters;
5. (Optional) It plots the results.

Function `implvol.m`

It computes BS implied volatilities for a set of different strikes and maturities by numerically inverting the Black–Scholes formula. Input prices can be either those quoted in the market and passed as input arguments or computed within a specific model.

1. It organizes data by grouping options according to the maturity;
2. It interpolates the LIBOR curve to calculate the rate for each maturity;
3. It selects the prices, for which the corresponding BS implied volatilities must be computed. These can be market prices (input arguments) or those calculated with a given model (the model and its parameters are optional arguments);
4. It computes BS implied volatilities by using the `fzero.m` built-in algorithm provided by the Matlab® Optimization toolbox;
5. It returns NaN or zero when no solutions are found.

Function `lossfunction.m`

It computes a given loss function given a vector of market data and their model counterparts.

14.4 Results and Comments

We have calibrated and tested models BS, HES, and HESJ using a large set of European options written on the FTSE 100 stock index. This index is based on the share prices of the 100 largest UK quoted companies. FTSE 100 Index Options are quoted on the Euronext LIFFE market as index points and have an assigned value of £10 for each index point variation. They are among the most actively traded European-style contracts in Europe. The analyzed sample period ranges from January 1, 2004 to April 23, 2004. The daily closing quotes for FTSE 100 options and FTSE index come from end-of-day quotations provided by LIFFE. We have also recorded annual continuous dividend yields on a daily basis. The interest rate curve refers to the

LIBOR market. Each day, interest rate for 1, 2, 3, 4, 6, and 12 months time horizons have been downloaded from the web site of the British Bankers' Association BBA (www.bba.org.uk). For intermediate maturities, rates have been obtained by linear interpolation.

In order to rule out possible biases in the estimates, we apply the following filters:

1. We exclude all options with less than six days to maturity, in order to avoid effects due to poor liquidity from the sample.
2. We drop all options that do not satisfy the no-arbitrage condition

$$c_T(\ln K) \geq \max\big(S(t)e^{-\delta(T-t)} - K P(t, T), 0\big).$$

3. We consider the moneyness range [0.9, 1.1] in order to exclude not liquid securities such as deeply out-of-the money or in-the-money options.
4. We dropped options whose price is less than £2.5 (£0.5 is the price unit) to reduce the impact of price discretization.

However, only a small percentage of the options have been excluded in this manner.[4] For each day in the sample period, we have calibrated the models BS, HES, and HESJ by minimizing six different loss functions, namely those corresponding to MAE, MSE, and PMSE on both option prices and implied volatilities. As an example, we report the calibration output on April 14, 2004. Option quotes and LIBOR rates are reported in Table 14.1 and Table 14.2, respectively. On this day, the closing FTSE 100 index was 4485.4 and the simply-compounded dividend yield 3.592%. Because dividend payments are usually concentrated in particular periods of the year, a more precise calibration requires an estimate of the term structure of dividend yields rather than their annual average.

Estimated parameters are shown in Table 14.3 for each model and loss function. The value of the different loss functions is listed in Table 14.4.

The BS model assumes a volatility parameter constant across all strikes and maturities. Therefore, the optimal value returned by the calibration of the BS model is a rough average of the market implied volatilities obtained inverting the Black–Scholes formula. A plot of implied volatilities obtained by interpolating market data is shown in Fig. 14.1. The implied-volatility surface exhibits a typical asymmetric pattern with a slope declining for longer maturities. Note that slightly higher values for the volatility parameter are obtained as we minimize the MAE/MSE functions using market and model prices. In this case, the main contribution to the loss function is due to in-the-money options, which usually exhibit higher implied volatilities. In contrast, using PMSE as a loss function or calibrating the model using implied volatilities, weights assigned to different options are more homogeneously spread over the entire moneyness range.

Calibration results for the HES model clearly indicate that volatility is not constant. The instantaneous volatility and the long run volatility are respectively lower

---

[4] For a more accurate study, one should disregard options whose last prices are quoted one hour before the closing time. In this case, both the underlying asset and the option prices are no more synchronous in that they are quoted at different times.

364   14 Fixing Volatile Volatility

Table 14.1. Option quotes on April 14, 2004

Strike	Moneyness	Maturity (days)				
		37	64	92	153	243
4125	0.9197	370.5	388.5	415.5	–	511.5
4175	0.9308	–	343.0	370.0	–	–
4225	0.9419	322.5	299.5	326.5	444.5	434.5
4275	0.9531	276.0	256.5	285.0	–	–
4325	0.9642	230.5	215.0	245.0	286.5	360.5
4375	0.9754	146.0	176.5	207.0	–	–
4425	0.9865	107.5	140.5	170.0	214.5	290.0
4475	0.9977	74.5	108.5	138.5	–	–
4525	1.0088	49.0	81.5	109.5	153.0	226.0
4575	1.0200	29.0	58.5	83.5	–	–
4625	1.0311	16.5	40.0	61.5	104.5	172.5
4675	1.0423	8.5	27.0	44.5	–	–
4725	1.0534	4.5	16.5	32.0	67.0	127.5
4775	1.0646	(2.5)	9.5	22.0	–	–
4825	1.0757	(1.5)	5.0	15.0	21.5	89.5
4875	1.0869	(1.5)	(2.5)	10.0	–	–
4925	1.0980	(0.5)	(1.5)	6.5	–	62.5
4975	1.1092	–	(0.5)	4.0	–	–

Table 14.2. LIBOR rates on April 14, 2004

Maturity (months)	O/N	1 m	2 m	3 m	4 m	6 m	9 m	12 m
LIBOR %	4.161	4.161	4.270	4.376	4.428	4.486	4.552	4.821

Table 14.3. Model parameters calibrated on April 14, 2004

Model	θ	Prices			Implied volatilities		
		MAE	MSE	PMSE	MAE	MSE	PMSE
BS	σ	0.1424	0.1503	0.1145	0.1405	0.1448	0.1351
HES	σ	0.1257	0.1208	0.1221	0.1255	0.1223	0.1181
	$\bar\sigma$	0.2086	0.2086	0.1988	0.2086	0.2086	0.2086
	α	3.2181	4.1118	3.6046	3.2074	3.2877	3.7844
	β	0.6638	0.8299	0.6879	0.6910	0.6482	0.7919
	ρ	−0.7456	−0.7473	−0.6562	−0.6864	−0.7213	−0.6628
HESJ	σ	0.1062	0.1047	0.0947	0.1086	0.1039	0.1064
	$\bar\sigma$	0.1591	0.1789	0.1098	0.1780	0.1851	0.1819
	α	1.3158	1.0437	3.0093	1.7365	2.4495	2.1548
	β	0.1699	0.2542	0.1713	0.2276	0.2893	0.3460
	ρ	−0.8426	−0.7273	−0.3108	−0.6498	−0.7210	−0.6352
	λ	0.0976	0.0910	0.2223	0.0933	0.1676	0.0719
	m_λ	−0.3000	−0.3000	−0.2605	−0.2583	−0.1212	−0.3000
	σ_λ	0.4755	0.5522	0.2254	0.3583	0.2219	0.4080

Table 14.4. Loss functions for different targets and models on April 14, 2004

Loss	Target	BS	HES	HES/BS	HESJ	HESJ/BS
MAE	Prices	1000.2706	224.5487	0.2245	162.6654	0.1626
MSE	Prices	25080.5662	1385.3014	0.0552	624.4359	0.0249
PMSE	Prices	1.9055	0.1333	0.0700	0.0894	0.0469
MAE	Vol	1.5132	0.3194	0.2111	0.2786	0.1841
MSE	Vol	0.0500	0.0032	0.0631	0.0022	0.0433
PMSE	Vol	2.1222	0.0790	0.0372	0.0790	0.0372

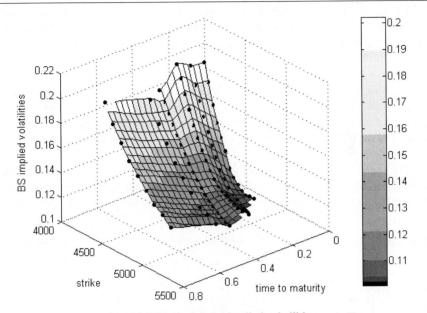

Fig. 14.1. Black–Scholes implied volatilities.

and higher than the corresponding BS parameter. This is consistent with a positively sloped term structure of volatilities, with the long run parameter approaching the implied volatility of long-term options. The vol–vol coefficient (β) assumes positive values in the interval [0.66, 0.83], depending on the chosen loss function. The mean reversion coefficient (α) varies in the range [3.2, 4.1]. Such values correspond to an average decay time toward the long run volatility of the order of 3 months. Finally, the correlation coefficient is remarkable negative as expected for an equity index and spans the range [−0.66, −0.75].

A sizeable value of (β) is needed to account for fatter tails in the underlying distribution, as implied by the curvature of the volatility surface. On the other hand, a too strong mean reversion parameter diminishes the impact of stochastic volatility and flattens the smile. As a consequence, the condition which guarantees strictly positive volatility, i.e. $2\alpha\overline{V}/\beta^2 \geq 1$, is not often satisfied by calibrated parameters. One can impose an additional constraint in the calibration but this generally worsens the quality of fit.[5] Therefore, HES model appears more compatible with the existence of scenarios where the volatility vanishes and stays low for long periods.

By allowing for stochastic volatility, smile patterns are nicely mimicked. The quality of fit is improved by almost one order of magnitude, as demonstrated by the decrease in the loss function. However, a closer inspection reveals that the HES model underestimates the slope and bowing across different moneyness, especially for short maturities. This behavior has been generally observed over the whole range of dates considered in this study.

By including jumps in the model, market prices/volatilities are more accurately reproduced as a result of the larger number of parameters. The volatility surface obtained by inverting the BS formula in order to match HESJ prices is shown in Fig. 14.2. The contribution of jumps seems justified by the value of the jump frequency (λ) and average size (m_λ). Estimated parameters indicate that there is a [10%, 18%] per year probability that a jump occurs, with an average amplitude of about -25%. The negative sign of (m_λ) indicates that the market considers extreme downward movements of the underlying more probable that upward ones. The dispersion of the jump size (σ_λ) is of the order 25–45%, depending on the choice of the loss function.

The HESJ model fits better also the implied volatility surface because the jump component increases the kurtosis of the distribution and enhances the volatility smile. Such effect is particularly evident for very short maturities where the curvature of

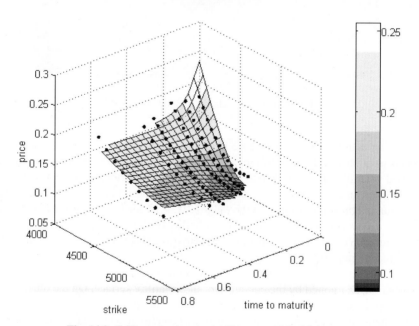

Fig. 14.2. Calibrated prices in the Heston model with jumps.

---

[5] In this setting, α is close to 3.5 and β to 0.55.

the volatility surface is very close to (or slightly overestimate) that implied by market data. Furthermore, the impact of jumps is to compensate for part of the diffusive volatility, whose dispersion parameter (β) assumes significantly lower values. We also note that the condition of strictly positive volatility is almost always met after unconstrained calibration. Finally, the correlation coefficient is unaffected or slightly higher.

If we look at the loss function returned by calibration, we confirm the impression that HESJ outperform HES, at least for in-sample data problem). Results are consistent with those found by Bates (1996) and Bakshi, Cao and Chen (1997), who examined other equity indexes. The drawback is that the estimation of parameters becomes more difficult and sensitive to the initial guess. In particular, the jump component plays a relevant role when calibrating short-maturity options.

We have also examined the stability over time of the estimated parameters. Descriptive statistics are reported in Table 14.5, and Fig. 14.3 illustrates their time behavior for the HES and HESJ models. Results vary for different choice of loss functions. In the figure, we plot results obtained by minimizing absolute errors on implied volatilities. Long run volatility, which mainly depends on the long-term options, does not change significantly in our range of dates and it is quite similar for the two models and different loss functions. On the contrary, the spot volatility is very "volatile", consistently with the hypothesis of the model, see panel (a). By direct comparison with the rescaled FTSE 100 return, we deduce that volatility is negatively correlated to underlying process, in agreement to correlation estimates. The mean reversion and vol–vol parameters are shown in panels (b) and (c). Their absolute values are usually lower in the HES model, because part of the volatility is explained by the jump

Table 14.5. Parameter statistics over the 3-month period considered in this study

Model	Par	% MSE prices				% MSE prices			
		Mean	Std	Max	Min	Mean	Std	Max	Min
BS	σ	0.144	0.017	0.123	0.186	0.134	0.019	0.110	0.179
	TLF	1.795				1.443			
HES	σ	0.126	0.025	0.090	0.197	0.119	0.027	0.078	0.192
	$\bar{\sigma}$	0.177	0.028	0.134	0.245	0.188	0.029	0.152	0.261
	α	2.930	1.049	1.000	5.000	3.341	0.289	2.700	4.625
	β	0.444	0.238	0.079	0.989	0.580	0.193	0.204	0.998
	ρ	−0.629	0.077	−0.778	−0.431	−0.564	0.110	−0.745	−0.346
	TLF	0.351				0.499			
HESJ	σ	0.114	0.022	0.085	0.180	0.107	0.024	0.072	0.181
	$\bar{\sigma}$	0.162	0.024	0.110	0.221	0.166	0.027	0.106	0.222
	α	2.187	0.809	1.013	4.000	2.338	0.682	1.000	4.000
	β	0.232	0.146	0.048	0.633	0.305	0.188	0.056	1.000
	ρ	−0.816	0.200	−1.000	−0.216	−0.738	0.154	−1.000	−0.396
	λ	0.192	0.090	0.007	0.423	0.180	0.066	0.049	0.324
	m_λ	−0.059	0.158	−0.300	0.125	−0.076	0.128	−0.299	0.112
	σ_λ	0.109	0.123	0.001	0.396	0.151	0.118	0.001	0.399
	TLF	0.250				0.417			

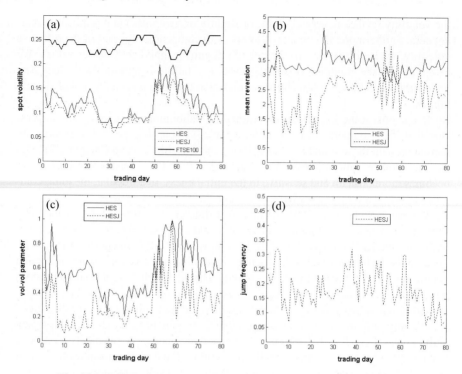

Fig. 14.3. Daily variation in the calibrated parameters of the Heston model.

component. However, the time series exhibits quite large variations, especially in the HESJ case. This can be due to the fact that calibration is not straightforward (at least for some dates) and global optimization should be preferred. Estimates also indicate that the jump component is not negligible. The frequency of jumps is rather stable and oscillating in the range [10–20%]. The average jump size is generally sizably negative [−25%] but it sometimes turns to small values around [0–5%] when the spot volatility is low.

Finally, we examine the effect of stochastic volatility on the option delta. In particular, sensitivities with respect to the spot price (such as delta, gamma) are easily computed by formula (14.14). In Fig. 14.4, we plot the delta function for a short and long maturity option, respectively. The delta computed in the HESJ model is significantly higher than the one stemming from the standard BS model. This would suggest that a higher amount of the underlying must be hold in order to hedge a short position in the call. However, we recall that this hedging strategy can only neutralize the uncertainty resulting from the stock randomness (at least, in a continuous trading framework). A comprehensive out-of-sample study of the hedging performance of various SV models has been performed by Bakshi, Cao and Chen (1997) on the S&P 500 index. These authors find that the Heston model outperforms the BS model and that inclusion of a jump component deteriorates the hedging performance, despite the good results obtained on the in-sample calibration. We conclude our treat-

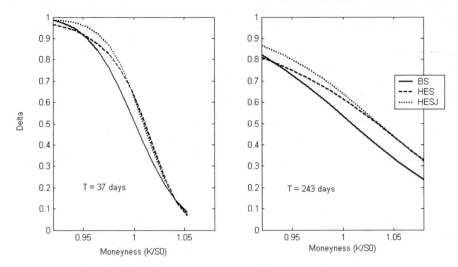

Fig. 14.4. Option delta for calibrated models.

ment by noticing that managing market risk affecting OTC derivatives requires for the adopted model to be consistent with quoted market information. Specifically, a model ought to fit liquid vanilla options to prevent the trader from arbitrage opportunities through vanilla option trading. In the general context of jump processes with stochastic volatility, implied calibration techniques achieving this goal have been studied by Galluccio and Le Cam (2006a, 2006b).

Exotic Derivatives

15

An Average Problem

Key words: Edgeworth expansions, Fourier and Laplace transforms, PDE, exotic options

In this chapter, we describe and compare alternative procedures for pricing Asian options. Asian options are written on an average. More precisely, prices of an underlying security (or index) are recorded on a set of dates during the lifetime of the contract. At the option's maturity, a pay-off is computed as a deterministic function of an average of these prices (see Carr and Schröder (2004)). As reported by Falloon and Turner (1999), the first contract linked to an average price was traded in 1987 by Bankers & Trust in Tokyo, hence the attribute "Asian".

Asian options are quite popular among derivative traders and risk managers. This is due to several reasons. Primarily, Asian options smooth possible market manipulations occurring near the expiry date. They also provide a suitable hedge for firms facing a stream of cashflows. This is the case, for instance, with commodity end-users that are financially exposed to average prices.

The standard Asian contract is written on the arithmetic average of an asset price or any financial index computed across weekly (or monthly) observations. It is a common practice to price this contract computing this average on a set of values recorded continuously over the option lifetime. In the standard Black–Scholes framework, the average depends on a sum of correlated lognormal variates. Unfortunately, the distribution of this sum does not admit a simple analytical expression. Consequently, numerical approximations need to be developed for the purpose of pricing arithmetic Asian options (Kat (2001)).

This case illustrates and compares the following methods:

(a) Approximation of the average distribution by fitting integer moments (Turnbull and Wakeman (1991), Levy (1992), Milevsky and Posner (1998) and Ju (2002)).
(b) Computation of lower and upper bounds for the price (Rogers and Shi (1992) and Thompson (1998)).

(c) A numerical solution of a rescaled version of the pricing partial differential equation (PDE) (Rogers and Shi (1992) and Vecer (2001)).
(d) Numerical inversion of the Laplace transform (Geman and Yor (1993)).
(e) Numerical inversion of the double transform (Fusai (2004) and Cai and Kou (2007)).

Other procedures presented in the literature, such as the eigenfunction method in Linetsky (2004) or the perturbation approach in Zhang (2001), are not considered here.

Our presentation is organized as follows: Sect. 15.1 introduces the Asian option pricing problem; Sect. 15.2 describes the five methods mentioned above. Sect. 15.3 details our implementation; Sect. 15.4 illustrates numerical results and provides a comparison among the examined methods.

15.1 Problem Statement

In the Black–Scholes framework, the risk-neutral process for the underlying asset is a geometric Brownian motion satisfying

$$\begin{cases} dS_t = rS\,dt + \sigma S\,dW_t, \\ S_0 = s_0, \end{cases} \tag{15.1}$$

where W_t is a standard Brownian motion, r is the continuously compounding rate of interest, and σ is the instantaneous percentage price volatility. The arithmetic average is computed over all prices spanning the period $[0, T]$. The pay-off of a continuously monitored Asian option is given by

$$(s_0 A_T / T - K)_+,$$

where K is a strike price and

$$\begin{cases} A_T = \int_0^T \exp\big((r - \sigma^2/2)t + \sigma W_t\big)\,dt, \\ A_0 \equiv 1. \end{cases}$$

The Asian option fair price is given by

$$e^{-rT}\widetilde{\mathbb{E}}_0\left(\frac{s_0 A_T}{T} - K\right)_+ = e^{-rT}\frac{s_0}{T}\widetilde{\mathbb{E}}_0(A_T - \overline{K})_+,$$

where $\widetilde{\mathbb{E}}_0$ denotes expectation under the risk-neutral probability measure and $\overline{K} := (K/s_0)T$. The pricing problem consists in finding the distribution function of A_T. In the next section we discuss several approaches.

15.2 Model and Solution Methodology

We present five methods for pricing Asian options. A first method derives a probability distribution sharing a number of moments with the distribution of the price average. A second approach aims at calculating tight upper and lower bounds for the exact option price. A third technique performs a numerical solution of the PDE satisfied by the Asian option price. The last two methods use integral transforms (i.e., Laplace and Fourier, respectively) to simplify the pricing problem.

15.2.1 Moment Matching

This is the most popular approach for pricing Asian options. The average price is assigned an arbitrary probability density function constrained to match a number of moments of A_T. Unfortunately, this method does not provide any assessment about the approximation error.

The first step is to derive a closed-form expression for the moments of A_T, as in Geman and Yor (1993):

$$\mu_n := \widetilde{\mathbb{E}}_0\big[A_T^n\big] = \frac{n!}{\lambda^{2n}} \left\{ \sum_{j=0}^{n} d_j^{(\gamma/\lambda)} \exp\left[\left(\frac{\lambda^2 j^2}{2} + \lambda j \gamma\right)T\right]\right\}, \qquad (15.2)$$

where

$$d_j^{(\beta)} = 2^n \prod_{\substack{0 \le i \le n \\ i \ne j}} \big[(\beta + j)^2 - (\beta + i)^2\big]^{-1},$$

$$\lambda = \sigma, \qquad \gamma = \frac{r - \sigma^2/2}{\sigma}. \qquad (15.3)$$

The next step is to choose and fit an arbitrary density function to a number of selected moments. Specifically, we considered lognormal, reciprocal gamma, and Edgeworth series approximations.

Lognormal Approximation

Turnbull and Wakeman (1991) and Levy (1992) assigned a lognormal distribution to the random variable A_T, i.e., $\ln A_T$ is normal with a mean m and variance v^2. Due to its simplicity, this approximation has gained popularity. Parameters m and v^2 match the mean and variance of A_T. The Asian call option price is given by the modified Black–Scholes formula

$$c_{\log} = s_0 e^{m + v^2/2 - rT} \mathcal{N}(d_1) - e^{-rT} \mathcal{N}(d_2), \qquad (15.4)$$

where

$$m = 2\log\mu_1 - \frac{1}{2}\log\mu_2 - \log T; \qquad v^2 = \log\mu_2 - 2\log\mu_1,$$

$$d_1 = \frac{\ln s_0/K + m + v^2}{v}, \qquad d_2 = d_1 - v. \qquad (15.5)$$

Edgeworth Series Approximation

The lognormal approximation only allows for fitting the mean and the variance of the average. In order to fit the third and fourth moment as well, i.e., skewness and kurtosis of the average, Turnbull and Wakeman (1991) proposed to adopt a fourth-order Edgeworth series expansion. The lognormal density $f_{\log}(y; m, v^2)$ with parameters m and v^2 is

$$f_{\log}\left(y; m, v^2\right) = \frac{1}{\sqrt{2\pi v^2}\,y}\,\exp\left(-\frac{(\ln y - m)^2}{2v^2}\right), \quad y > 0,$$

and then the fourth-order Edgeworth approximation $f_{\text{edg}}(y; m, v^2)$ is given by

$$f_{\text{edg}}\left(y; m, v^2\right) = f_{\log}\left(y; m, v^2\right) + \sum_{i=1}^{4} \frac{k_i}{i!} \frac{\partial^i f_{\log}(y; m, v^2)}{\partial y^i} + e(y), \qquad (15.6)$$

where k_i is the difference in the ith cumulant between the exact distribution and the approximate distribution, namely $k_i = \chi_i(f) - \chi_i(l)$, with

$$\chi_1(f) = \mu_1, \qquad\qquad\qquad \chi_2(f) = \widetilde{\mathbb{E}}_0(A_T - \mu_1)^2,$$
$$\chi_3(f) = \widetilde{\mathbb{E}}_0(A_T - \mu_1)^3, \qquad \chi_4(f) = \widetilde{\mathbb{E}}_0(A_T - \mu_1)^4 - 3\chi_2(f).$$

Once parameters m and v^2 have been set according to expression (15.5), then $k_1 = k_2 = 0$ in equation (15.6), and the approximate Asian option price is given by

$$c_{\text{edg}} = c_{\log} + e^{-rT} \frac{s_0}{T} \left[-\frac{k_3}{6} \frac{\partial f_{\log}(y; m, v^2)}{\partial y} + \frac{k_4}{24} \frac{\partial^2 f_{\log}(y; m, v^2)}{\partial y^2} \right]_{y=TK/s_0}, \quad (15.7)$$

where $c_{\log}$ is defined in formula (15.4).

The main problem of the Edgeworth series is that increasing the number of matched moments does not guarantee an improvement in the resulting approximation. Since the distribution of A_T is not univocally determined by its moments, the approximation (15.6) may even lead to a negative-valued density.[1] To solve this problem, Ju (2002) considers the Edgeworth series for approximating the distribution of $\ln A_T$ and he obtains a simple approximate price formula:

$$c_{\text{ju}} = c_{\log} + e^{-rT} K \left[z_1 n(y) + z_2 \frac{\partial n(y)}{\partial y} + z_3 \frac{\partial^2 n(y)}{\partial y^2} \right]_{y=\ln(K/s_0)}, \qquad (15.8)$$

where $c_{\log}$ is given in (15.4), $n(y) = n(y; m, v^2)$ is the density of the Gaussian distribution with mean m and variance v^2 (given in (15.5))

---

[1] Conditions under which the Edgeworth expansion is positive and unimodal are given in Barton and Dennis (1952). A discussion in the context of Asian option is provided in Ju (2002).

$$n(y; m, v^2) = \frac{1}{\sqrt{2\pi v^2}} \exp\left(-\frac{(y - m)^2}{2v^2}\right),$$

and derivatives are computed as

$$\frac{\partial n(y; m, v^2)}{\partial y} = -\frac{(y - m)}{v^2} n(y; m, v^2),$$

$$\frac{\partial^2 n(y; m, v^2)}{\partial y^2} = \frac{(m^2 - v^2 - 2my + y^2)}{v^4} n(y; m, v^2).$$

The remaining coefficients are as follows:

$$z_1 = -\sigma^4 T^2 \left(\frac{1}{45} + \frac{x}{180} - \frac{11x^2}{15120} - \frac{x^3}{2520} + \frac{x^4}{113400}\right)$$
$$- \sigma^6 T^3 \left(\frac{1}{11340} - \frac{13x}{30240} - \frac{17x^2}{226800} + \frac{23x^3}{453600} + \frac{59x^4}{5987520}\right),$$

$$z_2 = -\sigma^4 T^2 \left(\frac{1}{90} + \frac{x}{360} - \frac{11x^2}{30240} - \frac{x^3}{5040} + \frac{x^4}{226800}\right)$$
$$+ \sigma^6 T^3 \left(\frac{31}{22680} + \frac{11x}{60480} - \frac{37x^2}{151200} - \frac{19x^3}{302400} + \frac{953x^4}{59875200}\right),$$

$$z_3 = \sigma^6 T^3 \left(\frac{2}{2835} - \frac{x}{60480} - \frac{2x^2}{14175} - \frac{17x^3}{907200} + \frac{13x^4}{124700}\right),$$

$$x = rT.$$

Reciprocal Gamma Approximation

Milevsky and Posner (1998) proved that the stationary density for the arithmetic average of a geometric Brownian motion is given by a reciprocal gamma density. That is to say the reciprocal of the average A_T/T is gamma distributed. Indeed, these authors showed that

$$\frac{1}{\lim_{T \to \infty} A_T/T}$$

has a gamma distribution, provided that condition $r - \sigma^2/2 < 0$ is satisfied. Consequently, they suggest to approximate the distribution of A_T/T by a gamma density with parameters α and β. These parameters are chosen so as to match the first two moments of $s_0 A_T/T$

$$\alpha = \frac{s_0}{T} \frac{2\mu_2 - \mu_1^2}{\mu_2 - \mu_1^2}, \qquad \beta = \left(\frac{s_0}{T}\right)^2 \frac{\mu_2 - \mu_1^2}{\mu_2 \mu_1}.$$

The approximate Asian option price is given by

$$c_{\mathrm{rg}} = \frac{1 - e^{-rT}}{rT} s_0 G\left(\frac{1}{K}; \alpha - 1, \beta\right) - e^{-rT} KG\left(\frac{1}{K}; \alpha, \beta\right), \tag{15.9}$$

where $G(x; \alpha, \beta)$ is the gamma cumulative distribution

$$G(x; \alpha, \beta) = \frac{1}{\beta\Gamma(\alpha)} \int_0^x \left(\frac{u}{\beta}\right)^{\alpha-1} e^{-u/\beta} \, du,$$

and $\Gamma(z, a) = \int_a^{+\infty} t^{z-1}e^{-t} \, dt$ is the incomplete gamma function.

15.2.2 Upper and Lower Price Bounds

Rogers and Shi (1992) and Thompson (1998) derived lower and upper bounds for the Asian option price. For a lower bound, the idea is simple and powerful. Consider the random variable

$$X = \frac{s_0}{T} \int_0^T e^{(r-\sigma^2/2)s + \sigma W_s} \, ds - K.$$

The Asian option price is given by $\widetilde{\mathbb{E}}_0(X_+)$. Using the iterated rule for conditional expectations, the fact that $X_+ \geq X$, and the positiveness of X_+, we have

$$\widetilde{\mathbb{E}}_0(X_+) = \widetilde{\mathbb{E}}_0[\widetilde{\mathbb{E}}_0(X_+|Z)] \geq \widetilde{\mathbb{E}}_0[\widetilde{\mathbb{E}}_0(X|Z)_+],$$

for any conditioning variable Z. The accuracy of the lower bound $\widetilde{\mathbb{E}}_0[\widetilde{\mathbb{E}}_0(X|Z)_+]$ can be estimated using

$$0 \leq \widetilde{\mathbb{E}}_0(X_+) - \widetilde{\mathbb{E}}_0[\widetilde{\mathbb{E}}_0(X|Z)_+] \leq \frac{1}{2}\widetilde{\mathbb{E}}_0\left[\sqrt{\text{Var}(X|Z)}\right].$$

The idea is to select Z making the variance on the right-hand side in the expression above as small as possible. Rogers and Shi (1992) propose to set $Z = \int_0^T W_s \, ds$, and provide an analytical expression to the lower bound $c_{\text{low}} = \widetilde{\mathbb{E}}_0[\widetilde{\mathbb{E}}_0(X|Z)_+]$. Thompson (1998) obtained the same lower bound with the simpler expression

$$c \geq c_{\text{low}} = e^{-r} \int_0^1 s_0 e^{\alpha t + \sigma^2 t/2} \mathcal{N}\left(\frac{-\gamma^* + \sigma t(1 - t/2)}{1/\sqrt{3}}\right) dt - K\mathcal{N}\left(\frac{-\gamma^*}{1/\sqrt{3}}\right),$$
(15.10)

where the option maturity T has been standardized[2] to 1. Here $\mathcal{N}(x)$ denotes the standard normal cumulative function, and γ^* is the unique solution to the equation

$$\int_0^1 s_0 \exp\left(3\gamma^*\sigma t(1 - t/2) + \alpha t + \frac{1}{2}\sigma^2\left(t - 3t^2(1 - t/2)^2\right)\right) dt = K. \quad (15.11)$$

Computation of the value γ^* can be done rather quickly with standard root finder routines (e.g., the bisection method).

The upper bound c_{up} provided by Thompson (1998) is tighter than the one derived by Rogers and Shi (1992). It reads as

---

[2] For a general T, r and σ must be replaced by rT and $\sigma\sqrt{T}$, respectively.

$c \leq c_{\text{up}}$

$$\equiv e^{-r} \int_0^1 \int_{-\infty}^{+\infty} 2v\varphi(w)$$

$$\times \left[a(T,x) N\left(\frac{a(T,x)}{b(T,x)}\right) + b(T,x)\varphi\left(\frac{a(T,x)}{b(T,x)}\right) \right] dw\, dv, \quad (15.12)$$

where $\varphi(x)$ denotes the standard normal density function and the other parameters are given as follows:

$$a(T,x) = s_0 \exp(\sigma x + \alpha T) - K(\mu_T + \sigma x) + K\sigma(1 - T/2)x,$$

$$b(T,x) = K\sigma\sqrt{\frac{1}{3} - T(1 - T/2)^2},$$

$$\mu_T = \frac{1}{K}\left(s_0 \exp(\alpha T) + \gamma\sqrt{v_T}\right),$$

$$v_T = c_T^2 T + 2(K\sigma)c_T T(1 - T/2) + (K\sigma)^2/3,$$

$$c_T = s_0 \exp(\alpha T)\sigma - K\sigma,$$

$$\gamma = \left(K - s_0(e^\alpha - 1)/\alpha\right) \Big/ \int_0^1 \sqrt{v_T}\, dT,$$

$$v = \sqrt{T},$$

$$w = x/\sqrt{T}.$$

15.2.3 Numerical Solution of the Pricing PDE

The pricing problem for an Asian option can be formulated in terms of a two-state-variable PDE (Wilmott, Dewynne and Howison (1993)). Using an appropriate change of numéraire, Rogers and Shi (1992) and Vecer (2001) reduced the pricing problem to a single-state-variable PDE. If $\mathbb{E}_0^S$ denotes the expected value under the martingale measure $\mathbb{P}_S$ defined by

$$d\mathbb{P}_S = \frac{S_T}{s_0 e^{rT}}\, d\widetilde{\mathbb{P}},$$

the Girsanov theorem ensures that

$$W_t^S = W_t - \sigma t$$

is a Brownian motion under $\mathbb{P}_S$. The new measure corresponds to considering the stock price as the numéraire of reference. Therefore, we can write

$$e^{-rT}\widetilde{\mathbb{E}}_0\left[\left(\frac{s_0 A_T}{T} - K\right)_+\right] = s_0 \mathbb{E}_0^S\left[\frac{(s_0 A_T/T - K)_+}{S_T}\right]$$

$$= \frac{s_0}{T}\mathbb{E}_0^S\left[\left(\frac{s_0 A_T - KT}{S_T}\right)_+\right]$$

$$= \frac{s_0}{T}\mathbb{E}_0^S(Y_T)_+,$$

where

$$Y_t = \frac{s_0 A_t - KT}{S_t}, \qquad y \equiv Y_0 = -\frac{KT}{s_0}.$$

The expected value $\mathbb{E}_0^S(Y_T)_+$ can be derived through the following steps:

1. Write the stock price dynamics under the martingale measure $\mathbb{P}_S$:

$$dS = (r + \sigma^2)S \, dt + \sigma S_t \, dW_t^S. \qquad (15.13)$$

2. Compute the differential of the process $s_0 A_t$ as

$$d(s_0 A_t) = S_t \, dt.$$

3. Applying Itô's lemma to Y as a function of S and $s_0 A$ delivers

$$dY = (1 - rY) \, dt - Y\sigma \, dW_t^S.$$

4. As $g(t, y) = \mathbb{E}_0^S(Y_t)_+$ is a price relative to the numéraire S, it is a $\mathbb{P}_S$-martingale, therefore

$$\mathbb{E}_0^S[dg] = 0. \qquad (15.14)$$

5. Combining this equation with the drift of g, as computed by Itô's lemma, leads to

$$\left(\frac{\partial g(t, y)}{\partial t} + (1 - ry)\frac{\partial g(t, y)}{\partial y} + \frac{1}{2}\sigma^2 y^2 \frac{\partial^2 g(t, y)}{\partial y^2}\right) dt = 0,$$

which simplifies to

$$-\frac{\partial g(\tau, y)}{\partial \tau} + (1 - ry)\frac{\partial g(\tau, y)}{\partial y} + \frac{1}{2}\sigma^2 y^2 \frac{\partial^2 g(\tau, y)}{\partial y^2} = 0 \qquad (15.15)$$

using the time change $\tau = T - t$ and with an abuse of notation $g(\tau, y) = g(t + \tau, y)$.

Equation (15.15) has initial condition $g(0, y) = y_+$. Moreover, we can observe that if s_0 is relatively large compared to KT, the option will surely be exercised. This observation helps us to set the upper boundary condition. Indeed, we have

$$g(\tau, 0) = \lim_{y \to 0} \frac{1}{r} + e^{-rT}\left(y - 1 - \frac{1}{r}\right) = \frac{1}{r} - e^{-rT}\left(1 + \frac{1}{r}\right).$$

For $y \to -\infty$ (i.e., when $s_0 \to 0$), we have the lower boundary condition

$$g(T, y \to -\infty) = 0.$$

Consequently, the PDE must be solved for $y < 0$. When $y \geq 0$, the Asian option will be surely exercised at maturity, so that $g(\tau, y) = e^{-rT}(y - 1 - 1/r)$. Unfortunately the resulting PDE problem does not admit an analytical solution. It can be solved using numerical techniques, such as finite differences or transform methods (e.g., Fourier and Laplace).

We implemented a finite difference Crank–Nicolson scheme on a finite domain $[0, \tau] \times [Y_{\min}, 0]$. A first step consists of designing a grid with J time steps of length $\delta\tau$ and I space steps of length $\delta Y = -Y_{\min}/I$. The generic grid point (τ_j, Y_i) is defined by $\tau_j = j\delta\tau$ for $j = 1, \ldots, J$ and $Y_i = Y_{\min} + i\delta Y$ for $i = 1, \ldots, I$. The next step is to compute approximate values for the function g and its partial derivatives on the grid:

$$\frac{\partial g}{\partial \tau}(Y_i, \tau_j) \simeq \frac{g_{i,j+1} - g_{i,j}}{\delta\tau},$$

$$(1 - rY)\frac{\partial g}{\partial Y}(Y_i, \tau_j) \simeq \frac{(1 - rY_i)}{2}\left(\frac{g_{i+1,j+1} - g_{i-1,j+1}}{2\delta Y} + \frac{g_{i+1,j} - g_{i-1,j}}{2\delta Y}\right),$$

$$\frac{1}{2}\sigma^2 Y_i^2 \frac{\partial^2 g}{\partial Y^2}(Y_i, \tau_j)$$

$$\simeq \frac{\sigma^2 Y_i^2}{4}\left(\frac{g_{i+1,j+1} - 2g_{i,j+1} + g_{i-1,j+1}}{(\delta Y)^2} + \frac{g_{i+1,j} - 2g_{i,j} + g_{i-1,j}}{(\delta Y)^2}\right),$$

where $g_{ij} := g(Y_i, \tau_j)$. Substituting these expressions into equation (15.15) and denoting with h_{ij} the exact solution of the resulting finite difference scheme, we arrive at

$$0 = -\frac{h_{i,j+1} - h_{i,j}}{\delta\tau} + \frac{(1 - rY_i)}{2}\left(\frac{h_{i+1,j+1} - h_{i-1,j+1}}{2\delta Y} + \frac{h_{i+1,j} - h_{i-1,j}}{2\delta Y}\right)$$

$$+ \frac{\sigma^2 Y_i^2}{4}\left(\frac{h_{i+1,j+1} - 2h_{i,j+1} + h_{i-1,j+1}}{(\delta Y)^2} + \frac{h_{i+1,j} - 2h_{i,j} + h_{i-1,j}}{(\delta Y)^2}\right).$$

Collecting similar terms together and defining

$$A_i = -\frac{(1 - rY_i)}{4\delta Y} + \frac{\sigma^2 Y_i^2}{4(\delta Y)^2}, \qquad B_i = -\frac{1}{\delta\tau} - \frac{2\sigma^2 Y_i^2}{4(\delta Y)^2}, \qquad (15.16)$$

$$C_i = \frac{(1 - rY_i)}{4\delta Y} + \frac{\sigma^2 Y_i^2}{4(\delta Y)^2}, \qquad E_i = +\frac{1}{\delta\tau} - \frac{2\sigma^2 Y_i^2}{4(\delta Y)^2},$$

for $i = 1, \ldots, I - 1$, we obtain

$$A_i h_{i-1,j+1} + B_i h_{i,j+1} + C_i h_{i+1,j+1} = -A_i h_{i-1,j} - E_i h_{i,j} - C_i h_{i+1,j}.$$

This can be written in matrix notation as

$$\mathbf{M}\mathbf{h}_{j+1} = \mathbf{N}\mathbf{h}_j + \mathbf{b}_j, \qquad (15.17)$$

where $\mathbf{h}_j^\top = (h_{1,j}, \ldots, h_{I-1,j})$. Here matrices $\mathbf{M}$ and $\mathbf{N}$ are square tridiagonals of order $(I-1) \times (I-1)$, namely:

$$
\mathbf{M} = \begin{pmatrix}
B_1 & C_1 & 0 & \cdot & & \cdot \\
A_2 & B_2 & C_2 & \ddots & & \cdot \\
0 & A_3 & & \ddots & & 0 \\
\cdot & \ddots & \ddots & \ddots & C_{I-2} & \\
\cdot & \cdot & 0 & A_{I-1} & B_{I-1}
\end{pmatrix},
$$

$$
\mathbf{N} = \begin{pmatrix}
-E_1 & -C_1 & 0 & \cdot & & \cdot \\
-A_2 & -E_2 & -C_2 & \ddots & & \cdot \\
0 & -A_3 & & \ddots & & 0 \\
\cdot & \ddots & \ddots & \ddots & -C_{I-2} & \\
\cdot & \cdot & 0 & -A_{I-1} & -E_{I-1}
\end{pmatrix}.
$$

Vectors $\mathbf{b}_j$ take into account the boundary conditions at $y = Y_{\min}$ and $y = Y_{\max} = 0$:

$$
\begin{aligned}
\mathbf{b}_j^\top &= \left(-A_1 \times (h_{0,j} + h_{0,j+1}), 0, \ldots, 0, -C_{I-1} \times (h_{I,j} + h_{I,j+1})\right) \\
&= \left(0, \ldots, 0, -C_{I-1} \times \left[2r^{-1} + \left(e^{-rj\delta\tau} + e^{-r(j+1)\delta\tau}\right)\left(Y_{\max} - 1 - r^{-1}\right)\right]\right).
\end{aligned}
$$

The initial condition for the recursion (15.17) is $h_{i0} = Y_i^+$. At each time step the updated solution $\mathbf{h}_{j+1}$ is obtained through an LU decomposition implemented using a tridiagonal solver, as we explained in the chapter on Basic Numerical Methods for Partial Differential Equations.

15.2.4 Transform Approach

Geman and Yor (1993) and Fusai (2004) showed that it is possible to provide a simple expression for a suitable transformation of the option price. In particular, Geman and Yor obtained an analytical expression for the Laplace transform with respect to time-to-maturity of the option. Fusai (2004) in turn obtained a simple expression for a double transform (a Fourier transform with respect to the logarithmic strike, and a Laplace transform with respect to the option maturity) of the Asian option price.

Laplace Transform

Geman and Yor (1993) noticed that a geometric Brownian motion is a time-changed squared Bessel process. By using this property, these authors computed the Laplace transform of the Asian option price with respect to the time variable.[3] The Asian option price can be expressed as

---

[3] Lipton (1999) and Lewis (2002) obtain the same result by solving the PDE (15.15) using the Laplace transform with respect to τ.

$$c_{gy} = \frac{e^{-rT}}{T} \frac{4S}{\sigma^2} c(h, q). \tag{15.18}$$

The Laplace transform $C(\lambda, q)$ of $c(h, q)$ with respect to variable $h = \sigma^2 T/4$ can be computed as

$$
\begin{aligned}
C(\lambda, q) &= \int_0^{+\infty} e^{-\lambda h} c(h, q) \, dh \\
&= \frac{\int_0^{1/(2q)} e^{-x} x^{(\mu-v)/2-2} (1 - 2qx)^{(\mu+v)/2+1} \, dx}{\lambda(\lambda - 2 - 2v)\Gamma(1(\mu - v)/2 - 1)},
\end{aligned}
\tag{15.19}
$$

with $\lambda \in \mathbb{C}$ and

$$v = \frac{2r}{\sigma^2} - 1, \qquad q = \frac{\sigma^2}{4S} KT, \qquad \mu = \sqrt{2\lambda + v^2}.$$

As discussed in Chapter 7: "The Laplace Transform", the original function $c(h, q)$ is related to its Laplace transform through the Bromwich inversion integral

$$c(h, q) = \frac{1}{2\pi i} \int_{a-i\infty}^{a+i\infty} e^{\lambda h} C(\lambda, q) \, d\lambda, \tag{15.20}$$

with a greater than the right-most singularity of the Laplace transform.[4] This integral can be numerically evaluated by the Abate and Whitt (1992b) algorithm. First, the inversion formula is approximated by the trapezoidal rule

$$c(h, q) \approx \frac{e^{A/2}}{2h} \operatorname{Re}\left(C\left(\frac{A}{2h}, q\right)\right) + \frac{e^{A/2}}{h} \sum_{s=1}^{\infty} (-1)^s \operatorname{Re}\left(C\left(\frac{A + 2s\pi i}{2h}, q\right)\right), \tag{15.21}$$

where $\operatorname{Re}(C(\lambda, q))$ denotes the real part of function $C(\lambda, q)$, and $A = 2ah$. Next, the alternating series $\sum_{s=1}^{\infty}(-1)^s a_s$ appearing in the inversion formula (15.21) can be efficiently approximated by the Euler sum

$$E(m, n) = \sum_{s=0}^{m-1} \binom{m}{s} 2^{-m} s_{n+s},$$

with

$$s_n = \sum_{s=0}^{n} (-1)^s a_s.$$

The resulting (Euler) algorithm follows. First, we compute the first n terms in the series, i.e., s_n. The algorithm then extrapolates a highly accurate estimate of the series

---

[4] Given that the denominator of the Laplace transform in equation (15.19) is zero when $\lambda = 0$, or when $\lambda = 2 + 2v$, being $2 + 2v > 0$, a must be greater than $2 + 2v$. This figure is the right-most singularity of the Laplace transform.

by means of additional $m - 1$ terms (namely, $s_{n+1}, \ldots, s_{n+m-1}$). Consequently, the numerical inversion requires $n + m$ evaluations of the function a_s. It turns out that accurate results are obtained by using quite rather few terms. For example, in the numerical experiments we used $m = 35, n = 15$. On the contrary, a direct calculation of the sum of the series would require more than 100,000 evaluations of function a_s.

Our VBA® implementation of the numerical inversion algorithm provides accurate results provided that the volatility term $\sigma\sqrt{T}$ is relatively large, e.g. greater than 0.08, otherwise numerical problems arise during the computation of the integral term in expression (15.19).

Double Transform

Fusai (2004) provided an alternative solution based on integral transforms. The Asian option price can be written as

$$
e^{-rT} \widetilde{\mathbb{E}}_0 \left(\frac{s_0 A_T}{T} - K \right)^+ = \frac{4 s_0 e^{-rT + a_f k}}{\sigma^2 T} c(k, h; a_f) \Bigg|_{\substack{k=\ln(\frac{K}{s_0} \frac{\sigma^2 T}{4}) \\ h=\sigma^2 T/4}}. \tag{15.22}
$$

The double transform of $c(k, h; a_f)$ is available in closed-form. This is computed as the Fourier transform with respect to k, and the Laplace transform with respect to h,

$$
\mathcal{L}\big(\mathcal{F}(c(k, h; a_f))\big) \equiv \int_0^{+\infty} e^{-\lambda h} \int_{-\infty}^{+\infty} e^{i\gamma k} c(k, h; a_f) \, dk \, dh
$$
$$
= C(\gamma + i a_f, \lambda), \tag{15.23}
$$

where

$$
C(\gamma, \lambda) = \frac{4}{\sigma^2 \lambda 2^{1+i\gamma}} \frac{\Gamma(i\gamma)\Gamma(\frac{\mu+v}{2} + 1)\Gamma(\frac{\mu-v}{2} - 1 - i\gamma)}{\Gamma(\frac{\mu+v}{2} + 2 + i\gamma)\Gamma(\frac{\mu-v}{2})}, \tag{15.24}
$$

and $\mu = \sqrt{2\lambda + v^2}$. The complex gamma function can be computed by

$$
\Gamma(z + 1) = \left(z + \gamma + \frac{1}{2} \right)^{z+1/2} e^{-(z+\gamma+1/2)} \tag{15.25}
$$
$$
\times \sqrt{2\pi} \left[c_0 + \frac{c_1}{z+1} + \frac{c_2}{z+2} + \cdots + \frac{c_N}{z+N} + \epsilon \right]
$$

(see Press et al. (1992)). For $\gamma = 5$, $N = 6$, and suitable coefficients c_i's,[5] the error term is smaller than 2×10^{-10}.

Expression (15.24) is obtained exploiting the expressions for the real moments of A_T given in Yor (2001). See also Cai and Kou (2007).

---

[5] In particular, we have: $c_0 = 1.00000000019001$, $c_1 = 76.1800917294715$, $c_2 = -86.5053203294168$, $c_3 = 24.0140982408309$, $c_4 = -1.23173957245015$, $c_5 = 0.120865097386618 \times 10^{-2}$, $c_6 = -0.5395239384953 \times 10^{-5}$.

The original function $c(k, h; a_f)$ can be recovered using Laplace and Fourier inversion integrals. If we denote by $\mathcal{L}^{-1}$ and $\mathcal{F}^{-1}$, the formal Laplace and Fourier inverses respectively, we have

$$c(k, h; a_f) = \mathcal{L}^{-1}\left(\mathcal{F}^{-1}\left(C(\gamma + ia_f, \lambda); \gamma \to k\right); \lambda \to h\right)$$

$$= \frac{1}{2\pi i} \int_{a_l - i\infty}^{a_l + i\infty} \frac{e^{\lambda h}}{2\pi} \int_{-\infty}^{+\infty} e^{-i\gamma k} C(\gamma + ia_f, \lambda) \, d\gamma \, d\lambda, \quad (15.26)$$

where a_l is at the right of the largest singularity of the function $C(\gamma, \lambda)$. Therefore the function $c(k, h)$ is given by

$$c(k, h) = e^{a_f k} c(k, h; a_f).$$

The numerical computation of the double inversion integral in equation (15.26) can be performed by resorting to the multivariate version of the Fourier–Euler algorithm presented in Choudhury, Lucantoni and Whitt (1994). This is the iterated one-dimensional numerical inversion formula used to invert the Geman–Yor transform mentioned above. Given the double transform $C(\gamma, \lambda)$, we first numerically evaluate the Fourier inverse with respect to γ, and then invert the Laplace transform with respect to λ by using the numerical univariate inversion formula.

The Fourier inversion integral in (15.26) can be approximated by the trapezoidal rule, with a step size Δ_f

$$c(k, h) = e^{a_f k} \mathcal{L}^{-1}\left(\frac{1}{2\pi} \Delta_f \sum_{s=-\infty}^{+\infty} e^{-i\Delta_f s k} C(\Delta_f s + ia_f, \lambda)\right). \quad (15.27)$$

Setting $\Delta_f = \pi/k$ and $a_f = A_f/(2k)$, we have

$$c(k, h) = e^{A_f/2} \mathcal{L}^{-1}\left(\frac{1}{2k} \sum_{s=-\infty}^{+\infty} (-1)^s C\left(\frac{\pi}{k}s + i\frac{A_f}{2k}, \lambda\right)\right).$$

Applying the inversion integral for the Laplace transform leads to

$$c(k, h) = \frac{e^{A_f/2}}{4k\pi i} \int_{a_l - i\infty}^{a_l + i\infty} d\lambda \, e^{\lambda h} \sum_{s=-\infty}^{+\infty} (-1)^s C\left(\frac{\pi}{k}s + i\frac{A_f}{2k}, \lambda\right),$$

where a_l is at the right of the largest singularity of the function $C(\gamma, \lambda)$. By substituting $\lambda = a_l + i\omega$ in the last expression, we have

$$c(k, h) = \frac{e^{A_f/2 + a_l h}}{4hk} \int_{-\infty}^{+\infty} d\omega \, e^{i\omega h} \sum_{s=-\infty}^{+\infty} (-1)^s C\left(\frac{\pi}{k}s + i\frac{A_f}{2k}, a_l + i\omega\right).$$

The integral in this expression can be approximated using the trapezoidal rule with step size $\Delta_l = \pi/h$ and setting $a_l = A_l/(2h)$, with A_l such that a_l is greater than the right-most singularities. Finally, we obtain the desired expression

$$c(k, h) \approx \frac{e^{(A_f + A_l)/2}}{4hk} \sum_{m=-\infty}^{+\infty} (-1)^m \sum_{s=-\infty}^{+\infty} (-1)^s C\left(\frac{\pi}{k}s + i\frac{A_f}{2k}, \frac{A_l}{2h} + im\frac{\pi}{h}\right). \quad (15.28)$$

Choudhury, Lucantoni and Whitt (1994) discuss the sources of error in the multidimensional inversion algorithm described above, and propose methods of controlling the resulting bias. In particular, parameters A_f and A_l turn out to be important for controlling the discretization error. Numerical experiments not reported here suggest the following choice for these two parameters: $A_f = A_l = 22.4$. By splitting each of the above sums into a pair of sums with index ranging over the set of nonnegative integers, the inversion formula displays sums in the form $\sum_{s=1}^{\infty}(-1)^s a_s$, with complex-valued coefficients a_s. The numerical inversion requires the application of the Euler algorithm previously described twice, one for the Fourier inversion, the other for the Laplace inversion. This results in $(n_f + m_f)(n_l + m_l)$ evaluations of the double transform. Consequently, the computational cost of the inversion is directly related to this product. In our numerical experiments, we set $m_f = n_f + 15$ and $m_l = n_l + 15$, and the choice of n_f and n_l has been set according to the volatility level. If $\sigma\sqrt{T}$ is low (e.g., smaller than 0.1), we need to set high values for m_f and m_l, for example no smaller than 300. As $\sigma\sqrt{T}$ increases, we can reduce the value of m_f and m_l with no loss of accuracy.

15.3 Implementation and Algorithm

We now detail our VBA® implementation for all methods described in the previous section. Table 15.1 reports the module names containing all functions.

Algorithm (Moment Matching)

1. Compute the nth moments of the average by function moment_n (see equation (15.3)).
2. Compute the density approximation.
   2a. Lognormal approximation (15.4) → function AsianCallLOG.
   2b. Edgeworth approximation (15.6) → function AsianCallEdge. In particular, derivatives appearing in formula (15.6) can be computed using functions Der1Logdens and Der2Logdens.
   2c. Ju approximation (15.8) → function AsianCallJu.
   2d. Reciprocal gamma approximation in (15.9) → function AsianCallRG. Notice that the cumulative gamma function appearing in formula (15.9) is available in Excel, and could, in principle, be called for in a VBA® code. However, this function performs quite poorly. The user defined VBA® function DistributionGamma constitutes an efficient alternative. This function requires arguments A and B, both shape and scale parameters of the gamma distribution, and the number, n, of Gaussian points used to integrate the gamma density. We set n = 100.

Table 15.1. List of main VBA$^{\circledR}$ functions

Module name	Function	Formula	Sub-routines
mComplexFunctions	Operations between complex numbers		
mMomentsAverage	moment_n	(15.2)	djbeta
mLog_RGamma	AsianCallLOG	(15.4)	moment_n
			moment_n
mLog_RGamma	AsianCallRG	(15.9)	DensityRG
			DistributionRG logdens,
mEdgeworth	AsianCallEdge	(15.7)	Der1Logdens
			Der2Logdens
			momlog
mJu	AsianCallJu	(15.8)	moment_n
			GetGamma
			IntegrateGamma
mLowerBound	AsianLowerBound	(15.10)	FindGamma
			IntLowerBound
			phi, sqrtvarnt
mUpperBound	AsianUpperBound	(15.12)	IntegrateSqrtvarnt
			f_upper
			f_uppervw
			ltprice
mGemanYor	AsianCallGY	(15.21)	lt_gemanyor
			ltftasia
mDoubleTransform	AsianCallDFLT	(15.26)	transformasia
			infasia

Algorithm (Lower and Upper Price Bounds) The function `AsianLowerBound` (`Spot As Double, Strike As Double, rfT As Double, sigmaT As Double, n As Integer, gmin As Double, gmax As Double, gacc As Double`) computes c_{low} in equation (15.10). Note that the time to maturity T does not appear as explicit argument of this function. Indeed, the argument `rfT` is $r \times T$, and the argument `sigmaT` is $\sigma \times \sqrt{T}$.

1. Compute the integral in expression (15.10) by using Gaussian quadrature (Legendre). Therefore the integral is approximated by

$$\int_0^1 f(x)\,dx = \sum_{i=1}^{n} w_i f(x_i),$$

where the weights, w_i, and the abscissa, x_i, are computed by the function `gauleg(x1 as Double, x2 as Double, n as Integer)`. The arguments x_1 and x_2 are respectively the lower and upper limits of integration, and n (`n As Integer`), is the number of points used in the discretization of the integral. In the numerical experiments, we set $n = 50$.

2. In order to compute the integral (15.10) we need to find γ^*, the solution of equation (15.11). Again, the integral in expression (15.11) is computed using

Gaussian quadrature and γ^* has been found by a bisection algorithm with the function   FindGamma(..., gmin As Double, gmax As Double, gacc As Double). gmin and gmax are the values that bracket γ^* (in the numerical experiments, we set them equal to -7 and 7). gacc fixes the stopping criterion in the bisection algorithm (we set it equal to 0.0000001).

Computing the upper bound c_{up} in formula (15.12) is done using the function AsianUpperBound(..., n As Integer, m As Integer). Note that time to maturity, T, does not appear as an argument of this function. Indeed the argument rfT is $r \times T$, and the argument sigmaT is $\sigma \times \sqrt{T}$. The double integral in expression (15.12) is approximated, first reducing the infinite domain to the interval (L, U) (in the numerical experiments we set $L = -5$ and $U = 5$), and using Gaussian quadrature. Computing weights and abscissas again by the function gauleg, we get:

$$\int_0^1 \int_{-\infty}^{+\infty} f(x, y) \, dx \, dy \simeq \int_0^1 \int_L^U f(x, y) \, dx \, dy \simeq \sum_{j=1}^m u_j \sum_{i=1}^n w_i f(x_j, y_i).$$

In our numerical experiments we set $n = m = 75$.

Algorithm (Numerical Solution of the Pricing PDE) The numerical solution of the PDE (15.15) is implemented by the function AsiaPDECN(..., numspacestep As Integer, numtimestep As Integer), where numspacestep and numtimestep are the discretization points in space and time. (In the VBA®, Y_{min} is set equal to -6.)

1. Using expression (15.16) fill in the diagonals of the iteration matrices **M** and **N**, and set the initial condition.
2. Iterate for $j = 1, \ldots, J$ over time, and solve at each time step the linear system, (15.17), exploiting the tridiagonal structure of the matrix. The solution of the linear system is computed by the function tridag(LowDiag, diag, UpDiag, vecr, n As Integer), that takes the three diagonals (LowDiag, diag, UpDiag) as inputs, the vector (vecr) given by $\mathbf{N}\mathbf{h}_j + \mathbf{b}_j$, and n, the number of elements in diag.
3. The VBA® function returns a column vector with I elements, containing the solution at points y_i, corresponding to spot prices $S_i = -K/y_i$.[6]

In the numerical examples, we solved the PDE in the interval $[Y_{min} = -6, 0]$, and set $I = J = 3000$. Notice that the solution scales always with the square root of time. Therefore, in the numerical experiments it is convenient to set the time to maturity equal to 1, to replace the risk-free rate with rT, and the volatility coefficient with $\sigma\sqrt{T}$.

---

[6] Note that if we fix K we obtain solutions for different spot prices. Conversely, we can fix S and find the solution for different strike prices, $K_i = -Sy_i$.

Algorithm (Transform Functions) The numerical inversion of the transforms in expressions (15.20) and (15.24) requires the use of complex calculus. For this purpose, we have constructed the basic complex functions (available in the module mComplexFunctions in the Excel file). In particular, we defined a complex number as a Variant array, with the first component being the real part, and the second the imaginary part. As an example, the code for computing the difference of two complex numbers is given here:

```
Function Csub(A As Variant, B As Variant) As Variant
  Dim c As Variant
  ReDim c(2)
  c(1) = A(1) - B(1) 'Compute the real part of the difference
  c(2) = A(2) - B(2) 'Compute the imaginary part of the difference
  Csub = c 'Returns an array containing the real and imaginary parts
End Function
```

The arguments A and B are complex numbers represented by vectors with two elements, containing respectively the real and the imaginary part. Csub returns a vector with two elements, containing the real and the imaginary part of the difference. In a similar manner, we have defined the main operations among complex numbers, adapting the C routines available in Press et al. (1992), pp. 948–950.

Inverting the Laplace Transform

The numerical inversion of the Geman and Yor formula (15.19) is performed by the VBA® function AsianCallGY(..., aa As Double, terms As Integer, totterms As Integer, n As Integer). In particular:

1. In order to consider a constant integration domain in the integral (15.19), it is convenient to set $z = 2qx$ and to reduce the integral to the interval $(0, 1)$. The integrand appearing in formula is computed by the function ltprice.
2. The expression for the Laplace transform, formula (15.19), is given by the function lt_gemanyor. The integral has been computed using Gaussian quadrature (Legendre) with n points (argument n in AsianCallGY).
3. The numerical inversion, including the Euler algorithm, is done within the function AsianCallGY. The numerical inversion requires the specification of the free parameter A (aa as an argument of the VBA® function). The Euler algorithm requires assigning the parameters n (argument term), and $n + m$ (argument totterms).

In the numerical experiments we have used $n = 25$, $n + m = 36$, and $A = 18.4$.

Inverting the Fourier–Laplace Transform

The numerical inversion of the double transform is computed by the VBA® function AsianCallDFLT(..., aaf As Double, termsf As Integer, tottermsf As Integer, aa As Double, terms As Integer, totterms As Integer) As Double. In particular:

1. Complex function `ltftasia` computes the double transform (15.24) using the function `cgammln` (that returns the log-gamma complex function according to the Lanczos formula (15.25)).
2. Complex function `transformasia` returns formula (15.24).
3. Complex function `infasia` computes the inverse Fourier transform, i.e. expression (15.27).

In the numerical experiments we have used `aaf = 22.4`, `termsf = 315`, `tottermsf = 330`, and `aa = 22.4`, `terms = 55`, `totterms = 70`.

15.4 Results and Comments

We compared the effectiveness of the alternative procedures described so far. The first experiment was conducted under different sets of input parameters, as reported in Table 15.2. The results are displayed in Table 15.3, where we use prices obtained by the Monte Carlo simulation by Fu, Madan and Wang (1998) as our benchmark. The product $\sigma\sqrt{T}$ determines the accuracy of the method. For instance, Table 15.3 shows that moment matching methods deliver prices outside the band delimited by the upper and lower bounds, a bias that is particularly clear in the case of the reciprocal gamma approximation. Interestingly, the lower bound provides an exact approximation up to the third digit to the figures delivered by PDE, Laplace transform, and double transform methods. Case 4 is worth mentioning for the very narrow band selected by lower bound (0.0559859 versus an upper bound 0.055989). The PDE method delivers 0.055955, the Ju method provides 0.055984, the Laplace transform results in 0.055984, and the double transform yields 0.055986. We remark that the latter is the only value within the range of upper and lower bounds. Unfortunately, this attractive feature of the double transform method fades away as the number $\sigma\sqrt{T}$ increases beyond the threshold 0.7. This has been confirmed by experiment number 7.

Tables 15.4 and 15.5 provide a more precise comparison between alternative methods.[7] Specifically, we fix some parameters (e.g., option maturity = 1 year, spot

Table 15.2. Parameter set

Example	s_0	K	r	σ	T	$\sigma\sqrt{T}$
1	1.9	2	0.05	0.5	1	0.5
2	2	2	0.05	0.5	1	0.5
3	2.1	2	0.05	0.5	1	0.5
4	2	2	0.02	0.1	1	0.1
5	2	2	0.18	0.3	1	0.3
6	2	2	0.0125	0.25	2	0.3535
7	2	2	0.05	0.5	2	0.7071

[7] These tables can be generated using the VBA® macro `Comparison()` included in the VBA® module mComparison. An example is given in the sheet `Comparison` of the Excel file `PricingAsianOptions.xls`.

Table 15.3. Approximate prices for an Asian option under alternative numerical methods

Example	MC	LW	LG	RG	ED	PDE	JU	LT	LFT	UP
1	0.196	0.193	0.195	0.191	0.195	0.193	0.193	0.193	0.193	0.194
2	0.249	0.246	0.250	0.243	0.245	0.246	0.246	0.246	0.246	0.247
3	0.309	0.306	0.311	0.303	0.301	0.306	0.306	0.306	0.306	0.307
4	0.0565	0.056	0.056	0.056	0.056	0.056	0.056	0.056	0.056	0.056
5	0.22	0.218	0.220	0.217	0.217	0.218	0.218	0.218	0.218	0.218
6	0.172	0.172	0.173	0.171	0.174	0.172	0.172	0.172	0.172	0.172
7	0.348	0.350	0.359	0.342	0.364	0.350	0.350	0.350	0.370	0.353

MC = Monte Carlo, LW = Lower Bound, LG = Moment Matching (Lognormal Approximation), RG = Moment Matching (Reciprocal Gamma), ED = Moment Matching (Edgeworth Series Expansion), JU = Moment Matching (Normal Series Expansion), PDE = Numerical Solution of the Pricing Partial Differential Equation, LT = Laplace Transform Method, LFT = Laplace and Fourier Transform Method, UP = Upper Bound.

Table 15.4. Comparison of alternative Asian option pricing models (panel A)

K	σ	LW	LG	RG	ED	JU
90	0.05	13.72721	13.72721	13.72721	13.72721	13.72721
95	0.05	9.20317	9.20319	9.20303	9.20317	9.20317
100	0.05	4.72430	4.72553	4.72344	4.72430	4.72430
105	0.05	1.18748	1.18786	0.31599	1.18752	1.18762
110	0.05	0.07824	0.07680	0.00001	0.07829	0.07834
90	0.1	13.73246	13.73335	13.72723	13.73246	13.73244
95	0.1	9.28603	9.29086	9.22590	9.28603	9.28605
100	0.1	5.25444	5.26260	5.24864	5.25444	5.25451
105	0.1	2.29464	2.29627	1.64023	2.29464	2.29485
110	0.1	0.73012	0.72385	0.10461	0.73012	0.73060
90	0.3	15.23946	15.32305	14.67344	15.23946	15.24192
95	0.3	11.90191	11.98084	11.63274	11.90191	11.90290
100	0.3	9.05349	9.11390	9.00641	9.05349	9.05295
105	0.3	6.71327	6.74629	6.43541	6.71327	6.71197
110	0.3	4.85928	4.86279	4.06340	4.85928	4.85834
90	0.5	18.37067	18.62493	17.64425	18.37067	18.37986
95	0.5	15.62210	15.85178	15.28760	15.62210	15.62485
100	0.5	13.20052	13.39332	13.04198	13.20052	13.19738
105	0.5	11.09151	11.23974	10.80970	11.09151	11.08393
110	0.5	9.27337	9.37383	8.64116	9.27337	9.26323
MSE		0.00067	0.02200	0.10061	0.03173	0.00119
CPU time (s)		1	0	0	0	0
% Inside		100%	15%	0%	50%	55%

price $s_0 = 100$, interest rate $r = 10\%$), and vary both strike price K and percentage volatility σ (i.e., $K = 90, 95, 100, 105, 110$ and $\sigma = 5, 10, 30, 50$ percentage points). Visual inspection of the output values suggests a number of considerations.

Table 15.5. Comparison of alternative Asian option pricing models (panel B)

K	σ	PDE	LT	LFT	UP
90	0.05	13.72721	n.a.	13.72721	13.72721
95	0.05	9.20303	n.a.	9.20317	9.20319
100	0.05	4.71147	n.a.	4.72430	4.72445
105	0.05	1.18679	n.a.	1.18754	1.18765
110	0.05	0.09275	n.a.	0.07828	0.07851
90	0.1	13.73193	13.74067	13.73246	13.73321
95	0.1	9.28309	9.28468	9.28604	9.28719
100	0.1	5.24968	5.25514	5.25448	5.25483
105	0.1	2.29442	2.29452	2.29478	2.29511
110	0.1	0.73483	0.73040	0.73035	0.73106
90	0.3	15.24001	15.24058	15.24058	15.24972
95	0.3	11.90252	11.90301	11.90301	11.90843
100	0.3	9.05437	9.05468	9.05468	9.05914
105	0.3	6.71464	6.71471	6.71471	6.71942
110	0.3	4.86128	4.86111	4.86111	4.86690
90	0.5	18.37623	18.37642	18.37642	18.40774
95	0.5	15.62749	15.62764	15.62764	15.65638
100	0.5	13.20598	13.20608	13.20608	13.23400
105	0.5	11.09731	11.09735	11.09735	11.12574
110	0.5	9.27975	9.27973	9.27973	9.31044
MSE		0.00103	0.00042	0	0.00103
CPU time (s)			4	95	58
% Inside		55%	55%	100%	100%

First, the Laplace transform method is unable to provide accurate results for small values of $\sigma\sqrt{T}$ due to instability occurring on the numerical inversion of the transform. For instance, no value had been obtained with $\sigma = 5\%$. The double transform method allowed us to overcome this difficulty, and also delivers prices within the two bounds. However, to obtain this level of accuracy, we need to compute the Euler sum in the Fourier inversion with at least $n_f = n_l = 300$, for small values of $\sigma\sqrt{T}$. This requires a high computational cost. If $\sigma\sqrt{T} = 0.5$, we may set $n_f = n_l = 30$ and obtain the figures reported in Table 15.5. A similar problem arises while solving the pricing PDE numerically. For relatively small values of σ compared to the drift term, the PDE loses its parabolic characteristics, and a much finer discretization (e.g., 3000×3000) needs to be adopted to obtain a reliable value for the option price. Furthermore, the two transform methods and the PDE method deliver prices accurate up to the fourth digit. Finally, the reciprocal gamma approximation, for all values of price volatility, is unable to provide estimates inside the bounds. The MSE row reported in Tables 15.4 and 15.5 indicates the square root of the mean square error between the standing column and the column corresponding to the double transform method. The last row reports the CPU time[8] (as expressed in seconds), and the per-

---

[8] Numerical experiments were conducted on a PC equipped with a 1.70 GHz Pentium 4® CPU and 256 MB of RAM.

centage of times each model provides prices within the bounds. The MSE statistics confirm the excellent accuracy of the lower bound approximation together with the good quality of the PDE method and the upper bound. Edgeworth approximation performed quite poorly compared to the lognormal approximation, showing that raising the number of fitted moments does not necessarily provide a better approximation. Much better results (lower MSE, comparable to the PDE method) are obtained with the Ju approximation. However, it should be noted that the moment approximations require very simple implementation and can be run in the shortest amount of time.

16

Quasi-Monte Carlo: An Asian Bet[*]

Key words: Monte Carlo, quasi-Monte Carlo, Asian options, basket options

We consider the problem of pricing Asian options on a basket of underlying assets. Asian options are derivative contracts in which the underlying variable is the average price of given assets sampled over a period of time. Due to this structure, Asian options display a lower volatility and therefore are cheaper than their standard European counterparts. They are very useful in the financial industry and are one of the most popular path-dependent options today. The pricing of Asian options is computationally intensive and a great deal of literature explores this problem using various combinations of analytical methods and simulation techniques.

This paper is a survey of some recent enhancements to improve efficiency when pricing Asian options by Monte Carlo simulation. We present a comparison between the precision of the standard Monte Carlo method (MC) and the stratified Latin Hypercube Sampling (LHS). In particular, we discuss the use of low discrepancy sequences, also known as Quasi-Monte Carlo method (QMC),[1] and a randomized version of these sequences, known as Randomized Quasi-Monte Carlo (RQMC). The latter has proven a useful variance reduction technique for problems of up to 10 dimensions. We follow the pricing estimation procedures described by Imai and Tan (2002) and Dahl and Benth (2002) exploiting a numerical procedure introduced by Owen (2002). (See Benth et al. (2003) as well.)

Section 16.1 introduces the pricing problem in a Black–Scholes framework. Section 16.2 exposes the methodology. Pricing Asian options has a nominal dimension $d = M \times N$, where M is the number of underlying assets and N is the number of times price is recorded for the purpose of computing the price average. Both MC and LHS approaches provide root mean squared errors of order $\mathcal{O}(n^{-1/2})$, where n is the number of performed simulations, independently on the dimension d of the problem.

---

[*] by Piergiacomo Sabino.
[1] See, for instance, Niederreiter (1992) for a detailed account on this subject.

The RQMC method can produce errors with rates of order $\mathcal{O}(\ln^{d-1}(n)/n)$, which, for a low dimension d, is a smaller figure than the one obtained with standard MC. Section 16.3 explains the algorithm in great detail. We generate the trajectories of the underlying assets using the standard MC, the LHS and the RQMC approaches. For each technique, simulation is carried out by exploiting the Principal Component Analysis (PCA) and the Cholesky decomposition of the covariance matrix of the driving factors. These two approaches can be easily implemented with a low computational effort relying on the properties of the Kronecker product. Section 16.4 performs an experiment computing Asian option prices and reports the corresponding root mean squared errors obtained by the three simulation techniques, with each one implemented using either a PCA or a Cholesky decomposition. All computations are performed in MATLAB®.

16.1 Problem Statement

We introduce the problem of pricing Asian options in a Black–Scholes framework. Risky assets dynamics are assumed to be driven by a multivariate geometric Brownian motion under the risk-neutral probability, i.e.,

$$dS_i(t) = r S_i(t)\, dt + \sigma_i\, dW_i(t), \quad i = 1, \ldots, M. \tag{16.1}$$

Here $S_i(t)$ denotes the ith asset price at time t, σ_i represents the corresponding instantaneous return volatility, r is the risk-free continuously compounded interest rate, and $W(t) = (W_1(t), \ldots, W_M(t))$ is an M-dimensional Brownian motion. The $W_i(t)$'s satisfy the following properties:

$$\mathbb{E}[W_i(t)] = 0,$$
$$\mathrm{Var}[W_i(t)] = t,$$
$$\mathrm{Cov}[W_i(t), W_k(t)] = \rho_{ik} t,$$

where ρ_{ik} represents the constant instantaneous correlation between Brownian motions W_i and W_k.

Recall that the covariance matrix $\mathbf{R} = (\mathbb{E}[W(t_l)W(t_m)], l, m = 1, \ldots, N)$ of each Brownian motion in equation (16.1) is:

$$\mathbf{R} = \begin{pmatrix} t_1 & t_1 & \cdots & t_1 \\ t_1 & t_2 & \cdots & t_2 \\ \vdots & \vdots & \ddots & \vdots \\ t_1 & t_2 & \cdots & t_N \end{pmatrix}.$$

We let Σ denote the covariance matrix depending on the correlation among the Brownian motion whose elements are: $(\Sigma)_{i,k} = \rho_{ik}\sigma_i\sigma_k, i, k = 1, \ldots, M$, and define the global covariance matrix for the N-dim process Σ_{MN} as:

$$\Sigma_{MN} = \begin{pmatrix} t_1\Sigma & t_1\Sigma & \cdots & t_1\Sigma \\ t_1\Sigma & t_2\Sigma & \cdots & t_2\Sigma \\ \vdots & \vdots & \ddots & \vdots \\ t_1\Sigma & t_2\Sigma & \cdots & t_N\Sigma \end{pmatrix}.$$

Σ_{MN} is a block matrix obtained repeating Σ at each point of observation. This kind of mathematical operation is known as Kronecker product, denoted as $\otimes$. As such, Σ_{MN} can be identified as the Kronecker product between $\mathbf{R}$ and Σ, $\mathbf{R} \otimes \Sigma$. The Kronecker product reduces the computational complexity of the problem by enabling operations on a $(N \times M, N \times M)$ matrix using two smaller matrices which are $N \times N$, and $M \times M$ respectively (see Pitsianis and Van Loan (1993)). The Kronecker product offers many other properties, some of which are listed below:

(a) Trace factorization. $\mathrm{Tr}(A \otimes B) = \mathrm{Tr}(A)\,\mathrm{Tr}(B)$, where $\mathrm{Tr}(A)$ is the trace of the squared matrix A, i.e. the sum of its diagonal elements.
(b) Compatibility with Cholesky factorization. If we let A, B and D be semi-definite positive matrices, then $D = (C_A C_A^\top) \otimes (C_B C_B^\top) = (C_A \otimes C_B)(C_A \otimes C_B)^\top$. It follows that $(C_A \otimes C_B)$ is the Cholesky decomposition of D.
(c) Eigenvalues and eigenvectors. For A and B square, let λ_i be a member of the spectrum of A, $\sigma(A)$. Let x_A^i be a correspondent eigenvector of λ_i and let $\mu_j \in \sigma(B)$ such that x_B^j is a correspondent eigenvector. Then $\lambda_i \mu_j \in \sigma(A \otimes B)$ and $x_A^i \otimes x_B^j$ is the correspondent eigenvector of $A \otimes B$. To summarize, every eigenvalue of $A \otimes B$ arises as a product of eigenvalues of A and B.

Solving equation (16.1), asset prices at observation times are given by

$$S_i(t_j) = S_i(0) \exp\left[\left(r - \frac{\sigma_i^2}{2}\right)t_j + Z_i(t_j)\right], \quad i = 1, \ldots, M, \qquad (16.2)$$

where

$$\begin{pmatrix} Z_1(t_1) \\ \vdots \\ Z_M(t_1) \\ Z_1(t_2) \\ \vdots \\ Z_M(t_N) \end{pmatrix} = C \begin{pmatrix} \epsilon_1 \\ \vdots \\ \epsilon_M \\ \epsilon_{M+1} \\ \vdots \\ \epsilon_{M \times N} \end{pmatrix}.$$

C is a decomposed matrix of Σ_{MN} satisfying:

$$CC^\top = \Sigma_{MN}$$

and $\epsilon_1, \ldots, \epsilon_{MN}$ are independent standard normal variates.

Possible choices for C include the Cholesky decomposition of Σ_{MN} or a more versatile and efficient approach based on the Principal Component Analysis, PCA.

An Asian option is a European derivative contract written on a basket of assets. Its pay-off is given by the excess of a weighted average of historical prices of all assets comprised in the basket over a strike price K. This average covers values monitored on the time grid $t_1 < \cdots < t_N = T$ between the settlement day and the contract maturity. The pay-off reads as

$$a(T) = \max\left\{\sum_{i=1}^{M}\sum_{j=1}^{N} w_{ij} S_i(t_j) - K, 0\right\}, \qquad (16.3)$$

398   16 Quasi-Monte Carlo: An Asian Bet

where coefficients w_{ij} satisfy $\sum_{i,j} w_{ij} = 1$. European options with a pay-off function (16.3) are called Arithmetic Average options or Arithmetic Asian options. No closed-form solution exists for their arbitrage-free price.

To price the Asian options using simulation, it is necessary to simulate the multi-asset trajectories at each sampling time point as shown above. According to the formulas introduced, the pay-off at maturity of an Arithmetic Asian option can be written as:

$$a(T) = \max\{g(\epsilon) - K, 0\},$$

where

$$g(\epsilon) = \sum_{i=1}^{M \times N} \exp\left(\mu_i + \sum_{k=1}^{M \times N} C_{ik}\epsilon_k\right) \tag{16.4}$$

and

$$\mu_i = \ln(w_{i_1 i_2} S_i(0)) + \left(r - \frac{\sigma_{i_1}^2}{2}\right)t_{i_2}, \tag{16.5}$$

with $i_2 = [(i - 1)/M] + 1$, $i_1 = i - (i_2 - 1)M$, and $[x]$ denotes the greatest integer less than or equal to x.

The pricing of Asian options is thus, computationally cumbersome since the dimension of the problem is highly dependent on the number of underlying assets and the times of observation.

The general Monte Carlo method, as in the case for Asian options, is known to be useful when the dimensionality is large. In this case it is necessary to compute $N \times M$ realizations of the multidimensional process. However, the Monte Carlo method requires a very high computational effort, which must be controlled by suitable computations to reduce the number of simulations as well as variance reduction techniques to enhance its performance. In the following section we will illustrate some of most recent approaches to improve its precision such as the use of the Latin Hypercube Sampling and the low discrepancy sequences also known as Quasi-Monte Carlo method.

16.2 Solution Metodology

We aim to provide a fast and efficient technique which improves the precision of the general Monte Carlo method to price Asian option contracts. We recall that the crude Monte Carlo approach can be be seen as a method to compute the following integral:

$$I = \int_{[0,1)^d} f^*(x)\,dx. \tag{16.6}$$

According to the law of large numbers, the integral I, can be approximated by the following quantity:

$$\widehat{I} \simeq \widehat{I}_n = \frac{1}{n}\sum_{i=1}^{n} f^*(U_i), \tag{16.7}$$

where $(U_1, \ldots, U_n)$ is a sample of independent uniformly distributed random variables in the hypercube $[0, 1)^d$.

The actual problem consists of generating a sample of such random draws to uniformly cover the whole hyper cube $[0, 1)^d$. It is well known that the root mean squared error (RMSE) of the standard Monte Carlo method is $\mathcal{O}(1/\sqrt{n})$ and is independent on the dimension of the problem d. The cases of high values for d are especially interesting in finance because of their various applications, including the pricing of high-dimensional multi-factor path-dependent derivatives securities.

16.2.1 Stratification and Latin Hypercube Sampling

Stratified sampling is a variance reduction method for Monte Carlo estimates. It amounts to partitioning the hypercube $\mathcal{D} = [0, 1)^d$ into H disjoint strata $\mathcal{D}_h$, $(h = 1, \ldots, H)$, i.e., $\mathcal{D} = \bigcup_{i=1}^{H} \mathcal{D}_h$ where $\mathcal{D}_k \cap \mathcal{D}_j = \emptyset$ for all $j \neq k$, then estimating the integral over each set, and finally summing up these numbers (see Boyle, Broadie and Glasserman (1997) for more on this issue). Specifically, mutually independent uniform samples $x_1^h, \ldots, x_{n_h}^h$ are simulated within a stratum $\mathcal{D}_h$ and the resulting integrals are combined. The resulting stratified sampling estimator is unbiased. Indeed

$$\mathbb{E}[\widehat{I}_{\text{strat}}] = \sum_{h=1}^{H} \frac{|\mathcal{D}_h|}{n_h} \sum_{i=1}^{n_h} \mathbb{E}\left[f\left(x_i^h\right)\right]$$

$$= \sum_{h=1}^{H} |\mathcal{D}_h| \mu_h$$

$$= \sum_{h=1}^{H} \int_{\mathcal{D}_h} f(x) \, dx = I,$$

where $|\mathcal{D}_h|$ denotes the volume of stratum D_h. Moreover, this estimator displays a lower variance compared to a crude Monte Carlo estimation, i.e.,

$$\text{Var}[\widehat{I}_{\text{strat}}] \leq \frac{\sigma^2}{n}.$$

Stratified sampling transforms each uniformly distributed sequence $U_j = (U_{1j}, \ldots, U_{dj})$ in $\mathcal{D}$ into a new sequence $V_j = (V_{1j}, \ldots, V_{dj})$ according to the rule

$$V_j = \frac{U_j + (i_1, \ldots, i_d)}{n}, \quad j = 1, \ldots, n, i_k = 0, \ldots, n-1, k = 1, \ldots, d,$$

where $(i_1, \ldots, i_d)$ is a deterministic permutation of the integers 1 through d. This procedure ensures that one V_j lies in each of the n^d hypercubes defined by the stratification.

Latin Hypercube Sampling (LHS) can be seen as a way of randomly sampling n points of a stratified sampling while preserving the regularity from stratification (see, for instance, Boyle, Broadie and Glasserman (1997)). Let $\pi_1, \ldots, \pi_d$ be independent random permutations of the first n positive integers, each of them uniformly distributed over the $n!$ possible permutations. Set

$$T_{jk} = \frac{U_{jk} + \pi_k(j) - 1}{n}, \quad j = 1, \ldots, n, k = 1, \ldots, d,$$

where $\pi_k(j)$ represents the jth component of the permutation for the kth coordinate. Randomization ensures that each vector T_j is uniformly distributed over the d-dimensional hypercube. Moreover, all coordinates are perfectly stratified since there is exactly one sample point in each hypercube of volume $1/n$. For $d = 2$, there is only one point in the horizontal or vertical stripes of surface $1/n$ (see Fig. 16.1). The base and the height are $1/n$ and 1, respectively. For $d > 2$ it works in the same way. It can be proven that for all $n \geq 2, d \geq 1$ and squared integrable functions f, the error for the estimation with the Latin Hypercube Sampling is smaller than or equal to the error for the crude Monte Carlo (see Koehler and Owen (1996)):

$$\mathrm{Var}[\widehat{I}_{\mathrm{LHS}}] \leq \frac{\sigma^2}{n-1}.$$

Figure 16.1 shows the difference between the distribution of 32 points generated with the MATLAB® functions `lhsdesign` and `rand`. For the LHS method we notice that there is only 1 point (dotted points in Fig. 16.1) in each vertical or horizontal stripe whose base is 1 and height is 1/32: it means that there is only vertical and horizontal stratification.

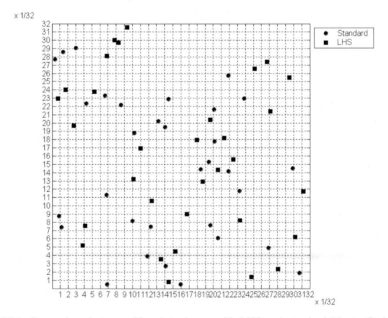

Fig. 16.1. Comparison between 32 points drawn with LHS and standard Monte Carlo approaches.

16.2.2 Low Discrepancy Sequences

As previously mentioned, the crude MC method is based on a completely random sampling of the hypercube $[0, 1)^d$ and its precision can be improved using stratification or Latin Hypercube sampling. These two methods ensure that there is only one point in each smaller hypercube fixed by the stratification as illustrated in Figure 16.1. At the same time, these techniques provide nothing more than the generation of uniform random variables in smaller sets.

A completely different way to approach the sampling problem is to obtain a deterministic sequence of points that uniformly covers the hypercube $[0, 1)^d$ and to run the estimation using this sequence. Obviously, there is no statistical quantity that may represent the uncertainty since the estimation always gives the same results. The Monte Carlo method implemented with the use of low-discrepancy sequences is called Quasi-Monte Carlo (QMC).

The mathematics involved in generating a low-discrepancy sequence is complex and requires the knowledge of number theory. In the following, only an overview of the fundamental results and properties is presented (for details see Owen (2002)).

We define the quantity $D_n^* = D_n^*(P_1, \ldots, P_n)$ as the star discrepancy. It is a measure of the uniformity of the sequence $\{P_n\}_{n \in \mathbb{N}^*} \in [0, 1)^d$ and it must be stressed that it is an analytical quantity and not a statistical one. For example, if we consider the uniform distribution in the hypercube $[0, 1)^d$, the probability of being in a subset of the hypercube is given by the volume of the subset. The discrepancy measures how the pseudo-random sequence is far from the idealized uniform case, i.e. it is a measure, with respect to the L_2 norm, of the inhomogeneity of the pseudo-random sequence.

A sequence $\{P_n\}_{n \in \mathbb{N}^*}$ is called low-discrepancy sequence if:

$$D_n^*(P_1, \ldots, P_n) = O\left(\frac{(\ln n)^d}{n}\right), \tag{16.8}$$

i.e. if its star discrepancy decreases as $(\ln n)^d / n$.

The following inequality, attributed to Koksma and Hlawka, provides an upper bound to estimation error of the unknown integral with the QMC method in terms of the star discrepancy:

$$|I - \widehat{I}| \leq D_n^* V_{\mathrm{HK}}(f). \tag{16.9}$$

$V_{\mathrm{HK}}(f)$ is the variation in the sense of Hardy and Krause. Consequently, if f has finite variation and n is large enough, the QMC approach gives an error smaller than the error obtained by the crude MC method for low dimensions d. However, the problem is difficult because of the complexity in estimating the Hardy–Krause variation, which depends on the particular integrand function.

In the following sections we briefly present digital nets and the well-known Sobol' sequence which is the most frequently used low-discrepancy sequence to run Quasi-Monte Carlo simulations in finance.

16.2.3 Digital Nets

Digital nets or sequences are obtained by number theory and owe their name to the fact that their properties can be recognized by their "digital representation" in base b. Many digital nets exist; the most often used and efficient are the Sobol' and the Niederreiter–Xing sequences.

The first and simplest digital sequence with $d = 1$ is due to Van der Corput and is called the radical inverse sequence. Given an integer $b \geq 2$, any non-negative number n can be written in base b as:

$$n = \sum_{k=1}^{\infty} n_k b^{k-1}.$$

The base b radical inverse function $\phi_b(n)$ is defined as:

$$\phi_b(n) = \sum_{k=1}^{\infty} n_k b^{-k} \in [0, 1),$$

where $n_k \in \{0, 1, \ldots, b - 1\}$ (Galois set).

By varying n, the Van der Corput sequences are constructed. Table 16.1 illustrates the first seven Van der Corput points for $b = 2$. Consecutive integers alternate between odd and even; these points alternate between values in $[0, 1/2)$ and $[1/2, 1)$. The peculiarity of this net is that any consecutive b^m points from the radical inverse sequence in base b are stratified with respect to b^m congruent intervals of length b^{-m}. This means that in each interval of length b^{-m} there is only one point.

Table 16.1 shows an important property which is exploited in order to generate digital nets since a computing machine can represent each number with a given precision, referred to as "machine epsilon". Let $z = 0.z_1z_2\ldots$ (base b) $\in [0, 1)$, define $\Psi(z) = (z_1, z_2, \ldots)$ the vector of the its digits, and truncate its digital expansion at the at the maximum allowed digit w: $z = \sum_{k=1}^{w} z_k b^{-k}$. Let $n = [b^w z] = \sum_{h=1}^{w} n_h b^{h-1} \in N^*$, where $[x]$ denotes the greatest integer less than or equal to x. It can be easily proven that:

Table 16.1. Van der Corput sequence

N	n base 2	$\phi_2(n$ base 2$)$	$\phi_2(n)$
0	000.	0.000	0.000
1	001.	0.100	0.500
2	010.	0.010	0.250
3	011.	0.110	0.750
4	100.	0.001	0.125
5	101.	0.101	0.625
6	110.	0.011	0.375
7	111.	0.111	0.875

$$n_h = z_{w-h+1}(z) \quad \forall h = 1, \ldots, w.$$

This means that the finite sequences $\{n_h\}_{h \in \{1, \ldots, w\}}$ and $\{z_k\}_{k \in \{1, \ldots, w\}}$ have the same elements in opposite order. For example, in Table 16.1 we allow only 3 digits; in order to find the digits of $\phi_2(1) = 0, 5$ we consider $\phi_2(1)2^3 = 4 = 0n_1 + n_2 0 + n_3 1$. The digits of $\phi_2(1)$ are then $(1, 0, 0)$ as shown in Table 16.1.

The peculiarity of the Van der Corput sequence is largely required in high dimensions where the contiguous intervals are replaced by multidimensional sets called the b-adic boxes.

Let $b \geq 2$, k_j, l_j with $0 \leq l_j \leq b^{k_j}$ be all integer numbers. The following set is called b-iadic box:

$$\prod_{j=1}^{d} \left[\frac{l_j}{b^{k_j}}, \frac{l_j + 1}{b^{k_j}} \right),$$

where the product represents the Cartesian product.

Let $t \leq m$ be a non-negative integer. A finite set of points from $[0, 1)^d$ is a (t, m, d)-net if every b-adic box of volume b^{-m+t} (bigger than b^{-m}) contains exactly b^t points. This means that cells that "should have" b^t points do have b^t points. However, considering the smaller portion of volume b^{-m}, it is not guaranteed that there is just one point.

A famous result of the theory of digital nets is that the integration over a (t, m, d) net can attain an accuracy of the order of $O(\ln^{d-1}(n)/n)$ while, restricting to (t, d) sequences, it raises slightly to $O(\ln^d(n)/n)$. The above results are true only for functions with bounded variation in the sense of Hardy–Krause.

16.2.4 The Sobol' Sequence

The Sobol' sequence is the first d-dimensional digital sequence ($b = 2$) ever realized. Its definition is complex and is covered only briefly in the following. Let $\{n_k\}_{k \in \mathbb{N}^*}$ be the digital representation in base $b = 2$ of any integer n; the nth element S_n of the Sobol' sequence is defined as:

$$S_n = \left(\sum_{k=1}^{+\infty} n_k \mathbf{V}_k 2^{-k} \right) \quad \mod 1,$$

where $\mathbf{V}_k \in [0, 1)^d$ are called directional numbers. In practice, the maximum number of digits, w, must be given. In Sobol's original method the ith number of the sequence S_{ij}, $i \in \mathbb{N}$, $j \in \{1, \ldots, d\}$, is generated by XORing (bitwise exclusive OR) together the set of V_{kj} satisfying the criterion on k: the kth bit of i is nonzero. Antonov and Saleev (1980) derived a faster algorithm by using the Grey code. Dropping the index j for simplicity, the new method allows us to compute the $(i + 1)$th Sobol' number from the ith by XORing it with a single V_k, namely with k, the position of the rightmost zero bit in i (see, for instance, Press et al. (1992)). Each different Sobol' sequence is based on a different primitive polynomial over the integers modulo 2, or

Table 16.2. Directional numbers

d	P	m	Principal polynomial	q
1	[1 1 1]	[1 3]	$x^2 + x + 1$	2
2	[1 0 1 1]	[1 1 5]	$x^3 + x + 1$	3
3	[1 1 0 1]	[1 3 7]	$x^3 + x^2 + 1$	3
4	[1 0 0 1 1]	[1 1 3 13]	$x^4 + x + 1$	4
5	[1 1 0 0 1]	[1 1 5 9]	$x^4 + x^3 + 1$	4
6	[1 0 0 1 0 1]	[1 1 5 13 17]	$x^5 + x^2 + 1$	5
7	[1 0 1 0 0 1]	[1 3 7 15 21]	$x^5 + x^3 + 1$	5
8	[1 0 1 1 1 1]	[1 3 3 11 25]	$x^5 + x^3 + x^2 + x + 1$	5
9	[1 1 0 1 1 1]	[1 3 5 15 27]	$x^5 + x^4 + x^2 + x + 1$	5
10	[1 1 1 0 1 1]	[1 3 5 15 31]	$x^5 + x^4 + x^3 + x + 1$	5

in other words, a polynomial whose coefficients are either 0 or 1. Suppose P is such a polynomial of degree q:

$$P = x^q + a_1 x^{q-1} + a_2 x^{q-2} + \cdots + a_{q-1} x + 1.$$

Define a sequence of integers M_k, by the qth term recurrence relation:

$$M_k = 2a_1 M_{k-1} \oplus 2^2 a_2 M_{k-2} \oplus \cdots \oplus 2^{q-1} M_{k-q+1} a_{q-1} \oplus \left(2^q M_{k-q} \oplus M_{k-q}\right).$$

Here $\oplus$ denotes the XOR operation. The starting values for the recurrence are $M_1, \ldots, M_q$ that are odd integers chosen arbitrarily and less than $2, \ldots, 2^q$, respectively. The directional numbers V_k are given by:

$$V_k = \frac{M_k}{2^k}, \quad k = 1, \ldots, w.$$

Table 16.2 shows the first ten primitive polynomials and the starting values used to generate the directional numbers for the 10-dimensional Sobol' sequence.

16.2.5 Scrambling Techniques

Digital nets are deterministic sequences. Their properties ensure good distribution in the hyper cube $[0, 1)^d$, enabling precise sampling of all random variables, even if they are very skewed. The main problem is the computation of the error in the estimation, since it is difficult to compute and depends on the chosen integrand function. To review, the crude MC provides an estimation with low convergence independent on d and the possibility to statistically evaluate the RMSE. On the other hand, the QMC method gives higher convergence, but there is no way to statistically calculate the error.

In order to estimate a statical measure of the error of the Quasi-Monte Carlo method we need to randomize a (t, m, d)-net and to try to obtain a new version of points such that it still is a (t, m, d)-net and has uniform distribution in $[0, 1)^d$.

This randomizing procedure is called scrambling. The scrambling technique permutes the digits of the digital sequence and returns a new sequence which has both the properties described above.

The scrambling technique we use is called Faure–Tezuka Scrambling (for a precise description see Owen (2002), Hong and Hickernell (2000)).

For any $z \in [0, 1)$ we define $\boldsymbol{\Psi}(z)$ as the $\infty \times 1$ vector of the digits of z.

Now let $\mathbf{L}_1, \ldots, \mathbf{L}_d$ be nonsingular lower triangular $\infty \times \infty$ matrices and let $\mathbf{e}_1, \ldots, \mathbf{e}_d$ be $\infty \times 1$ vectors. Only the diagonal elements of $\mathbf{L}_1, \ldots, \mathbf{L}_d$ are chosen randomly and uniformly in $\mathbf{Z}_b^* = \{1, \ldots, b\}$, while the other elements are chosen in $\mathbf{Z}_b = \{0, 1, \ldots, b\}$. $\mathbf{Y}$, the Faure–Tezuka scrambling version of $\mathbf{X}$, is defined as:

$$\boldsymbol{\Psi}(y_{ij}) = \left(\mathbf{L}_j \boldsymbol{\Psi}(x_{ij}) + \mathbf{e}_j\right) \quad \mathrm{mod}\, b.$$

All operations take place in the finite field $\mathbf{Z}_b$. Owen proved that, with his scrambling, it is possible to obtain (see Owen (2003)):

$$\mathrm{Var}[\widehat{I}] \le \frac{b^t}{n} \left[\frac{b+1}{b-1}\right]^d \sigma^2,$$

for any twice summable function in $[0, 1)^d$. These results state that for low dimension d, the randomized QMC (RQMC) provides better estimation with respect to Monte Carlo, at least for large n.

Figure 16.2 shows 32 points of the two-dimensional Sobol' sequence (between the first and the second coordinate) and its Faure–Tezuka scrambled version. It must

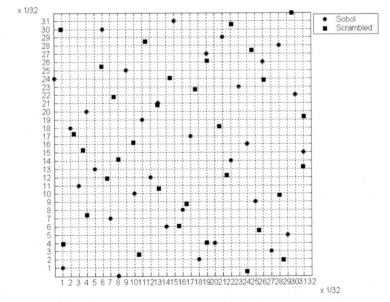

Fig. 16.2. Comparison between 32 Sobol' points and their Faure–Tezuka scrambled version.

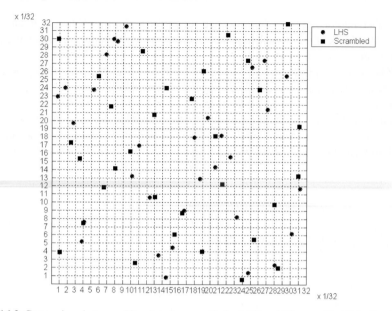

Fig. 16.3. Comparison between 32 points drawn with the LHS and 32 scrambled Sobol' points.

be noted that for both sequences there is only one point in each of all the subsets of measure $1/n$, and not only in the horizontal and vertical stripes as with the LHS. The most important observation is that the scrambled version is still a low discrepancy sequence. Figure 16.3 shows the difference between the former net and 32 points generated with the MATLAB® internal function lhsdesign.

16.3 Implementation and Algorithm

We illustrate the simulation procedure to compute the arithmetic Asian option price. It must be stressed that Quasi-Monte Carlo estimations are dramatically influenced by the problem's dimension, since the rate of convergence depends on the parameter d, as it can be seen in equations (16.8) and (16.9). Many studies and experiments suggest that Quasi-Monte Carlo methods only be used for problem dimensions up to 10 (see Boyle, Broadie and Glasserman (1997) for more on this issue). This condition translates into a relationship between the number M of underlying assets and the number N of monitoring times: $M \times N \leq 10$.

The pricing procedure consists of three main steps:

1. Random number generation by MC, LHS or RQMC.
2. Path generation.
   Monte Carlo estimation.

The first step can be performed as follows:

1.1. (MC) Run MATLAB® internal routine randn.m to sample independent standard normals.

1.2. (LHS) Run MATLAB® internal routine `lhsdesign.m` and sample independent uniforms and use the inverse transform method to generate normal extractions.

1.3. (RQMC) Run function `Sobol_Sequence.m` and obtain 10-dimensional Sobol' sequences; then, perform Faure–Tezuka scrambling by running function `scramble.m`. The file `Sobol_Sequence.m` relies on two MATLAB® functions: `GetDirNum.m` and `GetSob.m`. The former returns directional numbers as shown in Table 16.2; the latter delivers a Sobol' sequence. We suggest storing all generated points in order to reduce the computational effort.

The second step can be implemented by the following algorithm:

2.1. Define the parameters of the simulation.
2.2. Define the drift as in Eq. (16.5).
2.3. Create the $N \times N$ correlation matrix $(\mathbf{R})_{l,k} = (\min(t_l, t_k); l, k = 1, \ldots, N)$.
2.4. Define the correlation matrix $\boldsymbol{\Sigma}$ of M Brownian motions.
2.5. Perform either a PCA or the Cholesky decomposition on the global correlation matrix $\boldsymbol{\Sigma}_{MN}$. This matrix is built up by repeating the constant block of correlation $\boldsymbol{\Sigma}$ at all the times of observation.

The Cholesky decomposition of the global correlation matrix $\boldsymbol{\Sigma}_{MN}$ can be calculated by launching the MATLAB® code `chol_decom.m`. This function exploits property b of the Kronecker product and uses the MATLAB® internal routine `kron.m`. PCA decomposition can be achieved by running the MATLAB® routine `pca_decom.m` which evaluates the eigenvalues and eigenvectors of $\mathbf{R}$ and $\boldsymbol{\Sigma}$ and exploits property c of the Kronecker product.

The third step consists of running MATLAB® functions `asian_crude.m`, `asian_lhs.m` and `asian_rqmc.m`. This delivers two figures: (1) the expected value stating the Asian option price; (2) the RMSE of the estimated price.

As stratification introduces correlation among random drawings, the last two functions are based on the following "batch" method. The method consists of repeating N_B simulations for B times (batches) and computing the average Asian price for each batch, so the RMSE becomes:

$$RMSE = \sqrt{\frac{\sum_{b=1}^{B}(\bar{a}(0)_b - \bar{a}(0))^2}{B(B-1)}},$$

where $(\bar{a}(0)_1, \ldots, \bar{a}(0)_B)$ is a sample of the average present values of the Asian option generated in each batch.

16.4 Results and Comments

We perform a test of all the valuation procedures described in the previous section. Specifically, our experiments involve standard Monte Carlo, the Latin Hypercube Sampling and Randomized Quasi-Monte Carlo by the Faure–Tezuka scrambled version of the Sobol' sequences. Paths are simulated by using both PCA and the Cholesky decomposition as in Dahl and Benth (2002).

Table 16.3. Input parameters used in the simulation

$S_i(0) = 100$
$K = 100$
$r = 2\%$
$T = 1$
$\sigma_1 = 30\%$
$\sigma_2 = 40\%$
$\rho_{ij} = 0$ and 40% for $i, j = 1, 2$

Table 16.4. Correlation case. Estimated prices and standard errors

	Standard MC	LHS	RQMC
PCA	8.291 (0.053)	8.2868 (0.0073)	8.2831 (0.0016)
Cholesky	8.374 (0.055)	8.293 (0.026)	8.2807 (0.0064)

Table 16.5. Uncorrelation case. Estimated prices and standard errors

	Standard MC	LHS	RQMC
PCA	7.195 (0.016)	7.157 (0.013)	7.1696 (0.0017)
Cholesky	7.242 (0.047)	7.179 (0.022)	7.1689 (0.0071)

We consider an at-the-money arithmetic Asian option with strike $K = 100$, written on a basket of $M = 2$ underlying assets, expiring in $T = 1$ year and sampled $N = 5$ times during its lifetime. All results are obtained by using $S = 81920$ drawings. Table 16.3 reports input parameters for our test. The nominal dimension of the problem is $M \times N = 10$ which is equal to the number of rows and columns of the global correlation matrix Σ_{MN}. All the experiments can be performed by launching the MATLAB® file Pricing.m. Table 16.4 and Table 16.5 show results for the positive correlation and uncorrelated cases respectively. Simulated prices of the Asian basket options are in statistical accordance, while the estimated RMSE's depend on the sampling strategy adopted. Furthermore, from a financial perspective, it is normal to find a higher price in the positive correlation case than in the uncorrelated one.

Based on these results, we can make the following conclusions:

1. RQMC method and the use of the Faure–Tezuka scrambling technique provide the best estimation among all the implemented procedures for both the "Correlation" and "Zero Correlation" cases. The correspondent RMSE's are the smallest ones with a higher order of convergence with the same number of simulations.
2. The Kronecker product is a fast and efficient tool for generating multidimensional Brownian paths with low computational effort.
3. Relative to the standard Monte Carlo and LHS approaches, the use of scrambled low-discrepancy sequences provides more accurate results, particularly with PCA-based method.
4. The accuracy of the estimates is strongly dependent on the choice between the Cholesky or the PCA approach. In particular, independent of the simulation pro-

cedure (MC, LHS or RQMC), when using PCA decomposition the estimates are affected by a smaller sampling error (smaller standard error).

The methods presented can be viewed as a part of hierarchy of methods introducing an additional level of regularity in inputs at the expense of complicating the estimation errors. Some methods like stratified sampling fix the size of the sample, while others leave flexibility. The levels of this hierarchy are crude MC (completely random), LHS, QMC methods (completely deterministic). Based on the results reported we can conclude that the use of randomized low discrepancy sequences leads to dramatic improvements and enhances the precision of the numerical computation of the Asian option price. Furthermore the numerical results show that the performance of the RQMC depends on the choice of the matrix C. In particular the Asian option prices obtained with PCA-based path generations have smaller RMSE; however, its precision critically depends on the structure of the covariance matrix. From a computational point of view, the use of the Kronecker product provides a fast generation of the multidimensional path considering both the PCA and the Cholesky decomposition.

The order of convergence in the RQMC depends on the dimension d, and it gives the best results up to $d = 10$. Unfortunately, as d increases, it becomes necessary to consider a larger sample, limiting the benefit of the RQMC (see Boyle, Broadie and Glasserman (1997), p. 1298). To face high-dimensional simulations, a different approach, which goes beyond the purposes of this case, has been proposed by Owen (1998) based on the Latin Supercube Sampling method.

17

Lookback Options: A Discrete Problem[*]

Key words: PDE, Monte Carlo simulation, transform method, discrete monitoring, exotic option

The number of exotic options traded on the market has dramatically increased in the last decade. Correspondingly, a large demand has come about for the development of new, efficient, and fast methods for pricing these securities.

This case presents numerical and analytical methods for pricing discretely monitored lookback options in the Black–Scholes framework. Lookback options are path-dependent options. Their settlement is based on the minimum or the maximum value of an underlying index as registered during the lifetime of the option. At maturity, the holder can "lookback" and select the most favorable figure of the underlying as occurred during this period. Since this scheme guarantees the best possible result for the option holder, he will never regret the option payoff. As a consequence, a lookback option is more expensive than a vanilla option with similar payoff function. An important feature of this contract is the frequency of observation of the underlying assets for the purpose of identifying the best possible value for the holder. Discrete monitoring refers to updating the maximum/minimum price at fixed times (e.g., weekly or monthly). In general, a higher maximum/lower minimum occurs as long as the number n of monitoring dates increases. As noted by Heynen and Kat (1995), the discrepancy between option prices under continuous and discrete monitoring can largely be due to the slow convergence of the discrete scheme to the continuous one as the number n of monitoring dates increases. This figure is quantified in an order of proportionality of $1/\sqrt{n}$.

Closed-form solutions for continuous sampled lookback option prices have been obtained by Conze and Viswanathan (1991) and Goldman, Sosin and Gatto (1979). However, few papers have investigated the analytical pricing of discretely monitored lookback options, see for example Nahum (1998). In this case, we present three alter-

---

[*] with Matteo Bissiri.

native approaches to the pricing problem under discrete monitoring of the underlying index. The first method is analytical. It has been recently proposed by Atkinson and Fusai (2004), who cast the pricing problem in terms of an integral equation. This equation can be solved in closed-form. The second approach is numerical. It consists of using finite difference methods for solving the pricing partial differential equation (PDE). Solutions proposed in the literature (e.g., Wilmott, Dewynne and Howison (1993)) obtain a PDE with two state variables (the asset price and its standing maximum/minimum value) whose solution is the option price. We instead show how the pricing problem can be reduced to the computation of the distribution of the minimum (maximum) by numerically solving the Black–Scholes PDE. This distribution is then integrated, using for instance a quadrature rule, to obtain the lookback option price. As a third method, we consider the number computed by a Monte Carlo simulation and compare it to the results obtained using the other two methods.

In Sect. 17.1, we describe lookback options and the way the pricing problem can be formulated under discrete monitoring. In Sect. 17.2, we illustrate the first two solution methodologies. Section 17.3 details the algorithms and the MATLAB® implementation. In Sect. 17.4, we finally provide some numerical results and compare analytical to numerical approaches. We also empirically examine the convergence rate to the continuous monitoring case.

17.1 Problem Statement

A lookback option can be structured as a put or call. The strike can be either fixed or floating. We now consider two lookback options written on the minimum value achieved by the underlying index during a fixed time window:

- A *fixed strike lookback put*: The payoff is given by the difference, if positive, between the strike price and the minimum price over the monitoring period.
- A *floating strike lookback put*: The payoff is given by the difference between the asset price at the option maturity, which represents the floating strike, and the minimum price over the monitoring period.

Therefore floating strike options will always be exercised. We may formalize the problem and assume that the underlying asset return evolves according to an arithmetic Brownian motion

$$dX_t = \mu \, dt + \sigma \, dW_t,$$

starting at $X_0 = x$. Then, the stock price at time t is given by $S_t = S_0 \exp(X_t)$. We assume that the minimum is monitored at equally spaced dates $t_j = j\Delta$, $(j = 0, \ldots, n)$, with Δ denoting a fixed time period between consecutive monitoring dates. We define m_n as the minimum asset price return registered until time t_n:

$$m_n = \min_{s=\Delta, 2\Delta, \ldots, n\Delta} X_s.$$

The payoff of a fixed strike lookback option is given by:

$$\left(e^k - e^{m_n}\right)_+, \tag{17.1}$$

where k is the logarithm of the strike price. It is natural to assume that the initial spot price x is greater than k. The payoff for a floating strike lookback call option is

$$e^{x_n} - e^{m_n}. \tag{17.2}$$

In order to price lookback options, we need to compute the distribution law of m_n. This task requires computing the conditional expectation

$$\mathbb{P}_{0,x}(m_n > l) = \mathbb{E}_{0,x}(\mathbf{1}_{(X_0>l, X_1>l, \dots, X_n>l)}), \tag{17.3}$$

where

$$\mathbf{1}_{(x>l)} = \begin{cases} 1, & x \geq l, \\ 0, & x < l. \end{cases}$$

The price of the lookback put with fixed strike can be then computed by noting that, for $x > k$, we have

$$\mathbb{E}_{0,x}\left(e^k - e^{m_n}\right)_+ = \left(e^k - e^x\right)_+ \mathbb{P}_{0,x}(m_n = x) + \int_{-\infty}^k e^u \mathbb{P}_{0,x}(m_n \leq u)\,du$$

$$= \int_{-\infty}^k e^u \mathbb{P}_{0,x}(m_n \leq u)\,du. \tag{17.4}$$

Similarly the price of the floating strike call option requires the computation of the expected value $\mathbb{E}_{0,x}(e^{x_n} - e^{m_n})$. For a just issued lookback option we have $m_0 = x$, so that we can write

$$\mathbb{E}_{0,x}\left(e^{x_n} - e^{m_n}\right) = \mathbb{E}_{0,x}\left(e^{x_n}\right) - \mathbb{E}_{0,x}\left(e^{m_n}\right)$$
$$= e^{r_{t_n}} e^x - \mathbb{E}_{0,x}\left(e^{m_n}\right)$$
$$= e^{r_{t_n}} e^x - \left(e^x - \mathbb{E}_{0,x}\left(e^x - e^{m_n}\right)_+\right).$$

Consequently, a formula for pricing a floating strike call option is given once the price of a fixed strike lookback call is available. Lookback options on the maximum can be priced by exploiting the relation between maximum and minimum operators.

In order to evaluate the quantity $\mathbb{P}_{0,x}(m_n > l)$ in (17.3), it is convenient to rewrite it in terms of iterated (conditional) expectations

$$\mathbb{P}_{0,x}(m_n > l)$$
$$= \mathbb{E}_{0,x}(\mathbf{1}_{X_0>l,\dots,X_n>l})$$
$$= \mathbb{E}_{0,x}(\mathbf{1}_{(X_0>l, X_1>l,\dots,X_n>l)})$$
$$= \mathbf{1}_{(x>l)}\mathbb{E}_{0,x}\left(\mathbf{1}_{(X_1>l)} \cdots \mathbb{E}_{(n-2)\Delta, X_{n-2}}\left(\mathbf{1}_{(X_{n-1}>l)}\mathbb{E}_{(n-1)\Delta, X_{n-1}}(\mathbf{1}_{(X_n>l)})\right)\right). \tag{17.5}$$

In next section we illustrate how to compute $\mathbb{P}_{0,x}(m_n > l)$ by using two approaches.

17.2 Model and Solution Methodology

17.2.1 Analytical Approach

The main result in Atkinson and Fusai (2004) consists of an analytical representation for the distribution function of the discrete minimum of the arithmetic Brownian motion. These authors show that this quantity can be expressed as the inverse *z-transform* of the solution to an integral equation. The numerical valuation of this function is easy to perform. We now sketch the main steps involved in this computation.

Let us recursively define a function $h(x, t, j-1)$ according to the backward rule:

$$h(x, t, j-1) = \mathbb{E}_{t,x}\left[h\big(x + \mu\Delta + \sigma(W_{i\Delta} - W_{(j-1)\Delta}), t_j, j\big)\right]\mathbf{1}_{(x>l)}, \qquad (17.6)$$

for all $j = n, n-1, \dots, 1$. Here j refers to the number of monitoring dates that have already passed by. A terminal condition $h(x, t_n, n) = \mathbf{1}_{(x>l)}$ is also imposed. Then, $\mathbb{P}_{0,x}(m_n > l)$ is given by $h(x, 0, 0)$. In particular, if we assume t is a monitoring date and we define $h(x, j) = h(x, t_j, j)$, then we can look at $h(x, 0)$. If, instead, t is not a monitoring date, we can still use expression (17.7) as much as $n-1$ times. Notice that in the last step of the recursion we must replace Δ by $\tau = t_1 - t$.

By using the fact that the transition density of $\sigma(W_{i\Delta} - W_{(i-1)\Delta})$ is Gaussian with zero mean and variance $\sigma^2\Delta$ and that $h(x)$ is zero for $x < l$, we can write

$$\begin{aligned}
h(x, j-1) &= \mathbb{E}_{(j-1)\Delta,x}\left[h\big(x + \mu\Delta + \sigma(W_{j\Delta} - W_{(j-1)\Delta}), j\big)\right]\mathbf{1}_{(x>l)} \\
&= \mathbf{1}_{(x>l)} \int_{-\infty}^{+\infty} h(x + \mu\Delta + \sigma\xi, j)\frac{e^{-\xi^2/(2\Delta)}}{\sqrt{2\pi\Delta}}\,d\xi \\
&= \int_{\frac{l-x-\mu\Delta}{\sigma}}^{+\infty} h(x + \mu\Delta + \sigma\xi, j)\frac{e^{-\xi^2/(2\Delta)}}{\sqrt{2\pi\Delta}}\,d\xi, \qquad (17.7)
\end{aligned}$$

for $x > l$. We can further simplify this relation by defining $f(z, j) = h(z, n-j)$ and verify that this function satisfies a forward recursion

$$f(x, j+1) = \int_{\frac{l-x-\mu\Delta}{\sigma}}^{+\infty} f(x + \mu\Delta + \sigma\xi, j)\frac{e^{-\xi^2/(2\Delta)}}{\sqrt{2\pi\Delta}}\,d\xi, \qquad (17.8)$$

with a starting value $f(x, 0) = \mathbf{1}_{(x>l)}$. After changing variables according to the rule $x + \mu\Delta + \sigma\xi = y$, we obtain:

$$f(x, j+1) = \int_{l}^{+\infty} f(y, j)\frac{e^{-(y-\mu\Delta-x)^2/(2\sigma^2\Delta)}}{\sqrt{2\pi\sigma^2\Delta}}\,dy, \qquad (17.9)$$

i.e., $f(x, j)$ solves an integral-difference equation. We can eliminate the dependence on the index j by considering the *z-transform* of expression (17.9). This quantity is obtained by multiplying both sides of (17.9) by q^j, where $q \in \mathbb{C}$, and then summing over the index $j = 1, \dots$. Assuming that we are allowed to swap integral with infinite summation, we obtain

$$\sum_{j=1}^{\infty} q^j f(x, j) = q \int_l^{+\infty} K(x - \mu\Delta - y) \sum_{j=1}^{\infty} q^{j-1} f(y, j - 1) \, dy \quad (17.10)$$

$$= q \int_l^{+\infty} K(x - \mu\Delta - y) \sum_{j=0}^{\infty} q^j f(y, j) \, dy, \quad (17.11)$$

where $K(x) = \exp(-x^2/(2\sigma^2\Delta))/\sqrt{2\pi\sigma^2\Delta}$. By defining

$$F(x, q) = \sum_{j=0}^{+\infty} q^j f(x, j), \quad (17.12)$$

and adding $f(x, 0)$ to both sides of (17.11), we arrive at an integral equation for $F(x, q)$:

$$F(x, q) = q \int_l^{+\infty} K(x - \mu\Delta - y, \Delta) F(y, q) \, dy + \mathbf{1}_{(x>0)}. \quad (17.13)$$

This equation is to be solved on the interval $l < x < \infty$.

From a probabilistic viewpoint, considering the *z-transform* in (17.12) amounts to playing coin tossing, where q is the probability of getting tails. If the first heads comes out at the $(j + 1)$th tossing, the player wins the amount $f(x, j)$. $F(x, q)/(1-q)$ therefore represents the expected payoff of the game. In other words, this consists of randomizing the option maturity j according to the geometric distribution $(1 - q)q^j$. Compared to the case of fixed maturity, the randomized setting allows us to simplify the pricing problem. From a recursive integral equation satisfied by function $f(x, j)$, the problem is reduced to a more tractable integral equation satisfied by function $F(x, q)$: the dependence on parameter j has been eliminated. Once $F(x, q)$ has been computed, one can proceed backward to $f(x, j)$ by using the inversion formula for the *z-transform*:

$$f(x, n) = \mathcal{Z}^{-1}(F(x, q)) = \frac{1}{2\pi\rho^n} \int_0^{2\pi} F(x, \rho e^{iu}) e^{-inu} \, du, \quad (17.14)$$

defined for integer $n \geq 0$ and where i is the imaginary number $\sqrt{-1}$.

Atkinson and Fusai (2004) provide an analytical solution to the integral equation (17.13). The distribution of the minimum can be expressed in terms of the *z-transform* inverse of $F(x, q)$ as:

$$\mathbb{P}_{0,x}(m_n \leq l) = 1 - e^{\alpha(x-l)+\beta t_n} \mathcal{Z}^{-1}\left(F\left(\frac{x - l}{\sqrt{\sigma^2\Delta/2}}, q\right)\right), \quad (17.15)$$

for $l \leq x$, where

$$F(y_1, q) = \begin{cases} A e^{-\alpha_1 y_1} + g(\alpha_1) \sum_{n=-\infty}^{\infty} A_n e^{+i\mu_n y_1}, & y_1 > 0, \\ g(\alpha_1, q), & y_1 = 0, \end{cases} \quad (17.16)$$

and

$$A = \frac{1}{1 - qe^{\alpha_1^2}}, \tag{17.17}$$

$$A_n = \frac{L_+(\mu_n, q)}{2\mu_n(q)(\mu_n(q) - i\alpha_1)}, \tag{17.18}$$

$$g(\alpha_1, q) = \frac{1}{L_+(i\alpha_1, q)}\mathbf{1}_{(\alpha_1 \geq 0)} + \frac{L_+(-i\alpha_1, q)}{L(-i\alpha_1, q)}\mathbf{1}_{(\alpha_1 < 0)}, \tag{17.19}$$

$$\mu_n(q) = \begin{cases} \sqrt{\ln q + 2n\pi i} & \text{if } \Im\left(\sqrt{\ln q + 2n\pi i}\right) > 0, \\ -\sqrt{\ln q + 2n\pi i} & \text{if } \Im\left(\sqrt{\ln q + 2n\pi i}\right) < 0, \end{cases} \tag{17.20}$$

$$L_+(u, q) = \exp\left\{ \frac{u}{\pi i} \int_0^{+\infty} \frac{\ln(1 - qe^{-z^2})}{z^2 - u^2} dz \right\}, \quad \Im(u) > 0, \tag{17.21}$$

$$L(u, q) = 1 - qe^{-u^2}, \tag{17.22}$$

$$\alpha = -\frac{\mu}{\sigma^2}, \qquad \beta = \alpha\mu + \frac{\alpha^2\sigma^2}{2}, \tag{17.23}$$

$$\alpha_1 = \alpha\delta, \qquad \delta = \sqrt{\frac{\sigma^2}{2}\Delta}.$$

Here $\rho = |q| < 1$ and $\Im(q)$ denotes the imaginary part of q.

A numerical approximation to the *z-transform* inverse (17.14) can be found in Abate and Whitt (1992a). These authors approximate the inversion integral using a trapezoidal rule with a step size of π/n. The resulting figure is:

$$\begin{aligned} \mathcal{Z}^{-1}(F(x, q)) &\approx \tilde{f}(x, n) \\ &= \frac{1}{2n\rho^n} \sum_{j=1}^{2n} (-1)^j \Re\left(F\left(x, \rho e^{ji\pi/n}\right)\right) \\ &= \frac{1}{2n\rho^n} \left\{ F(x, \rho) + (-1)^n F(x, -\rho) + 2\sum_{j=1}^{n-1} (-1)^j \Re\left(F\left(x, \rho e^{ji\pi/n}\right)\right) \right\}, \end{aligned}$$

where $\Re(q)$ is the real part of q and the sum term vanishes for $n = 1$. Abate and Whitt (1992b) provide an upper bound to the discretization error

$$|f(z, n) - \tilde{f}(z, n)| \leq \frac{\rho^{2n}}{1 - \rho^{2n}}.$$

For practical purposes, one can assume this bound to be approximately equal to ρ^{2n}, at least for small values of ρ^{2n}. Hence, a computational accuracy of $10^{-\gamma}$ would require $\rho = 10^{-\gamma/(2n)}$.

The price of a floating strike put option (payoff (17.2)) requires evaluating the integral in (17.4). This task can be pursued by replacing the solution (17.15) into (17.4) and computing the integral. We obtain

$$e^{-rt_n} \mathbb{E}_{0,x} \left(e^k - e^{m_n} \right)^+ = e^{-rt_n} \left(e^k - e^{\alpha x + \beta t_n} \mathcal{Z}^{-1} (p(x, k, q, \alpha)) \right), \qquad (17.24)$$

where

$$p(x, k, q, \alpha)$$
$$= \begin{cases} A \exp(k - \alpha x) + g(\alpha_1, q) \\ \quad \times \sum_{n=-\infty}^{\infty} A_n \frac{\exp((1-\alpha-i\mu_n(q)/\delta)k + (i\mu_n(q)/\delta)x)}{1-\alpha-i\mu_n(q)/\delta}, & x > k, \qquad (17.25) \\ g(\alpha_1, q) g((1-\alpha)\delta, q) e^{(1-\alpha)x}, & x \leq k. \end{cases}$$

Finally, the pricing formula for a floating strike call option with payoff (17.1) is given by

$$e^{-rt_n} \mathbb{E}_{0,x} \left(e^{x_{t_n}} - e^{m_n} \right) = e^x \left(1 - e^{-rt_n} \right) + e^{-rt_n} \mathcal{Z}^{-1} (p(x, x, q, \alpha)), \qquad (17.26)$$

and the expression (17.25) can be used to price floating strike options.

17.2.2 Finite Difference Method

The second method we consider is a numerical solution of the PDE satisfied by the function h. According to the Feynman–Kac theorem, we can represent the iterated expectations in (17.5) as the solutions of a sequence of PDEs. In particular, if $t \in [t_{j-1}, t_j]$, then $h(x, t, j - 1)$ defined in (17.6) solves the backward PDE

$$\frac{\partial h(x, t, j - 1)}{\partial t} + \left(r - \frac{\sigma^2}{2} \right) \frac{\partial h(x, t, j - 1)}{\partial x} + \frac{1}{2} \sigma^2 \frac{\partial^2 h(x, t, j - 1)}{\partial x^2}$$
$$= 0, \qquad (17.27)$$

with updating conditions at the monitoring dates

$$h(x, t = t_j, j - 1) = h(x, t_j, j) \mathbf{1}_{(x>0)}, \quad j = n, n - 1, \ldots,$$
$$h(x, t_n, n) = \mathbf{1}_{(x>0)}.$$

Given the solution of the PDE, we have $h(x, 0, 0) = \mathbb{P}_{0,x-l}(m_n > 0)$ and then the distribution $\mathbb{P}_{0,x}(m_n > l) = h(x - l, 0, 0)$.

Between consecutive monitoring dates, we solve equation (17.27) using the Crank–Nicolson scheme. This transforms problem (17.27) into a sequence of systems of linear equations, each one solved through a tridiagonal method using the LU-decomposition. Details on this method are provided in Part I of this book. We hereby summarize the main steps of the procedure:

1. We restrict the domain of the PDE to the interval $[x_{\min}, x_{\max}] \times [0, n\Delta]$ and construct a grid with space step δx and time step $\delta \tau$; the generic grid points are $x_i = x_{\min} + i\delta x$, $\tau_l = l\delta \tau$, and the solution to the PDE at these points is denoted by $h_{il} = h(x_i, \tau_l)$, $i = 1, \ldots, 2I + 1, l = 1, \ldots, L$.

2. The partial derivatives are approximated by finite differences at the grid points

$$\frac{\partial h}{\partial t}(x_i, \tau_l) \simeq \frac{h_{i,l+1} - h_{i,l}}{\delta \tau},$$

$$\left(r - \frac{\sigma^2}{2}\right)\frac{\partial h}{\partial x}(x_i, \tau_l) \simeq \left(r - \frac{\sigma^2}{2}\right)\left(\frac{h_{i+1,l+1} - h_{i-1,l+1}}{2\delta x} + \frac{h_{i+1,l} - h_{i-1,l}}{2\delta x}\right),$$

$$\frac{1}{2}\sigma^2\frac{\partial^2 h}{\partial x^2}(x_i, \tau_l) \simeq \frac{\sigma^2}{4}\left(\frac{h_{i+1,l+1} - h_{i,l+1} + h_{i-1,l+1}}{(\delta x)^2}\right.$$
$$\left. + \frac{h_{i+1,l} - h_{i,l} + h_{i-1,l}}{(\delta x)^2}\right).$$

3. By inserting these expressions into equation (17.27) and denoting the exact solution of the finite difference scheme by $v(x_i, \tau_l)$, we have

$$-\frac{v_{i,l+1} - v_{i,l}}{\delta \tau} + \left(r - \frac{\sigma^2}{2}\right)\left(\frac{v_{i+1,l+1} - v_{i-1,l+1}}{2\delta x} + \frac{v_{i+1,l} - v_{i-1,l}}{2\delta x}\right)$$

$$+ \frac{\sigma^2}{4}\left(\frac{v_{i+1,l+1} - 2v_{i,l+1} + v_{i-1,l+1}}{(\delta x)^2} + \frac{v_{i+1,l} - 2v_{i,l} + v_{i-1,l}}{(\delta x)^2}\right) = 0.$$

4. After collecting terms appropriately, we get to:

$$+av_{i-1,l+1} + bv_{i,l+1} + cv_{i+1,l+1} = -av_{i-1,l} - (b-2)v_{i,l} - cv_{i+1,l},$$

where

$$a = -\frac{r - \sigma^2/2}{4\delta x} + \frac{\sigma^2}{4(\delta x)^2},$$

$$b = -\frac{1}{\delta \tau} - \frac{2\sigma^2}{4(\delta x)^2},$$

$$c = \frac{r - \sigma^2/2}{4\delta x} + \frac{\sigma^2}{4(\delta x)^2},$$

and $i = 2, \ldots, 2I$.

   a. If $i = 1$, or $i = 2I$, we apply boundary conditions $v_{1,l} = 1$ and $v_{2I,l} = 0$.

   b. If l corresponds to a monitoring date, we impose the updating condition

$$v_{i,l} = v_{i,l}\mathbf{1}_{(i\delta x > 0)}. \qquad (17.28)$$

Once the PDE is solved, the prices of lookback options are obtained through the integral representation (17.4). This figure can be computed by using for example a simple quadrature rule, such as the rectangular one

$$\mathbb{E}_{0,x}\left(e^k - e^{m_n}\right)_+ = e^{-rt_n}\int_{-\infty}^{k} e^u V(u, 0, n)\, du$$

$$\simeq \sum_{i=1}^{2I+1} e^{x_{min} + i\delta x} v(x_{min} + i\delta x, 0, 0)\delta x, \qquad (17.29)$$

where spacing δx corresponds to the spatial step used in the solution of the PDE.

17.2.3 Monte Carlo Simulation

Lookback options can be easily priced by standard Monte Carlo (MC) simulation. The underlying price is simulated at all monitoring dates using the discretized process

$$S_{i+1} = S_i e^{(r-0.5\sigma^2)\Delta + \sigma\sqrt{\Delta}\epsilon_i^{(j)}},$$

where $\epsilon_i^{(j)}$ is a standard normal random variate. We denote by $S_i^{(j)}$ the spot price at time $t_i = i\Delta$ as sampled in the jth simulation. The corresponding minimum price $mp_i^{(j)}$ over the interval until time t_i is updated at each monitoring date according to the rule

$$mp_i^{(j)} = \min\left[S_i^{(j)}, mp_{i-1}\right],$$

with a starting condition $m_0 = x$. The MC price for a lookback is given by the average of the discounted payoff computed over J simulated sample paths. For a lookback option with fixed strike K, the MC price is

$$e^{-rt_n}\frac{1}{J}\sum_{j=1}^{J}\left(K - mp_n^{(j)}\right)_+.$$

Similarly, for a floating strike lookback option, the MC price is

$$e^{-rt_n}\frac{1}{J}\sum_{j=1}^{J}\left(S_n^{(j)} - mp_n^{(j)}\right).$$

We have improved the accuracy of the computation by using standard antithetic variables.[1]

17.2.4 Continuous Monitoring Formula

When the minimum price is monitored continuously over time, i.e., the underlying variable is $\tilde{m}_t = \min_{0 \le s \le t} X_s$, Conze and Viswanathan (1991) and Goldman, Sosin and Gatto (1979) obtain the following pricing formulae for the floating strike option:

$$\mathbb{E}_{0,x}\left(e^{x_n} - e^{\tilde{m}_n}\right)$$
$$= S_0\mathcal{N}(d_2) - e^{-rt_n}mp_0\mathcal{N}\left(d_2 - \sigma\sqrt{t_n}\right)$$
$$+ e^{-rt_n}\left(\frac{\sigma^2}{2r}\right)S_0\left[\left(\frac{S_0}{mp_0}\right)^{-2r/\sigma^2}\mathcal{N}\left(-d_2 + 2r\sqrt{\frac{t_n}{\sigma^2}}\right) - e^{rt_n}\mathcal{N}(-d_2)\right],$$

and for the fixed strike option:

---

[1] MC prices and standard errors have been obtained by repeating J simulations M times. Thus, the effective number of simulations is $J \times M$, with $J = 100000$ and $M = 100$. This choice is due to the possibility of exploiting the vector structure in MATLAB® and reducing the computational time.

$$\mathbb{E}_{0,x}\left(e^k - e^{\tilde{m}_n}\right)_+$$

$$= \mathbf{1}_{(K < mp_0)}\left\{-S_0\mathcal{N}(-d) + e^{-rt_n}K\mathcal{N}\left(-d + \sigma\sqrt{t_n}\right)\right.$$

$$+ e^{-rt_n}\left(\frac{\sigma^2}{2r}\right)S_0\left[\left(\frac{S_0}{K}\right)^{-2r/\sigma^2}\mathcal{N}\left(-d + 2r\sqrt{\frac{t_n}{\sigma^2}}\right) - e^{rt_n}\mathcal{N}(-d)\right]\right\}$$

$$+ \mathbf{1}_{(K \geq mp_0)}\left\{e^{-rt_n}(K - mp_0) - S_0\mathcal{N}(-d_2) + e_0^{-rt_n}mp_0\mathcal{N}\left(-d_2 + \sigma\sqrt{t_n}\right)\right.$$

$$+ e^{-rt_n}\left(\frac{\sigma^2}{2r}\right)S_0\left[\left(\frac{S_0}{mp_0}\right)^{-2r/\sigma^2}\mathcal{N}\left(-d_2 + 2r\sqrt{\frac{t_n}{\sigma^2}}\right) - e^{rt_n}\mathcal{N}(-d_2)\right]\right\},$$

with

$$d_2 = \frac{1}{\sqrt{\sigma^2 t_n}}\left(\ln\left(\frac{S_0}{mp_0}\right) + rt_n + \frac{1}{2}\sigma^2 t_n\right),$$

$$d_2 = \frac{1}{\sqrt{\sigma^2 t_n}}\left(\ln\left(\frac{S_0}{mp_0}\right) + rt_n + \frac{1}{2}\sigma^2 t_n\right),$$

$$S_0 = e^x, \qquad K = e^k, \qquad mp_0 = e^{m_0}.$$

17.3 Implementation and Algorithm

MATLAB® codes for this study are stored in the folder PDElookback.
   The folder contains the following subfolders:

(1) **af**: contains analytical pricing functions and functions associated with the z-transform;
(2) **pde**: contains PDE pricing function and the tridiagonal system solver;
(3) **mc**: contains MC tools for simulating the underlying process and pricing lookback options;
(4) **analysis**: generates plots and tables.

We now detail the content of all functions included in the project.

Function af_lookbackcall_floatingstrike.m and function af_lookbackput_fixedstrike.m

These functions compute the price of a lookback option by using formulas (17.24) and (17.26), respectively. The inverse z-transform is inverted by numerical integration using the trapezoid rule, as described in the previous section. The function Ztlookback.m computes the z-transform expression given in (17.16). It calls the following auxiliary functions: An.m, Bn.m, mun.m, g.m, L.m and Lplus.m, defined in (17.17)–(17.23). In particular the Lplus.m function computes the integral in (17.21) by adaptive Lobatto quadrature (see MATLAB® help).

Function `af_Fmin.m` and function `af_Fplus.m`

These functions compute the analytical expression of the cumulative distribution of the minimum, formula (17.15).

Function `pde_lookbackcall_floatingstrike.m` and function `pde_lookbackput_fixedstrike.m`

These functions solve the PDE with a Crank–Nicholson scheme. The function `pde_tridiagonal_solver.m` solves the tridiagonal linear system using LU decomposition. The cumulative distribution as function of the initial spot price is returned by the function `pde_lookbackput_minimum.m`. The price of a lookback option is obtained by numerically integrating the cumulative distribution for the minimum, as in (17.29).

Function `mc_lookbackcall_floatingstrike.m` and function `mc_lookbackput_fixedstrike.m`

These two functions compute the price and standard error of a lookback option using Monte Carlo simulation. The function `mcsimulator.m` simulates the underlying spot and minimum price at a given set of dates. It repetitively calls the function `mcevolversolver.m` which evolves the state of the underlying between two dates.

The flowchart exhibited in Fig. 17.1 summarizes the structure of the project.

17.4 Results and Comments

In this section we discuss numerical results obtained by running the codes described above. First, we investigate the bias of discrete monitoring *vs.* continuous monitoring. Then, we compare the accuracy of the proposed numerical procedures. Finally, we study the effect of changing parameter values on the option price. Numerical experiments have been conducted under the following assumptions: (1) the infinite series in formula (17.24) has been computed using 51 terms, i.e., for n running from -25 to 25; (2) the PDE has been solved using a spatial discretization with 6000 points and a time discretization with 10,000 steps; (3) Monte Carlo simulation has been performed over 10^7 simulations with antithetic variates.

In Fig. 17.2 we represent the distribution of the minimum (Eq. (17.15)) for a varying number of monitoring dates. This illustrates how discrete monitoring can have an important effect on the distribution of the minimum. This is confirmed in Fig. 17.3, where we represent the price of a floating strike lookback option and vary the number of monitoring dates under the hypothesis that the remaining parameters have been set as follows: $r = 0.1$, $\sigma = 0.2$, $t = 1$, $S_0 = 100$. In this case, the continuous formula returns 19.6456. Assuming a year consists of 250 days and that 10 monitoring dates are available (i.e., monitoring occurs approximately once a

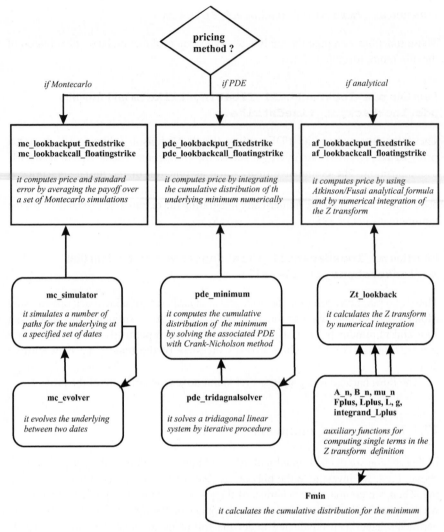

Fig. 17.1. Flow-chart of the MATLAB® functions.

month), the discrete formula gives 17.0007: a percentage difference about 15% with respect to the continuous case. Using 10,000 monitoring dates (i.e., monitoring occurs once every 36 minutes), the discrete formula returns 19.5523, a small but still appreciable difference in respect to the continuous case. Theoretical results discussed in Broadie, Glasserman and Kou (1999) show that the convergence to the continuous case is very slow, that is in the order of $C/\sqrt{n}$, for a suitable constant C.

In Table 17.1, we report values for fixed strike put options. Both analytical solution and MC simulation provide price estimates that agree up to the second-order decimal. The PDE solution is reliable, but it does not appear as accurate as the others

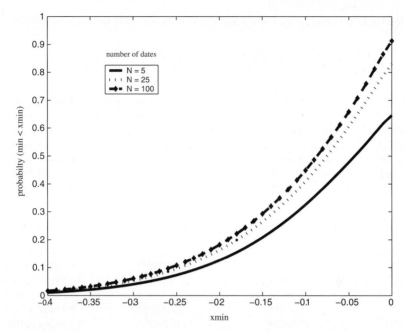

Fig. 17.2. Distribution function of the minimum for different monitoring dates.

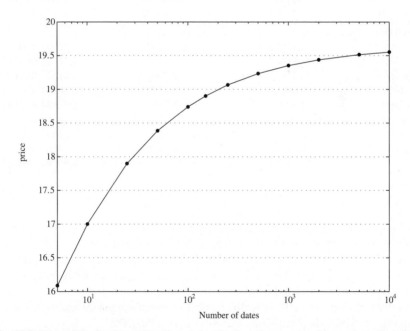

Fig. 17.3. Price of the floating strike discrete lookback option varying the number of monitoring dates. Parameters: $r = 0.1, \sigma = 0.2, t = 1, S_0 = 100$.

Table 17.1. Fixed strike put option: comparison of analytical, PDE and MC solutions. Payoff: $(e^k - e^{m_n})^+$

k	An. sol. (17.24)	PDE	MC (s.e.)
Fixed strike lookback call			
$n = 5$			
ln(90)	1.76898	1.76538	1.76780 (0.01027)
ln(92.5)	2.60123	2.59985	2.59989 (0.01175)
ln(95)	3.69009	3.68045	3.68881 (0.01296)
ln(97.5)	5.05627	5.04561	5.05497 (0.01339)
ln(100)	6.69778	6.69775	6.69641 (0.01313)
ln(102.5)	9.13606	9.11677	9.13469 (0.01313)
ln(105)	11.57433	11.59579	11.57296 (0.01313)
ln(107.5)	14.01261	14.03138	14.01124 (0.01313)
ln(110)	16.45088	16.47121	16.44951 (0.01313)
$n = 25$			
ln(90)	2.20212	2.19773	2.20282 (0.01026)
ln(92.5)	3.20671	3.20503	3.20793 (0.01104)
ln(95)	4.51040	4.49884	4.51154 (0.01117)
ln(97.5)	6.14766	6.13472	6.14826 (0.01078)
ln(100)	8.12410	8.12401	8.12437 (0.01043)
ln(102.5)	10.56238	10.54304	10.56264 (0.01043)
ln(105)	13.00065	13.02205	13.00091 (0.01043)
ln(107.5)	15.43893	15.45764	15.43919 (0.01043)
ln(110)	17.87720	17.89748	17.87746 (0.01043)
$n = 100$			
ln(90)	2.43091	2.42611	2.42997 (0.01116)
ln(92.5)	3.51807	3.51621	3.51718 (0.01179)
ln(95)	4.91891	4.90647	4.91807 (0.01185)
ln(97.5)	6.66689	6.65302	6.66617 (0.01157)
ln(100)	8.78048	8.78030	8.77984 (0.01129)
ln(102.5)	11.21876	11.19932	11.21812 (0.01129)
ln(105)	13.65703	13.67834	13.65639 (0.01129)
ln(107.5)	16.09531	16.11392	16.09467 (0.01129)
ln(110)	18.53358	18.55376	18.53294 (0.01129)

Parameters: $S_0 = 100$, $\sigma = 20\%$, $r = 5\%$, $t_n = 0.5$, $m_0 = \ln(S_0)$.

do. This is due to the time discretization, a problem that does not present itself for the other two methods, and to the presence of a discontinuity in the updating condition (17.28) of the PDE. As discussed in the chapter on Basic PDEs, the presence of discontinuities and the use of a Crank–Nicolson scheme can deteriorate the numerical solution in a neighborhood of the discontinuity point. The problem is clearly exacerbated as long as we increase the number of monitoring dates and therefore the frequency at which discontinuities are introduced in the PDE.

In Table 17.2 we price floating strike call options. Numerical results confirm the quality of the approximation, showing a perfect match up to the third digit, placing

Table 17.2. Floating strike put option: comparison of analytical, PDE and MC solutions. Payoff: $(e^{x_{t_n}} - e^{m_n})$

n	An. sol. (17.26)	PDE	MC
Floating strike lookback call			
$\sigma = 0.05$			
5	3.49103	3.49093	3.49071 (0.00353)
25	3.85405	3.85367	3.85435 (0.00324)
50	3.95357	3.95300	3.95349 (0.00334)
100	4.02721	4.02638	4.02703 (0.00329)
150	4.06071	4.05557	4.06097 (0.00331)
250	4.09485	4.09350	4.09452 (0.00318)
500	4.12976	4.12783	4.12970 (0.00382)
∞		4.21636	
$\sigma = 0.2$			
5	9.16679	9.16676	9.16522 (0.01875)
25	10.59311	10.59302	10.59317 (0.01535)
50	10.97201	10.97188	10.97095 (0.01721)
100	11.24949	11.24930	11.24846 (0.01676)
150	11.37495	11.33658	11.37632 (0.01624)
250	11.50235	11.50205	11.50181 (0.01630)
500	11.63214	11.63171	11.63338 (0.01834)
∞		11.95198	
$\sigma = 0.5$			
5	20.21359	20.21352	20.21004 (0.06078)
25	23.29912	23.29902	23.29767 (0.05099)
50	24.10732	24.10719	24.10148 (0.05641)
100	24.69628	24.69608	24.69325 (0.05603)
150	24.96177	24.86390	24.96601 (0.05214)
250	25.23083	25.22973	25.22985 (0.05285)
500	25.50445	25.50104	25.50941 (0.06009)
∞		26.17645	

Parameters: $S_0 = 100$, $r = 0.05$, $t_n = 0.5$ years, $m_0 = \ln(S_0)$.

this method between the analytical and the MC algorithms. Table 17.2 illustrates, for a fixed number of monitoring dates, the way rising volatility determines an increase in the lookback option price. While the expected value of the terminal spot price is unaffected by volatility, a higher volatility implies a higher probability of reaching a smaller minimum and thus a greater value of the floating strike lookback call option. In this table, we can also appreciate the importance of the discrete formula compared to the continuous time monitoring result case.

In Fig. 17.4, we show the impact of interest rate over the arbitrage price of a call option with floating strike and a put option with fixed strike ($K = 100$). The remaining parameters are set at $S_0 = 100$, $\sigma = 20\%$, $T = 1$, $N = 5$. A higher interest rate determines a larger positive drift and an increase in the probability that the minimum coincides with the starting value x of the process. At the same time, a higher interest rate determines a fall in the option price due to a discounting-related

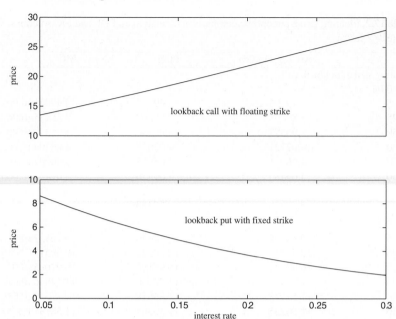

Fig. 17.4. Lookback option price vs. interest rate level.

Table 17.3. CPU times: WH and MC solutions. Put payoff: $(e^k - e^{m_n})_+$, $m_0 = \ln S_0$

Mon dates	Analytical solution		PDE		Monte Carlo	
	Time (s)	Price	Time (s)	Price	Time (s)	Price (s.e.)
10	1.14	7.41906	59.3	7.41901	25.8	7.41961 (0.01288)
50	1.76	8.50299	86.4	8.50287	119.3	8.50296 (0.01214)
100	3.71	8.78048	91.3	8.78030	235.7	8.77984 (0.01129)

Parameters: $S_0 = 100$, $\sigma = 0.2$, $r = 0.05$, $t_n = 0.5$.

effect. Given the structure of payoffs in (17.26) and (17.24), we see that the net result is a higher price for the floating option and a lower price for the fixed strike option, as is illustrated in Fig. 17.4.

Finally, Table 17.3 compares the CPU time for pricing fixed strike lookback options using the analytical solution, Monte Carlo simulation and numerical solution of the PDE.

18

Electrifying the Price of Power[*]

Key words: electricity prices, simulation, equilibrium models, AR(1) model

In this chapter we explore modeling forward prices for electricity. Power exchanges provide us with the one-day forward price for one megawatt-hour (MWh) of electricity for delivery at a specified location. The resulting figure is referred to as the electricity "day-ahead price".

Empirical evidence from most power markets across the globe show stylized patterns of day-ahead price dynamics. Incidentally, these features appear to have no direct counterpart in other financial security markets. Day-ahead price trajectories usually display periodical trends, recurrent spikes, sharp mean reversions, and time dependent volatility. It turns out that all these properties are a direct consequence of peculiar phenomena featuring the supply and demand sides in the energy commodity markets. In particular:

1. Almost all existing electricity markets reflect a chronic shortage of power for a spot delivery due to the limited number of power plants compared to the standing demand in the economies they serve.
2. The time requirement for increasing production capacity through new power plants, coupled with scarce incentives to bear the risks of the long term investments, make the supply curve approximately time invariant.
3. The demand is quite stiff to price variations, yet its dynamics are quite sensitive to weather conditions.

A typical market instance is depicted in Fig. 18.1.

From the graph, we can see that the equilibrium price can be extremely sensitive to the variations affecting the demand for immediate use of power. What makes this issue even more unique is that electricity cannot be stored, or it can be, but to a very limited extent and at a high cost. This prevents markets from clearing expected

---

[*] with Paolo Carta.

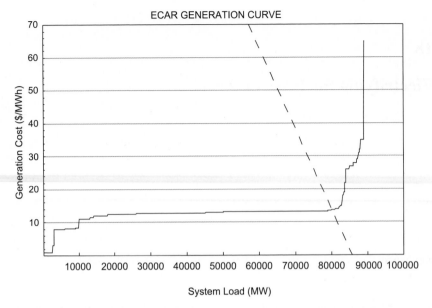

Fig. 18.1. Demand and supply stack function in the ECAR power market.

shocks through cash-and-carry strategies, a fact that explains the recurrence of the spiky behavior in the price dynamics.

Several papers have investigated power prices. Bhanot (2000), Knittel and Roberts (2001), Botterud, Bhattacharyya and Ilic (2002), among others, examine the empirical features of electricity price dynamics. Beamon (1998), Joskow and Kahn (2001), Stevenson (2001) investigate the formation process of equilibrium prices, while Manoliu and Tompaidis (2002) adopt a purely econometric approach based on the Kalman filter. Deng (1999), Barlow (2002), De Jong and Huisman (2002), Escribano, Peña and Villaplana (2002), Fiorenzani (2005), Huisman and Mahieu (2003), and Lucia and Schwartz (2002) propose reduced-form models for the risk neutral dynamics. In this respect, Roncoroni (2002) proposes a model reproducing both trajectorial and statistical properties of electricity prices in major US power markets.

In the present case, we examine the equilibrium model for the day-ahead price introduced by Hinz (2003), and provide an explicit numerical implementation. In our model, agents implement trading strategies involving a forward position to meet supply requirements. The goal is to determine a forward price compatible with the optimal trading behavior in the market.

Our development of the case is organized as follows: Sect. 18.1 introduces the model for both one-period and multi-period settings. Section 18.2 describes a particular instance of the multi-period model. Finally, Sect. 18.3 details the corresponding implementation and reports results of a few numerical experiments to evaluate the effect of demand predictability over the regime-switching behavior of price dynamics.

18.1 Problem Statement

18.1.1 The Demand Side

We consider a market where N agents act simultaneously as retailers as well as producers of electricity. In a one-period setting, we focus on two dates: T_0, which represents current time, and T_1, which indicates the delivery date. In this model, T_1 is set to the day following T_0. As seen from time T_0, the amount of energy $\widetilde{Q}_i$, that will be asked for from agent i at time T_1, is described by a random variable, whose distribution is assumed to be given exogenously. At time T_1, agent i sells each unit of energy at a fixed retail price p_i^r. As a result, the sale provides a random gross revenue

$$R(\widetilde{Q}_i) = p_i^r \, \widetilde{Q}_i. \tag{18.1}$$

18.1.2 The Bid Side

Agents can meet their demand in three ways. They may

(1) produce energy at time T_1;
(2) buy forward contracts at time T_0 for delivery one day later; or
(3) buy energy in the spot market at time T_1.

Let us examine each of these opportunities in detail.

Internal Production

Agent i owns power plants delivering an overall generation capacity of c_i MWh. The variable cost per MWh is p_i^v, whereas the general cost for operating the process is a constant P_i^f:

$$\text{Variable Unit Cost} = p_i^v,$$

$$\text{General Fixed Cost} = P_i^f.$$

Forward Trading

There is a forward market that quotes prices for one-day delivery of one MWh. Our goal is to provide a model for the equilibrium determination of the quoted day-ahead price p in a competitive power market. The model assumes that an equilibrium price p results from the sole interaction of the N market agents described. Moreover, speculators are not allowed to enter the day-ahead market. At time T_0, agent i takes a position in the day-ahead market. He can either buy or sell at the currently quoted forward price p. A positive (resp. negative) traded quantity q_i is interpreted as a net purchase (resp. sale). Naturally, this quantity cannot exceed the maximal production level in the rest of the economy $C_i := \sum_{j \neq i} c_j$, nor can it be lower than the opposite of his own production capacity c_i in the period (that is the maximum the agent can sell):

$$-c_i \leq q_i \leq C_i := \sum_{j \neq i} c_j,$$

$$\text{Forward Unit Cost} = p.$$

Spot Trading

At time T_1, agent i can either produce or buy in the spot market any demanded power units exceeding the forward purchase q_i. Production is preferred to buying in the market only if the spot price $\tilde{p}^s$ exceeds the overall production costs per MWh, namely $p_i^v + P_i^f / c_i$. Once production capacity is exhausted, accessing the spot market represents the only possibility to procure the residual capacity required by customers.

$$\text{Spot Unit Cost} = \tilde{p}^s.$$

However, sometimes the forward purchased capacity q_i exceeds the actual demand $\tilde{Q}_i$ at time T_1. In this case, the excess can be sold in the market at a "back-supply price" p^b, which is supposed have an upper bound defined as the smallest variable unit cost across all production units:

$$\text{Back-Supply Price} = p^b \leq p_i^v, \quad \text{for all } i = 1, \dots, N.$$

18.1.3 The Bid Cost Function

The overall cost P_i borne by agent i for meeting a demand level $\tilde{Q}_i$ is given by the sum of production, forward and spot purchase costs. Here we have developed three cases:

Case (a). If demand $\tilde{Q}_i$ is met by the forward provision q_i, i.e., $\tilde{Q}_i \leq q_i$, agent i bears the fixed production cost P_i^f, plus the forward purchase cost pq_i paid out at time T_0, minus the revenue from reselling the residual capacity $q_i - \tilde{Q}_i$ at the back-supply price p^b:

$$P_i(p, q_i, \tilde{Q}_i, \tilde{p}^s) = P_i^f + pq_i - p^b(q_i - \tilde{Q}_i), \quad \text{if } \tilde{Q}_i \leq q_i.$$

Despite the dependence on $\tilde{p}^s$ being fictitious, it is convenient to include it for consistency with the two cases below.

Case (b). If demand $\tilde{Q}_i$ exceeds the forward provision q_i, but can be met by capacity from internal production ($q_i < \tilde{Q}_i \leq q_i + c_i$), the agent then incurs the fixed production cost P_i^f, plus the forward purchase cost pq_i, plus the cost of the least expensive choice of either producing the remaining electricity internally or buying it on the market.[1] The final figure represents the minimum of either the variable production cost p_i^v or the spot price $\tilde{p}^s$, which is multiplied by the quota of the demand that has not been met by the forward purchase, i.e., $\tilde{Q}_i - q_i$. The resulting cost function reads as follows:

$$P(p, q_i, \tilde{Q}_i, \tilde{p}^s) = P_i^f + pq_i + \min\{p_i^v, \tilde{p}^s\}(\tilde{Q}_i - q_i), \quad \text{if } q_i < \tilde{Q}_i \le q_i + c_i.$$

Case (c). If demand $\tilde{Q}_i$ exceeds the forward provision q_i plus the maximal capacity c_i from internal production ($q_i + c_i < \tilde{Q}_i$), the agent bears the fixed production cost P_i^f, plus the cost of forward purchase done at time T_0, plus the minimum value between the variable production cost p^v and the spot purchase cost $\tilde{p}^s$ over the entire generating capacity c_i, plus the spot purchase cost $\tilde{p}^s$ over the residual requested capacity $\tilde{Q}_i - (q_i + c_i)$:

$$P(p, q_i, \tilde{Q}_i, \tilde{p}^s) = P_i^f + pq_i + \min\{p_i^v, \tilde{p}^s\}c_i + \tilde{p}^s(\tilde{Q}_i - q_i - c_i), \quad \text{if } q_i + c_i < \tilde{Q}_i.$$

The resulting overall cost function reads as

$$P(p, q_i, \tilde{Q}_i, \tilde{p}^s) = P_i^f + pq_i - p^b(q_i - \tilde{Q}_i)_+ + \min\{p_i^v, \tilde{p}^s\} \min\{(\tilde{Q}_i - q_i)_+, c_i\}$$
$$+ \tilde{p}^s(\tilde{Q}_i - q_i - c_i)_+ \tag{18.2}$$

for all $\tilde{Q}_i \ge 0$. Here $(x)_+$ denotes the maximum between x and 0. Combining formulae (18.1) and (18.2), we obtain an expression for the Profit and Loss (P&L) function for agent i:

$$G_i(p, q_i, \tilde{Q}_i, \tilde{p}^s) := R(\tilde{Q}_i) - P_i(p, q_i, \tilde{Q}_i, \tilde{p}^s)$$
$$= p_i^R \tilde{Q}_i + p^b(q_i - \tilde{Q}_i)_+ - P_i^f - pq_i$$
$$- \min\{p_i^v, \tilde{p}^s\} \min\{(\tilde{Q}_i - q_i)_+, c_i\} - \tilde{p}^s(\tilde{Q}_i - q_i - c_i)_+.$$

Notice that this quantity depends on four key variables:

(1) the day-ahead price p prevailing in the market today;
(2) the position q_i taken by agent i in the day-ahead market today;
(3) the demand of electricity $\tilde{Q}_i$ tomorrow; and
(4) the spot price $\tilde{p}^s$ of electricity tomorrow.

The first of these quantities is determined at time T_0 by the joint interaction of the market participants. The second term is a control variable for agent i, while the third and the fourth variables are random as seen from current time T_0. Notice that the gain function G is a concave function with respect to the control variable q_i, a key property for the agent's optimization program. Table 18.1 reports the set of model variables and parameters.

---

[1] If $p^v < \tilde{p}^s$, the internal production is preferred to a spot market purchase; if $\tilde{p}^s < p^v$, a procurement cost in the open market is lower than internal production and the residual capacity is procured in the spot market. Notice that general costs are borne by the agent in any case.

Table 18.1. Model variables and parameters

Symbol	Quantity	Nature
q_i	Forward purchase by agent i	Control
c_i	Production capacity of agent i	Fixed
$\tilde{Q}_i$	Demand for agent i	Random
p_i^r	Retail unit price	Fixed
p	Day-ahead forward price	Random
$\tilde{p}^s$	Spot market price	Random
p^b	Back-supply price	Fixed
p_i^v	Variable unit cost of production	Fixed
P_i^f	General cost of production	Fixed

18.1.4 The Bid Strategy

We summarize the model as follows. The market behavior results from the interaction of N retailer-producers. Each agent is characterized by a production capacity c_i, a fixed operational cost P_i^f, a variable unit cost of production p_i^v, and a retail price p_i^r. The next day, each agent faces a random demand $\tilde{Q}_i$ to be met by filling any gap with respect to the energy capacity procured in the day-ahead market one day before. The agent then makes a decision about the proper mix of internal power generation, forward market intervention, and trading in the spot market based on the random behavior of demand $\tilde{Q}_i$, the spot price $\tilde{p}^s$, the currently quoted day-ahead price p, and the other parameters fixed by the market. The agent's goal is to select q_i in such a way that the expected utility from the resulting net gain is maximized under market clearing conditions. Let us put this informal reasoning into a precise statement.

A market equilibrium is defined as a selection of day-ahead market purchases $(q_i^*)_{1 \leq i \leq N}$ for all agents, and a unique forward price p^* such that each agent i maximizes his expected utility

$$\mathcal{U}_i(p, q_i) = \mathbb{E}[U_i(G_i(p, q_i, \cdot, \cdot))]$$

under market clearing conditions

$$q_i^* \in [-c_i, C_i],$$

$$0 = \sum_{i=1}^{N} q_i^*.$$

Hinz (2003) shows that an equilibrium day-ahead price exists in the interval between the back-supply price p^b and the highest possible spot price defined as the supremum $\sup \tilde{p}^s < \infty$, provided that

$$\mathbb{P}\big(\tilde{Q}_i \in [q, \hat{q}], \tilde{p}^s > p_i^v\big) > 0, \quad \text{for all } q \in (0, \hat{q}) \text{ and all } i = 1, \dots, N.$$

The uniqueness of this equilibrium point is an open question to date.

18.1.5 A Multi-Period Extension

A multi-period model is a sequence of single period models. The random input is a $(1 + N)$-valued stochastic process, $\tilde{\pi}(n) = (\tilde{p}^s(n), \widetilde{Q}_1(n), \ldots, \widetilde{Q}_N(n))_{n \geq 1}$, where $\tilde{p}^s(n)$ is the spot price of energy at time n, and $\widetilde{Q}_i(n)$ is the capacity demanded to agent i at the same time. Let $\mathcal{F}^{\tilde{\pi}} := (\mathcal{F}_n^{\tilde{\pi}})_{n \geq 1}$ be the informational flow generated by observing the process $\tilde{\pi}$. An equilibrium is defined as an $\mathcal{F}^{\tilde{\pi}}$-adapted process $(p^*(n), q_1^*(n), \ldots, q_N^*(n))_{n \geq 1}$ of a day-ahead price and individual net purchases in zero net supply (i.e., $0 = \sum_{i=1}^N q_i^*(n)$ for $n \geq 1$), such that each pair $(p^*, q_i^*(n))$ maximizes the ith agent expected utility at all times $n \geq 1$ given all market information available one time step before:

$$\left(p^*, q_i^*(n)\right) = \arg \max_{p \geq 0, q_i \in [-c_i, C_i]} \mathbb{E}[U_i(G_i(p, q_i, \cdot, \cdot))|\tilde{\pi}(n-1)],$$

for $i = 1, \ldots, N$, $n \geq 1$. Maximization is intended path by path, i.e., for each sample ω.

Finally, we wish to underline the most restrictive feature of this multiperiod extension. The above model assumes that agents take a one-day horizon in their decision making process. In particular, market prices are irrespective of expectations by market participants that go beyond the delivery day of the traded security. A truly multiperiod extension ought to take a comprehensive view about the impact of daily decisions over the entire performance in the examined period. This would eventually lead to a much more complex stochastic optimal control problem. The present setting can be seen as an approximate solution of this issue.

18.2 Solution Methodology

We consider a multiperiod model where all agents are essentially the same and both the spot price and demand are statistically independent. For the sake of clarity, we shall proceed through steps.

- **Model setting** All agents share common variable cost p^v, fixed cost P^f, production capacity c, utility function U, and conditional distribution of electricity demand. These assumptions translate into the following set of conditions:

$$p_i^v = p_j^v = p^v,$$
$$P_i^f = P_j^f = P^f,$$
$$c_i = c_j = c,$$
$$U_i(x) = U_j(x) = U(x),$$
$$\widetilde{Q}_i(n)|\tilde{\pi}(n-1) = \widetilde{Q}_j(n)|\tilde{\pi}(n-1) = \widetilde{Q}(n)|\tilde{\pi}(n-1),$$

for all $i, j = 1, \ldots, N$. Notice that demands $\widetilde{Q}_i(n)$ and $\widetilde{Q}_j(n)$ may assume different values although their conditional distributions $\widetilde{Q}_i(n)|\tilde{\pi}(n-1)$ and

$\widetilde{Q}_j(n)|\tilde{\pi}(n-1)$ match. As long as the symbol $\widetilde{Q}(n)|\tilde{\pi}(n-1)$ denotes the common distribution, computations involving this latter are written with respect to a fictitious random variable $\widetilde{Q}(n)$ which may be identified with any of the two demand variables $\widetilde{Q}_i(n)$ and $\widetilde{Q}_j(n)$. The spot price $\tilde{p}^s$ is statistically independent of the power demand $\widetilde{Q}_i$, it exceeds the variable unit cost p^v with positive probability, and it is also bounded from above by a constant with unit probability. Finally, the conditional distribution $\mathbb{P}(\widetilde{Q}(n) \geq c|\tilde{\pi}(n-1))$ is normal.

- **Equilibrium price dynamics** It can be proved that the equilibrium price reads as

$$p^*(n) = p^v - \mathbb{E}\big[\big(p^v - \tilde{p}^s(n+1)\big)_+\big]$$
$$+ \mathbb{E}\big[\big(\tilde{p}^s(n+1) - p^v\big)_+\big]\mathbb{P}\big(\widetilde{Q}(n+1) \geq c|\tilde{\pi}(n)\big), \quad (18.3)$$

where $(y)_+ := \max\{0, y\}$. These dynamics write as

$$\frac{p^*(n) - \tilde{a}_n}{\tilde{b}_n - \tilde{a}_n} = \mathbb{P}\big(\widetilde{Q}(n+1) \geq c|\tilde{\pi}(n)\big), \quad (18.4)$$

where the random sequences $\tilde{a}_n$ and $\tilde{b}_n$ are defined by

$$\tilde{a}_n = p^v - \mathbb{E}\big[\big(p^v - \tilde{p}^s(n+1)\big)_+\big], \quad (18.5)$$
$$\tilde{b}_n = p^v - \mathbb{E}\big[\big(p^v - \tilde{p}^s(n+1)\big)_+\big] + \mathbb{E}\big[\big(\tilde{p}^s(n+1) - p^v\big)_+\big]. \quad (18.6)$$

Since $\mathbb{P}$ is a probability, it takes value between 0 and 1, and expression (18.4) shows that the day-ahead price $p^*(n)$ ranges in the interval $[\tilde{a}_n, \tilde{b}_n]$.

- **Price distribution** It is supposed that we can monotonically transform the unconditional and conditional demand distributions in a way that the resulting pair is jointly Gaussian. Specifically, we assume there is a strictly increasing function f defined on the positive real axis such that the pair

$$\big(f\big(\widetilde{Q}(n+1)\big), \mathbb{E}\big[f\big(\widetilde{Q}(n+1)\big)|\tilde{\pi}(n)\big]\big)$$

is normally distributed. It can be shown that the *relative* equilibrium day-ahead price $\bar{p}^*(n)$ in equation (18.4) reads as

$$\bar{p}^*(n) := \frac{p^*(n) - \tilde{a}_n}{\tilde{b}_n - \tilde{a}_n} = \Phi_{0,1}(M(n)), \quad (18.7)$$

where Φ_{μ,σ^2} is the cumulative normal distribution with a mean μ and variance σ^2, and $M(n)$ is a Gaussian variable defined by

$$M(n) = \frac{\mathbb{E}[f(\widetilde{Q}(n+1))|\tilde{\pi}(n)] - f(c)}{\sqrt{\text{Var}(\mathbb{E}[f(\widetilde{Q}(n+1))|\tilde{\pi}(n)] - f(\widetilde{Q}(n+1)))}}. \quad (18.8)$$

This quantity describes the power demand dynamics as rescaled according to the function f. Simple computations lead to the relative price distribution

$$f_{\tilde{p}^*(n)}(x) = \frac{\phi_{\mu_n,\sigma_n^2}(\Phi_{0,1}^{-1}(x))}{\phi_{0,1}(\Phi_{0,1}^{-1}(x))} \mathbf{1}_{(0,1)}(x), \qquad (18.9)$$

where ϕ_{μ,σ^2} is the normal probability density function with a mean μ and variance σ^2, $\mu_n = \mathbb{E}(M(n))$, and $\sigma_n^2 = \text{Var}(M(n))$.

Combining expressions (18.7) and (18.9), we see that the probability density function of the equilibrium day-ahead price p^* is

$$f_{p^*(n)}(x) = \frac{f_{\tilde{p}^*(n)}((x - \tilde{a}_n)/(\tilde{b}_n - \tilde{a}_n))}{\tilde{b}_n - \tilde{a}_n}, \qquad \text{for all } x \in [\tilde{a}_n, \tilde{b}_n].$$

The limit distribution as the equilibrium price approaches the boundary of the domain $[\tilde{a}_n, \tilde{b}_n]$, reads as

$$\lim_{x \to \tilde{a}_n^+} f_{p^*(n)}(x) = \lim_{x \to \tilde{b}_n^-} f_{p^*(n)}(x) = \begin{cases} \infty & \text{if } \sigma_n^2 > 1, \\ 0 & \text{if } \sigma_n^2 < 1. \end{cases}$$

This shows that the asymptotic regime switching is driven by the demand predictability.

The last ingredient for obtaining an explicit price distribution is a probability distribution for the electricity spot price $\tilde{p}^s(n+1)$.

18.3 Implementation and Experimental Results

To implement the price model (18.3), we need explicit dynamics for the spot price $\tilde{p}^s$ and the demand processes $\tilde{Q}_i$.

In order to have realistic spot price dynamics, we consider a simple mean reverting model

$$d\tilde{p}^s = \frac{d\mu(t)}{dt} + \theta(\mu(t) - \tilde{p}^s)\,dt + \sigma_p\,dW(t), \qquad (18.10)$$

where W is a standard Brownian motion, and the trend function is given by

$$\mu(t) = q - \cos(\omega t).$$

Here $q = 2.3$ is a constant value and $\omega = 2\pi/250$ is the trend frequency. In our experiment, the initial spot price was set to $\tilde{p}^s(0) = 0.8$, and the instantaneous volatility is $\sigma_p = 0.5$. The spot price dynamics (18.10) can be solved in closed form, and boundaries $\tilde{a}_n$ and $\tilde{b}_n$ can be computed through formulae (18.5) and (18.6). We consider a demand process $\tilde{Q}_i$ to be driven by the joint effect of a common state variable x, following a simple AR(1) process

$$x(n+1) = Ax(n) + Bv(n+1),$$

and a random walk $\tilde{w}_i$ which is idiosyncratic to the ith agent:

$$\tilde{Q}_i(n+1) = E \exp\bigl(Cx(n) + D\tilde{w}_i(n+1)\bigr).$$

In these formulae, both v and the w_i's are independent standard Gaussian variables, constants A, B, C, D, and E are positive, and $|A| < 1$. The latter condition ensures that the variance of the process is well defined as a positive real number. Letting $f = \ln$, we have that $f(\tilde{Q}_i(n+1)) = \ln E + Cx(n) + Dw_i(n+1)$ and $\mathbb{E}[f(\tilde{Q}_i(n+1))|\tilde{\pi}(n)] = \ln E + Cx(n)$.

Consequently, vector $(f(\tilde{Q}(n+1)), \mathbb{E}[f(\tilde{Q}(n+1))|\pi(n)])$ is Gaussian for any $\tilde{Q} = \tilde{Q}_i$. Plugging these expressions into formula (18.8), we get to

$$M(n) = \frac{\ln E + Cx(n) - \ln c}{\sqrt{\operatorname{Var}(-Dw_i(n+1))}} = \frac{Cx(n) + \ln(E/c)}{D}.$$

Process $M(n)$ can be interpreted as power demand in log scale. This process is stationary, with moments

$$\mu = \mathbb{E}[M(n)] = \frac{\log(E/c)}{D},$$

$$\sigma^2 = \operatorname{Var}(M(n)) = \frac{C^2 B^2}{D^2(1 - A^2)}.$$

Constants C and D drive the demand predictability: the greater (resp. lower) the value of D (resp. E), the more precise the demand forecast.

We performed numerical experiments under low demand predictability (i.e., $\sigma^2 > 1$) and high demand predictability (i.e., $\sigma^2 < 1$). Let constants $C = B = E = 1$ be fixed throughout, and set $c = 1.5 > E$, meaning that a large part of demand is met with in-house production.

In the low predictability case, we fixed $D = 2.5$, and obtained $\sigma^2 \simeq 1.64$. Figure 18.2 exhibits a sample log-scaled demand path $(M(n), n = 1, \ldots, 500)$, and the corresponding relative price path $(\Phi_{0,1}(M(n)), n = 1, \ldots, 500)$.

We see that prices vary rather wildly, and sometimes agents accept to pay prices near to the upper bound $\tilde{b}_n$. Indeed, low predictability can be interpreted as an expectation of power shortage. Notice that the relative price path displays the same properties as those exhibited by the log-scaled demand path. Figure 18.3 plots the (stationary) density function $g_n(x)$ in equation (18.9). In the high predictability case, we set $D = 5$ and got $\sigma^2 \simeq 0.41$. Figure 18.4 reports a sample log-scaled demand path $(M(n), n = 1, \ldots, 500)$, and the corresponding relative price path $(\Phi_{0,1}(M(n)), n = 1, \ldots, 500)$.

In contrast with the previous case, here there is an increased tendency for prices to jump upward. Again, both demand and price fluctuations are rather similar in their behavior. Figure 18.5 shows the density function $g_n(x)$ in this case. Finally, we sampled the spot price process according to equation (18.10). Combining paths $\tilde{p}^s(n), \tilde{a}(n), \tilde{b}(n), \Phi_{0,1}(M(n))$ in formula (18.3) we derive a forward price trajectory $(\tilde{p}^*(n), n = 1, \ldots, 500)$. Figure 18.6 exhibits these samples for the high and low predictable cases.

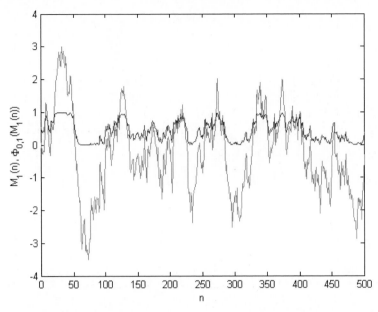

Fig. 18.2. Logarithmic demand path (red line) and the corresponding relative price path (blue line) under low predictability.

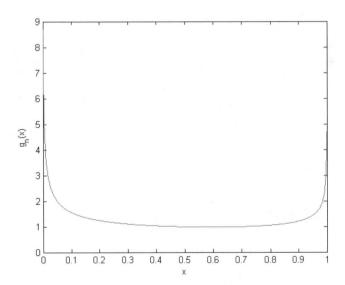

Fig. 18.3. Stationary density function.

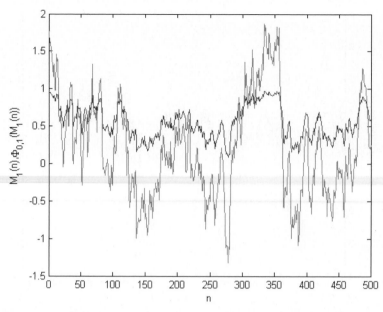

Fig. 18.4. Logarithmic demand path (red line) and the corresponding relative price path (blue line) under high predictability.

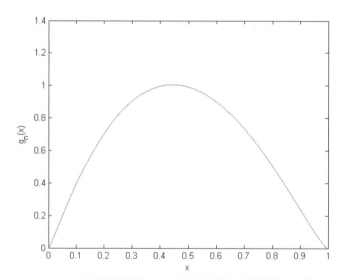

Fig. 18.5. Density function of g_n.

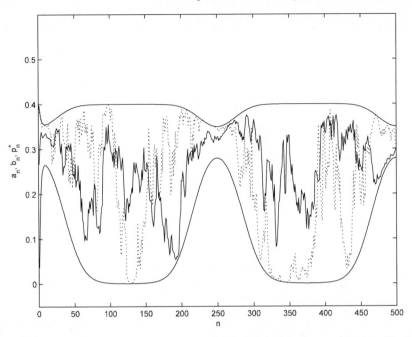

Fig. 18.6. Sample spot price dynamics under high (solid line) and low (dotted line) predictability. Upper and lower boundaries are depicted as smoothed solid lines.

19

A Sparkling Option[*]

Key words: energy prices, simulation, jump-diffusions, real options

19.1 Problem Statement

As a consequence of deregulation in the electricity market, a large number of utilities and producers in many European countries have become increasingly exposed to the risks of volatile energy prices. In this context, we consider the problem of dispatching a power generation unit on the basis of prevailing electricity and gas spot prices.

When quoted in open markets, electricity prices are often set hourly with price differences that can reach up to 40 times the opening level. These extreme fluctuations reflect the increasing marginal costs of production following demand peaks concentrated at some point in the day. Power markets are, in fact, day-ahead markets where producers and consumers place, respectively, their bids and orders. Equilibrium comes from the intersection between the electricity supply and demand. However, for different reasons such as seasonality, unexpected outages, problems in transmission and non-storability, there is a large probability of a short term disequilibrium between demand and supply sides. The demand curve is quite inelastic in the short-term, and shocks on the demand side (e.g., California during summer 2000, NordPool during the Lillehammer Winter Olympics, and Italy during summer 2003) or in the supply side can lead to huge upward price movements.

Nonstorability of electricity units does not allow a straightforward application of arbitrage pricing theory commonly used in stock and bond markets. Moreover, the characteristics of mean reversion, seasonality, and presence of spikes that commonly affect electricity and gas prices have forced researchers in the field to consider new

---

* with Mariano Biondelli, Enrico Michelotti and Marco Tarenghi.

stochastic processes for the purpose of describing the random dynamics of energy prices and evaluating energy derivatives.

This case-study describes and implements a method to value a power generation unit using a real options approach and the notion of a spark spread. This methodology has been adopted by Gardner and Zhuang (2000), Hsu (1998) and Deng, Johnson and Sogomonian (1999), among others. Benth and Saltyte-Benth (2006) study the approximation of spark spread option prices. The spark spread has developed in energy markets as an intermarket spread between electricity and natural gas. This spread is of interest because it determines the economic value of generation assets that are used to transform gas into electricity. The spark spread measures the difference between the costs of operating a gas-powered generation unit, given by the natural gas price, and the revenues from selling power at the market price. In day-to-day running, the plant operator generally consumes a particular gas unit only if the electricity spot price is greater than the cost of generating that unit. If the generation profit is negative, it would be unreasonable to burn a valuable commodity such as gas to obtain a low-valued product such as electricity. One would instead sell gas in the market, buy power, and stop running the plant. As it is clearly explained in Fiorenzani (2006c), the flexibility of turning the plant on and off, based on market prices, represents a real option for the asset owner.

The amount of natural gas that a gas-powered plant requires to generate a given amount of electricity depends on the asset's efficiency. This figure is represented by a heat rate Hr which is defined as the number of British Thermal Units (Btu) according to the Anglo-Saxon measurement standard (or cubic meters m^3 according to the decimal metric measurement system) required to produce one megawatthour MWh of electricity. If the heat rate is measured in Btu/kWh then a normal heat rate ranges between 8,000 and 12,000. The lower the heat rate, the more efficient the power plant.

Nowadays, the best heat rates are around 6,000. This performance can be achieved by means of Combined Cycle Gas Turbine (CCGT) plants or with cogeneration plants. CCGTs represent a relatively new standard for power plants. Turbines exhibit a series of major advantages with respect to previous plants, and are gradually replacing oil-based plants. Their primary feature is a high return per fuel unit (about 35%), although it can vary depending on the size of the generating unit. For instance, a unit dispatching 120 MWh has an efficiency rate around 32%, whereas a 250 MWh unit delivers a 38% return. Second, compared to the traditional oil burning plants, these new plants reduce the environmental pollution quite significantly. Finally, CCGT units are more effective in terms of ramp-up times (about 3 hours), and consequently can better manage peaks and congestion problems. Indeed, CCGT plants are rather insensitive to unexpected changes in the demand side and can be used with no interruption.

From the heat rate, we can infer whether a power plant is a baseload or a peakload unit, namely one that is exclusively employed to manage periods of high demand. A plant with a heat rate greater than 10 MMBtu/MWh is commonly considered a peakload unit. Moreover, a plant operator that uses natural gas should run the generation unit if it is worthwhile to do so, meaning that:

Power Spot Price > Operating Hr × Natural Gas Spot Price.

The spark spread is then given by:

Spark Spread = Power Spot Price − Operating Hr × Natural Gas Spot Price.

When the spark spread is positive, the plant is run, and an operating profit is earned. When the spark spread is negative, the plant is shut down. The plant owner then makes a profit by selling natural gas and buying power in the market (instead of burning valuable gas to produce low priced electricity). We model the profit of a CCGT as an option with a payoff

$$\pi = \max\{E - \text{Hr} \times G, 0\}, \tag{19.1}$$

where π denotes the profit per MWh, G is the spot price of gas, E represents the spot price of electricity, and Hr is the heat rate expressed in gas units per MWh.

We now introduce the concept of *Market Heat Rate* (MHr). This quantity is given by the ratio between the quoted prices of electricity and gas, namely:

$$\text{MHr} = \frac{E}{G}.$$

Equation (19.1) can be recast as follows:

$$\pi = \max\left\{\left(\frac{E}{G} - \text{Hr}\right) \times G, 0\right\} = \max\{(\text{MHr} - \text{Hr}) \times G, 0\}. \tag{19.2}$$

This expression sheds light on the financial interpretation of a plant as an option: the market heat rate represents the underlying asset and the operating heat rate is the strike price. When the power plant is in-the-money, the market heat rate is above the heat rate of the plant. An operator can take advantage of this situation and exercise the option by selling power and buying gas at the market spot price. Conversely, if the market heat rate goes below the operating heat rate, the operator shuts the plant down without incurring into any loss.

The analogy between a power plant and a financial option turns the problem of evaluating a power plant into the one of pricing a strip of spark spread options (Fiorenzani (2006a)). In other words, the gas power plant is seen as a portfolio of call options written on gas and electricity spot prices. Consequently, the power plant value can be computed as the expected discounted payoff of spark spread options. This approach ought to be preferred to the standard Net Present Value NPV, which systematically underestimates the value of the plant by skipping the underlying option.[1]

The remainder of the chapter is structured as follows. Section 19.2 presents the structure of the pricing procedure. In particular, we specify stochastic processes describing the random evolution of natural gas and electricity spot prices. Section 19.3 details the calibration procedure to fit models to market data. Section 19.4 evaluates a plant using Monte Carlo simulation over a ten-year horizon.

---

[1] We assume that the market is complete and refer to Fiorenzani (2006b) for a relaxation of this hypothesis in the present valuation context.

19.2 Model and Solution Methodology

The arbitrage valuation of spread options requires one to specify models for the random behavior of the underlying assets. In the case of a spark spread, these are the price processes for natural gas and electricity, i.e., the input and, respectively, the output of the plant.

Figure 19.1 offers a seven-year gas price path in the natural logarithmic scale as quoted in the New York Gate market at Nymex (source: Bloomberg). The period ranges from January 1996 to December 2002. Following Schwartz (1997), we model the logarithmic price dynamics using a mean reverting process with linear trend and constant volatility:

$$d(\ln G(t)) = \theta\big[\mu^g(t) - \ln G(t)\big]\,dt + \sigma\,dW^g(t). \tag{19.3}$$

Here $G(t)$ is the time t gas price, parameter θ measures the speed of reversion per unit of deviation from the linear trend $\mu^g(t) = \alpha + \beta t$, and σ represents the volatility of the instantaneous log-price variation. As it is customary with this kind of model, W^g is a standard Brownian motion driving unexpected price shocks.

The model can be estimated by maximum likelihood as in Greene (2002). The log-likelihood function can be computed by discretizing the process (19.3) through the Euler method and solving for the log-price:

$$\ln G(t + \Delta t) = \ln G(t) + \theta\big(\mu^g(t) - \ln G(t)\big)\Delta t + \sigma\sqrt{\Delta t}\,\varepsilon(t).$$

Here $\varepsilon(t)$ is a random draw from a standard normal distribution $\mathcal{N}(0, 1)$. The resulting variable has a normal conditional distribution given by

$$\ln G(t + \Delta t)|G(t) \curvearrowright \mathcal{N}(m(t), v(t)),$$

where

$$m(t) = \ln G(t) + \theta\big(\mu^g(t) - \ln G(t)\big)\Delta t,$$
$$v(t) = \sigma^2 \Delta t.$$

By setting $\Delta t = 1/250$, we obtain the following expression for the log-likelihood function with daily observations

$$\ln \mathcal{L} = -\frac{1}{2}\sum_{i=1}^{n}\left[\frac{(\ln G(t_i) - m(t_i))^2}{v(t_i)}\right] - \frac{1}{2}\ln(v(t_i)) - \frac{1}{2}\ln 2\pi. \tag{19.4}$$

Maximizing this quantity with respect to parameters α, β, θ and σ, we get to a statistically estimated model for the gas price dynamics.

Turning to electricity prices, any good model should reproduce three empirical features displayed by power price dynamics, namely:

- A mean reversion towards a seasonal trend;
- Unexpected price oscillations;
- Recurrent spikes.

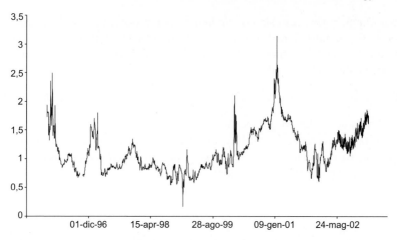

Fig. 19.1. The log-price of natural gas quoted at the NY Gate market during the period 1996–2002.

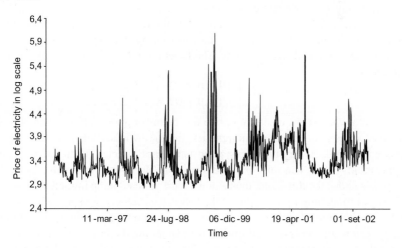

Fig. 19.2. Logarithmic spot prices of electricity recorded at the PJM power market during the period 1996–2002.

These latter are defined as large upward moves followed by a sudden recovery back to a normal price level. Figure 19.2 shows the historical path followed by electricity spot prices (in a natural logarithmic scale) between 1996 and 2002 in the PJM power market. All spikes cluster around the warm season. We remark that seasonality has been unaffected by the turbulent behavior displayed by the PJM market in more recent years. The price path for the PJM is quite similar to those observable at other US power markets.

The above mentioned features can be accommodated by using the spot price model introduced in Roncoroni (2002) and furthermore developed by Geman and Roncoroni (2006).[2] This is a Markov jump-diffusion process that has been proven to properly fit several US market data sets in terms of path properties and descriptive statistics. The model assumes that the natural logarithm of power price dynamics are described by a stochastic differential equation

$$dE(t) = \left[\mu^e(t)' + \vartheta_1\left(\mu^e(t) - E(t^-)\right)\right] dt + \sigma^e \, dW^e(t) + h(E(t^-)) \, dJ(t), \quad (19.5)$$

where μ^e denotes the price trend function, ϑ_1 is the mean reversion speed, and σ^e is a constant instantaneous volatility. Brownian motion W^e is assumed to be correlated to the Brownian motion W^g driving gas price dynamics according to a constant coefficient ρ, i.e., $\mathbb{E}(dW^e \, dW^g) = \rho \, dt$.

The price trend is represented by a mixed linear–sinusoidal function

$$\mu^e(t) = \mu(t; \alpha, \beta, \gamma, \varepsilon, \delta, \zeta) = \alpha + \beta t + \gamma \cos(\varepsilon + 2\pi t) + \delta \cos(\zeta + 4\pi t),$$

where α represents a fixed cost linked to the production of electricity, β takes into account a linear trend in time, and the other two terms indicate respectively an annual and a semi-annual seasonality component.

The last term in equation (19.5) represents the discontinuous part of the model featuring price spikes. This effect is characterized by three quantities defining occurrence, direction, and size of jumps. The function h defines a switch for the jump direction:

$$h(E(t^-)) = \begin{cases} +1 & \text{if } E(t) \leq \mu^e(t) + \Delta, \\ -1 & \text{if } E(t) > \mu^e(t) + \Delta, \end{cases}$$

according to whether the electricity price lies below or above a threshold defined as a constant spread Δ over the mean trend μ^e. In other words, if the price is below the threshold $\mu^e(t) + \Delta$, any upcoming jump is upward directed. If the price is above the threshold, jumps can only move downward. Moreover, we assume that when the price is above the threshold, then a downward directed jump immediately occurs. This way of modelling jumps is due to the fact that extremely high prices are singular and the market immediately reverts back to reasonable values. A process exibiting a strong mean reversion is not appropriate for explaining the fast negative movements from spiky to normal regime. The process J is a compound Poisson, defined as

$$J(t) = \sum_{i=1}^{N(t)} J_i.$$

The counting process N enumerates jumps occurring over any time interval $[0, T]$. An intensity function

$$\iota(t) = \vartheta_2 \times \left[\frac{2}{1 + |\sin[\pi(t - \tau)/k]|} - 1\right]$$

steers the jump occurrence once a year ($k = 1$) with a peak during the summertime ($\tau = 1/2$). The expected maximum number of jumps per year is represented by the

---

[2] See also Benth et al. (2007).

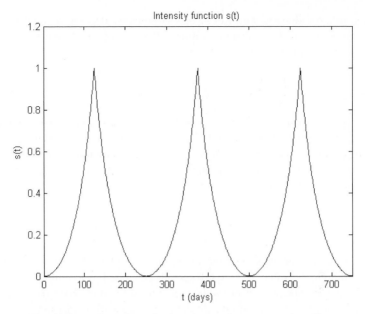

Fig. 19.3. The jump intensity function used to generate the spikes.

constant parameter ϑ_2, which is to be estimated. Figure 19.3 reports the graph of the selected intensity function. Jump sizes are modelled by a sequence of independent and identically distributed truncated exponential variables with density

$$p(J_i \in \mathrm{d}x; \vartheta_3, \psi) = \frac{\vartheta_3 e^{-\vartheta_3 x}}{1 - e^{-\vartheta_3 \psi}}$$

for $0 \leq x \leq \psi$. Here ϑ_3 is a constant parameter to be calibrated and ψ represents the maximum size of absolute price changes in the natural logarithmic scale.

In order to reflect the seasonal pattern followed by unexpected small market shocks, we introduce a slight modification to the model described above, and replace the constant volatility with the time dependent function.[3]

$$\sigma^2(t) = \sigma_0^2 + a \cos^2(\pi t + b). \tag{19.6}$$

Futhermore, a realistic evaluation of the plant should take into account the fact that in general the gas and electricity processes are correlated. We estimate the correlation between the standardized residuals in the two processes as obtained by filtering the trend out of the series of the log-prices (series without jumps). To separate the jump component from the diffusion part of the process, we introduce a threshold parameter Γ: a price variation larger than Γ is due to a jump, whereas a smaller variation is explained by the diffusion part of the electricity price process. From this analysis, we obtain a correlation estimate of $\rho = 0.1912$. Then, the two processes are

---

[3] Cartea and Figueroa (2005) provide an alternative method to estimate a time-varying volatility.

generated in the following way. Let ε_1 and ε_2 be two independent standard normal random variables. The Brownian innovations dW^e and dW^g in the electricity and gas processes are then obtained as:

$$\begin{cases} \sigma_1\, dW^e = \sigma_1\varepsilon_1\sqrt{dt}, \\ \sigma_2\, dW^g = \sigma_2\big(\rho\varepsilon_1 + \sqrt{1-\rho^2}\varepsilon_2\big)\sqrt{dt}, \end{cases} \tag{19.7}$$

so that $\sigma_1\, dW^e \sim \mathcal{N}(0, \sigma_1^2\, dt)$, $\sigma_2\, dW^g \sim \mathcal{N}(0, \sigma_2^2\, dt)$ and $\mathrm{Cor}(\sigma_1\, dW^e, \sigma_2\, dW^g) = \rho\sigma_1\sigma_2\, dt$, as is required. Equations (19.7) are the Cholesky decomposition generating correlated normal variates using two independent standard normal variables.

In summary, we can generate simulated paths for gas and electricity prices and then evaluate a power plant by formula:

$$\text{Power Unit Value} = \sum_{i=1}^{n} \mathbb{E}_0\big(e^{-r_i t_i} \pi_i\big), \tag{19.8}$$

where t_i runs over all days the plant is on and π_i is the profit (that is positive or zero according to the value of the spark spread) generated by a running plant at time t_i. In the equation above, r_i denotes the continuously compounded risk-free rate of interest for maturity t_i. For the sake of simplicity, we assume that interest rates are statistically independent from the process driving energy prices and write:

$$\text{Power Unit Value} = \sum_{i=1}^{n} \mathbb{E}_0\big(e^{-r_i t_i}\big)\mathbb{E}_0(\pi_i) = \sum_{i=1}^{n} P(0, t_i)\mathbb{E}_0(\pi_i), \tag{19.9}$$

where $P(0, t_i)$ denotes the current discount factor for a maturity t_i as can be obtained from the US interest rate curve. Quotations for certain maturities up to 10 years are recorded on the Bloomberg system, while the others can be inferred by interpolation between the observed numbers. Following a suggestion provided to us by a practitioner working in the energy sector, we adopt a conservative view and add a 4% premium to this structure.

In order to produce an even more realistic valuation of the plant, in a second simulation exercise we have taken into account the hourly profile in electricity spot price dynamics and the costs for turning the plant on and off. We now discuss about these aspects and the way we have included them in a simulation exercise.

The day-ahead market provides a schedule for electricity prices for each hour of the following day. We have built a hypothetical schedule curve as follows. First, we consider hourly prices over one year. Next, for each hour in the day, we compute the average price over the whole data set. Finally, we normalize these average prices by dividing each average value by the maximum price. The resulting number that corresponds to an hour i is denoted by p_i. Figure 19.4 displays the normalized averages for all 24 hours. We then simulate prices on a hourly basis as

$$\tilde{p}_i = p_i + \sigma_i\epsilon_i,$$

where σ_i is the standard deviation corresponding to the ith hour and ϵ_i denotes a standard normal sample. The hourly price schedule is obtained as

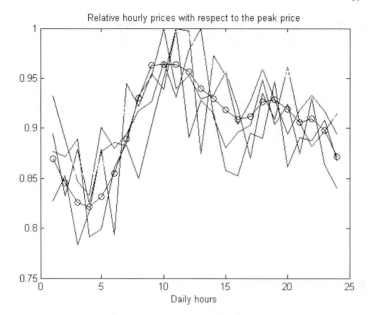

Fig. 19.4. Average normalized daily prices.

$$\hat{p}_i = \frac{\tilde{p}_i}{\max_{i=1,\dots,24}(\tilde{p}_i)} p_{\text{peak}},$$

for each hour i, where p_{peak} is a simulated peak price.

A further aspect in the evaluation process is related to the activation costs of the plant. The ramp-up time defines the required period for a power plant to producing electricity at its maximum level of efficiency. We assume that during the ramp-up time, which takes approximately one to two hours on average, the plant consumes a half of the gas required under normal conditions. We also include a fixed cost that does not affect the dispatch schedule, but enters into the valuation process of the plant.

The joint consideration of a spark spread and the activation/disconnection costs defines the opportunity to turn a plant on. Consider the following two cases:

(a) Turning the plant on is not a viable option in presence of an excessively low spark spread;
(b) A loss resulting from a negative spread can be smaller than the costs due to turning the plant off.

In the first case, the power plant should be left off even if the spark spread is positive. In the second case, the plant should be run even if the spread is negative. The problem of activating/disconnecting a plant should be tackled using a stochastic optimization model. To facilitate the implementation, we decided to simplify the problem by dividing the time period in alternating subperiods of positive and negative spark-spreads. For each subperiod, we compare the loss due to a negative spread to the resuming costs and select the lowest between the two. This number is further compared to the revenues that would be obtained by turning the turbine on whenever the spread is positive during the following period: if revenues are smaller, the plant is switched off; if the revenues are greater, the plant is turned on. Furthermore, if the activating costs are greater than the loss due to a negative spread, then the plant is kept running since one would prefer losing money by producing and selling electricity rather than stopping the production and restarting it later. This algorithm needs not be optimal, yet it provides a better solution than the straightforward strategy consisting of considering a spark spread option for each period separately.

19.3 Implementation and Algorithm

We calibrated the model on a data set including 1,750 daily observations obtained from the Bloomberg system. The period spans seven years, from January 1996 to December 2002. Data refer to electricity prices quoted at the PJM (Pennsylvania–New Jersey–Maryland) market and natural gas prices exhibited at the New York Gate, which is the largest natural gas utility for the area of New York City.

Calibration involves the:

1. Estimation of parameters in the gas price process by maximum likelihood (eq. (19.4));
2. Estimation of the electricity price process using a method proposed in Roncoroni (2002) and consisting of:
   (a) estimating the trend by Ordinary Least Squares (OLS);
   (b) disentangling normal and jump regimes in order to estimate Γ, ψ and ϑ_3;
   (c) estimating the mean reversion force ϑ_1 and volatility related parameters σ_0^2, a, b by maximum likelihood.

Estimates of the gas price process are reported in Table 19.1. The maximization of the likehood function has been performed using the Excel solver. As for the electricity price process, we first estimated the linear trend by OLS using EViews. This estimation is relative to the original data bounded from above by the 90-percentile of the sample price distribution. We filtered out of the series all the data exceeding this threshold, then pooled the residual data into a single 1-year period and averaged the seven numbers corresponding to each day. Finally, we estimated the periodic part of the trend using a Nonlinear Least Square procedure on this filtered data set. The trend component is shown in Fig. 19.5, together with the observed price path and

Table 19.1. Parameters for the natural gas price process

α	Average level	0.939152
β	Linear growth	0.063992
θ	Mean reversion force	4.4524
σ^2	Instantaneous variance	1.3150

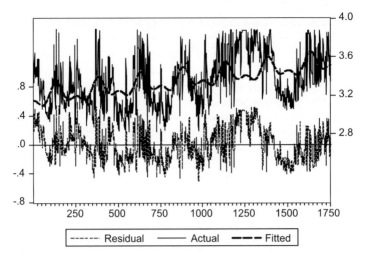

Fig. 19.5. Estimated electricity price trend.

the corresponding residuals. This graph clearly shows a small positive trend and a double periodic component. Moreover, there are two maxima during the year, one in summertime and a smaller one occurring in wintertime.

To disentangle normal and jump regimes, we separated each price variation into a continuous and a jump component as follows: all price variations greater than a threshold Γ are considered jumps; the residual variations are intended as stemming from the continuous part of the process. To select a proper Γ, we calibrated the model assuming different values for Γ (chosen as a percentile of the sample distribution of the daily price variations) and take the value that provides the best result in terms of moment matching and price properties of the resulting dynamics. In the present context, the 97-percentile is the optimal choice. This number corresponds to 7.57 jumps per year on average, which is a realistic figure in our view. We thus filtered the whole data set according to this choice and estimated the parameters related to the jump component. We detected the maximum jump size ψ and the average jump size and then used these two values to estimate the value of the parameter ϑ_3. Since the jump sizes are assumed to be i.i.d. random variates with a common distribution given by a truncated exponential, the expected value of such a distribution is

$$\mathbb{E}(J_i) = \frac{1}{1 - e^{-\vartheta_3 \psi}} \left[\frac{1}{\vartheta_3} - \left(\psi + \frac{1}{\vartheta_3} \right) e^{-\vartheta_3 \psi} \right]. \qquad (19.10)$$

This formula defines ϑ_3 implicitly. We used the Excel Solver to estimate ϑ_3 using the sample mean, which, for this distribution, coincides with the Maximum Likelihood estimate.

To find ϑ_2, we first computed the interarrival times between two jumps, then we found a mean interarrival time weighed according to the normalized intensity function $2(1 + |\sin[\pi(t - \tau)/k]|)^{-1} - 1$, and finally estimated ϑ_2 as the inverse of the resulting average interarrival time. The mean reversion force ϑ_1 and parameters

Table 19.2. Estimated parameters for the electricity price process

Γ	Jump size threshold	0.563521
α	Average level	3.167108
β	Linear growth	0.053377
γ	Yearly trend	−0.084373
ε	Yearly shift	−0.374401
δ	6-months trend	0.055531
ς	6-months shift	−0.390401
Δ	Jump-regime level	1.65
ψ	Maximum jump size	1.739186
K	Jump periodicity	1
τ	Jump time shift	0.5
σ_0^2	Constant variance	9.369858
a	Size of variance oscillations	−7.069798
b	Yearly variance shift	−0.278851
ϑ_1	Mean reversion force	32.04205
ϑ_2	Mean expected number of jumps	8.450
ϑ_3	1/average observed jump size	0.094438

σ_0^2, a, and b can be estimated using maximum likelihood on the data filtered by the 97-percentile. The conditional transition density function of the electricity price increment is normal, with mean and variance given by equation (19.6). We noticed that the phase in the cosine for the variance is fairly similar to the phase for the annual seasonality in the trend. This fact indicates that in periods of high electricity demand, daily price variations are larger than in quiet periods. Finally, in order to estimate Δ, we observe that higher values correspond to higher electricity price levels during high demand periods. Conversely, the smaller the value of Δ, the sooner a downward jump makes the price revert back to normal level. In order to reproduce these effects, we set Δ equal to one half of the range spanned by the log-prices data set. The estimates of the electricity process are reported in Table 19.2. Once the parameter set has been estimated, it is possible to perform the numerical simulation evaluating the power plant. We wrote a MATLAB® code which simulates a random path of the processes described in (19.3) and (19.5).

The simulation consists of the following steps:

1. Simulate a random path for electricity and gas prices and then, for a given heat rate, compute the spark spread for each day in the simulation period (the MATLAB® codes involved are el_path.m and gas_path.m).
2. Discount the daily power plant pay-off using an appropriate discount factor, eventually corrected by a market risk premium (MATLAB® function simul_ spark.m).
3. Repeat steps 1 and 2 about 10,000 times and compute mean value and the standard error of the simulated figures (MATLAB® code is plant_eval.m).

If instead we are interested in a valuation on a hourly basis, we may proceed as follows:

1. Steps 1 and 2 as above.
2. Determine the random hourly curve for electricity price by making a perturbation of the average hourly curve derived from market data (using the script shock.m).
3. Compute the hourly spark spread using the script simul_spark_hour.m (for the sake of simplicity, we assumed that the gas price remains constant during the day).
4. Detect the hours the plant must be turned on (or kept active) and those it must be switched off (or kept inactive) in order to maximize revenues (or minimize costs); to this aim, we have taken into account the cost structure of the plant (using the code activation.m).
5. Discount the positive (and sometimes negative) values of the spark spread using a discount factor corrected for the risk premium in the energy sector (running the script plant_eval_hour.m).

The remaining codes are called by the main codes cited above and need to be included for correct initialization and evaluation processes.

To simulate the jump component in (19.5), we sampled jump times from an exponential distribution with parameter ϑ_2; for each jump time candidate τ_k, we then applied an acceptance–rejection scheme consisting of drawing a uniformly distributed sample U_k and accepting τ_k provided that $U_k \leq \iota(\tau_k)$.

Figure 19.6 displays a sample trajectory for the gas price over a 7-year period in the natural log-scale. Figure 19.7 shows a sample path for the electricity price over the same period. Notice that both price paths present qualitative features similar to those exhibited by the empirical processes in the market.

19.4 Results and Comments

The expected value in (19.9) has been estimated by Monte Carlo over 10,000 simulations. Table 19.3 reports the value of the plant using different heat rates, namely 6, 8, 10, and 12. As one would expect, a soaring heat rate is followed by a falling proportion of the time spent under activity. The plant thus becomes peakload. The results we obtained are consistent with our expectations. In particular, by reducing plant efficiency, the value of the plant decreases. With a heat rate equal to 12, the unit becomes active only for one third of the time. However, in reality even a turbine with a heat rate equal to 6 is not activated for 82% of the time.

In order to produce an even more realistic valuation of the plant, a second simulation exercise takes into account the hourly profile in electricity spot prices and the costs of activating and disconnecting the power plant.

Table 19.4 reports the values of two plants with different heat rates and ramp-up times corresponding to one and two hours, respectively. For instance, if the heat rate equals 8, the plant value goes from $759,768 in the case of no ramp-up time

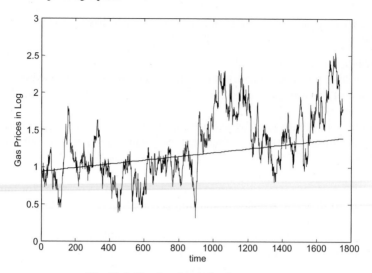

Fig. 19.6. Simulated path for the gas price.

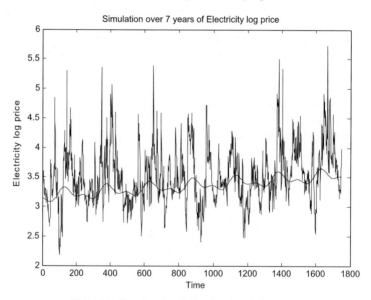

Fig. 19.7. Simulated path for the electricity price.

to $454,490 under the assumption of a 2-hour ramp-up time, that is to say, an adjustment of 40% of the initial value. Moreover, considering these costs results in an activation frequency more in line with actual percentages. Standard errors for the estimated plant values are indicated within round brackets.

The valuation procedure discussed above is computationally intensive. After devoting a particular attention to the matrix-based programming framework in

Table 19.3. Value of a power plant under varying heat rates

Heat rate (MWh/MMBtu)	Plant value ($/MWh)	Standard error	Frequency of activation (%)
6	1,064,232	11.06%	81.98
8	759,768	14.33%	63.39
10	545,664	16.84%	46.32
12	399,024	19.88%	33.19

Table 19.4. Values of a power plant under varying heat rates and ramp-up times

Heat rate (MWh/MMBtu)	Ramp-up time (h)	Plant value ($)	Frequency of activation (%)
8	0	759,768 (14.33%)	63.39
8	1	479,680 (21.25%)	54.46
8	2	454,490 (23.67%)	54.59
10	0	545,664 (16.84%)	46.32
10	1	295,330 (29.83%)	37.88
10	2	260,520 (30.04%)	37.88

MATLAB® (i.e., avoiding loops "for"), a PC equipped with a 1.6 GHz processor takes approximately one hour to run 10,000 simulations in the simplest case of a peak power unit. A large part of the computational time spent by the running process involves the generation of random numbers. In the simplest simulation, at least 125 million independent shocks are required to get to the final distribution of the plant price, whereas in the hourly-based case, at least 725 million shocks need to be generated. In this case, the computational time increases by up to 3 hours. The accuracy of our estimates can be easily improved through the adoption of suitable variance reduction techniques.

20

Swinging on a Tree[*]

Key words: interruptible contracts, energy prices, dynamic programming

A swing contract grants the holder a number of transaction rights on a given asset for a fixed strike price. Each right consists of the double option to select timing and quantity to be delivered under certain limitations. Transactions are specified by the contract structure and usually involve a supplementary right to choose between purchase and selling. Swing contracts are very popular in markets where delivery is linked to consumption or usage over time. This is the case of energy commodity markets such as oil, gas and electricity. There, swing features are usually embodied in a base-load contract providing a constant flow of the commodity for a fixed tariff.

A typical scheme involves a retailer selling gas to a final consumer. The contract contains an option to interrupt delivery for a predetermined number of times. From a financial viewpoint, the retailer is short one strip of forward contracts, one contract per delivery day, and long one swing option allowing for adjustments in the gas delivery according to contingent market conditions. The joint position is often referred to as a callable forward contract. Beyond side commitments existing between the two counterparts, the swing option exercise policy depends on standing market conditions such as the commodity spot price and availability. The net cash flow for the retailer is given by the forward price received upon delivery minus the option premium paid to the consumer and usually settled as a discount premium over the forward price; plus any cash flow stemming from exercising the option.

The purpose of this chapter is to evaluate a swing option with non-trivial constraints by means of dynamic programming. General treatments on swing contracts are discussed by Clewlow and Strickland (2000) and Eydeland and Wolyniec (2002). The seminal paper by Thompson (1995) tackles the issue of multiple exercise derivatives. Keppo (2004) proves that the optimal exercise policy in the case of no load penalty is "bang-bang" (i.e., an all-or-nothing clause) and derives explicit hedging

---

[*] with Michele Lanza and Valerio Zuccolo.

strategies involving standard derivatives such as forwards and vanilla options. Other papers devoted to the analysis of swing options include Baldick, Kolos and Tompaidis (2003), Barbieri and Garman (1996, 1997), Pilipovich and Wengler (1998), Clewlow and Strickland (2000), Clewlow, Strickland and Kaminski (2001), Carmona and Dayanik (2003), and Lund and Ollmar (2003). Cartea and Williams (2007) analyze the interplay between interruptible clauses and the price of risk in the gas market. Our development moves the main problem through a simplified argument and defer a rigorous treatment of the model and its solution methodology to the next section.

20.1 Problem Statement

The simplest swing option is defined by two input parameters: (1) a number n of transaction rights and (2) an exercise price K. The option holder, e.g., a gas retailer, manages the option exercise policy through control variables $\tau_1, \dots, \tau_n$ signalling the exercise times and $q_1, \dots, q_n$ indicating the transacted quantities ($q_i > 0$ for purchase and $q_i < 0$ for selling). The control signal is defined by

$$i(t) = \begin{cases} 1 & \text{if } t \text{ is an exercise time,} \\ 0 & \text{otherwise.} \end{cases}$$

The pair $(i(t), q(t))$ defines the control chosen by the option holder at time t. The option fair price is computed as the maximal expected pay-off obtainable by the holder over the contract lifetime $[0, T]$ through an admissible control policy $\{(i(t), q(t)), t \in [0, T]\}$.

Since markets randomly evolve, the optimal control cannot be fully determined at the contract inception. However, it is possible to specify a rule selecting the optimal control corresponding to any market instance. More precisely, we look for a number of variables whose knowledge at any time t uniquely determines the optimal control at that time. To detect the state variable underlying a swing option, we examine the spectrum of choices available at any time t. Suppose that no swing right is available anymore. Then, the only possible action is to continue the contract. The corresponding pay-off is given by the expected present value of the option price displayed one period later (i.e., at time $t + 1$). If instead a "swinging" gas load is still possible, then the holder decides to exercise the option whenever the resulting revenue is greater than the expected present value of the option price shown one period later. The revenue from exercising the option is given by three terms. First, the option pay-off $(G - K) \times q$, where G is the gas spot price and q is the stricken load. Second, a penalty $P(q)$ on the exercised quantity must be computed and subtracted to the previous term. Third, the present value $V(t + 1)$ of the residual contract, namely a swing option with the same features and a number of swing rights reduced by one. At first glance, this term may resemble the cash flow stemming from q forward positions on gas. In reality, this is not the case as long as both delivery time and load are optional. In standard financial jargon, one may call this term of the swing pay-off a "callable forward with flexible load". Figure 20.1 illustrates the described procedure.

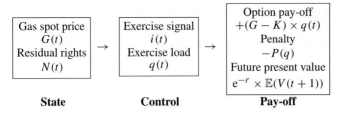

Fig. 20.1. Exercise decision scheme.

Notice that the time t optimal choice depends on two state variables, the gas price G and the number N of standing swing rights. Correspondingly, the swing option value is a function $V = V(t, G(t), N(t))$ of time, gas price and the number of residual swing rights.

Swing option valuation can be carried out through Dynamic Programming. The idea is to write a recursive algorithm for computing the swing option value as a function of its possible value one period later. For the sake of clarity, we consider a contract with swing rights to call for any quantity q under a penalty $P(q)$ affecting the option pay-off. The backward procedure starts at the contract maturity T. For all possible values of G and N, the value $V(T, G, N)$ is computed as the corresponding option pay-off:

$$V(T, G, n) = \begin{cases} 0 & \text{if } n = 0, \\ \max_q\{(G - K)_+ q - P(q)\} & \text{if } n > 0, \end{cases}$$

where $(x)_+ := \max\{0, x\}$. The general term in the backward recursion reads as

$$V(t - 1, G, n) = \overbrace{\mathrm{e}^{-r} \times \mathbb{E}_{t-1}\big(V(t, \cdot, n)\big)}^{\text{pay-off from continuation}} \qquad \text{if } n = 0,$$

$$V(t - 1, G, n) = \max_q\Big\{\underbrace{(G - K)_+ q - P(q) + \mathrm{e}^{-r} \times \mathbb{E}_{t-1}\big(V(t, \cdot, n - 1)\big)}_{\text{pay-off from exercise}},$$

$$\underbrace{\mathrm{e}^{-r}\mathbb{E}_{t-1}\big(V(t, \cdot, n)\big)}_{\text{pay-off from continuation}}\Big\} \quad \text{if } n > 0,$$

where r denotes the continuously compounded one-period risk-free rate of interest and $\mathbb{E}_{t-1}$ indicates the conditional expectation given the information available at time $t - 1$. The swing option value $V(0, G_0, n)$ depends on the number n of swing rights and the standing market situation as represented by the current gas price G_0.

In actuality, swing contracts include additional constraints beyond the simplified setting illustrated above. Exercise dates may be required to differ by more than a fixed refraction period ρ, i.e., $\tau_{i+1} - \tau_i \geq \rho$; delivery may be delayed; constraints may be strictly binding (e.g., $q_i(t) = q$) or floating ($a(t) \leq q_i(t) \leq b(t)$); the strike price may differ according to whether the holder decides to buy or sell gas upon exercise; its value can also be contingent upon other specified market variables.

An important feature of real-world swing contracts is the penalty function $P(q)$, which may be either local or global. A local penalty affects the revenue from exercising the option depending on the exercised quantity q, whereas a global penalty applies to the overall exercised quantity $Q = \sum_i q_i$ at the end of the contract.

We provide a framework for evaluating swing options to call or put gas under global penalties. Roncoroni and Zuccolo (2004) offer a deeper analysis of the optimal exercise policies for swing options under both local and global penalties.

20.2 Model and Solution Methodology

We consider a swing option with lifetime $[0, T]$ providing the holder with a number u_0 of upswing rights (i.e., call options) and a number d_0 of downswing rights (i.e., put options) on a gas load. The gas spot price is described by a stochastic process evolving over a discrete set of evenly spaced times. We assume these times constitute a refinement of the option lifetime, namely T/Δ is an integer for a time lag Δ. We allow for cash rolling-over at a rate r. Table 20.1 reports basic notation for the proposed model. The time t spot price of gas is denoted by $G(t)$. This variable evolves over time according to a stochastic process $(G(t), t \in \mathbb{T})$. The process is assumed to follow a trinomial discretization of a geometric Brownian motion. The exact specification is detailed in the implementation section below. We use symbols u and d to denote the standing number of upswing and downswing rights, respectively. Table 20.2 indicates the state variables of the swing valuation problem. At the outset, u_0 (resp. d_0) upswing (resp. downswing) rights are made available to the option holder. Each exercise consists of the delivery of a constant load q for a fixed price K. A final penalty P is applied to the gas load accrued over the lifetime of the contract, that is:

Table 20.1. Input parameters and basic definitions

Symbol	Quantity
$[0, T]$	Time horizon
Δ	Time period length
$n = [T/\Delta]$	Periods
$T = \{k\Delta, k = 0, \dots, n\}$	Horizon refinement
$G(t), t \in T$	Spot price process of gas
G_0	Initial spot price of gas
$\mathbb{G}_t$	Image space of $G(t), t \in T$
r	One-year risk-free rate of interest

Table 20.2. State variables

Symbol	Quantity
t	Time point
G	Spot gas price
u	Standing number of upswing options
d	Standing number of downswing options

Table 20.3. Contract features

Symbol	Quantity
$\mathbb{T}^{\mathcal{O}} \subseteq \mathbb{T}$	Set of optional dates
$n^{\mathcal{O}} = \#\mathbb{T}^{\mathcal{O}}$	Number of optional dates
$u_0 \in \mathbb{N}$	Initial number of upswing options
$d_0 \in \mathbb{N}$	Initial number of downswing options
q	Fixed callable/puttable load
K	Strike price
$Q(u, d)$	Overall load
$P(Q)$	Global penalty function

$$Q(u, d) = |q(u_0 - u) - q(d_0 - d)|.$$

We allow for the exercise times to be constrained within a subset $\mathbb{T}^{\mathcal{O}}$ of the time horizon $\mathbb{T}$ and denote by $n^{\mathcal{O}}$ the number of times available at the outset. Table 20.3 shows the contractual features defining the swing option under investigation. The problem can be cast as a maximization of the present value of the option future cash flow over a set of admissible exercise policies. The value function for this problem is denoted by V_G. Notice that if the swing load q is fixed, then each upswing right corresponds to an American call option and is therefore exercised at the last available time.

20.3 Implementation and Algorithm

20.3.1 Gas Price Tree

The benchmark model for spot price dynamics in continuous time is the geometric Brownian motion (GBM)

$$G(t) = \exp\left(\left(r - \frac{\sigma^2}{2}\right)t + \sigma W(t)\right), \tag{20.1}$$

where constants r and σ represent the instantaneous annualized risk-free rate of interest and price volatility, respectively. In energy markets, the use of this model is rather questionable. The main reason is that energy demand is driven by periodical trends and prices tend to revert back to a periodical mean level. Moreover, some markets display peculiar features such as stochastic volatility and spikes (see, e.g., Roncoroni (2002)). However, the popularity gained in the last thirty years by the GBM, mainly due to the Black and Scholes model for option pricing, suggests considering this specification as a theoretical reference. Also, this model is widely employed by energy traders for the purpose of obtaining benchmark values for exotic derivatives. Accordingly, we develop our implementation under this specification for the gas price process and leave the final user the option of selecting alternative dynamics.

For simplicity, we set an initial price $G_0 = 1$. The time horizon is refined into a set of $2n$ periods $\{k\delta, k = 1, \ldots, 2n\}$, where $\delta = \Delta/2$ (i.e., half the time period selected for the option lifetime discretization). On this "doubly refined" time horizon, we establish a binomial random walk

$$G^{\mathrm{B}}((k+1)\delta) = \begin{cases} G(k\delta) \times I & \text{with probability } p, \\ G(k\delta) \times D & \text{with probability } 1 - p. \end{cases}$$

Constants I and D represent the one-period percentage increase and decrease in the standing price and must satisfy the inequality $0 < D < 1 < I$. These figures are selected so that the two processes G and G^{B} match in their conditional mean and variance over each time step. To this aim, we set

$$\begin{aligned} \exp(r\delta)G(t) &= \mathbb{E}(G(t+\delta)|G(t)) \\ &= \mathbb{E}\big(G^{\mathrm{B}}(t+\delta)|G^{\mathrm{B}}(t)\big) \\ &= p \times I \times G(t) + (1-p) \times D \times G(t) \end{aligned}$$

and

$$\begin{aligned} e^{(2t+\sigma^2)\delta t} X_i^2 \exp(2t+\sigma^2)\delta \times G(t) &= \mathrm{Var}(G(t+\delta)|G(t)) \\ &= \mathrm{Var}\big(G^{\mathrm{B}}(t+\delta)|G^{\mathrm{B}}(t)\big) \\ &= \big(pI^2 + (1-p)D^2\big) \times G(t)^2 \\ &\quad - \big(pI \times G(t) + (1-p)D \times G(t)\big)^2, \end{aligned}$$

plus the symmetry condition

$$D = \frac{1}{I}$$

ensuring that the resulting tree does recombine (i.e., an upward move followed by a downward move has the same effect on the price quotation as a downward move followed by an upward move). Solving for I, D and p gives

$$\begin{aligned} I &= B + \sqrt{B^2 - 1}, \\ D &= B + \sqrt{B^2 + 1}, \\ p &= \frac{\exp(r\delta) - D}{I - D}, \end{aligned}$$

where $2B = \exp(-r\delta) + \exp[(r+\sigma^2)\delta]$.

From this binomial tree we build a trinomial tree with the desired time step Δt by merging all consecutive pairs of time periods into a single period $\Delta = 2\delta$: the resulting random walk reads as

$$G^{\mathrm{T}}((k+1)\Delta) = \begin{cases} G^{\mathrm{T}}(k\Delta) \times I^2 & \text{with probability } p^2, \\ G^{\mathrm{T}}(k\Delta) & \text{with probability } 2p(1-p), \\ G^{\mathrm{T}}(k\Delta) \times D^2 & \text{with probability } (1-p)^2, \end{cases}$$

where I, D and p are defined as above. Notice that the intermediate point is constant since $I \times D = 1$.

20.3.2 Backward Recursion

We now provide an algorithm for computing the option value at contract inception. In a general swing contract, the option holder has the right to choose delivery time and quantity. To avoid cumbersome algorithms, we simplify the context and suppose that upon exercise a *fixed* load q is delivered. Consequently, the option price is a function $V_G(0, G_0, u_0, d_0)$, where G_0 is the gas price prevailing in the market at time 0 and u_0 (resp. d_0) is the number of upswing (resp. downswing) rights specified as contractual clauses. This method considerably simplifies the valuation procedure in the case examined herein. Notice that under different contract specifications, e.g., a penalty function affecting each single exercise, the load constraint can be relaxed without incurring into the above mentioned complications.

The solution algorithm is a Dynamic Programming backward recursion.

Time T Value Function

The holder has the choice to maximize the profit from exercising either upswing or downswing rights provided that both of them are still available. An indicator function signals the availability of the corresponding right. In particular, if all option rights have been already exercised, the contract is worthless. The final value reads as

$$V_G(T, G, u, d) = \max\{q(G - K)_+ \mathbf{1}_{u \neq 0}, q(K - G)_+ \mathbf{1}_{d \neq 0}\} - P(Q(u, d)),$$

for all $G \in \mathbb{G}_T, u \leq u_0, d \leq d_0$. Here $\mathbb{G}_T$ denotes the set of all possible values taken by the gas price at time T (see Table 20.1) and $\mathbf{1}_{u \neq 0}$ denotes the indicator function of the set $\{u \neq 0\}$, i.e., $\mathbf{1}_{u \neq 0} = 1$ if $u \neq 0$ and 0 otherwise.

Time t Value Function

If $t \in \mathbb{T}$ is not an optional time, i.e., $t \notin \mathbb{T}^\mathcal{O}$, or neither upswing nor downswing rights are available anymore, i.e., $u = d = 0$, then the swing is worth the conditional expected value of its discounted price one time step forward:

$$\pi_N(t, G, u, d) = e^{-r\Delta} \mathbb{E}_t \left[V_G(t + \Delta, G(t + \Delta), u, d) \right].$$

Notice that expectation has been taken conditional to the information available at time t. If any swing right is available at time t (i.e., $t \in \mathbb{T}^\mathcal{O}$ and $u \neq 0$ or $d \neq 0$), the option holder compares the value π_N from continuing the contract (i.e., no exercise) to the pay-off from exercising a swing right. An upswing right exercise leads to an immediate cash flow $q(G - K)_+$ plus the expected value of its discounted price one time step later, namely the time $t + \Delta$ value of a swing contract with one upswing right less:

$$\pi_U(t, G, u, d) = q(G - K)_+ + e^{-r\Delta} \mathbb{E}_t \left[V_G(t + \Delta, G(t + \Delta), u - 1, d) \right].$$

A similar argument leads to the following value of a downswing exercise:

$$\pi_{\mathrm{D}}(t, G, u, d) = q(K - G)_+ + e^{-r\Delta}\mathbb{E}_t\big[V_G\big(t + \Delta, G(t + \Delta), u, d - 1\big)\big].$$

Combining these expressions, we obtain the following recursive relation for the time t value function:

$$V_G(t, G, u, d)$$
$$= \max\big\{\pi_{\mathrm{U}}(t, G, u, d)\mathbf{1}_{u\neq 0, t\in\mathbb{T}^{\mathcal{O}}}, \pi_{\mathrm{D}}(t, G, u, d)\mathbf{1}_{d\neq 0, t\in\mathbb{T}^{\mathcal{O}}}, \pi_{\mathrm{N}}(t, G, u, d)\big\},$$

where $t \in \mathbb{T} \setminus \{T\}$, $G \in \mathbb{G}_t$, $u \le u_0$, $d \le d_0$, and the indicator functions signal the availability of the corresponding swing rights. Notice that no penalty directly applies to any swing right exercise.

20.3.3 Code

Our code splits into three main parts. A first part computes the tree knots. A second part evaluates penalty. A final part calculates the option value. The penalty is computed as a fixed fare π for each unit of the net recourse to swing load (i.e., modulus of the number of exercised upswing minus the number of exercised downswing) over a threshold τ representing the minimal number of net exercises. For instance, if this level equals 10 and the number of stricken upswing (resp., downswing) rights is 20 (resp., 5), then the penalty paid-off at the contract expiration is $P = 3\max\{|20 - 5| - 10, 0\} = 15$.

Function $\text{Swing}(n, T, G_0, K, r, \sigma, n^{\mathcal{O}}, \rho, q, \tau, \pi)$ implements the valuation procedure. Here, n is the number of dates in the time horizon, T is the contract maturity, G_0 is the spot price of gas at the outset, K is the option strike price, r is the one-period interest rate (i.e., it equals $r\Delta$ in the previous section), σ is the annualized instantaneous volatility of gas price, $n^{\mathcal{O}}$ is the number of optional dates (which we assume to be uniformly spread over the contract lifetime), $\rho = u_0 = d_0$ is the number of upswing and downswing rights (which, for simplicity, are assumed to be equal), q is the deliverable quantity upon each exercise, τ is the penalty threshold, π is the unit penalty.

20.4 Results and Comments

We evaluate the swing option price across alternative parameter scenarios. In all cases, the option maturity is set to 1 year, the time step is approximately equal to one week, i.e., $1/49$, the exercise quantity is 5 MMBtu (Million British termal units), current spot price of gas is \$3/MMBtu, strike is set to \$2.90/MMBtu, and interest rate is 4% *per annum*. The spot price volatility is estimated at 60%. At contract maturity, a penalty π is charged for each delivered gas unit $Q(u, d)$ exceeding a threshold τ.

We perform a comparative study of swing option prices across different values for the number $n^{\mathcal{O}}$ of available exercise times, the number ρ of swing rights, the unit penalty π, and the penalty threshold τ.

Table 20.4. Option prices at varying unit penalties

Dates $n^{\mathcal{O}}$	Rights ρ	Threshold τ	Penalty π	Value
25	4	10	0	26.9024
25	4	10	1	23.4513
25	4	10	3	21.3629
25	4	10	5	20.9558
25	4	10	7	20.8667

Table 20.5. Option prices at varying penalty thresholds (high unit penalty)

Dates $n^{\mathcal{O}}$	Rights ρ	Threshold τ	Penalty π	Value
25	4	5	3	15.7126
25	4	10	3	21.3629
25	4	15	3	26.8605
25	4	20	3	26.9024

The first experiment considers 25 exercise dates, namely one exercise opportunity each two consecutive working days, a penalty threshold $\tau = 10$, four swing rights, each one for delivering 5 MMBtu. We examine the way the swing option fair price changes across varying overload penalties. The case $\pi = 0$ corresponds to an absence of penalty. Table 20.4 reports option values for the examined instances. Notice that swing prices sharply decrease for low level unit penalties, whereas they show steady behavior for high-level unit penalties. In particular, the price drop is rather significant moving from no penalty to a unit penalty.

Another way to affect the option price through penalties is to modify the penalty threshold as defined in Sec. 20.3.3. The lower this figure, the stronger the effect of a penalty on the option value. As Table 20.5 illustrates, if the threshold is set to 5, any exercise beyond the first one is subject to a penalty, whereas the level 20 makes the penalty totally ineffective for the examined case (i.e., 4 rights, each one delivering 5 MMBtu). Naturally, the impact of moving the penalty threshold depends on the unit penalty level. Table 20.6 shows that a sufficiently low unit penalty makes the option value quite insensitive to threshold resetting.

The next experiment shows the price sensitivity to a varying number of optional dates. Recall that an optional date is a point in time where a swing option can be exercised. We assume that these dates are evenly spread over the contract lifetime. Table 20.7 shows option prices corresponding to 7, 13, 25, and 49 optional dates. In general, the swing value is slightly sensitive to these dates as long as an exercise is allowed (i.e., $n^{\mathcal{O}} \geq \rho$).

Our last experiment involves price sensitivity to a varying number of swing rights. Each right affects the option value by modifying the quantity of gas that can be delivered and the effectiveness of a penalty constraint. Table 20.8 reports figures for all cases.

We observe an interesting behavior whenever the number of rights deviates from a relatively low figure. For instance, if rights go from 2 to 4, the price increases

Table 20.6. Option prices at varying penalty thresholds (low unit penalty)

Dates $n^{\mathcal{O}}$	Rights ρ	Threshold τ	Penalty π	Value
25	4	5	0.05	26.4352
25	4	10	0.05	26.6501
25	4	15	0.05	26.8618
25	4	20	0.05	26.9024

Table 20.7. Option prices at a varying number of optional dates

Dates $n^{\mathcal{O}}$	Rights ρ	Threshold τ	Penalty π	Value
7	4	10	3	19.5901
13	4	10	3	20.8096
25	4	10	3	21.3629
49	4	10	3	21.6317

Table 20.8. Option prices at a varying swing rights

Dates $n^{\mathcal{O}}$	Rights ρ	Threshold τ	Penalty π	Value
25	2	10	3	13.7417
25	4	10	3	21.3629
25	6	10	3	23.2434
25	10	10	3	26.3706
25	12	10	3	27.6269

by about 55%, whereas moving swing rights from 6 to 12, the option value only increases by about 19%. This is a typical scarcity item due to the relatively higher value of the former available swing rights compared to the latter ones.

It is interesting to explore the optimal exercise strategy by showing the distribution of the first few exercise times. These distributions are not available in closed-form. However, we may provide approximate versions by computing the exercise times over a large number of simulated paths of the underlying process. The procedure works as follows. First, the optimal exercise policy is computed: this is a rule applying to any point in the state variable domain. Next, several trajectories are sampled by simulation. For each simulation, the corresponding upswing and downswing times are calculated and stored. Finally, relative frequency functions are computed for each of the exercise times, starting with the first one, moving to the second one, and so on.

We consider a swing option with 2 downswing rights and 2 upswing rights. The resulting histograms are displayed in Fig. 20.2

These graphs display a clear picture of the time distribution of optimal exercises provided that the random evolution of the market is correctly described by the given random walk. As noted in Section 20.2, each upswing right corresponds to an American call option and is therefore exercised at the last available time.

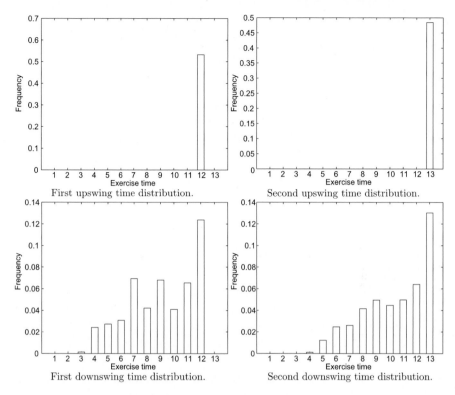

First upswing time distribution.   Second upswing time distribution.

First downswing time distribution.   Second downswing time distribution.

Fig. 20.2. Sample jump times distributions.

Interest-Rate and Credit Derivatives

21

Floating Mortgages[*]

Key words: mortgage prepayment policy, interest rates, dynamic programming, simulation

A mortgage is a loan secured on real estate. The borrowed amount, named principal, is repaid along a time horizon through installments. These payments are usually computed *pro-rata temporis*, meaning that each cash flow splits in two components: a first component represents a portion of the outstanding balance; a second component is the interest accrued since the last payment. This scheme guarantees that the loan is totally repaid by the end of the contract. Most of the residential mortgage loans offered to retail investors are expected to amortize through a "French" amortization scheme consisting of constant installments, usually paid off on monthly or quarterly bases.

In the US market, the vast majority of mortgages are collected into pools managed by government sponsored agencies. These institutions issue securitized notes backed by the cash flows generated by a specific pool of mortgages and sell these assets, known as mortgage-backed securities (MBS), to private investors, generally large investment funds, either directly of through dealers.

In principle, the value of an MBS is the value of a long-term annuity with fixed maturity. However, the borrower is allowed to pre-pay his debt back any time before the legal maturity of the loan. Prepayment may occur due to either exogenous reasons such as moving or any other personal issue, or endogenous reasons typically linked to the mortgagor's ability to enter a new mortgage at more favorable rate conditions. This usually occurs as market lending rates decrease and the borrower has the opportunity to get profit from such circumstances. In practice, market rates may fall and the standing mortgage is still continued. This is the case of a mortgagor whose credit status has deteriorated and who is thus required to have a greater spread over the market lending rate. Transaction fees applying to a mortgage prepayment constitute

---

[*] with Alessandro Moro.

a further incentive to continuing the standing contract while market rates decrease. In actuality, an MBS can be seen as a long position in a fixed-maturity annuity and a short position in a compound American-style call option.

The literature on the valuation of MBSs splits into two distinct frameworks. Early works by Dunn and McConnell (1981a, 1981b) and Brennan and Schwartz (1985) propose a rational model explaining how a mortgage borrower chooses to refinance his loan. They determine the fair value of an MBS by applying contingent claim valuation to the portfolio consisting of a long annuity and a short American-style option under the hypothesis that the option value is maximized. Notice that this is equivalent to assuming that the mortgagor minimizes the value of his standing mortgage position.

A second framework for valuing MBSs builds on the econometric identification of the prepayment behavior from historical data. Models within this setting have been proposed by Schwartz and Torous (1989, 1992) and Boudoukh et al. (1997) and now constitute the standard market practice for valuing MBSs. This setting suffers from a major drawback: they perform quite badly in out-of-sample predictions. This is mainly due to a strong dependence on market conditions producing the historical data on which econometric analysis is conducted.

The present case-study considers the mortgage refinancing problem as seen from the mortgagors' viewpoint, much in the spirit of the first line of research mentioned above. We show the way to determine the optimal prepayment rule for a borrower minimizing the value of his mortgage position by choosing the time the mortgage is to be refinanced. Following the theoretical framework introduced in Roncoroni (2000), Roncoroni and Moro (2006) add a constraint on the number of refinancing options. This feature accommodates the possibility that the lender gives the borrower the option to refinance his mortgage internally at smaller additional costs than the ones presented upon repaying the principal and entering into a new mortgage with another lender (e.g., transaction costs and credit spread variations). This option represents a way to attract the customer's fidelity to the lending institution and let it save all costs required to search for alternative investment opportunities. The optimal recursive determination of prepayment policies has recently been investigated by Longstaff (2002) and Gocharov and Pliska (2003) under unconstrained refinancing opportunities. Stanton (1995) develops an alternative model combining elements from the econometric and the rational prepayment approaches.

Roncoroni and Moro (2006) assume that refinancing is subject to small fees or transaction costs. They focus on refinancing decisions exclusively steered by better market conditions for the borrower. The inclusion of exogenous elements driving the refinancing policy does not pose any particular problem to the proposed setting. It suffices to allow the time horizon be dependent on the occurrence of the random events triggering the contract expiration. The number N of refinancing opportunities over the time horizon is fixed at the outset. At each point in time the mortgagor sets the debt rate to either the current rate or to newly available floating lending rate. Over the contract lifetime, this option can be exercised N times at most.

From a financial viewpoint, the mortgagor's decision problem takes the form of a multiple compound American-style option. This allowed for the use of stochastic dynamic programming methods to tackle the determination of the optimal prepayment policy defined as the one minimizing the value of the contract.

This case-study is organized as follows. Section 21.1 describes the contract feature of fixed-rate and floating rate mortgages under limited refinancing options. Section 21.2 details the solution methodology and extends the analysis to the case including transaction costs. Section 21.3 illustrates empirical results obtained by performing experiments under alternative scenarios. A final section presents a few concluding remarks.

21.1 Problem Statement and Solution Method

21.1.1 Fixed-Rate Mortgage

We consider a finite time horizon $\mathbb{T}_{0,T} := (0, 1, \ldots, T)$. At time 0, an individual borrows 1 Euro. This generates a balance due $B(0) = 1$. The borrower is required to pay back this amount together with interest over the horizon according to a constant installment scheme. For each period $[t - 1, t]$ an interest $I(t - 1)$ is calculated on the outstanding balance $B(t - 1)$ at a rate equal to $r(t - 1)$. The outstanding exposure $B'(t - 1)$ of the debtor over the period $[t - 1, t]$ is defined by the due interest $I(t - 1)$ plus the standing balance $B(t - 1)$. At the end of the period, i.e., time t, the borrower pays a fraction $f(t)$ which is written down depending on the exposure $B'(t - 1)$. This number is computed as a proportion of the residual contract lifetime, namely $f(t) = 1/(T - (t - 1))$. Let us denote this payment by $P(t - 1)$. The explicit dependence on the starting day $t - 1$ of the period underlines that this amount is set at this time, though paid off at the end of the period. After performing this payment, the new outstanding balance for the debtor becomes $B(t) = B'(t - 1) - P(t - 1)$. Table 21.1 summarizes the steps involved in this payment scheme. It is clear that the new balance due depends on the initial balance and on the debt rate process $r = (r(t), t = 0, \ldots, T - 1)$. Since by hypothesis the initial debt position is $B(0) = 1$, the only exogenous ingredient is the the debt rate process r.

Notice that the borrowed capital is totally repaid by the end of the time horizon, i.e., $B(T) = 0$. This can be proved by showing that the last payment $P(T - 1)$ matches the standing balance at time $T - 1$:

Table 21.1. The amortization scheme

Quantity	Symbol	Formula
Standing debt balance at $t - 1$	$B(t - 1)$	Given by induction
Debt rate for period $[t - 1, t]$	$r(t - 1)$	Random
Interest accrued on $[t - 1, t]$	$I(t - 1)$	$r(t - 1) \times B(t - 1)$
Standing exposure on $[t - 1, t]$	$B'(t - 1)$	$I(t - 1) + B(t - 1)$
Constant installment coefficient	$f(t)$	$1/(T - (t - 1))$
Payment to lender at time t	$P(t - 1)$	$f(t) \times B'(t - 1)$
Standing debt balance at time t	$B(t)$	$B'(t - 1) - P(t - 1)$

$$P(T-1) = B'(T-1)/(T-(T-1))$$
$$= B'(T-1).$$

The *cost to go* associated to a given debt rate process r is defined as the sum of all cash flows stemming from the payment scheme just described, i.e., $\sum_{t=0}^{T-1} P(t)$. We now move to the description of the debt rate process r.

21.1.2 Flexible-Rate Mortgage

Let us assume that the market quotes an interest rate R and suppose this number follows a time-homogeneous Markov chain $R = (R(t), t = 0, \ldots, T)$ with finite state space $S = \{s_{min} + k\Delta s, k = 0, 1, \ldots, K\}$. Here the minimum rate s_{min}, the interest lag Δs and the cardinality $K + 1$ are all fixed. Let $S' = \{s_{min} + k\Delta s, k = 1, \ldots, K-1\}$. The transition probabilities of the process R are assigned as follows:

$$p(x, y) = \begin{cases} \frac{1}{3} & \text{if } (x, y) \in S' \times \{x, x \pm \Delta s\}, \\ \frac{1}{2} & \text{if } (x, y) \in \{s_{min}\} \times \{s_{min}, s_{min} + \Delta s\} \\ & \quad \cup \{s_{max}\} \times \{s_{max}, s_{max} - \Delta s\}, \\ 0 & \text{otherwise.} \end{cases} \quad (21.1)$$

This function, coupled with an initial market rate $R(0)$, induces a probability measure $\mathbb{P}_{R(0)}$ on the path space S^{T+1}.

Turning to the debt rate process r, we suppose its initial value $r(0)$ matches the market rate $R(0)$ at the same time. At any date t, the debtor is faced with the choice of continuing the mortgage at the standing debt rate $r(t-1)$ or to repay the entire capital due by using the proceeds from entering a new mortgage at the currently available conditions expressed by the prevailing market rate $R(t)$. In this case we say that the mortgage has been refinanced.

We study the case where this option can be exercised a maximum number N of times over the horizon $(0, \ldots, T-1)$. If N is equal to the number $T-1$ of setting times, the optimal strategy for the borrower is trivial: he exercises the option each time the market rate R goes down. Consequently, we suppose that the number N of refinancing opportunities is strictly smaller than the number of dates. Table 21.2 indicates the input parameters and dynamic processes defining the contract provisions. Turning to the mortgage refinancing decision, this can be described by a process α specifying at each time $t = 0, \ldots, T-1$ whether the mortgagor continues his position at the standing conditions ($\alpha = 0$) or, if possible, refinances his debt ($\alpha = 1$). The control process is described in Table 21.3.

Of course, the chosen control policy affects the state variables dynamics featuring the borrower's position over time. These variables are as follows: (1) the number $n^\alpha(t)$ of available refinancing opportunities left for future exercise; (2) the current interest rate $r^\alpha(t)$; (3) the installment cash flow $P^\alpha(t)$ for the current period; (4) the resulting outstanding balance $B^\alpha(t)$. Table 21.4 reports all processes subject to a control. It is clear that choosing to refinance or not at a give time t has has effect on $n^\alpha(t+1)$ and $r^\alpha(t+1)$. The link to the other two quantities is clarified in Table 21.5.

Table 21.2. Input variables and parameters

Quantity	Symbol	Formula
Time horizon	$T_{0,T}$	$\{0, 1, \ldots, T\}$
Market rate range	S	$\{s_{\min} + k\Delta s, k = 0, 1, \ldots, K\}$
Initial market rate	$R(0)$	Given
Market rate dynamics	R	$(R(t), t = 0, \ldots, T) \sim (21.1)$
Number of options	$n(0)$	$N < T$
Initial debt rate	$r(0)$	$R(0)$
Balance due	$B(0)$	1

Table 21.3. Control variables

Quantity	Symbol	Formula
Control policy process	α	$(\alpha(t), t = 0, \ldots, T - 1)$
Control policy domain	$\mathcal{D}(\alpha(t))$	$\begin{cases} \{0, 1\} & \text{if } \sum_{i=1}^{t} \alpha(i) < n(0) \\ \{0\} & \text{otherwise} \end{cases}$
Control policy at time t	$\alpha(t)$	$\begin{cases} 0 & \text{continuation} \\ 1 & \text{refinancing} \end{cases}$

Table 21.4. Controlled system

Quantity	Symbol	Formula
Refinancing options process	n^{α}	$(n^{\alpha}(t), t = 0, \ldots, T - 1)$
Refinancing opportunities at t	$n^{\alpha}(t)$	
Controlled debt rate process	r^{α}	$(r^{\alpha}, t = 0, \ldots, T - 1)$
Debt rate at time t	$r^{\alpha}(t)$	$\begin{cases} R(t) & \text{if } \alpha(t) = 1 \\ r^{\alpha(t-1)}(t-1) & \text{if } \alpha(t) = 0 \end{cases}$
Pro-rata payment process	P^{α}	$(P^{\alpha}(t), t = 0, \ldots, T - 1)$
Standing debt balance process	B^{α}	$(B^{\alpha}(t), t = 0, \ldots, T - 1)$

Table 21.5. Debt process

$B^{\alpha}(t)$	Outstanding debt balance at time t
$B'(t) := B^{\alpha}(t)(1 + r^{\alpha}(t)/52)$	Capital plus interest on $[t, t+1]$
$P^{\alpha}(t) := B'(t)/(T - t)$	Pro-rata payment for $[t, t+1]$
$B^{\alpha}(t+1) := B'(t) - P^{\alpha}(t)$	Outstanding debt balance at time $t+1$

The borrower wishes to select a control policy that minimizes the cost associated with the entire repayment stream $(P^{\alpha}(t), t = 0, \ldots, T - 1)$. For the purpose of illustrating this issue, we consider the sum of all these payments as a raw measure of this cost. Notice that we do not consider discounting for evaluation purposes. The problem is qualified as follows:

$$\min_{\alpha \in \mathcal{A}} \mathbb{E}\left(\sum_{j=0}^{T-1} P^{\alpha}(j)\right), \tag{21.2}$$

where $\mathcal{A}$ denotes the class of admissible control policies $\alpha = \{\alpha_0, \ldots, \alpha_{T-1}\}$ determining the controlled payment process $(P^\alpha(t), 0 \le t < T)$.[1]

We note that any control policy α defines a multivariate stopping rule $(\tau_1, \ldots, \tau_n)$, $n \le N$, which is recursively defined by

$$\tau_1 = \inf\{t \ge 0 \colon \alpha(t) = 1\},$$
$$\tau_{k+1} = \inf\{t > \tau_k \colon \alpha(t) = 1\}.$$

Our task is to build a model for describing the financial structure illustrated above, then provide an algorithm delivering the minimum value in (21.2), and finally determine the corresponding optimal control policy

$$\alpha^* = \big(\alpha^*(0), \ldots, \alpha^*(T-1)\big).$$

This latter is equivalent to the multivariate optimal stopping rule $(\tau_1^*, \ldots, \tau_N^*)$. We end this section by deriving an expression for the time $t + 1$ outstanding balance in terms of the adopted control policy and the debt level recorded one time-step before:

$$B(t+1) = B'(t) - P(t) = B'(t) - \frac{B'(t)}{T-t} = c\big(r^{\alpha(t)}(t)\big)B(t)\mathcal{T}(t),$$

where $c(x) = 1 + x/52$ denotes the accrual factor corresponding to rate x and $\mathcal{T}(t)$ is the *pro-rata* coefficient defined by $1 - 1/(T - t)$.

21.2 Implementation and Algorithm

21.2.1 Markov Control Policies

Recall that the state variable of a Markov control problem is defined as the information upon which the control policy is chosen at any time. More precisely, this is the set of observable variables upon which $\alpha(t)$ can be determined at a given point in time.

We observe that the time t standing balance $B(t)$ and debt rate $r(t)$ fully determine both the updated capital-plus-interest $B'(t)$ and the constant installment payment scheme $P(t)$, which in turn goes into the objective functional (21.2). Since the target is affected by both $B(t)$ and $r(t)$, the time t control $\alpha(t)$ should depend on these variables. The dependence on $B(t)$ is straightforward. If any refinancing opportunity is still available, i.e., $0 < n(t) \le N$, then the debt rate is set to the best performing one between its previous value $r(t - 1)$ and the current market rate $R(t)$ observed in the market place at the same time. If instead all refinancing opportunities have already been exhausted, i.e., $n(t) = 0$, then the new current debt rate $r(t)$ must

---

[1] Specifically, the repayment process $(P^\alpha(t), 0 \le t < T)$ is defined on the probability space $(\mathcal{S}^T, \mathcal{B}(\mathcal{S}^T), \mathbb{P}_{R(0)})$ and the expectation is performed under the probability measure $\mathbb{P}_{R(0)}$ induced on the space of paths by the transition probabilities (21.1) and the initial condition $R(0)$.

agree with its previous value $r(t-1)$. In general, $r(t)$, and thus a Markov control policy $\alpha(t)$, depends on the triplet $(R(t), r(t-1), n(t))$.

These considerations lead to consider control policies whose time t value depends on the outstanding balance $B(t)$, the market rate $R(t)$, the one-period-ahead debt rate $r(t-1)$ and number $n(t)$ of available refinancing options:

$$\alpha(t) = F\big(t, \big(B(t), R(t), r(t-1), n(t)\big)\big). \tag{21.3}$$

If B, R, r^-, n represent possible values taken by $B(t), R(t), r(t)$, and $n(t)$, respectively, the 4-uple (B, R, r^-, n) is a candidate state variable.

We adopt the dynamic programming principle for the purpose of computing the value function and the optimal exercise policy over the contract lifetime. Since the state variable $B(t)$ is continuous, a direct application of this principle is prevented. However, the value function is homogeneous of degree one in this variable and the value function can be written as

$$V(t, (y, x, r, n)) := y \times V^*(t, (x, r, n)). \tag{21.4}$$

We refer to V^* as the "Unitary Value Function" because it represents the value function per unit of outstanding balance. Since y is non-negative, the optimal policy for V coincides with the optimal policy for V^*. As a consequence, admissible stopping rules are independent of y. We therefore consider control variables of the form

$$\alpha(t) = F\big(t, \big(R(t), r(t-1), n(t)\big)\big). \tag{21.5}$$

The control variable so defined is $\mathcal{F}_t$-measurable, implying that the control policy (21.5) is admissible. The state variable of our problem is the triple (R, r^-, n) defined on a subset of $\mathcal{X} = S \times S \times \{0, \dots, N\}$. For instance, if $t = 1, r \in \{r_0, r_0 + \Delta s, r_0 - \Delta s\} \subset S$. It turns out that the loss in terms of computational complexity for restricting numerical calculations to the exact domain of the state variable highly overcomes the gain resulting from reducing the number of computations. Consequently, we decide to skip considering the domain constraints while performing the optimization algorithm and compute the value function over the entire domain $\mathcal{X}$.

21.2.2 Dynamic Programming Algorithm

Having identified the state variable, we now turn to the recursive computation of the value function V^* and the determination of the optimal refinancing strategy consisting of an N-uple of $\mathcal{T}$-valued strictly increasing stopping times.

Time T Value Function

At time T the debtor pays off the time $T-1$ standing capital plus the interest accrued between $T-1$ and T. Then the contract extinguishes and becomes worthless:

$$V(T, (y, x, r, n)) = 0,$$

for all admissible y, x, r, and n.

Time $T - 1$ Value Function

If no refinancing option is available anymore, i.e. $n = 0$, the debtor incurs a payment equal to time $T - 1$ outstanding debt y plus the interest accrued between $T - 1$ and T according to the rate r, that is

$$CNR^0(T - 1, y, x, r) = y \times c(r) \times T^*(T - 1),$$

where $c(r) := (1 + r/52)$, $T^*(t) := 1/(T - t)$ is the complement of the coefficient $T(t)$ to the unit. If any refinancing option is still available, i.e., $n > 0$, we compare the cost CR resulting from refinancing, i.e., $\alpha = 1$, to the one CNR from continuing under the standing conditions, i.e., $\alpha = 0$. The former is the sum of the interest $CP_{[T-1,T]}(y, x)$ accrued on the outstanding debt y over the period between $T - 1$ and T according to the new market rate x and the discounted expected value function $DEC(T, y, x, r, n - 1)$ calculated one time step later, i.e., at time $(T - 1) + 1 = T$, corresponding to one refinancing opportunity less, that is

$$CR(T - 1, y, x, r, n) = CP_{[T-1,T]}(y, x) + DEC(T, y, x, r, n - 1)$$
$$= y \times c(x) \times T^*(T - 1) + 0.$$

The latter is defined similarly, except for the new debt rate which now equals the current debt rate r:

$$CNR(T - 1, y, x, r, n) = CP_{[T-1,T]}(y, r) + DEC(T, y, x, r, n)$$
$$= y \times c(r) \times T^*(T - 1) + 0.$$

The time $T - 1$ value function is thus:

$$V\left(T - 1, (y, x, r, n)\right) \tag{21.6}$$
$$= \begin{cases} \min\{CR(T - 1, y, x, r, n), CNR(T - 1, y, x, r, n)\}, & n \geq 1, \\ CNR^0(T - 1, y, x, r), & n = 0 \end{cases}$$
$$= \begin{cases} \min\{y \times c(x)T^*(T - 1), y \times c(r)T^*(T - 1)\}, & n \geq 1 \\ yc(r)T^*(T - 1), & n = 0. \end{cases}$$

Since $c(\cdot)$ is an increasing function and both y and $T^*(T - 1)$ are positive, if $n > 0$, the optimal decision is to refinance the mortgage provided that the new market rate x is lower than the standing debt rate r. The time $T - 1$ optimal control policy reads as

$$\alpha(T - 1)(y, x, r, n) = \alpha(T - 1)(x, r, n) \tag{21.7}$$
$$= \begin{cases} 1 & \text{if } n > 0 \text{ and } x < r, \\ 0 & \text{if } n = 0 \text{ or } r \leq x. \end{cases}$$

Time t Value Function

The value function at time t is to be computed as in formula (21.6) by replacing $T - 1$ is replaced with t:

$$V(t, (y, x, r, n))$$
$$= \begin{cases} \min\{CR(t, y, x, r, n), CNR(t, y, x, r, n)\} & \text{if } n = 1, \ldots, N, \\ CNR^0(t, y, x, r) & \text{if } n = 0. \end{cases} \quad (21.8)$$

The main difference here is that more intensive computations are needed.

$$CR(t, y, x, r, n) = CP_{[t,t+1]}(y, x) + DEC(t + 1, y, x, r, n - 1)$$
$$= yc(x)T^*(t) + \mathbb{E}^{\mathbb{P}_x}\big(V\big(t, \big(B(t + 1), \cdot, x, n - 1\big)\big)\big), \quad (21.9)$$

where $T^*(t) := 1/(T - t)$ is the complement to one of the coefficient $T(t)$ and the symbol "$\cdot$" indicates the argument with respect to which expectation is to be computed. Noting that (1) the time $t + 1$ standing balance is equal to $B(t + 1) = c(r(t))B(t)T(t)$, (2) the expected value is computed over the possible values taken by the market rate $R(t + 1)$, and (3) V factors in the product $y \times V^*$, the cost for refinancing becomes

$$CR = yc(x)T^*(t) + \mathbb{E}^{\mathbb{P}_x}\big(V\big(t + 1, \big(c(x)yT(t), R(t + 1)(\cdot), x, n - 1\big)\big)\big)$$
$$= yc(x)T^*(t) + c(x)T(t)y\mathbb{E}^{\mathbb{P}_x}\big(V^*\big(t + 1, \big(R(t + 1)(\cdot), x, n - 1\big)\big)\big).$$

Analogous computations for the cost of continuation lead to

$$CNR = CP_{[t,t+1]}(y, r) + DEC(t + 1, y, x, r, n - 1)$$
$$= yc(r)T^*(t) + \mathbb{E}^{\mathbb{P}_x}\big(V\big(t + 1, \big(B(t + 1), \cdot, r, n\big)\big)\big)$$
$$= yc(r)T^*(t) + c(r)T(t)y\mathbb{E}^{\mathbb{P}_x}\big(V^*\big(t + 1, \big(R(t + 1)(\cdot), r, n\big)\big)\big).$$

The same cost of continuation when all options have been exercised, i.e., $n = 0$, reads as

$$CNR^0(t, y, x, r) = CNR(t, y, x, r, 0).$$

By gathering these expression altogether and dividing by the outstanding balance y we come up to a formula for the time t unitary value function:

$$V^* = \begin{cases} \min\big\{c(x)T^*(t) \\ \quad + c(x)T(t)\mathbb{E}^{\mathbb{P}_x}\big(V^*\big(t + 1, \big(R(t + 1)(\cdot), x, n - 1\big)\big)\big), \\ \quad c(r)T^*(t) + c(r)T(t)\mathbb{E}^{\mathbb{P}_x}\big(V^*\big(t + 1, \big(R(t + 1)(\cdot), r, n\big)\big)\big)\big\}, & n \geq 1, \\ c(r)T^*(t) \\ \quad + c(r)T(t)\mathbb{E}^{\mathbb{P}_x}\big(V^*\big(t + 1, \big(R(t + 1)(\cdot), r, 0\big)\big)\big), & n = 0. \end{cases}$$

The optimal stopping rule at time t is not as simple as in formula (21.7) because explicit minimization is required to decide whether to refinance or not. We finally remark that the probability under which expectations are taken is determined by (21.1) through the rule $P_x(\{y\}) := P(x, y)$ in all cases.

Time 0 Value Function

At time $t = 1$ the unitary value function V^* becomes independent of variables r and n. Indeed, this is the first time the debtor is allowed to refinance and all the N options are available. Note that the debt rate on $[0, 1]$ is r_0 and $R(1) \in \{r_0, r_0 + \Delta s, r_0 - \Delta s\}$. The value function is given by

$$V = V^*(0, (r_0, r_0, N)) = c(r_0)\mathcal{T}^*(0) + c(r_0)\mathcal{T}(0) \times \mathbb{E}^{\mathbb{P}_{r_0}}\big(V^*(1, (\cdot, r_0, N))\big).$$

This is the fair price of the mortgage contract at time 0.

21.2.3 Transaction Costs

To provide an effective disincentive many mortgages prescribe a fee to apply upon prepayment: the mortgagor wishing to extinguish the contract is required to pay an additional fee usually ranging from 0.5 to 2 percentage points over the outstanding balance. A slight modification is required to accommodate this feature with our framework. Again, the value function factorizes as

$$V(t, (y, x, r, n)) := y \cdot V^*(t, (x, r, n)). \tag{21.10}$$

Since transaction costs are proportional to the outstanding balance, they can be included into the value function as a constant adding to the function CR. Let ϕ represent the outstanding balance quota that is singled out as a prepayment fee. In general, the time t value function $V^* = V^*(t, (x, r, n))$ per unit balance is computed as

$$V^* = \begin{cases} \min\{c(x)\mathcal{T}^*(t) \\ \quad + c(x)\mathcal{T}(t)\mathbb{E}^{\mathbb{P}_x}\big(V^*(t+1, (R(t+1)(\cdot), x, n-1))\big) + \phi, \\ \quad c(r)\mathcal{T}^*(t) + c(r)\mathcal{T}(t)\mathbb{E}^{\mathbb{P}_x}\big(V^*(t+1, (R(t+1)(\cdot), r, n))\big)\}, & n \geq 1, \\ c(r)\mathcal{T}^*(t) + c(r)\mathcal{T}(t)\mathbb{E}^{\mathbb{P}_x}\big(V^*(t+1, (R(t+1)(\cdot), r, 0))\big), & n = 0. \end{cases}$$

21.2.4 Code

The program described above has been implemented in three Matlab® codes. These modules perform the following operations:

1. Path simulation of the Markov chain driving market rate dynamics.
2. Backward induction determining optimal control policy.
3. Scenario generation and graphics plotting.

Code I

1. The first code is called by the main program contained in the third code. We define some global variables.
   T = time horizon = 249 (weeks);
   $R0$ = initial interest rate = $R(0) = 0.05$;
   Ds = weekly absolute variation of market rates = $\Delta s = 0.002$ ($K = 41$);
   $ALPHA$ = initial number of allotted refinancing opportunities = $n(0) = 4$.

2. Each knot in the tree accounts for the following information: time t, current market rate x, outstanding debt rate one r, and remaining number of refinancing opportunities.

3. The output of `gen_traj` is the T-sized vector y representing the weekly movement of market interest rate according to the following rule:
   - $y(t) = +1 =$ Soaring by 20 b.p. (basis points) in the market rate between week $t - 1$ and week t.
   - $y(t) = -1 =$ Falling by 20 b.p. in the market rate between week $t - 1$ and week t.
   - $y(t) = 0 =$ Standing steady between week $t - 1$ and week t.

Code II

The main aim of Code II is to evaluate the Value Function as a value of the different input-parameters.

4. We first evaluate the value function at time $t = T - 1 = 249$. Since Matlab® requires positive integer-valued indexes, some quantity needs to be rescaled and/or translated. For instance the Value Function V depends on t, ri, xj, n, all positive, where:

   t is time in weeks;

   ri denotes standing debt rate;

   xi represents current market rate;

   n is the number of standing refinancing opportunities (plus one).

5. The state-space has been rescaled and translated so that ri and xj take values between 1 and 41, where

   $+1$ stands for $xi = 0.01$, the lower limit;

   $+41$ stands for $xi = 0.09$, the upper limit;

   $+21$ stands for $xi = 0.05$, the starting reference rate.

6. Index n is between 1 and 5. Value 1 means that no refinancing opportunity is available anymore, while value 5 means that all refinancing opportunities are still available. If refinancing is still possible (i.e. $n > 1$), the $(T - 1)$-Value Function results from comparing the current market rate to the standing debt rate. Finally, the Value Function can be computed as
   `V(t,ri,xj,n)= min(Vrif,VNrif).`

Code III

This code is divided into three parts: the first part contains all global variables; the second part evaluates probability distribution and control on the last trajectory; the third part plots results as well as the control corresponding to a sample trajectory (see Chapter 6 for the distinction between "control policy" and "control").

7. We set a seed for the uniform random generator to its initial default value.

```
rand('seed',0);
rand('state',0);
```

8. Stat is a (4×179)-matrix whose $(i.j)$ entry contains the probability of the ith refinancing at time j. Last is a vector to be used for plotting sample paths.
9. A sample scenario is generated by calling function gen_traj, with a starting rate $r = 0.05$.

```
for k = 1 : N
  y = gen_traj;
  n = ALPHA+1;
  ri = 21;  \%-Kmin+1
  xj = 21;  \%-Kmin+1
```

10. At each knot in a given scenario, we check for the value taken by C. If $C = 1$, the debt rate is set to the current market rate and n decreases by 1. For each of the N sample paths ($N = 10E^4$), function Stat stores the exact point where the nth refinancing opportunity has been used. Each time an opportunity is accepted, Stat increases by 1 in the appropriate cell.

```
if C(t,ri,xj,n)==1
  if (k==N)
  graphx(ALPHA+2-n) = t;
  graphy(ALPHA+2-n) = 0.05+(xj-21)*0.002;
  end
  ri = xj;
  n = n - 1;
  Stat(ALPHA+1-n,t) = Stat(ALPHA+1-n,t)+1;
```

11. Functions last, graphx andgraphy display the last sample trajectory and the corresponding stochastic control.

21.3 Results and Comments

Roncoroni and Moro (2006) computed the fair value of the contract $V(0, (1, r_0, N))$ corresponding to a standing rate $r_0 = 5\%$ an a number $N = 4$ of refinancing opportunities. This is the minimal cost for the mortgagor, namely the one resulting from the optimal control policy. They obtain an optimal cost equal to 1.4404.

The optimal refinancing policy is expressed as a rule transforming any possible configuration of the state variable into an action consisting of either continuing or refinancing the mortgage. To have a concrete idea about the optimal policy, we plot one sample path of the market rate process and indicate the four optimal refinancing times (see Fig. 21.1).

Notice that the first refinancing option is exercised after a few weeks compared to the repayment horizon. While market rates increase for about 75 weeks, no refinancing takes place. In the following weeks, market rates steadily decrease and a new refinancing option is exercised as soon as they go below the standing debt rate. The remaining two options are exercised in a relatively narrow time window, always

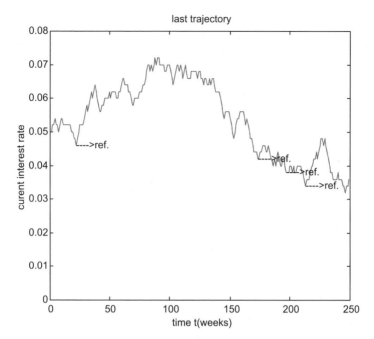

Fig. 21.1. Sample trajectory of the market rate of interest. Expression "→ ref" indicates the four optimal refinancing times.

at points of minimum along the rate path. Naturally, this behavior is specific to the examined path and other samples may give rise to different control actions.

A clearer picture of the optimal control policy is obtained by computing the distributions of the four optimal refinancing times. Analytical expressions for these distributions are not available. However, we may compute a sample estimate of their shape by simulating a large number of trajectories, then storing the corresponding optimal stopping times, and finally computing the relative frequency histogram of the four refinancing times. Figure 21.2 shows a sample probability distribution for the first two optimal refinancing times as obtained through 10,000 sample paths of the market rate process.

The first option tends to be exercised within the first two weeks, whereas the second stopping time seems to span a wide time period. It is interesting to note that a refinancing option is never exercised during the first period. The first refinancing time displays a spiky distribution with a maximum attained at time $t = 2$. It seems that if the market rate decreases during the first period, the high proximity to the initial period tends to offset the advantage stemming from refinancing. On the contrary, if the market rate decreases in the second period, this effect reverses and the mortgage is refinanced.

Figure 21.3 shows the last two refinancing times.

These times cluster on the farthest end of the horizon spectrum. However, they also display a marked spiky behavior. We also computed the empirical relative fre-

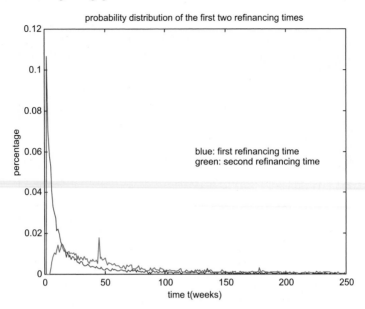

Fig. 21.2. Sample probability distribution of the first two refinancing times over $n = 10,000$ sample paths.

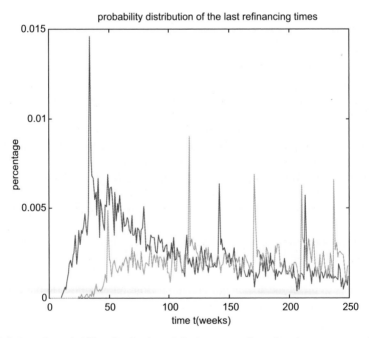

Fig. 21.3. Sample probability distribution of the last two refinancing times over $n = 10,000$ sample paths.

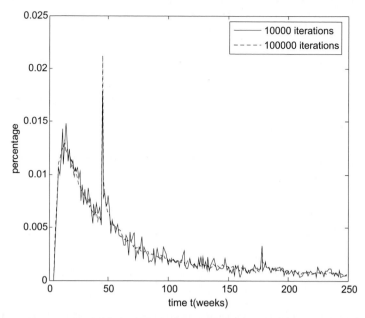

Fig. 21.4. First optimal refinancing time sample distributions computed over alternative sample sizes. Case 1: (plain line) sample size = 10,000. Case 2: (dotted line) sample size = 100,000.

quency of the first refinancing time over 10,000 and 100,000 sample paths. Figure 21.4 displays the two distributions obtained by independent path generations.

The two graphs exhibit similar properties, showing that convergence to the exact distribution of the first refinancing time is fairly quick. In particular, the large spike occurring in about 50 weeks seems to be an intrinsic feature of the control policy. Our final experiment checks for the behavior of the mortgage value as a function of the entry level mortgage rate r_0 and an additional fee is applied to any refinancing decision. Interest rates vary from 1 to 9 percentage points. Prepayment fees range from 0 to 2 percentage points over the standing capital. The contract is supposed to expire in 180 months and 4 refinancing opportunities are allotted at the outset. Figure 21.5 reports a plot of the corresponding two-dimensional surface.

It is clear that higher entry rates or larger fees have a negative effect on the minimal cost attainable by the mortgage holder. This is reflected by an increase in the time 0 value function as shown by the graph.

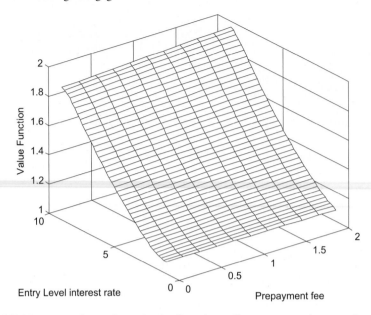

Fig. 21.5. Mortgage value under optimal refinancing policy across varying entry level mortgage rates and prepayment fees.

Basket Default Swaps[*]

Key words: credit derivatives, defaultable bonds, simulation

22.1 Problem Statement

Credit derivatives are financial contracts allowing the transfer of credit risk from one market participant to another. This feature improves the efficient allocation of credit risk amongst market participants. For this reason, in recent years credit derivatives have become the main tool for transferring and hedging risk.

Credit derivative markets have rapidly grown in volume and variety of traded products. Credit-linked securities have experienced various applications, ranging from hedging default risk, to freeing up credit lines, to reducing regulator capital requirements, to hedging dynamic credit exposure driven by market variables and to diversifying financial portfolios by gaining access to otherwise unavailable credits. The outstanding balance of credit derivatives contracts has increased from an estimated USD 50 billion in 1996 to almost USD 500 billion at the end of the year 2000. According to the 2003 ISDA survey the volume of the credit derivatives market has reached an outstanding notional of more than USD 2 trillion in February 2003. The exact size of the global credit derivatives markets, however, is difficult to estimate, given the potential for overcounting when contracts involve more than one counterpart. Moreover, the notional amounts outstanding considerably overstate the net exposure associated with those contracts.

Credit default swaps (CDS) constitute the fundamental class of credit derivatives (Schonbucher (2003)). These instruments allow banks to hedge credit risk underlying loans and other interest rate positions. A more complicated credit-linked product is the basket default swap (BDS), whose pay-off is contingent on the default occurring within a basket of bonds. More precisely, a basket default swap is a bilateral contract whereby one counterpart, named the "protection seller", agrees to make

---

[*] with Giacomo Le Pera and Davide Meneguzzo.

a specific compensation payment to the other counterpart, named the "protection buyer", whenever a specific credit event occurs. This event is defined with respect to a basket of obligors. A 1st-to-default basket provides for a payment upon default of any of the obligors within the basket. Settlement may involve cash only, e.g., a (single name) credit default swap, or it may require a physical delivery of the defaulted asset in return for a cash payment of the par amount. In return for protection against the 1st-to-default, the protection buyer pays a basket spread to the protection seller defined as a sequence of cash flows. These payments terminate upon the first credit event. Similarly, other credit products may depend on one or several default events. An nth-to-default basket triggers a credit event after n (or more) obligors have defaulted. A last-to-default basket refers to the case where all assets in the basket are required to default.

Modeling the joint default dependency is the main issue one has to investigate for the purpose of pricing these securities and monitoring their risk level. Since the Merton (1974) model, several approaches have been suggested for modeling the dependence structure among different assets. Existing approaches can be grouped into two classes of models, known as structural models and reduced-form models.

Structural models detect a default event whenever the firm's value (as modelled using a suitable stochastic process) breaks a certain lower bound meant to signal company liquidation or restructuring. Assuming that the firm's value can be described through a geometric Brownian motion, the default event probability distribution can be determined in analogy to the pricing of barrier options. The underlying is the firm value and the strike is given by the firm's debt.

Reduced-form models are based on the concept of default time. This approach treats default as a jump process described through a corresponding intensity process specifying the time evolution of the expected number of default per unit time. Depending on the complexity of the model, the intensity process is assumed to be either deterministic or random. The resulting framework provides tractable models which can be easily calibrated to market data.

We hereby aim at illustrating three alternative methods for pricing multi-name credit derivatives. In particular, we compare the Monte Carlo approach with two semi-analytical techniques, each one providing an appreciable improvement in terms of computational efficiency. In particular, we show that the Monte Carlo setting suffers from the fact that an extremely significant number of simulations are required in order to achieve an acceptable convergence. Moreover, a large number of obligors, usually more than 100, are involved in the basket and the resulting procedure becomes computationally inefficient, if not totally unfeasible. Finally, the pairwise correlation matrix for N obligors has dimension $N \cdot (N-1)/2$. For instance, 150 credits require specifying 11,175 parameters! This problem urges the final user to develop alternative valuation methods such as the ones we present below.

22.2 Models and Solution Methodologies

22.2.1 Pricing nth-to-default Homogeneous Basket Swaps

In this section we present a methodology for pricing basket default swaps. We consider a basket that pays a spread s at dates $\{t_1, t_2, \ldots, t_{i-1}, \ldots, T_M\}$. The periodic payments of the protection buyer to the protection seller are denoted as the premium leg (PL); the payment of the protection seller to the protection buyer in case of default of the underlying credit is called default leg (DL). Δ denotes the year fraction representing the period between payments (i.e., $1/4$ for quarterly payments, 1 for annual frequency) and with $B(0, t_i)$ the nonstochastic discount factor for the maturity t_i. For simplicity, we do not take into account the accrued premium until the default date.

Given the definition of counting process $N(t) := \sum_{i=1}^{N} \mathbf{1}_{\tau_i \leq t}$, where $\mathbf{1}_{\tau_n > t}$ is the survival indicator of the nth default time (see Chapter "Dynamic Monte Carlo"), the nth-to-default basket premium leg is:

$$PL = s\Delta \sum_{i=1}^{M} \left[B(0, t_i) \mathbb{P}\big(N(t_i) < n\big) \right] \tag{22.1}$$

$$= s\Delta \sum_{i=1}^{M} \left[e^{-rt_i} \sum_{j=0}^{n-1} \mathbb{P}\big(N(t_i) = j\big) \right],$$

where $\mathbb{P}(N(t_i) < n)$ is the probability to have less than n defaults no later than time t_i. The above expression means that the protection buyer pays the spread s until n names have defaulted, where s is quoted *per annum*.

If the nth default occurs before the maturity T, the protection seller has to pay the difference between the par value and the recovery rate (R) of the reference entity. Considering the same recovery rate for each obligor in the basket,[1] the default leg (DL) is given by

$$DL = (1 - R)\mathbb{E}[B(0, \tau_n) \cdot \mathbf{1}_{\tau_n \leq T}].$$

If we assume that default can occur continuously over time, we may define $F_n(t) = \mathbb{P}(N(t) \geq n)$ and we rewrite the above expression as:

$$DL = (1 - R) \int_0^T B(0, t)\, \mathrm{d}F_n(t).$$

Under the hypothesis of constant interest rate and integrating by parts, we arrive at:

$$DL = (1 - R)\left\{ e^{-rT} F_n(T) + r \int_0^T F_n(t)e^{-rt}\, \mathrm{d}t \right\}. \tag{22.2}$$

---

[1] This assumption greatly simplifies the analysis that follows. A more realistic viewpoint ought to consider the special features exhibited by empirical recovery rates as it has been illustrated, e.g., in Altman, Resti and Sironi (2004, 2005) and Titman, Tompaidis and Tsyplakov (2004).

Imposing that $PL = DL$, we obtain the equilibrium spread

$$s^* = \frac{(1-R)\{e^{-rT} F_n(T) + r \int_0^T F_n(t)e^{-rt}\, dt\}}{\Delta \sum_{i=1}^{T}[e^{-rt_i} \sum_{j=0}^{n-1} \mathbb{P}(N(t_i) = j)]}. \tag{22.3}$$

Pricing a basket default swap requires estimating the probability distribution of default times. This can be modeled in the following two ways:

1. Sampling a large number of random scenarios from the joint distribution;
2. Using a factor model to deliver tractable semi-analytic computations.

We provide an implementation of the first solution using copula functions and Monte Carlo simulations. We describe the second methodology adopting two different factor models: one based on the use of the Fast Fourier Transform, the other on a recursive algorithm.

22.2.2 Modelling Default Times

Default probabilities p_i can be obtained either directly or implicitly. In the first case, information stems either from the market via the observation of credit default swaps spreads, or from rating agencies via advertised default rating matrices. In the second case, they are derived from an assessment of the so-called hazard rate h_i as implied from market prices of credit derivatives. We now shortly illustrate a hybrid version of this approach and refer to Chapter "Dynamic Monte Carlo" for a self-contained introduction to jump processes and their simulation algorithms.

Consider an increasing sequence of stopping times $\tau_1 < \cdots < \tau_n < \cdots$, and define a counting process as

$$N(t) := \sum_{i=1}^{N} \mathbf{1}_{\tau_i \le t}. \tag{22.4}$$

A (homogeneous) Poisson process with intensity $\lambda > 0$ is a counting process whose increments are independent and the probability that n jumps occur in any interval $[s, t]$ is given by

$$\mathbb{P}[N(t) - N(s) = n] = \frac{1}{n!}(t-s)^n \lambda^n e^{-(t-s)\lambda}.$$

In other words, $N(t)$ has independent increments and $N(t) - N(s)$ is a Poisson variable with parameter $\lambda(t-s)$. The time of default τ is defined as the first jump time of the process N, that is

$$\tau = \inf\{t > 0 \colon N(t) > 0\}.$$

Intensity based models assume that an intensity function is given, then build a counting process with that intensity, and finally derive the corresponding default time distribution. In real applications, the intensity function is bootstrapped from bond prices and credit default swap spreads.

22.2.3 Monte Carlo Method

The Gaussian Copula model introduced by Lì (2000) is one of the most popular methods for pricing financial products whose payout depends on default times $\tau_1, \ldots, \tau_N$. From here on, we assume the recovery rate R_i paid off in case of default of the ith underlying asset is a fraction of the nominal.

The basic assumptions of this framework are as follows:

1. The default time τ_i is modeled as the first jump time of an inhomogeneous Poisson process.
2. The dependence structure of default times is assigned through a Gaussian copula (see (22.5)).
3. Under the risk-neutral probability $\mathbb{P}^*$, the default-free interest rate dynamics are independent of default times.
4. The fraction of the nominal which is paid in case of default is given by the recovery rate, which we assume to be the same for all obligors.

Under these assumptions, the default time for the ith defaultable product in the basket has a cumulative distribution function given by

$$F_i(t) = 1 - \exp\left(-\int_t^T \lambda_i(s)\,ds\right).$$

Moreover, using a Gaussian copula leads to the joint distribution of default times as

$$F_{\tau_1,\ldots,\tau_n}(t_1, t_2, \ldots, t_n) = \Phi_n\big(\Phi^{-1}(F_1(t_1)), \Phi^{-1}(F_2(t_2)), \ldots, \Phi^{-1}(F_n(t_n))\big),$$

$$(22.5)$$

where Φ_n denotes the n-dimensional normal cumulative function with correlation matrix Σ. This matrix is just the correlation matrix among asset price returns and may be estimated from the observation of equity indices.

Correlated default times can be simulated by the following algorithm.

Algorithm (Simulating Correlated Events via Gaussian Copula)

1. Sample $Y_1, \ldots, Y_n$ from a multivariate normal $\mathcal{N}(\mathbf{0}, \Sigma)$.
2. Return $\tau_i = F_i^{-1}(\Phi(Y_i)), i = 1, \ldots, n$.

This method can be easily proved by noting that

$$Y_i = \Phi^{-1}(F_i(t)), \quad \text{for all } i = 1, \ldots, n.$$

22.2.4 A One-Factor Gaussian Model

According to the classical structural framework proposed by Merton (1974), the one-factor model postulates that a company defaults when the firm value falls below a certain threshold. Let the firm value be driven by two components, namely a common factor Y representing a source of systematic risk and a noise term ε_i that is idiosyncratic to the firm. The systematic risk component can be viewed as an indicator of the state of the business cycle (e.g., stock index, GDP, interest rates). The idiosyncratic part is a firm-specified indicator and depends on events strictly linked to the firm (e.g., the quality of the management, the market share, the position with respect to competitors). We assume that Y and ε_i are independent and normally distributed variables with zero mean and unit variance.

The dynamics of firm value i is given by

$$V_i(t) = \rho_i Y + \sqrt{1 - \rho_i^2}\,\varepsilon_i, \tag{22.6}$$

where factor $\rho_i \in [0, 1]$. If $\rho_i = 0$ then the firm value does not depend on the general state of the economy; if $\rho_i = 1$, the randomness affecting the firm value is exclusively due to macroeconomic events. Furthermore, we assume that the correlation between two firm values is $\mathrm{cov}(\varepsilon_i, \varepsilon_j) = \rho_i \rho_j$.

The default time distribution can be determined by defining a default event time for a firm i as the first date for which $V_i(t) \leq \Phi^{-1}(F_i(t))$, i.e.,

$$\varepsilon_i \leq \frac{\Phi^{-1}(F_i(t)) - \rho_i \cdot Y}{\sqrt{1 - \rho_i^2}}. \tag{22.7}$$

Knowing $\Phi^{-1}(F_i(t))$ and ρ_i, the conditional default probability $p_{i|y}$ that credit i will default at time t, conditional on Y is given by:

$$
\begin{aligned}
p_{i|y}(t) &= \mathbb{P}\left[V_i(t) \leq \Phi^{-1}(F_i(t))|Y = y\right] \\
&= \mathbb{P}\left[\rho_i \cdot Y + \sqrt{1 - \rho_i^2} \cdot \varepsilon_i \leq \Phi^{-1}(F_i(t))|Y = y\right] \\
&= \mathbb{P}\left[\varepsilon_i \leq \frac{\Phi^{-1}(F_i(t)) - \rho_i \cdot Y}{\sqrt{1 - \rho_i^2}}\bigg|Y = y\right] \\
&= \Phi\left(\frac{\Phi^{-1}(F_i(t)) - \rho_i y}{\sqrt{1 - \rho_i^2}}\right),
\end{aligned}
\tag{22.8}
$$

for all $i = 1, \ldots, n$. The independence property leads to the joint default probability distribution for the N obligors as:

$$F(t) = \mathbb{P}(\tau_1 \leq t, \tau_2 \leq t, \ldots, \tau_N \leq t) = \int_{-\infty}^{+\infty} \prod_{i=1}^{N} p_{i|y}(t)\phi(y)\,\mathrm{d}y, \tag{22.9}$$

where ϕ_Y is the factor density function. The corresponding joint survival distribution reads as:

$$S(t) = \mathbb{P}(\tau_1 > t, \tau_2 > t, \ldots, \tau_N > t) = \int_{-\infty}^{+\infty} \prod_{i=1}^{N} q_{i|y}(t)\phi(y)\,\mathrm{d}y,$$

where $q_{i|y}(t) = 1 - p_{i|y}(t)$.

22.2.5 Convolutions, Characteristic Functions and Fourier Transforms

In some cases, it is possible to derive analytical formulas for the value of the default distribution or to compute the value of the default distribution with Fourier methods. These results can be used to check the outcome of Monte Carlo simulations.

We now show how to use the Fast Fourier Transform to get to the distribution function of a counting process $N(t)$. We start with a few definitions.

Definition (Convolution Product) If f and g are measurable functions, we define their convolution as:

$$(f \otimes g)(u) = \int_{-\infty}^{\infty} f(x)g(u - x) \, dx.$$

If X and Y are independent random variables with probability density f_X and f_Y, the distribution function of the sum $X + Y$ is given by $f_X \otimes f_Y$.

Definition (Characteristic Function) The characteristic function (CF) of a continuous r.v. X with d.f. f_X is defined as $\Gamma(t) = \mathbb{E}(e^{itX})$, i.e.,

$$\Gamma(t) = \int_{-\infty}^{+\infty} f_X(x)e^{itx} \, dx, \tag{22.10}$$

where $i = \sqrt{-1}$ denotes the imaginary unit.

Definition (Fourier Transform) The Fourier Transform (FT) of a function f is defined by:

$$F(t) = \int_{-\infty}^{\infty} f(x)e^{itx} \, dx$$

if this integral exists. The function

$$f(x) = \frac{1}{\sqrt{2\pi}} \int_{-\infty}^{\infty} F(w)e^{iwx} \, dw \tag{22.11}$$

is called the Inverse Fourier Transform (IFT), if it exists.

In what follows, we use the Fast Fourier Tranform (FFT) algorithm to implement discrete Fourier transforms and the Inverse Fast Fourier Transform (IFFT) to implement discrete IFT. Given a finite set of data, e.g., a periodic sample taken from a real-world signal, the FFT expresses these data in terms of their component frequencies. It also solves the essentially identical inverse problem of reconstructing a signal from the frequency data.

The moment generating function of a counting process $N(t)$ is given by

$$\Psi_{N(t)}(u) = \mathbb{E}(u^{N(t)}) = \sum_{k=0}^{N} \mathbb{P}(N(t) = k)u^k. \tag{22.12}$$

Given the observation of factor Y, we can write

$$\Psi_{N_i(t)|Y}(u) = \mathbb{E}\left[u^{N_i(t)}|Y\right]$$
$$= u^1 p_{i|y}(t) + u^0\left(1 - p_{i|y}(t)\right) = up_{i|y}(t) + \left(1 - p_{i|y}(t)\right),$$

where $p_{i|y}(t)$ is the marginal default probability as defined in (22.8). By using the iterated expectation theorem and the independence of $N_i(t)$, we get to the unconditional moment generating function:

$$\Psi_{N(t)}(u) = \mathbb{E}\left(\prod_{i=1}^{n} \Psi_{N_i(t)|Y}(u)\right) = \int \prod_{i=1}^{n}\left[up_{i|y}(t)+\left(1-p_{i|y}(t)\right)\right]\phi(y)\,dy, \quad (22.13)$$

where $\phi(y)$ denotes the factor's density function. The probability $\mathbb{P}(N(t) = k)$ can be determined by calculating the coefficient of the term u^k from expression (22.12) and (22.13).

This formal expansion approach is well-suited for small values of N, whereas for large values we can use the FFT approach. The idea is to compute the characteristic function of $N(t)$ by replacing variable u in (22.12) with e^{iu}. The characteristic function has the following property: if N and K are independent, the characteristic function of the sum $N + K$ is the product of the characteristic functions of N and K. With this in mind, we obtain:

$$\Gamma_{N(t)}(u) = \mathbb{E}\left(\prod_{i=1}^{n} \Gamma_{N_i(t)|Y}(u)\right) = \int \prod_{i=1}^{n}\left[e^{iu}p_{i|y}(t) + \left(1 - p_{i|y}(t)\right)\right]\phi(y)\,dy,$$

where $\Gamma_{N_i(t)}(u)$ is the characteristic function of $N_i(t)$. Under certain technical conditions, the FFT of the probability density function is exactly its characteristic function. Therefore, we can write:

$$\Gamma_{N(t)|Y}(u) = \mathbb{E}\left[e^{iuN(t)}|Y\right] = \prod_{i=1}^{n} FFT(f_{i|y}(t)),$$

where $f_{i|y}(t)$ is the distribution of $N_i(t)$ that is a Bernoulli distributed random variable with probability $p_{i|y}(t)$. By inverting the expression above through the Inverse Fast Fourier Transform (IFFT) defined as

$$f_y(t) = IFFT\left\{\prod_{i=1}^{n} FFT(f_{i|y}(t))\right\}, \quad (22.14)$$

we can obtain the distribution of $N(t)$ given factor Y. Finally, the unconditional distribution is given by integration over the range spanned by variable y, namely:

$$f(t) = \int f_y(t)\phi(y)\,dy.$$

One of the greatest advantages of this approach is that the CF uniquely characterizes the distribution of a random variable.

22.2.6 The Hull and White Recursion

Recall that $q_{i|y}(t)$ denotes the survival probability $\mathbb{P}(\tau_i > t)$; then, the conditional probability of no-default is

$$P_{0|y}(t) = \prod_{i=1}^{N} q_{i|y}(t), \tag{22.15}$$

while the probability of one default

$$
\begin{aligned}
P_{1|y}(t) &= \sum_{i=1}^{N} \left\{ [1 - q_{i|y}(t)] \prod_{j=1, j\neq i}^{N} q_{j|y}(t) \right\} \\
&= \sum_{i=1}^{N} \left\{ [1 - q_{i|y}(t)] \prod_{j=1}^{N} \frac{q_{j|y}(t)}{q_{i|y}(t)} \right\}.
\end{aligned}
\tag{22.16}
$$

By using expression (22.15) in formula (22.16), we obtain

$$P_{1|y}(t) = P_{0|y}(t) \sum_{i=1}^{N} w_i(t|Y),$$

where $w_i(t|Y) = (1 - q_{i|y}(t))/q_{i|y}(t)$. Hull and White (2003) prove that the conditional probability of n names being in default by time t is:

$$S_{n|y}(t) = S_{0|y}(t) \sum w_{z(1)} w_{z(2)} \cdots w_{z(n)}. \tag{22.17}$$

If $z(1), z(2), \ldots, z(n)$ denote n distinct positive integers lower than $N+1$, the above sum is computed over $\binom{N}{n} = N!/(n!(N-n)!)$ possible selections. The unconditional probability that there will be exactly n defaults by time t can be determined using numerical integration over the distributions of factor Y, i.e.,

$$P_n(t) = \int_{-\infty}^{+\infty} P_{0|y}(t) \sum w_{z(1)} w_{z(2)} \cdots w_{z(n)} \phi(y) \, dy. \tag{22.18}$$

22.3 Implementation and Algorithm

We have developed MatLab® routines for pricing the nth to default basket swap using Monte Carlo simulation, the FFT approach and the Hull and White algorithm for the purpose of computing the probability of having n defaults by time t, $F_n(t)$. Let us consider a basket consisting of n assets; the algorithm requires a preliminary identification of the following variables:

(1) Contract maturity T;

(2) Correlation ρ among the underlying assets;
(3) Hazard rate h_i for each obligor;
(4) Common recovery rate R;
(5) Instantaneous interest rate r;
(6) Number of payments *per annum* f.

Then a contract type is selected (e.g., 1st-to-default, ..., nth-to-default) and a method for computing the joint probability distribution of default times is adopted (e.g., MC, FFT, or HW). All hazard rates are supposed to be constant; consequently, we are implicitly assuming that the jump dynamic of the defaults is driven by a Poisson process and the distribution of default times is exponential.

22.3.1 Monte Carlo Method

The Monte Carlo approach is based on the generation of random numbers from a Gaussian Copula. Given a basket with N references with default time correlation matrix Σ, the steps are as follows:

1. Find the Cholesky decomposition A of Σ;
2. Simulate n independent random variates $\mathbf{z} = (z_1, z_2, \ldots, z_n)^\top$ from $\mathcal{N}(0, 1)$.
3. Compute $\mathbf{x} = A\mathbf{z}$;
4. Put $u_i = \Phi(x_i)$ with $i = 1, 2, \ldots, n$ and where Φ denotes the univariate standard normal cumulative distribution function.
5. Return $(\tau_1, \ldots, \tau_N) = (F_1^{-1}(u_1), \ldots, F_N^{-1}(u_N))$, where F_i denotes the ith marginal.

With the present value of the default and premium leg for all scenarios the spread of the basket default swap is determined as quotient of the sum of the present values of the default legs divided by the sum of the present value of the premium legs (22.3).

22.3.2 Fast Fourier Transform

The FFT method is implemented by the routine `convolution.m`. This function delivers a vector containing the probabilities of having $0, 1, 2, \ldots, N$ defaults conditional on the factor Y. Starting with (22.8), the steps are as follows:

1. Define N vectors of size $(N + 1)$ as

$$\tilde{p}_{i|y}(t) := [\, 1 - p_{i|y}(t), \quad 0, \quad \ldots \quad 0, \quad p_{i|y}(t) \,],$$

for $i = 1, \ldots, N$.
2. Apply the FFT to each vector $\tilde{p}_{i|y}(t)$;
3. Set $\tilde{p}_y(t) = \prod_{i=1}^{N} FFT[\tilde{p}_{i|y}(t)]$;
4. Apply the IFFT to vector $\tilde{p}_y(t)$.

This procedure delivers a vector of conditional probabilities as a convolution of N input vectors.

22.3.3 Hull–White Recursion

Hull and White (2003) describe a method for calculating the expression (22.17). Calling $A_k = \sum [w_{z(1)} w_{z(2)} \cdots w_{z(k)}]$ with $k < N$ and $B_j = \sum_{j=1}^{N} w_i^j$, they prove the following recurrence relation:

$$A_1 = B_1,$$
$$2A_1 = B_1 A_1 - B_2,$$
$$3A_3 = B_1 A_2 - B_2 A_1 + B_3,$$
$$\vdots \qquad \vdots$$
$$k A_k = B_1 A_{k-1} + B_2 A_{k-2} - \cdots + (-1)^k B_{k-1} A_1 + (-1)^{k+1} B_k.$$

The function hw.m provides the related conditional probability vector. We have improved the code efficiency by writing all routines in vector notation. This means that all for and while loops have been replaced by vector or matrix operations. For the purpose of showing the computational burden stemming for non-vectorial coding, we provide routine hw_uv.m which performs the same algorithm as hw.m in a traditional loop-like programming style.

22.3.4 Code

For all the types of models the code is composed of a first common part that computes the present value of the default leg and the premium leg. While in the Monte Carlo algorithm we only have to generate the default times using the function GCdef.m, in the FFT procedure or the Hull and White method the code splits into three main parts. The first one evaluates the unconditional default probabilities with the chosen model. The second part calculates the unconditional probability by integrating over the factor's distribution. The third function gives the probability of having at least n defaults (i.e., the second summation in Eq. (22.1)).

The main function is ntd(model,T,rho,h,r,R,k,f), where model can assume values MC, FFT, or HW; T is the contract maturity; rho is the N-dimensional correlation vector as defined in (22.6); h is the N-dimensional hazard rate vector; r is the instantaneous rate of interest; R is the common recovery rate; k is the specification of the type of contract (i.e., for a 1st-to-default basket this value is set to 1); and f is the number of annual payments.

22.4 Results and Comments

We evaluate the fair spread of an nth-to-default basket swap under different scenarios. We run the function ntd(model,T,rho,h,r,R,k,f), with T = 5, rho ranging over [0, 1] with step 0.1, h = 0.01, r = 0.05, R = 0.5, k = {1, 2, 5}, and f = 1. The Monte Carlo algorithm has been run over 100,000 simulations.

Table 22.1. Fair spread (bps) for 5 names using FFT

ρ	1st t.d.	CPU time	2nd t.d.	CPU time	5th t.d.	CPU time
0	262.917	0.281	21.442	0.234	$\simeq 0$	0.234
0.1	261.775	0.281	22.209	0.281	$\simeq 0$	0.281
0.2	259.201	0.281	24.529	0.281	0.001	0.281
0.3	251.882	0.281	28.439	0.281	0.005	0.281
0.4	242.354	0.281	33.875	0.293	0.024	0.281
0.5	229.104	0.344	40.579	0.343	0.108	0.281
0.6	211.612	0.328	48.036	0.328	0.412	0.281
0.7	189.287	0.328	55.462	0.328	1.351	0.281
0.8	161.162	0.328	61.740	0.328	3.955	0.281
0.9	124.693	0.328	64.919	0.36	11.067	0.269
1	50	$\simeq 0$	50	$\simeq 0$	50	$\simeq 0$

Table 22.2. Fair spread (bps) for 5 names using HW recursion

ρ	1st t.d.	CPU time	2nd t.d.	CPU time	5th t.d.	CPU time
0	262.917	0.422	21.442	0.421	$\simeq 0$	0.422
0.1	261.775	0.5	22.209	0.0516	$\simeq 0$	0.5
0.2	259.201	0.5	24.529	0.5	0.001	0.5
0.3	251.882	0.516	28.439	0.531	0.005	0.516
0.4	242.354	0.593	33.875	0.531	0.024	0.515
0.5	229.104	0.593	40.579	0.625	0.108	0.516
0.6	211.612	0.594	48.036	0.601	0.412	0.516
0.7	189.287	0.594	55.462	0.594	1.351	0.5
0.8	161.162	0.656	61.740	0.594	3.955	0.5
0.9	124.693	0.656	64.919	0.687	11.067	0.547
1	50	$\simeq 0$	50	$\simeq 0$	50	$\simeq 0$

Tables 22.1, 22.2 and 22.3 report results showing the computational speed improvement over the standard Monte Carlo method. Indeed, it took 0.9320 seconds for the 1st-to-default under the assumption of $\rho = 0$. There are no remarkable differences between the Fast Fourier Transform method and the Hull–White recursion in terms of fair spread, while the former requires a shorter computational time. Figure 22.1 shows the proportion (in percentage points) of Monte Carlo paths that result in a default payoff for a test basket with correlation 0.5. We see that the estimated probability that the nth-default occurs before maturity is not stable. Figure 22.2 shows that the Monte Carlo estimator is unstable and converges at a slow speed. This experiment can be performed by running the script graph.m with correlation 0.5 and the same parameters as those reported above. Even if basket credit default swaps typically involve a limited number of reference entities ($n < 20$), we test whether the FFT and HW approach can be used for pricing Collateralized Debt Obligations (CDOs) where the number of obligors is typically greater than one

Table 22.3. Fair spread (bps) for 5 names using 100,000 Monte Carlo simulations

ρ	1st t.d.	CPU time	2nd t.d.	CPU time	5th t.d.	CPU time
0	264.439	0.923	21.566	1.011	$\simeq 0$	1.016
0.1	263.107	0.937	22.437	0.937	$\simeq 0$	0.984
0.2	259.953	0.953	24.657	0.937	0.009	0.921
0.3	253.329	0.968	28.279	0.968	0.018	0.953
0.4	244.084	0.937	33.587	0.953	0.038	0.968
0.5	230.825	0.953	40.001	0.953	0.086	0.968
0.6	212.805	0.968	47.987	0.968	0.339	0.953
0.7	190.428	0.953	56.082	0.968	1.126	0.968
0.8	162.561	0.984	62.741	0.968	3.650	0.953
0.9	126.004	0.953	66.072	0.953	10.591	0.953
1	50	$\simeq 0$	50	$\simeq 0$	50	$\simeq 0$

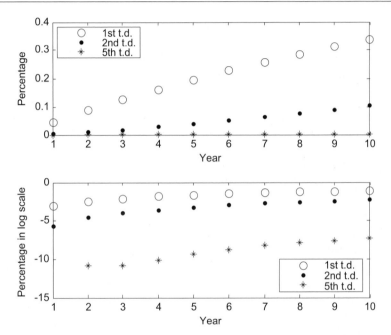

Fig. 22.1. Percentage of paths in default for a first, second and last to default basket. The results have been obtained running 10,000 Monte Carlo simulations.

hundred.[2] Due to the extremely burdensome computational time requirements, we decided not to run Monte Carlo. For a 2nd-to-default basket swap with correlation 0.5, we need about 9.5 seconds. We price a 2nd-to-default basket swap with 50 names

---

[2] A CDO comprises a pool of underlying instruments (called "collateral") against which notes of debt are issued with varying cashflow priority. These notes vary in credit quality depending on the subordination level. The collateral of a CDO is typically a portfolio of corporate bonds or bank loans or other type of financial facilities (residential or commercial mortgages, leasing, lending, revolving facilities, even other credit derivatives, to name few).

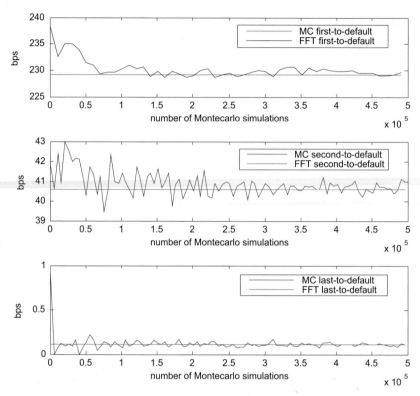

Fig. 22.2. Fair spread vs. number of Monte Carlo simulations.

and the usual parameter set. Table 22.3 shows that the FFT method works quite well and performs significantly faster than the Hull–White recursion (see Table 22.4). Moreover the HW algorithm does not converge for rho greater than 0.8 (see Table 22.5), due to the instability of the quadrature method adopted by the MatLab® function quadv.m. Indeed this function makes use of the recursive adaptive Simpson quadrature, whereas implementing a Gauss–Legendre quadrature would lead to convergence of the required integral (22.11). Hull and White (2003) use the recursion just to price nth-to-default basket swaps that involve a smaller number of obligors; for pricing correlation products that have a large number of underlings, they suggest a different algorithm. We propose a different implementation of the Hull–White recursion, using cycles instead of matrix syntax. Editing file prob.m with "@hw_un" substituting "@hw", we notice that the CPU time grows quite remarkably. For instance, in the case of a basket with 50 names and $\rho = 0.5$ the 1st-to-default premium is 1397.8 bps. However, the FFT method takes 1.95 seconds to run, whereas the Hull–White recursion requires 21.15 seconds in the vectorial code and 195.23 seconds in the loop-based routine.

Hence, a CDO consists of a set of assets (its collateral portfolio) and a set of liabilities (the issued notes).

Table 22.4. Fair spread (bps) for 50 names using FFT

ρ	2nd t.d.	CPU time
0	1205.799	1.25
0.1	1168.722	1.62
0.2	1071.935	1.71
0.3	944.151	1.76
0.4	807.688	1.78
0.5	674.218	1.76
0.6	548.140	1.95
0.7	429.898	2.03
0.8	317.615	1.67
0.9	206.42033	2.12
1	50	$\simeq 0$

Table 22.5. Fair spread (bps) for 50 names using HW

ρ	2nd t.d.	CPU time
0	1205.799	13.062
0.1	1168.722	17.328
0.2	1071.935	18.265
0.3	944.1515	18.422
0.4	807.688	19.047
0.5	674.218	19.329
0.6	548.140	20.672
0.7	429.901	22.327
0.8	–	–
0.9	–	–
1	50	$\simeq 0$

All the results highlight that the 1st-to-default premium decreases for higher values of correlation. The higher the correlation, the greater the advantage the contract offers to the protection buyer. The paid basket spread is smaller if compared to buying n single name credit default swaps. Since this contract only offers partial protection, we would expect that the upper bound ($\rho = 0$) for the premium is the sum of the individual default swap premiums, $\sum_{i=1}^{N} s_i$.

Similarly, the lower bound for the premium is the maximum of the individual default swap premiums, $\max(s_1, s_2, \ldots, s_N)$, which corresponds to the case in which $\rho = 1$ (since here the riskiest credit will always be the 1st-to-default): $\max(s_1, s_2, \ldots, s_N) \leq s \leq \sum_{i=1}^{N} s_i$ (see Fig. 22.3). Up to a certain correlation value, the premium for a 2nd-to-default increases as n does so. Over a certain level ρ, the propensity to jointly default implies that the protection offered to the contract holder decreases. For this reason the 2nd-to-default basket behaves like a first-to-default basket: the premium starts to decrease. The premium also depends on the value of the recovery rate R. We fix the value of the hazard rates and evaluate different contracts under varying correlation and recovery rate. As the recovery rate increases, the contract value becomes smaller and smaller since the protection it offers is less valu-

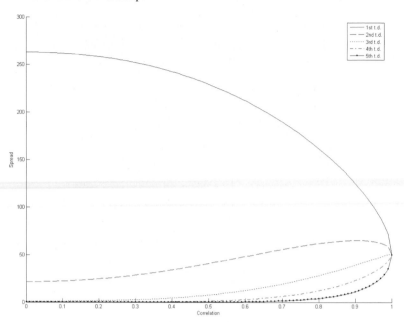

Fig. 22.3. Premium vs. correlation.

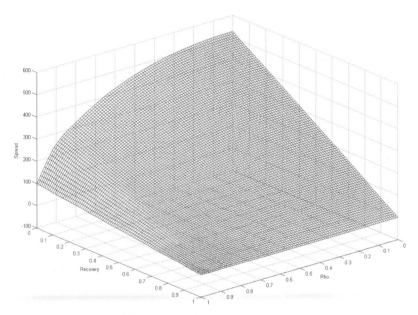

Fig. 22.4. First-to-default premium.

able. The resulting charts are displayed in Figs. 22.4 and 22.5. We finally analyze the
relation between the hazard rate h and the contract premium. As shown in Fig. 22.6,

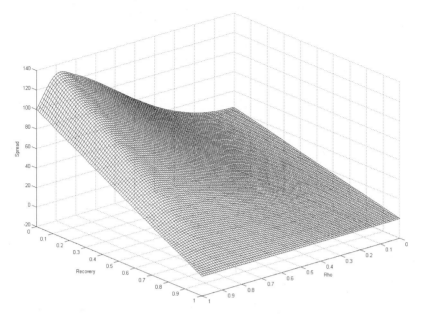

Fig. 22.5. Second-to-default premium.

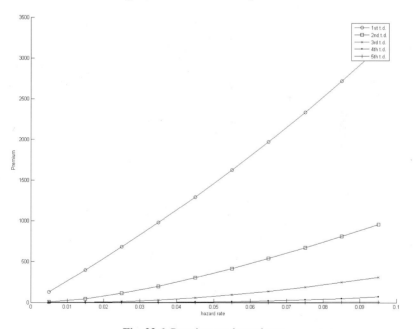

Fig. 22.6. Premium vs. hazard rate.

as h increases, the probability of default increases and the contract becomes more and more expensive.

Scenario Simulation Using Principal Components[*]

Key words: principal components analysis, Monte Carlo simulation, risk analysis, term structure of interest rates, forward curve

Portfolio managers, hedge funds, and derivative traders, among others, are confronted with the valuation and risk management of financial portfolios involving an increasingly large number of securities. The heterogeneity of asset pools makes these tasks very challenging and arduous to deal with using standard valuation tools, such as naive Monte Carlo methods.

Principal Components Analysis (PCA) and scenario simulation are complexity reduction techniques that can be applied to portfolios involving correlated assets.

The PCA method, as introduced in Steeley (1990) for the purpose of financial applications, consists of identifying statistically independent factors affecting the random evolution of a given pool of assets, sorting them by order of relative importance, and selecting the most significant ones. This task can be undertaken through the computation of a historical cross-asset covariance matrix and the identification of principal components representing independent factors underlying the market risk borne by the portfolio.

Jolliffe (1986) is the standard reference for PCA. Avellaneda and Scherer (2002) and D'Ecclesia and Zenios (1994) apply this technique to interest rate markets and portfolio selection. Gabbi and Sironi (2005) provide as extensive factor analysis in the Eurobond market. PCA has been applied to measure Value-at-Risk figures in Roncoroni (2004).[1] Based on early results by Roncoroni (1997, 1999), a functional version of PCA has been proposed by Guiotto and Roncoroni (2001), Roncoroni, Galluccio and Guiotto (2003), and Galluccio and Roncoroni (2006). These authors develop a method allowing to detect "Shape Factors" affecting cross-sectional de-

---

[*] with Alessandro Moro.

[1] Gatti et al. (2006) and Maspero and Saita (2005) provide key applications of Value-at-Risk to asset management and project finance transactions.

formations and perform a study on the empirical performance of hedging strategies based on these factors compared to the traditional PCA components.

The scenario simulation method, as introduced in Jamshidian and Zhu (1997), consists of replacing the joint probability distributions of the significant factors selected using PCA by appropriate discrete-valued approximations and then evaluating the distribution of the market value of an asset portfolio as a function of these approximated distributions.

The main hypothesis underlying these methods is that factors are jointly normally distributed. Correspondingly, the performance of the resulting reduced-form model depends on four elements, namely

(1) the similarity of actual asset price distributions to normal variates,
(2) the correlation level among assets in the portfolio,
(3) the number of retained factors,
(4) the number of points in the approximating discrete distribution.

As these values become larger, the quality of the resulting model improves.

This case-study is organized as follows. Here the factor detection problem and the complexity reduction issue are introduced. Section 23.1 details the solution methodology from a theoretical viewpoint, and provides an application to interest rate markets. Section 23.2 describes an implementation algorithm and the corresponding code. Section 23.3 concludes with some numerical experiments aimed at testing the quality of the method.

23.1 Problem Statement and Solution Methodology

Simulating cross-sectional data, such a term structure of interest rates, a commodity forward price curve, or a volatility smile, is a challenging task due to the high number of variables involved in the process. PCA aims at identifying and reducing the number of mutually independent factors driving the random evolution of a given cross-section. Scenario simulation improves PCA results by reducing the number of possible outcomes that need to be simulated for each selected factor.

We introduce the main issue through a simple example. Consider a portfolio π involving interest-rate dependent securities only. The portfolio value V_π depends of the term structure of interest rate expressed as a continuously compounded yield curve y_x, where $x \geq 0$ denotes a time-to-maturity. This can be seen as a prototypical instance of a cross-section. We assume that π contains positions depending on a number n of yields and that each yield may take one among a set of k values within the portfolio time horizon. Evaluating the risk underlying the portfolio value by Monte Carlo requires several simulations of possible scenarios for the n yields. In general, there are k^n possible outcomes, though a reasonably well-approximated portfolio distribution can be obtained by fewer Monte Carlo simulations. However, the computational burden remains quite important even for small portfolios.

There are two sources of reducible complexity in this procedure. First, different yields may be correlated to a certain extent, so simulating their outcomes independently of each other introduces a loss of efficiency. PCA proposes a solution to this issue by replacing the n yields with a number $m \ll n$ of mutually independent factors $w_1, \ldots, w_m$. Second, sampling the distribution of V_π by simulating samples from factor distributions is highly time consuming due to the high number of possible states taken by each factor. For instance, a four-factor model ($m = 4$) with $M = 100$ states for each factor leads to $N = 100^4$ scenarios and a same number of sample portfolio values V_π. Monte Carlo simulation usually requires a high number, though smaller than N, of runs to come up with a reasonable assessment of the portfolio distribution. Scenario simulation suggests replacing the underlying distributions by discrete ones involving a very small number of states, say from 3 to 9 in most applications. Let us explore the two techniques to a greater depth.

Empirical analysis of interest rate dynamics shows that the yield curve $y = \{y_x, x \geq 0\}$ evolves over time according to a wide variety of possible deformations. For instance, Central Bank may decide to decrease the short-term rate while the market is quoting a higher 10-year bond yield; as a result, the yield curve is expected to steepen. This phenomenon can be described as follows.

We consider n points on the yield curve. On a given time interval, all points may vary quite heterogeneously, giving the impression that n factors are actually driving the yield curve evolution. A moment's reflection suggests that pairs of yields corresponding to neighboring times-to-maturity are likely to move together whereas far-away yields are likely to have a lower degree of mutual dependence in their behavior. Qualitatively, we may say that a perturbation affecting a single yield spreads over the entire term structure according to the degree of proximity to the shocked point. This observation suggests that PCA should be performed on the empirical covariance (or correlation) matrix of absolute (or relative) yield returns.

Slightly relaxing our notation, we let $y_i(t)$ denote the time t quoted yield corresponding to time-to-maturity x_i ($i = 1, \ldots, n$). In a standard diffusion setting, we may assume yield return dynamics

$$
\frac{dy_1(t)}{y_1(t)} = \mu_1 \, dt + dW_1(t),
$$

$$
\frac{dy_2(t)}{y_2(t)} = \mu_2 \, dt + dW_2(t),
$$

$$
\vdots \qquad \vdots \tag{23.1}
$$

$$
\frac{dy_n(t)}{y_n(t)} = \mu_n \, dt + dW_n(t),
$$

where the W_i's are *correlated* Brownian motions. Principal components analysis delivers dynamics for y_i's in terms of *independent* Brownian shocks $\widetilde{W}_i(t)$ ($i = 1, \ldots, n$). These processes are obtained as linear combinations of the original Brownian motion in that $d\widetilde{W}_i = \sum_{k=1}^{n} \alpha_k \, dW_k$ for suitable coefficients $\alpha_1, \ldots, \alpha_n$. Moreover, they are ranked according to their relative importance in reproducing the underlying volatility. Indeed, if λ_k denotes the instantaneous variance per time unit of the process $\widetilde{W}_k$, namely $\lambda_k \, dt = \mathrm{Var}(d\widetilde{W}_i)$, then $\lambda_i \geq \lambda_j$, whenever $i > j$.

Complexity reduction can be achieved by setting a number $\rho \in (0, 1]$ representing the proportion of the overall market volatility that needs to be reproduced by a reduced-form yield curve model and defining m as the smallest number of Brownian motions required for generating this figure, i.e.,

$$m := \min\left\{ i \geq 1 : \frac{\sum_{k=1}^{i} \lambda_k}{\sum_{k=1}^{n} \lambda_k} \right\} \geq \rho.$$

We then consider reduced-form dynamics

$$\frac{dy_1(t)}{y_1(t)} = \mu_1 \, dt + \sum_{k=1}^{m} \beta_{1k} \, d\widetilde{W}_k(t),$$

$$\frac{dy_2(t)}{y_2(t)} = \mu_2 \, dt + \sum_{k=1}^{m} \beta_{2k} \, d\widetilde{W}_k(t),$$

$$\vdots \qquad \vdots \tag{23.2}$$

$$\frac{dy_n(t)}{y_n(t)} = \mu_n \, dt + \sum_{k=1}^{m} \beta_{nk} \, d\widetilde{W}_k(t).$$

Once this representation is established, a scenario simulation can be performed on these dynamics. In most applications m is about 8–10 times lower than n.

Standard Monte Carlo simulation requires sampling among hundreds of possible outcomes for the underlying Brownian motions. Scenario simulation is a reduction technique aimed at simplifying the burden of generating paths from the exact noise distribution. The idea is to select a very limited number of key states (called scenarios) of the noise driver, to assign them probabilities consistent with the initial probability distribution, and finally to sample from this reduced-form model.

Jamshidian and Zhu (1997) suggest selecting 9 states for the first factor, 7 for the second, 5 for the third, and 3 for the last driver. The scenario probability distribution is taken as a multinomial approximation of the underlying multivariate normal distribution. If m denotes the number of possible outcomes for each of the four factors, then the probability of a state i is given by

$$\mathbb{P}(i) = 2^{-m} \frac{m!}{i!(m-i)!}.$$

The distance between two different states is $2/\sqrt{m}$ standard deviations and the distance between the center of the distribution and the farthest state is $\frac{1}{2}m\frac{2}{\sqrt{m}} = \sqrt{m}$.

23.2 Implementation and Algorithm

23.2.1 Principal Components Analysis

Consider a vector-valued diffusion process with constant coefficients

$$\frac{d\mathbf{y}(t)}{\mathbf{y}(t)} = \boldsymbol{\mu}\, dt + d\mathbf{W}(t), \tag{23.3}$$

where $\mathbf{W}$ is a correlated n-dimensional Brownian motion. Here division between vectors has to be interpreted componentwise, e.g., $\frac{\mathbf{a}}{\mathbf{b}} = (\frac{a_1}{b_1}, \ldots, \frac{a_n}{b_n})$. A time series of cross-sectional data will be denoted by $\{\mathbf{y}(t), t = \delta, \ldots, N\delta\}$, where $\mathbf{y}(t) = (y_1(t), \ldots, y_n(t))$.

Algorithm (Principal Components Analysis on a Diffusion Process)

1. *Data setting.* Consider the annualized relative yield variations $[\mathbf{y}(t + \delta) - \mathbf{y}(t)]/(\delta \mathbf{y}(t))$ for all dates $t = \delta, \ldots, N\delta$. The sample mean of these data is given by

$$\boldsymbol{\mu} := (\delta N)^{-1} \sum_{k=1}^{N} \left[\frac{\mathbf{y}(k\delta) - \mathbf{y}((k-1)\delta)}{\mathbf{y}((k-1)\delta)} \right]. \tag{23.4}$$

If we subtract this average from each variation observed in the market, we obtain a set of annualized centered relative yield variations

$$\boldsymbol{\Delta}(t) := \delta^{-1} \left[\frac{\mathbf{y}(t + \delta) - \mathbf{y}(t)}{\mathbf{y}(t)} \right] - \boldsymbol{\mu}, \quad t = \delta, \ldots, N\delta. \tag{23.5}$$

The main hypothesis at this stage is that these vectors are all independent samples from a common multivariate normal distribution.

2. *Descriptive statistical analysis.* Compute the sample covariance matrix

$$C := \mathrm{Cov}(\boldsymbol{\Delta}) = N^{-1} \sum_{t=1}^{N} \boldsymbol{\Delta}(t)\boldsymbol{\Delta}(t)^{\top}$$

$$= \left(N^{-1} \sum_{t=1}^{n} \Delta_i(t)\Delta_j(t) \right)_{i,j=1,\ldots,n}.$$

3. *Diagonalization.* Decompose C as $C = U^{\top}\Lambda U$, where

- Λ is the $n \times n$ diagonal matrix gathering the eigenvalues of C in a decreasing order, i.e.,

$$\Lambda = \mathrm{Diag}(\lambda_1, \ldots, \lambda_n), \quad \text{with } \lambda_1 > \cdots > \lambda_n,$$

- U is the $n \times n$ matrix assembling the corresponding normalized eigenvectors column by column, i.e.,

$$U = (\mathbf{u}^1 | \ldots | \mathbf{u}^n), \quad \text{with } \lambda_i \mathbf{u}^i = C\mathbf{u}^i \text{ for all } i.$$

4. *Principal components.* For each $i = 1, \ldots, n$, define the ith principal component f_i as the variance normalized linear combination of annualized variations with weights given by the elements $u^i_1, \ldots, u^i_n$ of the ith eigenvector $\mathbf{u}^i$:

$$\mathbf{f} = \text{Diag}\left(\sqrt{\lambda_1^{-1}}, \ldots, \sqrt{\lambda_n^{-1}}\right) U^\top \boldsymbol{\Delta}.$$

In matrix-like form, this expression reads as

$$\begin{pmatrix} f_1 \\ \vdots \\ f_n \end{pmatrix} = \begin{pmatrix} \sum_{k=1}^n \sqrt{\lambda_1^{-1}} u_k^1 \Delta_k \\ \vdots \\ \sum_{k=1}^n \sqrt{\lambda_n^{-1}} u_k^n \Delta_k \end{pmatrix},$$

where Δ_k is the kth entry of vector $\boldsymbol{\Delta}$. As a linear combination of normal variables with zero mean, vector $\mathbf{f}$ is also normal with mean equal to the zero vector $\mathbf{0} = (0, \ldots, 0)^\top$. Moreover, its covariance is given by the $n \times n$ identity matrix, that is,

$$\begin{aligned}
\text{Cov}(\mathbf{f}) &= \text{Cov}\left(\text{Diag}\left(\sqrt{\lambda_1^{-1}}, \ldots, \sqrt{\lambda_n^{-1}}\right) U^\top \boldsymbol{\Delta}\right) \\
&= \text{Diag}\left(\lambda_1^{-1}, \ldots, \lambda_n^{-1}\right) U \, \text{Cov}(\boldsymbol{\Delta}) U^\top \\
&= \text{Diag}\left(\lambda_1^{-1}, \ldots, \lambda_n^{-1}\right) U C U^\top \\
&= \text{Diag}\left(\lambda_1^{-1}, \ldots, \lambda_n^{-1}\right) U U^\top \Lambda U U^\top \\
&= I_n,
\end{aligned}$$

where we applied the property that the transpose of an orthogonal matrix is equal to its inverse, e.g., $U^\top = U^{-1}$. In short, $\mathbf{f} \sim \mathcal{N}_n(\mathbf{0}, I_n)$. Since $\lambda_i > \lambda_{i+1}$, f_i can be interpreted as the ith most important component in explaining the cross-sectional risk C.

5. *Diffusion coefficients identification.* By setting $dt = \delta$ and $d\mathbf{W}(t) = \boldsymbol{\Delta}$ in equation (23.3), and by noticing that

$$\boldsymbol{\Delta} = \text{Diag}\left(\sqrt{\lambda_1}, \ldots, \sqrt{\lambda_n}\right) U \mathbf{f} = \begin{pmatrix} \sum_{j=1}^n \sqrt{\lambda_1} u_1^j f_j \\ \vdots \\ \sum_{j=1}^n \sqrt{\lambda_n} u_n^j f_j \end{pmatrix}, \qquad (23.6)$$

we come up to the following set of equations:

$$\begin{aligned}
\frac{dy_1(t)}{y_1(t)} &= \mu_1 \, dt + \sum_{k=1}^n \sqrt{\lambda_1} u_1^k \, d\widetilde{W}_k(t), \\
\frac{dy_2(t)}{y_2(t)} &= \mu_2 \, dt + \sum_{k=1}^n \sqrt{\lambda_2} u_2^k \, d\widetilde{W}_k(t), \\
&\vdots \qquad \vdots \\
\frac{dy_n(t)}{y_n(t)} &= \mu_n \, dt + \sum_{k=1}^n \sqrt{\lambda_n} u_n^k \, d\widetilde{W}_k(t).
\end{aligned} \qquad (23.7)$$

where $d\widetilde{W}_j := f_j$. By reducing the number of factors to $m < n$, the resulting dynamic system reads as in equation (23.2), where $\beta_{ik} := \sqrt{\lambda_i} u_i^k$.

The entries of each eigenvector $\mathbf{u}^i$ are referred to as "factor loadings". Indeed, formula (23.6) implies that a unit perturbation in the component $\sqrt{\lambda_i} f_i$ generates a shock u_1^i on the yield y_1, a shock u_2^i on the yield y_2, and so on. Usually, the entries of the first eigenvector display a common order of magnitude and share the same sign. This means that a unit shock in the first component is reflected in a similar movement of all points on the curve, that is a parallel shift. Similar interpretations can be attributed to some of the remaining factors, as will be clear at the end of our analysis.

The first applications of this decomposition in finance go back to Steeley (1990) and Litterman and Scheinkman (1991). These authors studied the yield curve dynamics of the US interest rate market. Their main conclusion was that the first three components $\mathbf{f}^1$, $\mathbf{f}^2$ and $\mathbf{f}^3$ explain more than 90% of the historical market volatility during the eighties. Moreover, these components can be respectively interpreted as a parallel shift, a change in slope, and a convexity adjustment in the shape of the term structure. Although this method has been criticized for the strong assumption made about the normality of the distribution of centered yield returns, it has now become a market standard in the banking industry for the purpose of modelling cross-sectional risk.

23.2.2 Code

Our implementation involves two codes, one for the scenario simulation model and another for the Monte Carlo simulation. Centered yield variations are calculated for a set of times-to-maturity. Function `cov` performs and delivers the variance–covariance matrix of the time series of these quantities. Function `eig` returns the set of decreasingly ordered eigenvalues and the corresponding normalized eigenvectors. Once PCA has been performed, the four most significant components are singled out and sample yield curves are generated.

The two codes differ in the way these curves are obtained. The scenario simulation code uses a finite set of possible outcomes for each of the four components. The Monte Carlo code generates pseudo-random samples for these figures. Once outcomes for the selected components have been obtained, their value is plugged into formula (23.7) and a set of sample yields, one per time-to-maturity, is returned as an output.

23.3 Results and Comments

We implemented and tested the simulation algorithm using a time series of yield curves in the US Treasury bond market. Our data span the period from March 1, 2000, to February 28, 2002, and the cross-section includes eleven yields, namely those referring to yearly times-to-maturity for years one to ten, plus the six-month yield.

Recall that the jth factor is given by a constant multiplied by the linear combination of yield increments $\Delta_1, \ldots, \Delta_{11}$ defined in (23.5) with loadings $u_1^j, \ldots, u_{11}^j$,

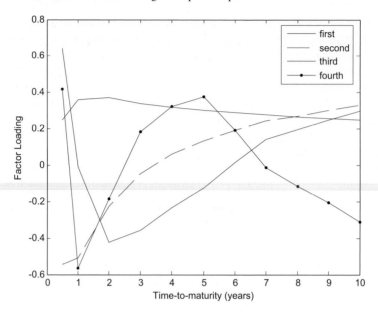

Fig. 23.1. Factor loadings of the most significant four principal components.

i.e., the entries of the normalized eigenvector corresponding to the jth greatest eigenvalue of the yield increments covariance matrix. We may figure out the shape of a given factor j by visually inspecting the graph of all corresponding factor loadings plotted against the time-to-maturity variable, i.e., $\{(T_i, u_i^j), i = 1, \ldots, 11\}$. Figure 23.1 shows a graph of factor loadings defining the most significant four factors. We see that the first factor assigns quite uniform loading coefficients to all yield increments. This feature suggests that the most important component driving the yield curve risk is a parallel shift in the curve shape. The second factor loadings display an increasing path. Correspondingly, this factor can be interpreted as a change in the yield curve slope. Finally, the third factor can be identified with a curvature change in the yield curve shape.

Table 23.1 reports instantaneous annualized standard deviations for the 11 principal components. The overall variance is given by the sum of the diagonal entries in the covariance matrix R or, equivalently, by the sum of all its eigenvalues[2]

$$\text{Var}_{\text{Tot}} = 0.153370 + 0.018219 + \cdots + 0.000003 = 0.174838.$$

---

[2] Since the trace operator $\text{Tr}(A) := \sum_i A_{ii}$ is invariant under orthogonal transformations and commutation among arguments, we have:

$$\sum_i R_{ii} \triangleq \text{Tr}(R) = \text{Tr}(\Sigma \Lambda \Sigma^\top) = \text{Tr}(\Sigma \Sigma^\top \Lambda) \stackrel{\Sigma^\top = \Sigma^{-1}}{=} \text{Tr}(\Lambda) \triangleq \sum_i \lambda_i.$$

Table 23.1. Instantaneous annualized volatilities reproduced by the eleven principal components

Factor	Label	Variance, λ_j	Std. dev., $\sqrt{\lambda_j}$	% Var., $\lambda_j / \sum_{k=1}^{11} \lambda_k$	% Cumul., $\sum_{k=1}^{j} \lambda_k / \sum_{k=1}^{11} \lambda_k$
1	$\widetilde{W}_1$	0.153370	3.916%	87.721%	87.721%
2	$\widetilde{W}_2$	0.018219	1.350%	10.420%	98.141%
3	$\widetilde{W}_3$	0.002309	0.481%	1.321%	99.462%
4	$\widetilde{W}_4$	0.000585	0.242%	0.335%	99.796%
5	$\widetilde{W}_5$	0.000181	0.135%	0.104%	99.900%
6	$\widetilde{W}_6$	0.000137	0.117%	0.078%	99.978%
7	$\widetilde{W}_7$	0.000017	0.041%	0.010%	99.988%
8	$\widetilde{W}_8$	0.000008	0.028%	0.005%	99.992%
9	$\widetilde{W}_9$	0.000005	0.023%	0.003%	99.995%
10	$\widetilde{W}_{10}$	0.000004	0.019%	0.002%	99.997%
11	$\widetilde{W}_{11}$	0.000003	0.018%	0.002%	100.000%

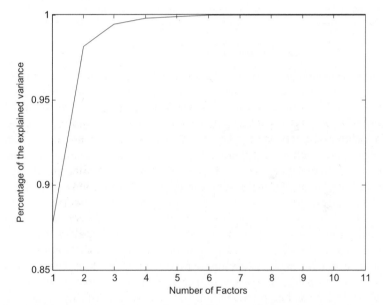

Fig. 23.2. Quota of the sample variance reported as a function of the cumulative number of ranked principal components (factors).

In particular, the first component accounts for $0.153370^2/0.174838 \cong 87.721\%$ of the total variance. Figure 23.2 displays the cumulated variance reproduced by an increasing number of principal components. It is clear from this picture that the most significant four components represent almost the entire sample variance in the market under investigation. Accordingly, we select $\widetilde{W}_1$, $\widetilde{W}_2$, $\widetilde{W}_3$ and $\widetilde{W}_4$ for the purpose of generating sample yield curves.

Our next experiment involves the yield curve as recorded on February 28, 2002, that is the last available datum in our set. We then generate eleven yields defining a sample yield curve as will be quoted in three months from the current date. This task is accomplished through standard Monte Carlo and scenario simulation. To put this program into action, let t_0 and $(y_1(t_0), \ldots, y_{11}(t_0))$ denote the starting date and the quoted term structure, respectively. If the simulation horizon is $t = t_0 + 3$ months, a standard Monte Carlo sample $(y_1(t), \ldots, y_{11}(t))$ is obtained as

$$y_i(t) = y_i(t_0) \exp\left[1 + \mu_i(t - t_0) + \sum_{k=1}^{4} \sqrt{\lambda_i} u_i^k \sqrt{t - t_0} \times \mathcal{N}^k(0, 1) \right],$$

where μ_i is the ith entry of vector $\boldsymbol{\mu}$ defined in (23.4) and $\mathcal{N}^1(0, 1), \ldots, \mathcal{N}^4(0, 1)$ are independent samples from a standard normal distribution. A sample performed using the scenario simulation model can be obtained by assuming that each $\mathcal{N}^k$ is a state of a discrete distribution approximating the normal density. We consider a discrete set of states $w_1^k, \ldots, w_m^k$ together with a distribution function defined by

$$\mathbb{P}(w_i^k) = 2^{-m} \frac{m!}{i!(m-i)!}, \quad i = 1, \ldots, m, k = 1, \ldots, 4.$$

We choose the number m_k of states depending on factor k according to the rule $m_k = 2(5 - k) + 1$. So we have nine states for factor 1, seven states for factor 2, five states for factor 3, and three states for factor 4. These are centered at zero and differ by a multiple of $2/\sqrt{m_k}$. Table 23.2 reports the exact figures for all states. We plotted 105 sample yield curves obtained by the two methods on a common pair of Cartesian axes. Figure 23.3 highlights the stronger uniformity among samples generated using the scenario simulation model as compared to those obtained using the standard Monte Carlo technique. Deeper insight can be gained by examining descriptive statistics over a larger sample of yield curves. We generated 945 sample curves. Each curve included eleven times-to-maturity. We then computed maximum, minimum, mean, and variance of sample yields corresponding to each maturity. Results are reported in Table 23.3 for the Monte Carlo simulation, and Table 23.4 for the scenario simulation model. Scenario simulation performs as well as Monte Carlo in terms of variety of samples paths. However, it outperforms standard Monte Carlo in terms of computational cost, as long as the number of states is dramatically reduced and no simulation is required at all.

It is important to underline that the quality of our results highly is highly dependent on the assumption that the underlying model is Gaussian. Whenever the underlying security prices display non-normal distributions, this hypothesis may result in a significant underestimation of the risk borne by a standing portfolio, and consequently reduce the effectiveness of hedging strategies based on Value-at-Risk measures. Gibson and Pristsker (2000) provide a few improvements of the Jamshidian and Shu (1997) method as applied to the determination of Value-at-Risk figures using PCA.

In all cases, PCA and scenario simulations constitute important tools for all risk managers for at least three reasons. First, these techniques are easy to understand and

Table 23.2. Scenario simulation approximation: states and probabilities for factors 1–4

Factor 1, $k = 1$								
w_i^k:	w_1^1	w_2^1	w_3^1	w_4^1	w_5^1	w_6^1	w_7^1	w_8^1	w_9^1
	−2.828	2.121	−1.414	−0.707	0.000	0.707	1.414	2.121	2.828
pw_i^k:	0.004	0.031	0.109	0.219	0.273	0.219	0.109	0.031	0.004

Factor 2, $k = 2$							
w_i^k:	w_1^2	w_2^2	w_3^2	w_4^2	w_5^2	w_6^2	w_7^2
	−2.450	−1.630	−0.820	0.000	0.820	1.630	2.450
pw_i^k:	0.016	0.094	0.234	0.313	0.234	0.094	0.016

Factor 3, $k = 3$				
w_i^k:	w_1^3	w_2^3	w_3^3	w_4^3	w_5^3
	−2.000	−1.000	0.000	1.000	2.000
pw_i^k:	0.062	0.250	0.370	0.250	0.062

Factor 4, $k = 4$		
w_i^k:	w_1^4	w_2^4	w_3^4
	−1.410	0.000	1.410
pw_i^k:	0.250	0.500	0.250

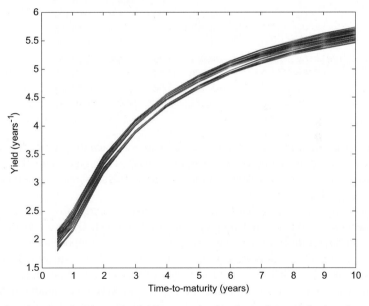

Fig. 23.3. Joint plot of 105 sample yield curves simulated on a 3-month horizon by standard Monte Carlo.

Table 23.3. Monte Carlo simulation, 4 factors

Time-to-maturity	Min	Mean	Max	St. dev.
6 m	1.785	2.061	2.245	0.01113
1 y	2.140	2.445	2.652	0.01438
2 y	3.155	3.414	3.600	0.01351
3 y	3.865	4.071	4.224	0.01211
4 y	4.319	4.511	4.662	0.01135
5 y	4.639	4.827	4.984	0.01086
6 y	4.903	5.080	5.234	0.01046
7 y	5.090	5.265	5.426	0.01024
8 y	5.328	5.414	5.582	0.01002
9 y	5.350	5.530	5.706	0.00989
10 y	5.452	5.634	5.819	0.00977

Table 23.4. Scenario simulation, 4 factors

Time-to-maturity	Min	Mean	Max	St. dev.
6 m	1.649	2.077	2.506	0.01220
1 y	1.995	2.469	2.943	0.01505
2 y	2.997	3.427	3.857	0.01421
3 y	3.725	4.075	4.425	0.01275
4 y	4.180	4.509	4.838	0.01191
5 y	4.496	4.822	5.148	0.01138
6 y	4.757	5.073	5.389	0.01099
7 y	4.932	5.257	5.582	0.01081
8 y	5.076	5.406	5.736	0.01060
9 y	5.184	5.521	5.859	0.01051
10 y	5.281	5.625	5.969	0.01042

to implement. Secondly, their limits are well known, which may not be the case for alternative and more sophisticated methods. Finally, PCA constitutes the standard market practice. Thus, no serious comparative risk analysis should avoid this setting as a benchmark.

It is worth mentioning that traditional PCA may be extended to the analysis of cross-sectional shifts at a functional level. In this respect, Guiotto and Roncoroni (2001) propose a theoretical framework where principal components are identified in the stylized deformations observed on a time-series of cross-sectional data. The study conducted by Galluccio and Roncoroni (2006) supported this approach from an empirical viewpoint. In particular, these authors showed that factors underlying cross-sectional shifts reproduce the yield curve risk more accurately than the traditional factors derived through standard PCA. Moreover, hedging a simple liability against functional risk is more effective than hedging against the PCA factors in terms of descriptive statistics of P&L distributions.

Financial Econometrics

24

Parametric Estimation of Jump-Diffusions[*]

Key words: maximum likelihood estimation, jump-diffusions, Monte Carlo simulation, term structure of interest rates

We consider the problem of estimating the coefficients of diffusion processes with jumps. These processes have been widely employed for describing sudden changes both in stock price and in interest rate dynamics and reproducing high-order moments of sample data. In most cases, the lack of analytical expressions for transition densities makes the use of maximum likelihood methods a very challenging task.

This case describes and tests the simulated maximum likelihood (SML) method introduced by Pedersen (1995) and Brandt and Santa Clara (2002). This technique allows for the estimation of a wide variety of diffusion processes, including those which lack closed-form expressions for the transition density. We illustrate the SML method by developing algorithms for the estimation of both continuous and mixed-jump-diffusion processes. The latter process has been employed in interest rate modelling by Piazzesi (2001).

The main idea underlying SML is to numerically evaluate the transition probabilities of the process corresponding to all pairs of values taken by the state variable at consecutive times. If a discretization of the time–space axes is properly refined, the resulting transition density approaches a Gaussian distribution. The likelihood estimator becomes a reliable approximation of the exact maximum likelihood estimator, namely the one stemming from the exact, yet unknown, transition density of the process.

Sections 24.1 and 24.2 introduce the estimation problem and illustrate the general methodology, respectively. Section 24.3 details the algorithm, while Sect. 24.4 provides two applications. First, we consider the Cox, Ingersoll and Ross (1985) model (CIR) for short-term interest rate dynamics. Since the transition density for

---

[*] with Piergiacomo Sabino.

this model is known in closed-form, parameters can be estimated using the exact maximum likelihood method. Consequently, we can use this model as a benchmark against which we evaluate the relative performance of SML. Second, we adapt SML to jump-diffusion processes and implement the resulting algorithm to the general case of time-dependent jump intensity.

24.1 Problem Statement

We consider the problem of estimating the parameters of a continuous time diffusion process Y satisfying a stochastic differential equation:

$$dY(t) = \mu(Y(t), t; \boldsymbol{\theta}) \, dt + \sigma(Y(t), t; \boldsymbol{\theta}) \, dW(t). \tag{24.1}$$

Here W is a standard Brownian motion and $\boldsymbol{\theta}$ is a vector of unknown parameters. Assume that vector $\widehat{\mathbf{Y}} = (\widehat{Y}_0, \widehat{Y}_1, \ldots, \widehat{Y}_N)$ gathers a set of observations of the process as recorded at consecutive times $t_0, \ldots, t_N$. If the exact transition density function $p_Y = p_Y(y, s, x, t; \boldsymbol{\theta})$ is available in analytic form for any pair of times $t < s$ and states x, y, the likelihood function of the process is defined as the joint density computed at the observed value $\widehat{\mathbf{Y}}$, i.e.,

$$\mathcal{L}(\boldsymbol{\theta}; \widehat{\mathbf{Y}}) := \prod_{n=0}^{N-1} \hat{p}_n(\boldsymbol{\theta}) = \prod_{n=0}^{N-1} p(\widehat{Y}_{n+1}, t_{n+1}, \widehat{Y}_n, t_n; \boldsymbol{\theta}). \tag{24.2}$$

The maximum likelihood estimator determines the parameter value maximizing the likelihood function computed on a sample $\widehat{\mathbf{Y}}$:

$$\hat{\boldsymbol{\theta}}_{\mathrm{ML}} : \widehat{\mathbf{Y}} \to \arg\max_{\theta} \mathcal{L}(\boldsymbol{\theta}; \widehat{\mathbf{Y}}).$$

In many instances $\mathcal{L}$ is an exponential function. If this is the case, we may find it more convenient to maximize the natural logarithm of $\mathcal{L}$. The theory of stochastic processes estimation using maximum likelihood is largely developed (see, e.g., Prakasa-Rao (1999)). However, the method requires an analytical expression for the transition densities of the process. The SML technique approximates transition function values using Monte Carlo simulation. This technique provides an answer whenever other approximating methods fail. However, it suffers from all the typical inconveniences of simulation-based techniques, including intensive time consumption and some difficulty arising whenever approximation errors need to be assessed.

24.2 Solution Methodology

Our goal is to provide an approximation for the transition function $\hat{p}_n(\boldsymbol{\theta})$ between consecutive times t_n and t_{n+1}. For the sake of simplicity, we assume that sampling occurs on an evenly spread time grid $t_0, \ldots, t_N$ (i.e., $t_{n+1} - t_n = \Delta$). Each interval $[t_n, t_{n+1}]$ is first split into M subintervals of length $\delta = \Delta/M$. Next, the transition

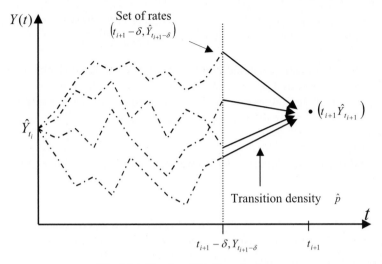

Fig. 24.1. Simulated likelihood scheme.

$\hat{p}_n(\boldsymbol{\theta})$ is represented as a convolution of densities over the two consecutive intervals $[t_n, t_{n+1} - \delta]$ and $[t_{n+1} - \delta, t_{n+1}]$, resulting in the following expression:

$$\hat{p}_n(\boldsymbol{\theta}) = p(\widehat{Y}_{n+1}, t_{n+1}, \widehat{Y}_n, t_n; \boldsymbol{\theta})$$
$$= \int_{\mathbb{R}} p(\widehat{Y}_{n+1}, t_{n+1}, y, t_{n+1} - \delta; \boldsymbol{\theta}) \times p(y, t_{n+1} - \delta, \widehat{Y}_n, t_n; \boldsymbol{\theta}) \, \mathrm{d}y.$$

Figure 24.1 illustrates these points.

This expression can be interpreted as the expected value of the transition probability $p(\widehat{Y}_{n+1}, t_{n+1}, y, t_{n+1} - \delta; \boldsymbol{\theta})$ as a function of y, with respect to the distribution of the process Y at time $t_{n+1} - \delta$ (the underlined quantity in the expression below) given that $Y(t_n) = \widehat{Y}_n$, i.e.,

$$\hat{p}_n(\boldsymbol{\theta}) = \mathbb{E}_{t_n, \widehat{Y}_n}\left(p\left(\widehat{Y}_{n+1}, t_{n+1}, \underline{Y(t_{n+1} - \delta)}, t_{n+1} - \delta; \boldsymbol{\theta}\right)\right). \tag{24.3}$$

To compute this expectation, we need information about two quantities: (1) an expression for p and (2) the distribution of process Y at time $t_{n+1} - \delta$ given $Y(t_n) = \widehat{Y}_n$. Both quantities are computed using numerical approximation. To this aim, we discretize Eq. (24.1) according to the Euler scheme over the refined time grid:

$$t_n,$$
$$t_n + \delta,$$
$$\vdots$$
$$t_n + (M - 1)\delta = t_{n+1} - \delta,$$
$$t_n + M\delta = t_{n+1}.$$

This method produces the following discrete time process:

$$Y_{t_n+(m+1)\delta}^M = Y_{t_n+m\delta}^M + \mu\left(Y_{t_n+m\delta}^M, t_n + m\delta; \boldsymbol{\theta}\right) \times \delta \tag{24.4}$$
$$+ \sigma\left(Y_{t_n+m\delta}^M, t_n + m\delta; \boldsymbol{\theta}\right)\sqrt{\delta} \times \varepsilon_{n,m}.$$

In the above equation, $\varepsilon_{n,m}$ are independent standard Gaussian variables defined for all discretization indices $n = 0, \ldots, N-1$ and all refining indices $m = 0, \ldots, M-1$. Since δ is "small" in comparison to Δ, we approximate the transition function in the expectation (24.3) by the transition of the discrete process (24.4) on the same interval:

$$p(\widehat{Y}_{n+1}, t_{n+1}, y, t_{n+1} - \delta; \boldsymbol{\theta}) \simeq p_M(y) := \phi_{\mathcal{N}}\left(\widehat{Y}_{n+1}; \mu_n(y), \sigma_n(y)\right),$$

where:

- $\mu_n(y) = y + \mu(y, t_n + (M-1)\delta; \boldsymbol{\theta})\delta$ denotes a trend term,
- $\sigma_n(y) = \sigma(y, t_n + (M-1)\delta; \boldsymbol{\theta})\sqrt{\delta}$ represents a volatility component, and
- $\phi_{\mathcal{N}}(\cdot; \mu, \sigma)$ is a normal density with mean μ and standard deviation σ.

The first approximation reads as

$$\hat{p}_n(\boldsymbol{\theta}) \simeq \mathbb{E}_{t_n, \widehat{Y}_n}\left(p_M\left(Y(t_{n+1} - \delta)\right)\right). \tag{24.5}$$

This expression is the expected value of the *approximated transition* density p_M from a random starting time-state $(t_{n+1} - \delta, Y(t_{n+1} - \delta))$ to the known target time-state $(t_{n+1}, \widehat{Y}_{n+1})$, with respect to the *exact distribution* p of $Y(t_{n+1} - \delta)$ conditional upon $Y(t_n) = \widehat{Y}_n$.

The second approximation involves the distribution p. The expectation in expression (24.5) is computed by running a Monte Carlo simulation of the discrete time process Y^M on the set $t_n, t_n + \delta, \ldots, t_n + (M-1)\delta$, starting at $Y_{t_n}^M = \widehat{Y}_n$. The estimate resulting from K sample paths is as follows:

$$\hat{p}_n(\boldsymbol{\theta}) \simeq \tilde{p}_n(\boldsymbol{\theta}) := \frac{1}{K}\sum_{i=1}^{K} p_M\left(\tilde{y}^{(i)}\right)$$

$$= \frac{1}{K}\sum_{i=1}^{K} \phi_{\mathcal{N}}\left(\widehat{Y}_{n+1}; \mu_n\left(\tilde{y}^{(i)}\right), \sigma_n\left(\tilde{y}^{(i)}\right)\right), \tag{24.6}$$

where $\tilde{y}^{(i)}$ are i.i.d. samples of Y^M at time $t_n + (M-1)\delta$ given $Y_{t_n}^M = \widehat{Y}_n$. Each sample is calculated by iterating the recursive relation (24.4) over the index m ($m = 0, \ldots, M-1$), starting with $Y_{t_n}^M$.

Once $\hat{p}_n(\boldsymbol{\theta})$ is computed for all n, the likelihood function can be evaluated by formula (24.2) by substituting $\tilde{p}_n$ for $\hat{p}_n$. Maximization of this numerical function of the parameter θ can be performed by standard numerical optimization routines.

24.3 Implementation and Algorithm

We can summarize the previous considerations into the following algorithm.

Algorithm (Simulated Maximum Likelihood)

1. Fix $\boldsymbol{\theta}$.
2. For all $n = 0, \ldots, N - 1$, compute the simulated transition using (24.6).
   Generation of $\tilde{y}^{(i)}$:
   2.1. Set $Y_{t_n}^M = \widehat{Y}_n$.
   2.2. Generate $\varepsilon_{n,m} \overset{\text{i.i.d.}}{\sim} \mathcal{N}(0, 1)$ for $m = 0, \ldots, M - 2$.
   2.3. Compute $Y_{t_n+(m+1)\delta}^M$ using formula (24.4).
   2.4. Return $\tilde{y}^{(i)} := Y_{t_n+(M-1)\delta}^M$.
3. Compute the approximated likelihood as $\widetilde{\mathcal{L}}(\boldsymbol{\theta}; \widehat{\mathbf{Y}}) := \prod_{n=0}^{N-1} \tilde{\hat{p}}_n(\boldsymbol{\theta})$.
4. Return $\tilde{\boldsymbol{\theta}}_{\text{ML}} := \arg\max_{\boldsymbol{\theta}} \widetilde{\mathcal{L}}(\boldsymbol{\theta}; \widehat{\mathbf{Y}})$.

Pedersen (1995) and Brandt and Santa Clara (2002) prove that the simulated transition density converges $\mathbb{P}$-almost surely to the exact transition density. Consequently, both the SML estimator and the maximum likelihood estimator share common asymptotic properties. Unfortunately, the rate of convergence of the SML scheme is the same as the one for the crude Monte Carlo method, so the algorithm runs quite slowly. This is not surprising, as the power of a Monte Carlo-based approach usually implies a sharp increase in the computational burden.

24.3.1 The Continuous Square-Root Model

The Cox, Ingersoll and Ross (1985) model is a popular device for modelling short-term interest rate dynamics. We adopt this model as a benchmark for the purpose of evaluating the relative quality of estimations delivered by SML.

The diffusion process is driven by a stochastic differential equation

$$dY(t) = \alpha\big(\mu - Y(t)\big)\,dt + \sigma\sqrt{Y(t)}\,dW(t). \tag{24.7}$$

This defines a mean-reverting process Y with reversion frequency α, long run mean μ, and[1] local variance σ. We perform SML estimation on data generated by simulation. Equation (24.7) is accordingly discretized using a Euler scheme

$$Y_{n+1} = Y_n + \alpha(\mu - Y_n) \times \Delta t + \sigma\sqrt{Y_n} \times \sqrt{\Delta t} \times \varepsilon_{n+1}. \tag{24.8}$$

Here $\Delta t = t_{n+1} - t_n$ and $\varepsilon_{n+1} \overset{\text{i.i.d.}}{\sim} \mathcal{N}(0, 1)$. While this is not the best way to simulate this process (a more efficient scheme is detailed in Chapter "Dynamic Monte Carlo"), it can be easily adapted to extended CIR dynamics with jumps, as is done in the next paragraph. Consequently, we have adopted this framework as a main reference. In the present context, n ranges from 1 to $N = 250$, and the step is $\Delta t = 1/250$. The file `CIR.m` implements the simulation algorithm for the discretization scheme (24.8) corresponding to the parameter choice

---

[1] The parameter α usually refers to the mean reversion speed of the process Y. However, this terminology does not appear to be correct because α has the dimension of a frequency. The mean reversion speed of Y at time t is actually given by the product $\alpha \times (\mu - Y(t))$.

$$\Theta = \{\alpha = 1.5, \mu = 0.035, \sigma = 0.1\}.$$

Notice that these values satisfy the restriction

$$2\alpha\mu \geq \sigma^2,$$

which is required for the continuous time process $Y(t)$ to be strictly positive-valued. Figure 24.2 shows a simulated trajectory produced by code `cir.m`. The SML estimation of model parameters can be split in three steps:

1. Computation of simulated likelihood for a given set of parameters.
2. Maximization of the simulated likelihood using a routine based on the Simplex method.
3. Evaluation of the estimated parameters Root Mean Square Error (RMSE) by numerically computing the inverse negative Hessian matrix of the log-likelihood function.

The starting set is arbitrarily chosen. Code `likelihood.m` computes the simulated log-likelihood by implementing formulae (24.3) and (24.6) in the previous section. The N observations $\widehat{Y}_0, \ldots, \widehat{Y}_{N-1}$ are given as an input to the routine which simulates the log-likelihood function for K different scenarios. For the purpose of optimizing the simulated likelihood we adopt the internal function `fminsearch.m`, which can handle discontinuities in the target functional. This routine is based on the Simplex search method consisting of a non-gradient-based search procedure providing local minima. At each step, we reset the seed of the random number generator to its original value in order to ensure that the normally distributed innovations remain unchanged, while maximizing the simulated likelihood. Code `SML1.m` defines

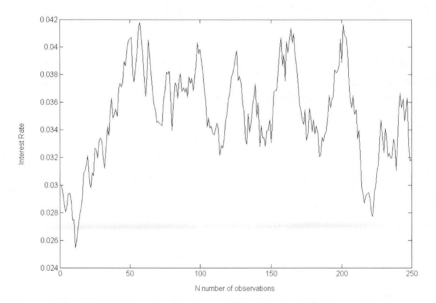

Fig. 24.2. Sample path of continuous CIR dynamics (250 time units).

the maximization procedure and the computation of RMSEs. All variables are defined within this routine and observations are generated by running code `CIR.m`. The maximization tolerance is set to the default value 10^{-4}.

24.3.2 The Mixed-Jump Square-Root Model

The CIR model can be generalized to include jumps. This extension allows us to capture effects that have been observed in several markets. For example, sudden shifts in the short-term interest rate are observed as a direct effect of Central Bank announcements about monetary policy regime changes. This setting can also be used for modelling electricity price dynamics, where periodic shocks occur during warmer seasons due to unexpected unbalances between supply and demand. The underlying process follows a stochastic differential equation

$$dY(t) = \alpha(\mu - Y(t)) \, dt + \sigma \sqrt{Y(t)} \, dW(t) + dJ(t). \tag{24.9}$$

Here the additional term $dJ(t) := J(t) - \lim_{s \uparrow t} J(s)$ represents the differential of a compound jump process defined as

$$J(t) = \sum_{i=1}^{N(t)} X_i. \tag{24.10}$$

In the above expression, N is a counting process representing the number of jumps occurred by time t:

$$N(t) = \text{number of jumps on } [0, t].$$

This process can be univocally assigned by specifying an intensity process $\lambda(t)$ stating the frequency of jump per time unit (i.e., the year) at each date. In other words, we may interpret $\lambda(t) \times \Delta t$ as the annualized conditional expected number of jumps occurring on the interval $[t, t + \Delta t]$. In most instances, λ is a deterministic function of time.

Once $N(t)$ is known, the net effect of all jumps is calculated. We suppose that jump sizes are described through independent and normally distributed random variables X_i $(i = 1, \ldots)$:

$$X_i \overset{\text{i.i.d.}}{\sim} \mathcal{N}(\mu_{\text{jump}}, \sigma_{\text{jump}}),$$

and that these are all independent of the Brownian motion W.

For the purpose of illustrating the SML method in a reasonably general case, we consider a time-varying intensity function. An instance of particular interest is provided by

$$\lambda(t) = \theta \times \frac{2}{1 + |\sin(\pi \frac{(t-\tau)}{k})|}.$$

Notice that the jump occurrence exhibits peak levels at multiples of k years, starting at time τ with a maximum. Consequently, the specification above can be adopted in all instances where jumps occur on a periodic basis. This is the case in interest

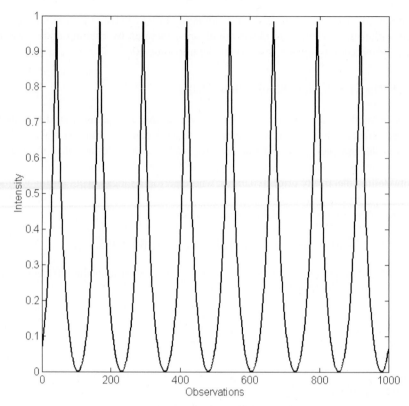

Fig. 24.3. Intensity function of an inhomogeneous Poisson process.

rate markets where balance requirements are periodically monitored (e.g. the Euro money market) and also in energy markets where price shocks display a seasonal pattern. Setting $k = 1/2$ (half a year) and $\tau = 1/6$ (two months), the jump intensity shows two peaks per year, starting on March 1. Parameter θ represents the maximum value attained by the intensity function. Figure 24.3 reports the graph for the selected intensity.

We implement the model with an initial rate $Y(t_0) = 0.03$ and a parameter set

$$\Theta = \{\mu_{\text{jump}} = 0.01, \sigma_{\text{jump}} = 0.002, \theta_{\text{jump}} = 10, \alpha = 1.5, \mu = 0.035, \sigma = 0.1\}.$$

These values ensure a very small probability of obtaining negative values in the process. The parametric choice above also ensures that the probability of negative jumps, i.e., $P[X_i \leq 0] = \phi_{\mathcal{N}}(0; \mu_{\text{jump}}, \sigma_{\text{jump}})$, is 2.8665×10^{-7}.

Following the experimental strategy adopted in the continuous case, these parameters are used to generate sample paths of the process. These paths are the input for the SML procedure. Estimated parameters generated by SML are then compared to the true values as a benchmark. Sample path generation is implemented in `true-jump.m`. First, we generate a path with no jump. Next, we run `position.m` and get a sample set of jump times. Then, we simulate a normally distributed random sample with size equal to the number of sample jumps. Finally, we overlay these jumps onto the continuous path at the sample jump times, which results in a sample path of the jump-diffusion process Y.

Figure 24.4 displays a sample generated by the described procedure. The main computation is given by formula (24.6), which we reproduce here:

$$\hat{p}_n(\boldsymbol{\theta}) \simeq \tilde{\hat{p}}_n(\boldsymbol{\theta}) := \frac{1}{K} \sum_{i=1}^{K} p\left(\widehat{Y}_{n+1}, t_{n+1}, \tilde{y}^{(i)}, t_{n+1} - \delta; \boldsymbol{\theta}\right).$$

Here $\widehat{Y}_{n+1}$ is the time t_{n+1} observation, δ is a "small" time lag, and $\boldsymbol{\theta}$ is the unknown vector of parameters. More importantly, p is the transition of the discretized process on the interval $[t_{n+1} - \delta, t_{n+1}]$ and the $\tilde{y}^{(i)}$'s $(i = 1, \ldots, K)$ are the time $t_n + (M - 1)\delta = t_{n+1} - \delta$ realizations of mutually independent sample paths of the process Y drawn over the interval $[t_n, t_{n+1} - \delta]$. A simulation is performed using the Euler iterative formula

$$Y_{n+1} = Y_n + \alpha(\mu - Y_n) \times \delta t + \sigma \sqrt{Y_n} \times \sqrt{\delta t} \times \varepsilon_{n+1} + I_{\{t_{n+1}\}} X_{n+1}.$$

Here jump occurrence is described by the indicator function

$$I_{\{t_{n+1}\}} = \begin{cases} 1 & \text{if } t_{n+1} \text{ is a jump time,} \\ 0 & \text{otherwise.} \end{cases}$$

Since jump sizes are normally distributed, the transition density of the discrete process between t_i and t_{i+1} is a mixture of normal variables and the approximated transition p on time interval $[t_{n+1} - \delta, t_{n+1}]$ reads as follows:

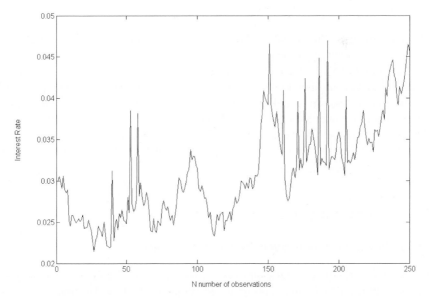

Fig. 24.4. Sample path of CIR dynamics with jumps (250 time units).

$$p\big(\widehat{Y}_{n+1}, t_{n+1}, \tilde{y}^{(i)}, t_{n+1} - \delta; \boldsymbol{\theta}\big) = p_0^{(i)} \phi_{\mathcal{N}}\big[\widehat{Y}_{n+1}; \mu^{\mathrm{c}}, \sigma^{\mathrm{c}}\big] + p_1^{(i)} \phi_{\mathcal{N}}\big[\widehat{Y}_{n+1}; \mu^{\mathrm{d}}, \sigma^{\mathrm{d}}\big],$$

where

$$\mu^{\mathrm{c}} = \tilde{y}^{(i)} + \alpha\big(\mu - \tilde{y}^{(i)}\big)\delta t, \qquad\qquad \sigma^{\mathrm{c}} = \sigma\sqrt{\tilde{y}^{(i)}}\sqrt{\delta},$$

$$\mu^{\mathrm{d}} = \tilde{y}^{(i)} + \alpha\big(\mu - \tilde{y}^{(i)}\big)\delta t + \mu_{\mathrm{jump}}, \qquad \sigma^{\mathrm{d}} = \sqrt{\sigma\tilde{y}^{(i)}\delta + \sigma_{\mathrm{jump}}^2}.$$

In this expression, $p_0^{(i)}$ is the probability of no jump between t_i and $t_{i+1} = t_i + \delta$. This is computed through the jump intensity process as

$$p_0^{(i)} = \mathrm{e}^{-\int_{t_i}^{t_i + \delta t} \lambda(s)\, \mathrm{d}s} \simeq \mathrm{e}^{-\lambda(t_i)\delta t}.$$

We assume that no more than one jump can occur on each interval of length δ in the refinement of the time axis.

For the purpose of computing the likelihood function in this model, we implement the program `likelihoodj.m`. This program generates a pair of independent innovation variables, one for the continuous process and one for the jump component. The latter is sampled by running `genindex.m`, which iteratively calls subroutine `subposition.m` (a modified version of `position.m`).

We estimate the parameters of the jump component using a model where all other parameters have been maintained fixed. Notice that optimization is performed with respect to the jump parameters under fixed random innovations. In other words, we simulate a standard normal variable Z and define the random jump size for any pair of possible parameters μ_{jump} and σ_{jump} by

$$X = \mu_{\mathrm{jump}} + \sigma_{\mathrm{jump}} Z.$$

The maximization of the simulated likelihood and the estimation of the parameters of the model under study are performed in a way similar to that of the process without jumps. Program `SML2.m` launches function `likelihoodj.m` for the computation of the simulated likelihood, and maximizes the returned value to produce a set of optimal parameters. We avoid running the estimation for the six parameters of the jump-diffusion model simultaneously. Rather, we fix the diffusion parameters on the "true" values set for the generation of the sample path ($\alpha = 1.5$, $\mu = 0.035$, $\sigma = 0.2$) and then run the maximization program `SML2.m` delivering the estimated set of values $\{\mu_{\mathrm{jump}}, \sigma_{\mathrm{jump}}, \theta_{\mathrm{jump}}\}$.

24.4 Results and Comments

24.4.1 Estimating a Continuous Square-Root Model

We perform an experimental estimation of the original CIR model. For the test, we use 100 simulated paths, 250 observed values, and $M = 10$ simulation steps within each of the $N - 1$ intervals. Using a personal computer equipped with a

Table 24.1. Estimated parameters using simulated maximum likelihood (continuous CIR dynamics)

Parameter	Actual value	Estimated value	RMSE
α	1.5	3.720	0.029
μ	0.035	0.0273	0.0092
σ	0.1	0.1126	0.0035

1.8 GHz processor and 448 megabytes of random access memory, the simulation takes about 4.5 hours (1 hour for the optimization and 3.5 for the computation of the RMSE's), and renders values for the fitting parameters and their RMSE's as reported in Table 24.1. The estimated parameters μ and σ are quite close to the true values, whereas α_{max} is about two times greater than expected. Better results may be obtained by increasing the number of simulations to enhance estimation effectiveness. Furthermore, improvements may also be made by enlarging the set of rates, i.e., N larger than 250, and imposing a finer structure between consecutive observations, i.e., M larger than 10. Notice that our choice has been constrained by computational time considerations. However, we believe that the bias in the estimation of parameter α is symptomatic of the slow performance of the SML methodology. In our opinion, SML represents a viable solution as long as alternative methods are not available. This is actually the case for several jump-diffusion processes.

Before examining the jump-diffusion case, we review the performance of the available alternative methods in the present context. Since transition densities for CIR dynamics are known in closed-form, we can run a maximum likelihood estimation. We set $\Delta = t - s > 0$, $c = 2\alpha[\sigma^2(1 - e^{\alpha\Delta}]$, $q = 2\alpha\mu\sigma^2 - 1$ and verify that the transition is a noncentral chi-square distribution

$$p_Y(y, t, x, s) = c e^{u-v}(v/u)^{q/2} I_q\left(2\sqrt{uv}\right),$$

where $u = cxe^{\alpha\Delta}$, $v = cy$ and I_q is the order q modified Bessel function of the first kind.[2] Samples from this distribution can be drawn by means of the Matlab® internal procedure ncx2pdf.m as detailed in Brigo and Mercurio (2006). Code likelihood.m computes the exact negative log-likelihood function as described above and routine MLE_true.m returns estimated parameters and corresponding RMSE as reported in Table 24.2. These figures are quite close to those we obtain using the simulated maximum likelihood estimator. Furthermore, while the former estimations have a higher RMSE due to the small size of observed data, the complete procedure only requires about half an hour. Even in this case the estimation for α is not in perfect accordance with theoretical values. In general, the simulated maximum-likelihood approach proved to be reasonably effective, whereas its efficiency needs to be improved.

---

[2] A Bessel function $y(z; n)$ of order n is a solution of the ordinary differential equation

$$z^2 y'' + zy' - (z^2 + n^2)y = 0.$$

Table 24.2. Estimated parameters using maximum likelihood (continuous CIR dynamics)

Parameter	Actual value	Estimated value	RMSE
α	1.5	4.51	0.55
μ	0.035	0.029	0.020
σ	0.1	0.0926	0.0042

Table 24.3. Estimated parameters using simulated maximum likelihood (mixed-jump CIR dynamics)

Parameter	Actual value	Estimated value
σ_{jump}	0.002	0.002
μ_{jump}	0.01	0.011
θ_{jump}	10	10.525

24.4.2 Estimating a Mixed-Jump Square-Root Model

We estimate a CIR model with jumps by performing 1000 simulated paths, each consisting of 250 values. The period between consecutive times is split into $M = 10$ subperiods. Due to the extremely high computational time required for each run, we do not estimate the RMSEs in this case. Our complete simulation requires about 2.5 hours and returns estimated values as indicated in Table 24.3. Compared to the continuous CIR model, estimating this process requires more time; however, it still provides estimations quite close to the actual values. This comparison should not be overvalued since the estimation in the two cases concern two distinct sets of parameters.

We finally suggest the final user to consider the following two issues before adopting the SML method:

(1) the algorithm has a quite slow convergence order;
(2) the target function is highly irregular and the optimization tool must be selected with care.

25

Nonparametric Estimation of Jump-Diffusions[*]

Key words: kernel methods, financial econometrics, estimation of jump-diffusions, simulation

We consider the problem of estimating diffusion processes with jumps. These processes have been widely used for describing the random evolution of financial figures such as stock prices, market indices and interest rates.

This case describes and tests a nonparametric procedure introduced in Stanton (1997) in a continuous path setting and then extended to mixed-jump-diffusions in Johannes (1999, 2004) and Bandi and Nguyen (1999, 2003), who provide a rigorous treatment of the underlying statistical theory. The technique we illustrate allows for the estimation of a wide variety of homogeneous diffusion processes, including those for which transition densities are not available in analytic form. We illustrate the method by developing algorithms for the estimation of both continuous and mixed-jump diffusion processes.

We organize the presentation as follows. Section 25.1 introduces the main issue of estimating diffusion processes by finite sample data. Section 25.2 describes the estimation technique by specifying a finite sample version of the conditional moments featuring continuous and mixed-jump models. Section 25.3 details a step-by-step implementation procedure for all model specifications. Section 25.4 describes our code and details an estimation experiment based on a time series of the 3-month EURI-BOR, which is a reference short-term rate of interest in the European money market. Results are described together with an assessment of the quality of the proposed procedure. An important by-product of our analysis is that continuous diffusion models are proved to be unable to capture important features displayed by market data.

---

* with Gianna Figà-Talamanca.

25.1 Problem Statement

We consider a one-dimensional time-homogeneous diffusion process X defined as the unique solution of a stochastic differential equation

$$dX(t) = \mu(X(t))\, dt + \sigma(X(t))\, dW(t). \qquad (25.1)$$

Here W is a one-dimensional standard Brownian motion and μ and σ are regular functions ensuring existence and uniqueness of a weak solution to the equation (25.1) (see, e.g., Karatzas and Shreve (1997) and Kloeden and Platen (2000)). Incidentally, we remark that the methods and results to follow can be easily extended to a multi-dimensional setting.

It is common practice in financial literature to identify a model by specifying particular parametric forms for the drift and the diffusion functions. Focusing on interest rate models, the drift can be an affine function of the process level. For instance, we may consider a linear drift $\mu(x) = \delta x$ or a function $\mu(x) = \alpha(\beta - x)$ generating mean reversion to a constant level β. In the former case, the constant δ is the sole drift parameter; in the latter instance, parameter α represents a mean reversion force, while parameter β defines a trend which the process X reverts to in the long run. As for the diffusion function, several specifications have been proposed: Vasicek (1977) takes a simple constant value σ; Cox, Ingersoll and Ross (1985) suppose a level dependence as modeled by the function $\sigma(x) = \sigma\sqrt{x}$; Chan et al. (1992) generalize these models by assuming a function of two parameters σ and γ, namely $\sigma(x) = \sigma x^{\gamma}$. These models are estimated by parametric methods aimed at assigning a number to each of the parameters specifying the process.

Assessing a form for both drift and diffusion coefficient should be in agreement with their statistical meaning. It can be shown that the drift coefficient represents the instantaneous average speed of the process at each point in time, conditional to its value, i.e.,

$$\mu(x) = \lim_{\Delta t \to 0} \frac{1}{\Delta t} \mathbb{E}[X(t + \Delta t) - X(t)|X(t) = x].$$

Similarly, the diffusion coefficient represents the instantaneous average quadratic speed, or centered moment of order two, of the process at each point in time, conditional to its value, namely

$$\sigma^2(x) = \lim_{\Delta t \to 0} \frac{1}{\Delta t} \mathbb{E}\big[\big(X(t + \Delta t) - X(t)\big)^2|X(t) = x\big].$$

Moreover, all continuous diffusion processes, namely those exhibiting continuous paths, display no higher-order moments:

$$\lim_{\Delta t \to 0} \frac{1}{\Delta t} \mathbb{E}\big[\big(X(t + \Delta t) - X(t)\big)^j|X(t) = x\big] = 0, \quad j > 2. \qquad (25.2)$$

Several studies have shown that continuous diffusion models suffer from relevant shortcomings in explaining the dynamics of interest rate time series. For instance, empirical moments computed on market proxies for the short rate often violates condition (25.2). This observation seems to confirm that observed data is not generated by a continuous diffusion model.

This problem can be overcome by modeling interest rate dynamics through a mixed-jump diffusion model defined by

$$dX_t = \mu(X(t^-)) \, dt + \sigma(X(t^-)) \, dW(t) + dJ(t), \qquad (25.3)$$

$$J(t) = \sum_{i=1}^{N(t)} Y_i. \qquad (25.4)$$

Here J is a compensated compound jump process with intensity $\lambda(x)$, i.e., $J(t) - \lambda(X(t))$ a martingale, and $Y_i \overset{\text{i.i.d.}}{\sim} Y$ represent the random jump size whose common distribution p_Y is assumed to be independent of the Brownian noise W and the counting process N. For these dynamics, the corresponding relations between model coefficients and the descriptive statistics of the process read as

$$\mu(x) = \lim_{\Delta t \to 0} \frac{1}{\Delta t} \mathbb{E}[X(t + \Delta t) - X(t) | X(t) = x], \qquad (25.5)$$

$$\sigma^2(x) + \lambda(x) \mathbb{E}_Y[Y^2] = \lim_{\Delta t \to 0} \frac{1}{\Delta t} \mathbb{E}\left[(X(t + \Delta t) - X(t))^2 | X(t) = x\right], \quad (25.6)$$

$$\lambda(x) \mathbb{E}_Y[Y^j] = \lim_{\Delta t \to 0} \frac{1}{\Delta t} \mathbb{E}\left[(X(t + \Delta t) - X(t))^j | X(t) = x\right], \quad \text{for all } j > 2.$$
$$(25.7)$$

The key point in these expressions is that jumps allow for the modeling of time series displaying high-order moments (i.e., $j > 2$) in the instantaneous increments.

Statistical estimation of diffusion processes such as (25.1) and (25.3) aims at determining coefficients $\mu(x)$, $\sigma(x)$, $\lambda(x)$, and $p_Y(y)$ compatible to observed data in some sense to be specified. Parametric estimation assumes that the diffusion coefficients belong to a parametric family of deterministic functions of the system state x. An estimator is a function assigning a parameter selecting diffusion coefficients within the assumed family of functions to a set of empirical observations of the process. Nonparametric estimation instead aims at determining $\mu(x)$, $\sigma(x)$, and $\lambda(x)$ pointwise, i.e., for each value x chosen in a suitable interval of the real line, without any reference to parametric families. Of course, these functions are required to satisfy a number of regularity conditions to ensure that the resulting process is well defined as the unique solution of the corresponding stochastic differential equation (see, e.g., Oksendal and Sulem (2004) and Protter (2005)).

25.2 Solution Methodology

A possible method for nonparametrically estimating a diffusion process is to define its coefficient by means of finite sample versions of the expressions on the right-hand side of formulae (25.5), (25.6), and (25.7). Three logical steps are required:

(1) observations $x_1, x_2, \ldots, x_n$ are collected at a possibly large time frequency Δ;
(2) conditional moments are estimated nonparametrically using their finite sample counterparts through, e.g., kernel methods;
(3) coefficients are singled out of the estimated conditional moments.

We now turn this program into a concrete estimation procedure for the EURIBOR. Assume that n discrete observations $r_1, \ldots, r_n$ of the short rate process are recorded with time frequency Δ. Finite sample estimates for the first two conditional moments can be obtained by kernel convolutions as

$$M_1(r) = \frac{\sum_{t=1}^{n} K(\frac{r_t - r}{h})(r_{t+1} - r_t)}{\Delta \sum_{t=1}^{n} K(\frac{r_t - r}{h})}, \tag{25.8}$$

$$M_2(r) = \frac{\sum_{t=1}^{n} K(\frac{r_t - r}{h})(r_{t+1} - r_t)^2}{\Delta \sum_{t=1}^{n} K(\frac{r_t - r}{h})}. \tag{25.9}$$

In our setting, the *kernel function* K is a symmetric probability density on the real axis expressing the influence a point at zero has on all other points in its domain. Parameter h is a *bandwidth* driving the smoothing behavior of the kernel. For instance, in the above expressions, the mean variation of the process starting at a level r is obtained by weighing all sample variations $r_{t+1} - r_t$ recorded in the past. The kernel function provides loadings depending on the level r. If we take K to be a standard Gaussian density, the larger the distance between r_t and r, the lower the impact of sample increment $r_{t+1} - r_t$ on the instantaneous average increment of the process starting at level r. Conversely, the closer r_t is to the actual level r, the higher the importance of the sample increment $r_{t+1} - r_t$ will be in explaining the average increment from r. It is clear from this remark that the influence of sample increments is driven by the kernel function and the bandwidth level.[1]

For a continuous diffusion process, M_1 and M_2 are estimates of the drift and the squared diffusion coefficient, respectively. For a mixed-jump diffusion, M_2 is the sum of the instantaneous squared diffusion term and the product between the second-order moment of the jump size distribution and the jump intensity as is reported in expression (25.6). Naturally, in the case of mixed-jump-diffusions, high-order moments are estimated for the purpose of completely specifying the estimated process. Indeed, high-order moments can be computed through the following finite sample statistics:

$$M_j(r) = \frac{\sum_{t=1}^{n} K(\frac{r_t - r}{h})(r_{t+1} - r_t)^j}{\Delta \sum_{t=1}^{n} K(\frac{r_t - r}{h})}, \quad j > 2. \tag{25.10}$$

These statistics may be employed for the purpose of testing the hypothesis that data comes from a diffusion process with continuous paths.

Notice that, if we define kernel weights

---

[1] Silverman (1986) offers a detailed account on kernel estimation methods, whereas James and Webber (2000) provide a quick overview from a financial modeling perspective.

$$\omega_t(r) = \frac{K(\frac{r_t - r}{h})}{\sum_{t=1}^{n} K(\frac{r_i - r}{h})}, \quad t = 1, 2, \ldots, n,$$

then equations (25.8), (25.9) and (25.10) read as

$$M_j(r) = \sum_{t=1}^{n} \frac{(r_{t+1} - r_t)^j}{\Delta} \omega_t(r), \quad j \geq 1.$$

Consequently, we may interpret these moments as weighing averages of an appropriate power of sample increments.

25.3 Implementation and Algorithm

We implement an estimation scheme aimed at simultaneously detecting whether sample data is generated by a continuous or a diffusion model and identifying the functional forms of the selected dynamics. Our procedure is carried out through the following:

Algorithm (Nonparametric Estimation)

1. Compute $M_1(r)$, $M_2(r)$, $M_3(r)$, and $M_4(r)$ for all values r in the interval $[r_{min}, r_{max}]$, where r_{min} and r_{max} denote the minimum and the maximum values for the input data, respectively.

2. Simulate m paths of the continuous diffusion process defined by Eq. (25.1) with $\mu(r) := M_1(r)$ and $\sigma^2(r) := M_2(r)$.

3. For each path labeled by $j = 1, 2, \ldots, m$, compute moments $M_1^{(j)}(r)$, $M_2^{(j)}(r)$, $M_3^{(j)}(r)$, and $M_4^{(j)}(r)$ by formulae (25.8), (25.9), and (25.10), respectively.

4. For each r in the interval $[r_{min}, r_{max}]$, sample moments $M_1(r)$, $M_2(r)$, $M_3(r)$, and $M_4(r)$ computed at step 1 are then compared to the median value and to the 10th and 90th percentiles of the m values obtained for $M_1^{(j)}(r)$, $M_2^{(j)}(r)$, $M_3^{(j)}(r)$, and $M_4^{(j)}(r)$. The resulting figures serve as confidence bands for a diffusion. If the sample estimates lie inside the confidence bands for all $r \in [r_{min}, r_{max}]$, the null hypothesis stating a continuous diffusion behavior for interest rate dynamics is not rejected and the estimation procedure is terminated with a final assessment

$$\hat{\mu}(r) = M_1(r),$$
$$\hat{\sigma}^2(r) = M_2(r).$$

If the sample estimates lie outside the confidence bands for some values of r, move to the next step.

5. Compute the higher-order conditional moment $M_6(r)$ by formula (25.10) and use this figure, together with $M_4(r)$, to identify both jump intensity and jump size distributions.

The last step deserves a few comments. In our experiments, we assume that the jump size Y is normally distributed with zero mean and variance $\sigma_Y^2 > 0$. Formulae (25.6), (25.7), (25.9), and (25.10) lead to the following three approximated relations

$$M_2(r) \simeq \sigma^2(r) + \lambda(r)\sigma_Y^2, \tag{25.11}$$

$$M_4(r) \simeq 3\lambda(r)\sigma_Y^4, \tag{25.12}$$

$$M_6(r) \simeq 15\lambda(r)\sigma_Y^6, \tag{25.13}$$

where $\sigma^2(r)$, $\lambda(r)$, and σ_Y^2 are values of the true coefficient functions. Estimates are obtained by setting an exact equality in these relations. Using these formulae and the definition of drift, we finally come up to nonparametric estimates for the coefficients defining the mixed-jump diffusion process compatible with the sample data. More precisely, by expressions (25.5) and (25.8) we have

$$\hat{\mu}(r) = M_1(r).$$

Taking the ratio between the sixth and the fourth weighed moments and using relations (25.13) and (25.12) (with a strict equality replacing "$\simeq$"), we arrive at $\hat{\sigma}_Y^2 = \frac{M_6(r)}{5M_4(r)}$, which is rate dependent; as a first-order estimate for the constant jump size variance we take the sample average of this quantity, namely

$$\hat{\sigma}_Y^2 = \frac{1}{n}\sum_{t=1}^{n}\frac{M_6(r_t)}{5M_4(r_t)}. \tag{25.14}$$

Given this estimate, the jump intensity directly follows from expression (25.12) as

$$\hat{\lambda}(r) = \frac{M_4(r_t)}{3\hat{\sigma}_Y^4}.$$

The diffusion term in formula (25.9) is obtained as

$$\hat{\sigma}^2(r) = M_2(r) - \hat{\lambda}(r)\sigma_Y^2.$$

Under technical assumptions reported in Bandi and Nguyen (2003), these estimators can be proven to be consistent and asymptotically normal.

In our experiment we adopt a Gaussian kernel defined as

$$K(x) = \frac{1}{\sqrt{2\pi}}\exp\left(-\frac{x^2}{2}\right),$$

to derive moments estimates (25.8)–(25.10). We remark that, to the best of our knowledge, no theoretical result driving the selection process of a bandwidth value is available to date. Silverman (1986) suggests to use $h = \sigma n^{-1/5}$, where n is the sample length and σ is the standard deviation of the data set. A practical solution sometimes adopted in the existing literature is a cross-validation over a range of selected values for h. Following Johannes (1999, 2004), we select a bandwidth value

Table 25.1. Bandwidth values, where s is the annualized standard deviation of the Euribor daily changes

Moment, M_k	Bandwidth, h
M_1	1.25s
M_2	0.4s
$M_{3,4,5,6}$	0.75s

for each of the moments to compute. These values are reported in Table 25.1 and are expressed in terms of the empirical standard deviation s of EURIBOR daily changes. This choice is consistent with the empirical findings in Chapman and Pearson (2000) suggesting to oversmooth the drift function with respect to the diffusion coefficient. The functional estimates are obtained point-by-point by considering m values for the interest rate r.

Implementation is performed by routines `kernelgauss.m`, `kernelweights.m` and `kernelmoments.m`. We also compute confidence intervals for the estimates by simulation. This is carried out by running code `simul.m` where a large number of sample paths are generated and computing the empirical moments over each sample path using function `kernelmomsim.m`. Notice that this routine differs from `kernelmoments.m` in that it takes kernel weights as input. A shortcut aimed at avoiding an excessively intensive computation is to evaluate kernel weights once for all paths, e.g., by using the whole data set. In the case of a mixed-jump diffusion process, functional forms of drift, diffusion coefficients and jump intensity are obtained through code `Estimates.m`. Empirical moments are computed by running `kernelmoments.m` on the interest rate data set. Script `figuremom.m` prints out all graphs presented below. Script `total.m` gathers the entire implementation into a singe routine. Estimation results and graphical representations require a five-minute running time on a 2.8 GHz processor.

25.4 Results and Comments

We report results obtained by applying the procedure described above to a time series of 3-month EURIBOR rates. Our data set can be obtained by Datastream. It involves 1382 daily observations from January 4, 1999 to April 20, 2004. Figure 25.1 displays graphs of the time series of rates and the corresponding daily variations. Figures 25.2 and 25.3 show the first and the second empirical moments.

We jointly plot the corresponding 10% and 90% confidence intervals. Circles refer to the median of simulated values. The second moment function is split in two graphs, one for $r < 3.5\%$ and the other for $r > 3.5\%$. Notice that the estimated first moment is always inside the corresponding confidence interval and does not differ much from the median of simulated values. Unfortunately, the estimated second moment lies inside the relative confidence interval only for central values of the interest rate. Figures 25.4 and 25.5 display graphs for the third and fourth empirical moments as a function of rate r. We also indicate the median obtained through simulated paths, the 10% and the 90% confidence intervals.

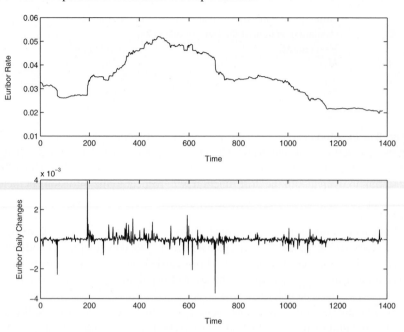

Fig. 25.1. The 3-month EURIBOR from January 1999 to April 2004.

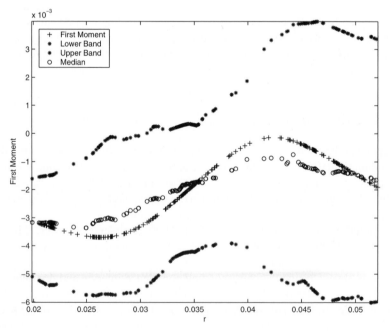

Fig. 25.2. Estimated first-order moment for the 3-month EURIBOR (+); 10–90th confidence bands obtained by simulation (∗); median value (o).

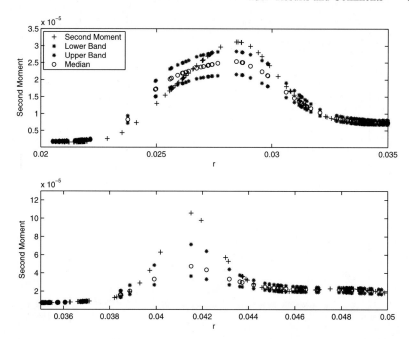

Fig. 25.3. Estimated second-order moment for the 3-month EURIBOR ($+$); 10–90th confidence bands obtained by simulation ($*$); median value ($\circ$).

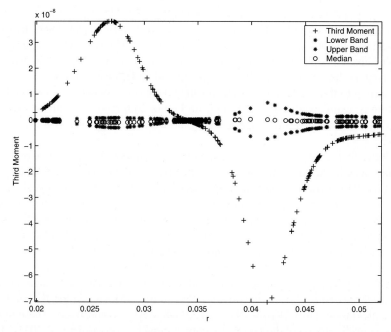

Fig. 25.4. Estimated third-order moment for the 3-month EURIBOR ($+$); 10–90th confidence bands obtained by simulation ($*$); median value ($\circ$).

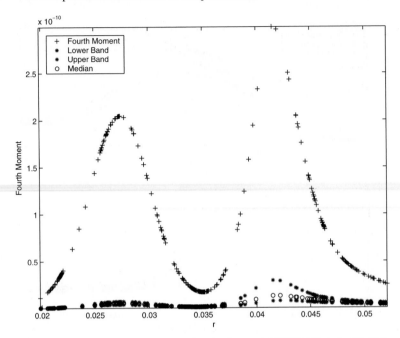

Fig. 25.5. Estimated fourth-order moment for the 3-month EURIBOR (+); 10–90th confidence bands obtained by simulation (∗); median value (○).

Notice that both moments lie outside confidence bands. In particular, the fourth moment is well above the 90% confidence upper bound. These findings suggest that a simple univariate diffusion model is not capable of reproducing the random dynamics displayed by the 3-month EURIBOR in the period under investigation.

Turning to the estimation of a mixed-jump diffusion process, the estimated drift function is again equal to the first moment. Figure 25.6 reports estimated drift (top panel), jump intensity (mid panel), and volatility (bottom panel) as a function of rate r.

Notice that jump intensity exhibits two modes and three local minima, one at the median value in the data set, that is at 3%, the others at the two endpoints of the sample range. The diffusion coefficient is instead rather stable, showing a peak around 4.2%. As for the jump size variance, the issue is quite subtle. Recall that our estimator for this figure is defined as the ratio between the sixth moment and five times the fourth moment (formula (25.14)). This quantity varies with the level r of interest rates. In our experiments we have computed 1382 values for this function. We argue that the lower the variation of this sample curve, the better the estimation of the jump size volatility using the average of these values. In our case, the 1382 sampled numbers display a standard deviation about 6.0151×10^{-7}. Moreover, the 10th and 90th percentiles amount to 1.2816×10^{-6} and 2.9668×10^{-6}, respectively. These figures prove a low dispersion of the sample function around its average $\hat{\sigma}_Y^2 = 2.3031 \times 10^{-6}$

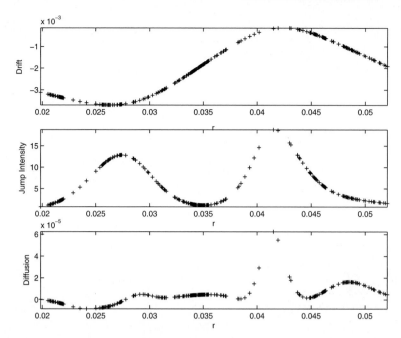

Fig. 25.6. Estimation of the drift function, the jump intensity and the diffusion function for a jump-diffusion with centered and normal jumps.

(which corresponds to a standard deviation $\hat{\sigma}_Y = 0.0015$). As a consequence, this number seems to be a reliable estimation of the jump size volatility.

A Smiling GARCH[*]

Key words: econometrics, Monte Carlo simulation, implied volatility, GARCH model

26.1 Problem Statement

In equity markets, empirical option prices document a well-known anomaly of the Black and Scholes formula. Volatility as implied by traded premia changes with respect to the exercise price and the maturity of an option. As shown in Fig. 26.1, once plotted against different exercise prices and maturities, implied volatility – the volatility value deduced from feeding observed option premia to the Black and Scholes formula – presents a characteristic convex shape which resembles a smirk (the *smile* effect). If we define option moneyness as the ratio between spot price and strike price,[1] in the stock option market we often observe that *deep-in-the-money* call options (that are call options with moneyness significantly larger than one), or *deep-out-of-the-money* put options, typically present an implied volatility greater than *at-the-money* options. Conversely, *deep-out-of-the-money* call options or *deep-in-the-money* put options admittedly show smaller implied volatility.

The immediate explanation of this phenomenon relates to the assumptions on stock price distribution embedded in the Black and Scholes formula. More precisely, since the latter assumes equity prices to be lognormally distributed, convexities in implied volatility arise because observed equity prices distribute with a thinner right tail and a fatter left tail than the canonical lognormal form. In fact, if the theoretical

---

[*] with Mariano Biondelli, Enrico Michelotti and Marco Tarenghi.

[1] Moneyness equals one if an option is *at-the-money* (ATM); whilst it is greater or lower than one if a call option is *in-* or *out-of-the-money* (ITM or OTM), respectively.

Garch estimated volatility surface against MktVol surface

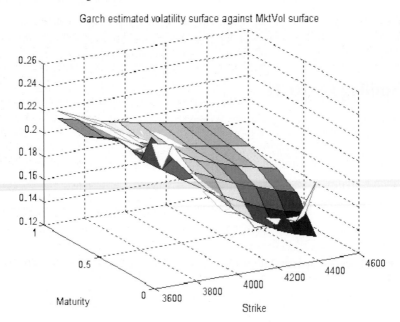

Fig. 26.1. Garch estimated versus market implied volatility surface.

assumption of lognormal distribution were to be perfectly confirmed in the market-place, the cross-section of implied volatilities against moneyness would be flat by construction.

This factual explanation, however, simply shifts the focus to price distributions: why do they do not confirm the Black–Scholes assumptions? Several arguments have been proposed. The most important is defined as the *leverage effect*. According to this argument, when stock prices move downward, the leverage of a company increases and its stocks become more risky. As a result, the volatility of prices increases to reflect a potentially more risky corporate profile. Moreover, particularly negative price performances which occurred in the past (for instance, during the 1987 crash) may influence the reactions of market players during a period of price downturn, which may have additional bearing on price volatility. A fat left tail (higher frequency in low values) in the distribution of prices (hence, a smile in implied volatilities) may thus be an attempt to factor the impact of extremely negative events into option prices. For example, Jackwerth and Rubinstein (1996) remark that the stock crash of October 1987 – under normality assumption – should occur only once in 14,756 years. Nonetheless, from that event onward, implied volatilities for the S&P 500 Index options have exhibited the characteristic smile effect.

Leverage or stock crash effects, however, may be sensible to explain the smile curve, but are not very useful to clarify the term structure of implied volatility (Fig. 26.1). As far as the time dimension is concerned, different aspects need be taken into consideration. In particular, empirical evidence shows that volatility is mean reverting. So, if the current level is below its long term trend, then implied

volatility plotted against time-to-maturity should be positively sloped (this constitutes a proxy for the expectation of a return to normal levels over time). Conversely,
it should be negatively sloped, in periods in which its recent past value has been ab-
normally high. These observations clearly question the assumption of constancy of
the Black and Scholes framework and carry relevant implications for practitioners. In
fact market agents price new options using the implied volatilities from quoted and
liquid market prices. Hence, they are exposed to a serious risk of mispricing if their
pricing models do not carefully fit the volatility surface and its expected behavior.

Different remedies have therefore been proposed (see Hull (2005)). In these
pages we focus on the approach proposed by Duan (1995, 1996) based on a GARCH
option pricing model (see also Campbell, Lo and MacKinlay (1997)). The story goes
as follows. In Sect. 26.1, we describe the mathematical model, based on a GARCH
specification for stock volatility. We then show how to calibrate the model parameters, by trying to fit the volatility surface on empirical observations. To this end, we
resort to a Monte Carlo simulation coupled with a nonlinear least squares procedure,
as described in Sect. 26.3. Subsequently, in Sect. 26.4, we provide a few numerical
results and comments about our main findings. Finally, in Sect. 5, we describe all the
functions implemented in the code.

26.2 Model and Solution Methodology

The trade-off between return volatility and asset returns is well documented in finance. However, as introduced above, equity data also exhibit a leverage effect, since
negative innovations in prices show greater positive impact on stock volatility than
positive shocks on prices have in the opposite direction. Hence, a model in which
asset prices and volatility are negatively and asymmetrically correlated seems to be
appropriate. Engle and Ng (1993) propose to this end the Nonlinear Asymmetric
GARCH$(p; q)$, or NGARCH$(p; q)$ model. The most important aspect of this model
is the asymmetric effect that news have on volatility. Indeed, a parameter θ is added
to the normal GARCH specification to control the size and direction of the shock. If
this parameter θ is positive, negative shocks in the underlying price process have a
larger impact on volatility than positive ones.

Therefore, in a discrete time setting, under the physical probability measure, one
period logarithmic returns are assumed to be conditionally normally distributed, and
volatility depends on itself and on past shocks in an asymmetric way:

$$\ln \frac{S(t+1)}{S(t)} = r + \lambda \sigma(t) - \frac{1}{2}\sigma^2(t) + \sigma(t) \times \varepsilon(t+1), \tag{26.1}$$

$$\sigma^2(t+1) = \beta_0 + \beta_1 \times \sigma^2(t) + \beta_2 \times \sigma^2(t) \times \big(\varepsilon(t) - \theta\big)^2, \tag{26.2}$$

$$\varepsilon(t+1)|\mathcal{F}_t \sim \mathcal{N}(0,1). \tag{26.3}$$

In the above statement, $S(t)$ is the asset price at time t, $\varepsilon(t)$ is a standard normal
random variable, r is the one-period continuously compounded risk-free rate, λ is a

risk premium and θ the parameter controlling the leverage effect. This model guarantees a positive conditional volatility, if the additional constraints $\beta_0, \beta_1, \beta_2 \geq 0$ and $\beta_1 + \beta_2 \leq 1$ are satisfied.

The above specification has the advantage of generating a skew in the volatility structure. In fact, by looking at θ, it may be observed that this leverage parameter drives an asymmetry on the information coming from the market. Suppose θ is positive: negative shocks in the underlying price process have a larger impact on volatility than positive ones. This creates an asymmetric impact on volatility. As a consequence the parameter θ allows for a skew (or a smirk) in the volatility structure: positive θ means higher volatility for low stock prices, negative θ means higher volatility for high stock prices.

Now, in order to price derivative contracts, we need the stock price process under the risk-neutral probability measure. Duan (1995) has indeed shown how to risk-neutralize the process described by (26.1)–(26.1). In particular, the model under the risk-neutral probability measure $\mathbb{P}^*$ becomes:

$$\ln \frac{S(t+1)}{S(t)} = r - \frac{1}{2}\sigma^2(t) + \sigma(t) \times \xi(t+1), \tag{26.4}$$

$$\sigma^2(t+1) = \beta_0 + \beta_1 \times \sigma^2(t) + \beta_2 \times \sigma^2(t) \times \left(\xi(t) - \theta - \lambda\right)^2, \tag{26.5}$$

with $\xi(t) = \varepsilon(t) + \lambda$ a standard normal random variable under $\mathbb{P}^*$.

As usual, the current price of an option contract can be seen as the average of its discounted random payoff. For example, for a plain vanilla European call option, with time-to-maturity τ and strike X, we have

$$c_t(\tau, X) = \mathbb{E}^*\left[e^{-r\tau} \max\{S(t+\tau) - X, 0\}|\mathcal{F}_t\right], \tag{26.6}$$

where $\mathbb{E}^*$ is the expectation operator under the risk-neutral probability measure $\mathbb{P}^*$, $S(t+\tau)$ is the asset price at the option maturity $t+\tau$, X denotes the strike price and $\mathcal{F}_t$ is the σ-algebra representing the information set available at time t. Notice that expression (26.6) requires a numerical valuation by appropriate methods, such as Monte Carlo simulation. Running a simulation of the above NGARCH process and discounting, we can estimate the option price (26.6) by the following computation

$$c_t^{(n)}(\tau, X) = \frac{1}{n}e^{-r\tau} \sum_{i=1}^{n} \max\{S_i(t+\tau) - X, 0\}, \tag{26.7}$$

where $S_i(t+\tau)$ is the ith simulated spot price at the option maturity.

Unfortunately, a crude Monte Carlo simulation is unable to guarantee that the price process satisfies the martingale restriction. This fact obviously has a negative impact on the quality of the estimate and the discounted average of a European call option payoff computed from a Monte Carlo simulation, that is $c_t^{(n)}(\tau, X)$, may not be a reliable estimate of $c_t(\tau, X)$.

To overcome this problem we follow the Empirical Martingale Simulation (EMS) procedure proposed by Duan and Simonato (1998). This simple correction of the

Monte Carlo procedure imposes a constraint on the simulated sample paths of the underlying prices, in order to ensure that these paths are martingales in an empirical sense. The procedure modifies simulated asset prices at different time points $t_1, t_2, t_3, \ldots, t_m = t + \tau$ of the simulation, so as to get a new estimate for the price $S_i^*(t_j, n)$ that respects martingality:

$$S_i^*(t_j, n) = S(t) \frac{Z_i(t_j, n)}{Z_0(t_j, n)},$$

for all $i = 1, 2, \ldots, n$ and $j = 1, 2, \ldots, m$. The quantities $Z_i(t_j, n)$ are recursively defined by

$$Z_i(t_j, n) = S_i^*(t_{j-1}, n) \frac{\widetilde{S}_i(t_j)}{\widetilde{S}_i(t_{j-1})}, \qquad (26.8)$$

$$Z_0(t_j, n) = \frac{1}{n} e^{-rt_j} \sum_{i=1}^{n} Z_i(t_j, n), \qquad (26.9)$$

for $i = 1, 2, \ldots, n$ and $j = 1, 2, \ldots, m$. In practice EMS uses a simple transformation of the simulated asset price $\widetilde{S}_i(t_j)$, to guarantee the martingale restriction. Indeed, EMS can be described through the following chain that transforms the simulated price process into a new one satisfying the martingale condition:

$$\frac{\widetilde{S}_i(t_j)}{\widetilde{S}_i(t_{j-1})} \rightarrow Z_i(t_j, n) \rightarrow Z_0(t_j, n) \rightarrow S(t) \frac{Z_i(t_j, n)}{Z_0(t_j, n)} = S_i^*(t_j, n). \qquad (26.10)$$

The attractive feature of this adjustment is that the EMS option price estimate will reduce the Monte Carlo simulation error. For deep-in-the-money options, this error will even be negligible. Indeed, for these options, the option price will be approximately $\widehat{S}(t) - e^{-rt_n} X$, where $\widehat{S}(t)$ is discounted sample average asset price. Using the EMS, this quantity coincides, by construction, with $S(t)$, so that the EMS option price will have a negligible variance if the option is deep-in-the-money. EMS has one drawback however. Because of the dependency among sample paths created by the EMS adjustment, the standard error of the price estimate is not readily available from using one simulation sample. The EMS accuracy is studied in detail in Duan, Gauthier and Simonato (2001).

26.3 Implementation and Algorithm

Several approaches have been proposed in order to estimate the NGARCH model. A simple solution is to estimate the parameters by directly using a time series of the underlying asset returns; for example, by maximum likelihood estimation. However, in this manner one implicitly assumes that physical and risk-neutral probability measures coincide, bypassing the risk-neutralization of the NGARCH process proposed above. Therefore, Duan (1996) proposes to estimate the NGARCH(1, 1) parameters $\beta_0, \beta_1, \beta_2$ and the sum $\beta_3 = \theta + \lambda$, by extracting implied volatilities from simulated option prices and subsequently minimizing the sum of squared errors with respect to market implied volatilities.

The resulting procedure can be summarized in the Algorithm below. In particular, we calibrate the NGARCH model on July 21, 2003, using FTSE 100 index options, with strikes ranging from 3625 to 4425 and maturities, beginning on August 2003 and ending on June 2004.

Algorithm

1. Extract from option market quotes, using the put-call parity, implied estimates of the underlying index level and of the discounting factor. This procedure avoids problems related to the synchrony of the stock and option markets;
2. Fix arbitrary NGARCH parameters, e.g., estimate them by using a time series approach, and simulate the price of the underlying using (26.4) and applying the EMS correction methodology;
3. Compute the simulated option prices (26.7) and extract the implied volatilities on simulated prices by using the Black and Scholes formula;
4. Compute the sum of squared residuals between estimated implied volatilities at the previous step and market implied volatilities on observed market prices;
5. Repeat Steps 2–5 by using a new set of GARCH parameters until the sum of squared residuals has been minimized.

We tackle the first step by following Shimko (1993). This author exploits the put-call parity theorem in order to compute the underlying asset value adjusted for dividends and the discounting factor from option market prices. Indeed, we have

$$c_t(\tau, X) - p_t(\tau, X) = S(t)e^{-q\tau} - Xe^{-r\tau}, \qquad (26.11)$$

where q is the continuous dividend yield on stock prices. Let us include in the above relationship an error term η in order to take into account slight violations of the put-call parity due for example to a bid-ask spread:

$$Y = \alpha + \beta X + \tilde{\eta}, \qquad (26.12)$$

where we relate $Y = c_t(\tau, X) - p_t(\tau, X)$ linearly to X and then we have an equation suitable for ordinary least square (OLS) regression. Therefore, having observations of call and put options for different strikes X, we obtain, for each time-to-maturity τ, the OLS estimates $\hat{\alpha}$ and $\hat{\beta}$. Then, we can estimate implied spot prices adjusted for dividends through $S(t)e^{-q\tau}$ by $\hat{\alpha}$, and the implied interest rate r through $-\ln(-\hat{\beta})/\tau$.

In Table 26.1 we report the results of the OLS Regression (26.12). Subsequently, we can feed these estimates to the Black and Scholes formula in order to obtain the market implied volatilities σ_{mkt}. To do this, we extract implied volatility from out-of-the-money options. Therefore, if estimated stock prices $S(t)$ are greater than X, we use implied put-volatility, otherwise implied call-volatility.[2] Finally, by using the

---

[2] Note that for the put-call parity, we have:

Table 26.1. Estimated values, according to the Shimko methodology, for the spot price including dividends and the risk free rate for different time to maturities

	Time-to-maturity (in days)					
	25	60	88	151	242	336
S_0	4064.01	4071.41	4033.90	4044.72	3962.59	3927.41
r	−0.0125	−0.0217	0.0282	0.0367	0.0346	0.0351

risk-neutral GARCH specification, (26.4) and (26.5), and the EMS correction (26.8), we run 5,000 Monte Carlo simulations for each maturity. The starting values for the GARCH parameters $\beta_0, \beta_1, \beta_2, \lambda + \theta$ and σ_1 have been arbitrarily set. In particular, we set σ_1 equal to the ATM implied volatility or the GARCH equilibrium volatility:

$$\frac{\beta_0}{1 - \beta_1 - \beta_2}.$$

Then, the option prices are obtained by replacing $S_i(t)$ with $S_i^*(t_j)$ in expression (26.7):

$$c_t^{(n)}(\tau, X) = \frac{1}{n} e^{-r\tau} \sum_{i=1}^{n} \max\{S_i^*(t + \tau) - X, 0\}.$$

From simulated option prices, for each strike X and time-to-maturity τ, we compute the GARCH implied volatilities $\sigma_{\text{garch}}(X, \tau)$, by numerically inverting the Black and Scholes formula. Then, given market implied volatilities $\sigma_{\text{mkt}}(X, \tau)$, we estimate the parameters of the NGARCH model by minimizing the sum of squared differences. The problem reads as a non-linear least squares minimization

$$\min_{\beta_0, \beta_1, \beta_2, \lambda + \theta} \sum_{i=1}^{k} \sum_{j=1}^{m} \left(\sigma_{\text{mkt}}(X_i, \tau_j) - \sigma_{\text{garch}}(X_i, \tau_j)\right)^2,$$

$$p_{\text{B\&S}} + S_t e^{-q\tau} = c_{\text{B\&S}} + X e^{-r\tau}, \tag{26.13}$$

where $p_{\text{B\&S}}$ and $c_{\text{B\&S}}$ are the option premiums evaluated using Black and Scholes formula. The same relationship is valid for market prices:

$$p_{\text{m}} + S_t e^{-q\tau} = c_{\text{m}} + X e^{-r\tau}, \tag{26.14}$$

where p_{m} and c_{m} are the market prices of the options with the same strike (X) and time to maturity (τ) considered before. Subtracting (26.13) from (26.14) we get:

$$p_{\text{B\&S}} - p_{\text{m}} = c_{\text{B\&S}} - c_{\text{m}}. \tag{26.15}$$

This means that the error we commit using the Black and Scholes formula is the same when evaluating call and put options with the same strike and maturity, and so the implied volatility is not dependent on the kind of option we are dealing with, of course *ceteris paribus*. For this reason, we extract implied volatility from the most liquid options, i.e. OTM puts and OTM calls.

under the natural constraints

$$\beta_0 > 0,$$
$$\beta_1 \geqq 0, \qquad \beta_1 + \beta_2 < 1,$$
$$\beta_2 \geqq 0, \qquad \lambda + \theta > 0.$$

The results of the minimization procedure are reported in Table 26.2. In particular, notice the behavior of the GARCH parameter $\beta_3 = \lambda + \theta$. This parameter controls the leverage effect under the risk-neutral measure and, thus, the skewness of the risk-neutral density. We observe that β_3 increases as the option maturity shortens, reflecting the fact that for short maturities risk-neutral densities appear more skewed. A comment on the optimization procedure is necessary. Our aim is to find the NGARCH parameters such that the above loss function, defining the distance between market implied volatility smirk and the simulated one, is minimized. In theory, a standard minimization procedure could be run. However sampling errors in the simulation can create problems in the optimization procedure. In other words, we may end up maximizing the sampling error! In order to overcome this problem, we suggest the use of a grid minimization method. In practice we define a reasonable parameter domain by taking the parameter values estimated on a time series basis as reference points. Then, we divide the range for each parameter in a grid with ten bins and we minimize the objective function across that multi-dimensional grid, with a tolerance of 10^{-3}. However, we still have to manage $3 \times 10^4 = 30,000$ Monte Carlo simulations, each one consisting of 5,000 iterations, for every maturity. This implies a huge computational effort. In order to limit this figure, we resort to some additional simplifications:

1. We separately estimate the leverage effect parameter $(\lambda + \theta)$ and the other three parameters.
2. We set the equilibrium variance of the GARCH model equal to the long term market ATM variance:

$$\beta_0 = \sigma_{\mathrm{ATM}}^2 (1 - \beta_1 - \beta_2).$$

These simplifications afford a reduction in the number of simulations to $3 \times (10 + 10^2) = 330$ and, as a result, in the computational time effort (nonetheless, still remarkably high: at least 10 hours with a 2.0 GHz processor). Finally, using a grid

Table 26.2. The GARCH parameters estimated via the Duan methodology

Time-to-maturity (days)	NGARCH parameters			
	β_0	β_1	β_2	β_3
25	0.00135	0.80551	0.10924	0.86
60	0.00138	0.80551	0.10654	0.8
88	0.00139	0.80233	0.10685	0.8
151	0.0012	0.805	0.10612	0.81
242	0.0011	0.80321	0.10924	0.78
336	0.00117	0.80351	0.10824	0.78

search we guaranteed a good approximate solution to our problem. Using an optimization routine like `fminsearch` in Matlab®, the minimization procedure does not produce reliable results.

As an alternative, we also consider a two-step procedure. In the first step, the three parameters β_0, β_1 and β_2 are found by a search over a three-dimensional grid, given an arbitrary starting value for β_3. In the second step, given the values of the three parameters β_0, β_1 and β_2, the value of β_3 is optimized using a one-dimensional minimization routine. We repeat this two-stage procedure only once, so that no optimality is ensured. In particular, parameter estimates strongly depend on the initial (arbitrary, but hopefully meaningful) value for β_3.

26.3.1 Code Description

Function `initial`

Provides the workplace of global variables which are necessary to run the following functions. In this case, the M-File Initial contains: the initial values of the GARCH parameters, the number of Monte Carlo simulations and all Market data (strikes, maturities, implied volatility surface, underlying asset prices and interest rates).

Function `impl_vol`

1. Sets an initial guess for the implied volatility:

$$\sigma = \sqrt[2]{\left| \left[\ln\left(\frac{S(t)}{X}\right) + r\tau \right] \frac{2}{\tau} \right|}.$$

2. Decides if the option is Call or Put;
3. Starts a "for cycle" to determine the implied volatility using the Newton method;
   a. Computes d_1, d_2, Vega, $\mathcal{N}(d_1)$, and $\mathcal{N}(d_2)$, at every step;
   b. Compares the theoretical premium to the real one and if the difference is less than the tolerance level, then the implied volatility is accepted;
   c. If the previous requirement is not fulfilled, it adjusts it in the following manner (Newton recursion):

$$\sigma = \sigma - (\text{Premium}_{\text{theo}} - \text{Premium}_{\text{mkt}})/\text{Vega};$$

4. Sets the volatility value coming from the minimization $\notin \mathbb{R}^+$ equal to zero. In other words, volatility is forced to be non-negative.

Function `vol_surface`

1. Sets an external cycle for each maturity available on the market;

2. For every maturity runs a Monte Carlo simulation in order to obtain n paths of the underlying asset, supposing that the underlying geometric Brownian motion has a time varying variance described by the NGARCH equation:

$$\sigma^2(t+1) = \beta_0 + \beta_1\sigma^2(t) + \beta_2\sigma^2(t)\big(\xi(t) - \theta - \lambda\big)^2;$$

3. Once a Monte Carlo path is simulated for each maturity, the code applies the EMS procedure;
4. Using the corrected paths, the values of out-of-the-money call and put options are computed for all strikes available on the market;
5. The implied volatility of these options is computed;
6. The value of the loss function (sum of squared residuals between simulated implied volatilities and market implied volatilities) is then computed.

Function `optimization`

1. Sets a "for cycle" across maturities;
2. Retrieves the ATM volatility corresponding to the selected maturity;
3. Launches an optimization function (`opt_garch`) to fit the three GARCH parameters (excluding the leverage effect, minimized apart from these), supposing that the equilibrium variance of the GARCH model is always equal to the ATM variance;
4. A double entry grid of 10×10 bins is used to make the screening of parameters β_1 and β_2, with a tolerance level of 10^{-3} for both;
5. The parameter β_0 is automatically determined given the following relation:

$$\beta_0 = \sigma^2_{ATM}(1 - \beta_1 - \beta_2);$$

6. Given the first three optimized parameters, considers another optimization (`opt_leverage` routine) to find the best leverage parameter such that the distance between the market implied volatility surface and the GARCH simulated implied volatility surface is minimized. Also in this case a grid approach is used with ten bins and a tolerance of order 10^{-3};
7. The optimal parameters are stored in a matrix.

Function `opt_garch`

1. Defines the maximum and minimum values for parameter β_1 (the GARCH effect), computes the range, and the medium plus initial value;
2. Start the first "for cycle" across the different values of β_1 (grid);
3. For each value of β_1 defines the maximum and minimum value of β_2 (ARCH effect parameter) in such a way that $\beta_1 + \beta_2 < 1$, i.e. $\max[\beta_2] = 1 - \beta_1 - 0.01$;
4. Starts the second internal cycle "`for`" across the different and admissible values for β_2 (grid);

5. For each admissible pair (β_1, β_2) runs the function `garch_loss` in order to obtain estimated implied volatilities and the loss function with respect to the market implied volatility;

6. A matrix of losses is obtained according to the two cycles "`for`" and the minimum of this matrix identifies the optimal (β_1, β_2) pair;

7. A refining procedure is applied to reach the tolerance level of 10^{-3} for the optimal parameter values: it repeats Steps 1–6 twice, with a range equal to 10^{-2} and 10^{-3} respectively, in an interval around the optimal values found.

Function `opt_leverage`

1. Given the values for the first three parameters, defines the minimum and maximum value for the leverage effect parameter $(\theta + \lambda)$ and computes the mean and initial value;

2. Starts a cycle "`for`" across the different values of $(\theta + \lambda)$ (grid method), from the minimum to the maximum value;

3. For each of them computes the loss function as a distance between market implied volatility and GARCH estimated volatility (running `garch_loss` code);

4. Refines the search to get the required tolerance approximation (10^{-3} as for the other parameters), repeating Steps 2 and 3.

Function `garch_loss`

1. Defines a vector of initial values for time dependent volatility using the equilibrium value of GARCH variance,

$$\sigma_{ATM}^2 = \beta_0/(1 - \beta_1 - \beta_2);$$

2. Defines a matrix of standardized normal shocks (rows equal to the number of Monte Carlo simulation and columns equal to the number of steps in each simulation);

3. Starts a "`for`" loop across time to get the path of the underlying asset. Starting with S_0, the function generates, according to a Geometric Brownian Motion with a GARCH volatility, n paths for the underlying asset using the recursion:

$$S_{t+dt} = S_t \exp\left((r - 0.5\sigma_t^2)\,dt + \sigma_t\sqrt{dt}\,\varepsilon_{t+dt}\right);$$

4. At each step the EMS procedure is applied to guarantee martingality and to reduce Monte Carlo variance. At each step, GARCH variance is also updated according to the normal shock which affects prices in the previous step:

$$\sigma_t^2 = \beta_0 + \beta_1\sigma_{t-1}^2 + \beta_2\sigma_{t-1}^2(\xi_{t-1} - \theta - \lambda)^2;$$

5. The final vector of underlying prices (corrected with EMS procedure) is used to obtain the Monte Carlo estimate of the option price:

$$c_t^{(n)}(\tau, X) = \frac{1}{n}e^{-r\tau}\sum_{i=1}^{n}\max\left[S_i^*(t+\tau) - X, 0\right];$$

6. A cylce "for" across strikes is used to obtain the out-of-the-money call or out-of-the-money put option prices;
7. For each option the implied volatility is extracted running the impl_vol routine;
8. The loss function value is computed as the sum of squared residuals between market implied volatilities and estimated volatilities.

26.4 Results and Comments

The results of the estimation procedure are given in Tables 26.3 and 26.4, where we compare market implied volatilities (Table 26.3) and the ones computed according to the Duan procedure (Table 26.4).

We observe that, for short maturities, the model undervalues implied volatilities. On the contrary, for longer maturities, GARCH based implied volatilities are higher. It seems that GARCH based implied volatility estimates tends to flatten out the peaks

Table 26.3. Market implied volatilities on FTSE 100 index on July 21st 2003

Strike	Time-to-maturity (in days)					
	25	60	88	151	242	336
3625	0.24145	0.23058	0.22579	0.22854	0.22654	0.22261
3725	0.23678	0.209931	0.21986	0.21863	0.21997	0.21544
3825	0.20892	0.204449	0.20754	0.20875	0.21064	0.20858
3925	0.19238	0.184878	0.19745	0.19745	0.20221	0.20172
4025	0.17504	0.16508	0.18424	0.18123	0.19539	0.19529
4125	0.15254	0.15915	0.17881	0.18193	0.18844	0.18888
4225	0.15278	0.153475	0.17097	0.17153	0.18374	0.18309
4325	0.15194	0.149329	0.16374	0.16271	0.17264	0.17746
4425	0.18368	0.150364	0.15303	0.16115	0.17078	0.17227

Table 26.4. Estimated implied volatilities on FTSE 100 index on July 21st 2003

Strike	Time-to-maturity (in days)					
	25	60	88	151	242	336
3625	0.22471	0.20661	0.19768	0.19497	0.20169	0.19686
3725	0.21496	0.20048	0.19589	0.19384	0.20104	0.19677
3825	0.20431	0.19567	0.19449	0.19819	0.20078	0.19629
3925	0.19735	0.19017	0.19277	0.19204	0.20061	0.19583
4025	0.19227	0.18601	0.19134	0.19144	0.20031	0.19529
4125	0.18762	0.18342	0.19079	0.19112	0.19986	0.19489
4225	0.18291	0.18115	0.19075	0.19083	0.19938	0.19451
4325	0.17953	0.17938	0.19046	0.19065	0.19865	0.19418
4425	0.17668	0.17694	0.19072	0.19057	0.19994	0.19379

of the volatility surface, and to produce volatility data less affected by shocks than observed implied volatilities.

In addition, the estimation procedure shows two main features:

(1) The decreasing trend of the leverage effect, from 0.86 for the shortest maturity to 0.78 for the longest one. This is mirrored in market data, where far-out-of-the-money puts with a very short life present a higher volatility premium than the far-out-of-the-money put with a longer residual life.
(2) While the smirk effect is captured by the leverage parameter θ, the other GARCH parameters remain almost constant across maturities. This can reflect the stability of the underlying process which is common for all analyzed options.

Finally, from Figure 26.1, where the market quoted volatility is plotted against the estimated volatility, we note that the GARCH estimation produces a kind of smoothing effect of the surface compared to what is obtained by original market prices.

Appendix A: Proof of the Thinning Algorithm

Consider the problem of generating samples of a counting process $N(t) = \sum_{i \geq 1} \mathbf{1}_{\{\tau_i \leq t\}}$ with (possibly random) intensity $\lambda(s)$. This amounts to sampling jump times τ_i, $i \geq 1$, from the knowledge of $\lambda(s)$. The thinning method requires to (1) select λ^* as an upper bound for $\lambda(s)$ on its domain; (2) sample jump times τ_i^* of a Poisson process N^* with constant intensity λ^*; (3) perform an acceptance–rejection test on each τ_i^*, which consists of sampling the random variable $\mathbf{1}_{\{\lambda^* U_i \leq \lambda(\tau_i^*)\}}$, where $U_i \overset{\text{i.i.d.}}{\sim} \mathcal{U}[0, 1]$; (4) accept τ_i^* as a jump time of N provided that the test succeeds, namely the independent uniform variable $\lambda^* U_i$ on $[0, \lambda^*]$ is smaller than $\lambda(\tau_i^*)$. It is assumed that λ, N^* and the U_i's are all statistically independent.

To prove this, we need to check that the process the counting process of the accepted jump times $N(t) = \sum_{i \,:\, 0 < \tau_i^* \leq t} \mathbf{1}_{\{\lambda^* U_i \leq \lambda(\tau_i^*)\}}$ has intensity λ, that is:

$$\mathbb{E}\big(N(t) - N(s)|\mathcal{F}_s^N\big) = \mathbb{E}\bigg(\int_s^t \lambda(u)\,\mathrm{d}u\,\bigg|\,\mathcal{F}_s^N\bigg), \tag{A.1}$$

where $\mathcal{F}_s^N$ is the completed filtration generated by N. This formula has a clear interpretation in the deterministic case: the expected number of jump times occurring over an interval $(s, t]$ is given by accruing the jump intensity λ over the same interval. For the sake of clarity, we assume that the function λ is superiorly bounded and set $\lambda^* = \sup \lambda(t, \omega)$. It is left to the reader to prove that the intensity process λ is $\mathcal{F}^N$-adapted. Using the linearity property and the rule of iterated conditional expectation, we have

$$\mathbb{E}\big(N(t) - N(s)|\mathcal{F}_s^N\big)$$

$$= \mathbb{E}\bigg(\sum_{i:s<\tau_i^*\leq t} \mathbf{1}_{\{\lambda^* U_i \leq \lambda(\tau_i^*)\}}\,\bigg|\,\mathcal{F}_s^N\bigg)$$

$$= \mathbb{E}\bigg(\sum_{i:s<\tau_i^*\leq t} \mathbb{E}\big(\mathbf{1}_{\{\lambda^* U_i \leq \lambda(\tau_i^*)\}}|\mathcal{F}_{\tau_i^*}^N\big)\,\bigg|\,\mathcal{F}_s^N\bigg) \qquad \text{as } \mathcal{F}_{\tau_i^*}^N \supset \mathcal{F}_s^N$$

$$= \mathbb{E}\left(\sum_{i:s<\tau_i^*\leq t} \mathbb{P}\big(U_i \leq \lambda(\tau_i^*)/\lambda^* | \mathcal{F}_{\tau_i^*}^N\big) \Big| \mathcal{F}_s^N \right) \qquad \text{as } \mathbb{P}(\mathcal{U}[0, 1] \leq X | X = u) = u$$

$$= \frac{1}{\lambda^*} \mathbb{E}\left(\sum_{i=N^*(s)+1}^{N^*(t)} \lambda(\tau_i^*) \Big| \mathcal{F}_s^N \right) \qquad \text{as } \lambda(\tau_i^*) \text{ is } \mathcal{F}_{\tau_i^*}^N\text{-adapted}$$

$$= \frac{1}{\lambda^*} \mathbb{E}\left(\mathbb{E}\left(\sum_{i=N^*(s)+1}^{N^*(t)} \lambda(\tau_i^*) \Big| \mathcal{F}_t^N \right) \Big| \mathcal{F}_s^N \right) \qquad \text{as } t > s.$$

Recall that the jump times are uniformly distributed on the interval $(s, t]$ conditional on the number of jumps having occurred over the same interval, that is $\tau_{N^*(s)+1}, \ldots, \tau_{N^*(t)} | \mathcal{F}_t \overset{\text{i.i.d.}}{\sim} \tau \sim \mathcal{U}(s, t]$. The Wald theorem leads to:

$$\mathbb{E}\big(N(t) - N(s) | \mathcal{F}_s^N\big) = \frac{1}{\lambda^*} \mathbb{E}\big(\mathbb{E}[N^*(t) - N^*(s) | \mathcal{F}_t^N] \times \mathbb{E}[\lambda(\tau) | \mathcal{F}_t^N] | \mathcal{F}_s^N\big).$$

Given $\mathcal{F}_s^N$, the increment $N^*(t) - N^*(s)$ and the random variable τ are mutually independent. Moreover, τ is uniformly distributed in $[s, t]$ given $\mathcal{F}_t^N$. The last term then becomes equal to:

$$\frac{1}{\lambda^*} \mathbb{E}\big(N^*(t) - N^*(s) | \mathcal{F}_s^N\big) \times \mathbb{E}\big(\lambda(\tau, \omega) | \mathcal{F}_s^N\big)$$

$$= \frac{1}{\lambda^*} \int_s^t \lambda^* \, du \, \mathbb{E}\left(\int_s^t \lambda(u) \frac{du}{t-s} \Big| \mathcal{F}_s^N \right)$$

$$= \mathbb{E}\left(\int_s^t \lambda(u) \, du \Big| \mathcal{F}_s^N \right),$$

which is the claim to be proved.

Appendix B: Sample Problems for Monte Carlo

1. Let the state variable X be the instantaneous short rate r. Fix three increasing times $t < T_1 < T_2$ and let P_{T_2} be the process describing the value of a default free zero-coupon bond maturing at time T_2. We choose $T_2 \geq T_1$ because the asset needs to exist at time T_1. We consider a call option on P_{T_2}, expiring at T_1 and strike at K Euros. We want to design a procedure to evaluate the fair value of this option at time t by making use of Monte Carlo methods. This option is a T_1-maturing contingent claim with pay-off function $(x - K)_+$ on a T_2-maturing contingent claim with pay-off function equal to 1 in all states of the sample world:

$$
V(t) = \mathbb{E}_t^* \left(e^{- \int_t^{T_1} r(s)\, ds} \left(P_{T_2}(T_1) - K \right)_+ \right)
$$
$$
= \mathbb{E}_t^* \left(e^{- \int_t^{T_1} r(s)\, ds} \left(\mathbb{E}_{T_1}^* \left(e^{- \int_{T_1}^{T_2} r(s)\, ds} \right) - K \right)_+ \right).
$$

To simulate this value, we may follow two routes. The *first algorithm* is naive. We set a small Δt and evolve r over time points $t, t + \Delta t, t + 2\Delta t, \ldots, T_1 - \Delta t, T_1, T_1 + \Delta t, \ldots, T_2 - \Delta t$. This is done by discretizing the s.d.e. for r over that partition. For each simulated interest rate path, we compute the option pay-off. Summing up the resulting payoff over n sampled paths and dividing by n, results in the following approximated value:

$$
C(t) \approx \frac{1}{n} \sum_{i=1}^{n} \left[e^{- \sum_{t=0}^{T_1 - \Delta t} r^{(i)}(t)\Delta t} \left(\left(e^{- \sum_{t=T_1}^{T_2 - \Delta t} r^{(i)}(t)\Delta t} \right) - K \right)_+ \right].
$$

The *second algorithm* is more effective. For each $i = 1, \ldots, N$:

(1) generate a path $(r^{(i)}(t + \Delta t), \ldots, r^{(i)}(T_1 - \Delta t))$ up to time $T_1 - \Delta t$;
(2) generate M "continuations":

$$
\left\{ \left(r^{(i,k)}(T_1), r^{(i,k)}(T_1 + \Delta t), \ldots, r^{(i,k)}(T_2 - \Delta t) \right), k = 1, \ldots, M \right\},
$$

after time T_1, until time $T_2 - \Delta t$ is reached;

560

(3) use these continuations to obtain a Monte Carlo estimate for the time T_1 value of the bond:

$$P_{T_2}^{(i)}(T_1) \approx \frac{1}{M} \sum_{k=1}^{M} \left[e^{-\sum_{t=T_1}^{T_2-\Delta t} r^{(i,k)}(t)\Delta t} \right];$$

(4) plug this value into the call option pay-off;
(5) discount between T_1 and t using the path $r^{(i)}(t+\Delta t), \ldots, r^{(i)}(T_1-\Delta t)$ generated at the first step; finally
(6) sum all these terms up and divide by n. The corresponding Monte Carlo estimate is:

$$C(t) \approx \frac{1}{n} \sum_{i=1}^{n} \left[e^{-\sum_{t=0}^{T_1-\Delta t} r^{(i)}(t)\Delta t} \left(\left(\frac{1}{M} \sum_{k=1}^{M} \left[e^{-\sum_{t=T_1}^{T_2-\Delta t} r^{(i,k)}(t)\Delta t} \right] \right) - K \right)_+ \right].$$

Write and run a computer code implementing the two procedures above. Elaborate and apply a test to compare their performance.

2. Suppose a model for the underlying factor dynamics X depends on a parameter. For instance, in the Black–Scholes model, X is the underlying security and we may take the volatility σ as a parameter. We wish to evaluate the rate of variation of the current fair value $V(t)$ of a given contingent claim resulting from a small change in the parameter δ. In the Black–Scholes example, we look for the Vega of an option. We may perform two Monte Carlo estimations: one under dynamics for the underlying factor corresponding to a parameter value $\delta + \varepsilon$; the other for a parameter value $\delta - \varepsilon$, where ε is a small positive constant. To reduce computations, in both cases we may use the same sequence of drawn r.v.'s needed to generate a path (this is one of the variance reduction techniques we will develop in the last section). We come up with estimations $V(t; \delta + \varepsilon)$ and $V(t; \delta - \varepsilon)$. The sensitivity can be approximated by a first-order difference $\partial_\delta V(t) \approx (V(t; \delta + \varepsilon) - V(t; \delta - \varepsilon))/(2\varepsilon)$. This method lets us compute hedge ratios too: indeed, the call option hedger is required to take a Δ-position in the underlying asset S, where Δ is just the sensitivity $\partial_S V(t, S(t))$. Write and run a computer code for computing sensitivities of the options detailed in the examples of Section 1 with respect to the parameters involved in a Black–Scholes model.

3. We wish to generate a simulated random path for a short rate model so that a given set of observed discount bond prices is perfectly matched. Suppose time t discount function is given by N observed zero-coupon bond prices $P_{T_1}(t) = p_1, \ldots, P_{T_N}(t) = p_N$, where $T_k = t + k\Delta t$. Recall the theoretical value for a discount bond price is: $P_T(t) = \mathbb{E}_t^*(\exp(-\int_t^T r(s)\,ds))$. Let $r(t) = r_0$. If we simulate a number n of discrete trajectories

$$\left\{ r_{tT}^{(i)} = \left(r^{(i)}(T_1), \ldots, r^{(i)}(T_{N-1}) \right), i = 1, \ldots, n \right\}$$

for the risk-neutral dynamics of the short rate process r, the theoretical value of any discount bond $P_{T_i}(t)$ need not match the corresponding observed price p_i:

$$p_i = P_{T_i}(t) \neq P_{T_i}^{\text{sampled}}(t) = \frac{1}{n}\sum_{i=1}^{n} e^{-\Delta t \sum_{k=1}^{N-1} r^{(i)}(T_k)}.$$

We want to bias each simulated path $r_{tT}^{(i)}$ so as to obtain a new path $\hat{r}_{tT}^{(i)}$ which is compatible with observed prices $p_1, \ldots, p_N$ in that:

$$p_i = \frac{1}{n}\sum_{i=1}^{n} e^{-\Delta t \sum_{k=1}^{N-1} \hat{r}^{(i)}(T_k)}.$$

To each sample $r^{(i)}(T_k)$, we

(1) add the continuously compounded forward rate spanning $[T_k, T_{k+1}]$ computed from the simulated path $r_{tT}^{(i)}$ by:

$$\{r_{tT}^{(1)}, \ldots, r_{tT}^{(n)}\} \rightarrow f_{T_k T_{k+1}}(t) = \frac{1}{\Delta t} \lg\left(\frac{P_{T_k}(t)}{P_{T_{k+1}}(t)}\right)$$

$$= \frac{1}{\Delta t} \lg\left(\frac{\frac{1}{n}\sum_{i=1}^{n} e^{-\Delta t \sum_{j=1}^{k-1} r^{(i)}(T_j)}}{\frac{1}{n}\sum_{i=1}^{n} e^{-\Delta t \sum_{j=1}^{k} r^{(i)}(T_j)}}\right),$$

and
(2) subtract the continuously compounded forward rate spanning $[T_k, T_{k+1}]$ implied in the observed discount function by:

$$\{p_1, \ldots, p_n\} \rightarrow \hat{f}_{T_k T_{k+1}}(t) = \frac{1}{\Delta t} \lg\left(\frac{p_k}{p_{k+1}}\right).$$

Then the resulting paths

$$\{\hat{r}_{tT}^{(i)} = (\hat{r}^{(i)}(T_1), \ldots, \hat{r}^{(i)}(T_{N-1})), i = 1, \ldots, n\}, \tag{B.1}$$

defined by

$$\hat{r}^{(i)}(T_k) = r^{(i)}(T_k) + f_{T_k T_{k+1}}(t) - \hat{f}_{T_k T_{k+1}}(t),$$

satisfy the matching property. That is bond prices estimated by Monte Carlo over the modified random paths in (B.1) exactly equal the observed prices $p_1, \ldots, p_N$:

$$p_i = P_{T_i}^{\text{sampled}}(t) = \frac{1}{n}\sum_{i=1}^{n} e^{-\Delta t \sum_{k=1}^{N-1} \hat{r}^{(i)}(T_k)}.$$

Write a program to implement this fitting procedure.

4. Give a formal proof for generating jump times by using the algorithm detailed in Sect. 2.3.4.

5. The recursive formula for simulating r in the Vasicek model is:

$$r(t_{i+1}) = \mu(t_{i+1}; t_i, r(t_i)) + \sigma(t_{i+1}; t_i, r(t_i))\sqrt{t_{i+1} - t_i}\mathcal{N}(0, 1)$$

starting at $(0, r_0)$. For the CIR short rate model we have $dr = \alpha(b - r)\,dt + \sigma\sqrt{r}\,dW(t)$ and:

$$r(t_{i+1}) = r(t_i) + \frac{\sigma^2}{4\alpha}\left(1 - e^{-\alpha\Delta t}\right)\chi^2\left(\frac{4\alpha\beta}{\sigma^2}, \frac{\sigma^2}{\alpha}\left(1 - e^{-\alpha\Delta t}\right)e^{-\alpha\Delta t}r(t_i)\right),$$

where $\chi^2(d, c)$ denotes a random sample from a non-central chi-square distribution with d degrees of freedom and non-centrality parameter c. See Johnson and Kotz (1995) for details on this family of distributions. For each of these models, compare Monte Carlo simulated European bond option values over increasing samples to the theoretical values given by closed form formula.

6. Prove that the approximating process $X^{(n)}$ in the Algorithm "Sampling theorem method for stationary Gaussian processes" is not stationary by verifying that its covariance function $c^{(n)}(t, s) = \mathbb{E}(X^{(n)}(t)X^{(n)}(s))$ does not depend on t and s through $t - s$. By applying the sampling theorem stated in the same section, show that $c^{(n)}(t, s)$ converges to a function of $t - s$ as $n \to \infty$. This justifies the approximation made in the above-mentioned algorithm.

Appendix C: The Matlab® Solver

Here we present the PDE solver implemented in a Matlab® environment. This tool aims at solving initial-boundary value problems for systems of parabolic and elliptic partial differential equations (PDEs) in one space variable. The syntax to be used in the command windows is as follows:

```
sol = pdepe(m,pdefun,icfun,bcfun,xmesh,tspan).
```

Here, m is parameter corresponding to the symmetry of the problem and can be set equal to 0, 1, or 2; pdefun defines the PDE; icfun sets up initial conditions; bcfun states boundary conditions; xmesh is a vector [x0,x1,...,xn] of increasing entries specifying the points at which a numerical solution is to be returned for each time in tspan=[t0,t1,...,tf]. Entries xmesh(1) and xmesh(end) are equal to z_L and z_U, respectively. Moreover, the dimensions n and f of the vectors xmesh and tspan must be greater or equal to 3. Entries tspan(1) and tspan(end) are the starting and final maturities, respectively. Function pdepe performs the time integration with an ODE solver that selects the time step in a dynamic manner.[1] Second-order approximations to the solution of the PDE are made on the mesh specified in xmesh. Notice that the function pdepe does not select the input mesh automatically. Therefore, it must be provided by the final user, who may decide to use an unevenly spaced grid. Therefore, it is a good practice to stagger mesh points by smaller amounts on the region of the domain where the solution changes rapidly. It is usually convenient to refine the mesh near the strike price and to use a sparse grid away from this region. The computational cost of the resulting routine strongly depends on the length of xmesh.

Function pdepe solves PDEs of the following type:

$$c(z, \tau, u, \partial_z u)\partial_\tau u = z^{-m}\partial_z(z^m f(z, \tau, u, \partial_z u)) + s(z, \tau, u, \partial_z u). \qquad (C.1)$$

The initial condition at $\tau = \tau_0$ is $u(\tau_0, z) = u_0(z)$. For all τ and either $z = z_L$, or $z = z_U$, the solution components satisfy a boundary condition:

---

[1] For reference, see Shampine and Reichelt (1997) and Skeel and Berzins (1990).

$$p(z, \tau, u) + q(z, \tau) f(z, \tau, u, \partial_z u) = 0. \tag{C.2}$$

Elements in $q(z, \tau)$ are either all zero or none of them is null. Nonzero values for q are associated to Neumann and Robin boundary conditions, whereas Dirichlet boundary conditions lead to a vanishing $q(z, \tau)$. Note that boundary conditions are expressed in terms of the function $f(z, \tau, u, \partial_z u)$ rather than $\partial_z u$. Notice that $p(z, \tau, u)$ depends on u whereas $q(z, \tau)$ is independent of it.

For the sake of clarity, let us examine a concrete example of utilization of the function pdepe. We consider the initial value problem illustrated in Chapter 4 "Finite Difference Methods". On the unit interval [0, 1], we consider the heat equation:

$$-\partial_\tau u(\tau, z) + \partial_{zz} u(\tau, z) = 0, \tag{C.3}$$

with initial condition:

$$u(0, z) = \begin{cases} 2z, & 0 \le z \le \frac{1}{2}, \\ 2(1 - z), & \frac{1}{2} \le z \le 1, \end{cases} \tag{C.4}$$

and boundary conditions:

$$u(\tau, 1) = u(\tau, 0) = 0. \tag{C.5}$$

The analytical solution of this *initial value problem* is:

$$u(\tau, z) = \frac{8}{\pi^2} \sum_{n=1}^{\infty} \frac{1}{n^2} \sin\left(\frac{n\pi}{2}\right) \sin(n\pi z) e^{-n^2 \pi^2 \tau}. \tag{C.6}$$

Equation (C.3) can be reduced to the form (C.1) by setting

$$\begin{aligned} m &= 0, & c(z, \tau, u, \partial_z u) &= 1, \\ f(z, \tau, u, \partial_z u) &= \partial_z u, & s(z, \tau, u, \partial_z u) &= 0. \end{aligned} \tag{C.7}$$

The boundary conditions are (C.5) and can be written in the form (C.2) using

$$p(z, \tau, u) = u(\tau, z), \qquad q(z, \tau) = 0,$$

where z can be either 0 or 1.[2] The arguments of function pdepe can be built as follows.

Let us examine a Matlab® formulation of this example.

_____

[2] If, instead, we have opted for boundary condition (C.5), we would have set $z_L = 0, z_U = 1,$ and

$$\begin{aligned} p(z_U, \tau, u) &= u(\tau, z_U), & q(z_U, \tau) &= 0, \\ p(z_L, \tau, u) &= u(\tau, z_L), & q(z_U, \tau) &= 0. \end{aligned}$$

- pdefun is a function that returns the terms c, f, and u. Input variables are defined by scalars x and t and vectors u and DuDx that approximate the solution and its partial derivative with respect to x. Comparing this to formula (C.7), the Matlab® code reads as follows:

```
function [c,f,s] = pdefun(x,t,u,DuDx)
c = 1;
f = DuDx;
s = 0;
```

- icfun evaluates the initial conditions. It has the form u = icfun(x). When called with an argument x, icfun evaluates and returns the initial values of the solution components at x in the column vector u. With reference to the initial condition assigned in (C.4), the Matlab® code reads as follows:

```
function u0 = pdex1ic(x)
u0 = 2*(x>=1/2).*(1-x)+2*(x<1/2).*x;
```

- bcfun evaluates terms p and q on boundary conditions (C.5) and returns the vector [pl,ql,pr,qr], where pl and ql are scalars corresponding to p and q evaluated at xL, similarly pr and qr correspond to xr. Therefore bcfun assumes form [pl,ql,pr,qr] = bcfun(xl,ul,xr,ur,t), where ul is the solution at the left boundary xL and uR is the solution at the right boundary xR. With reference to our example, the Matlab® formulation given in (C.2) becomes:

```
function [pl,ql,pr,qr] = pdex1bcfun(xl,ul,xr,ur,t)
pl = ul;
ql = 0;
pr = ur;
qr = 0;
```

Below, we use subfunctions to place all the functions required by pdepe in a single M-file that is named pdeexample.m. This file can be run from the Matlab® command window. The proposed function provides a solution for the example discussed above using, for illustrative purposes, 10 points for the xmesh and 24 points for the t-mesh. It calls function pdepe using the command sol = pdepe(m,@pdefun,@pdex1ic,@pdex1bcfun,x,t). The resulting numerical solution, generated by the line surf(x,t,sol), is illustrated in Fig. C.1.

```
function pdeExample
m = 0;
```

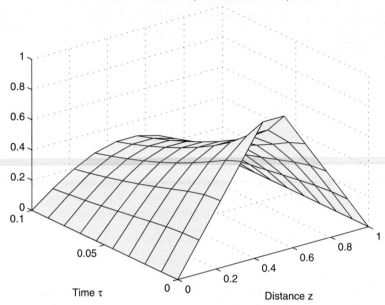

Fig. C.1. Numerical solution of the PDE given in the example.

```
x = linspace(0,1,10);
t = linspace(0,0.1,10);
%computes the numerical solution of the pde
sol = pdepe(m,@pdefun,@pdex1ic,@pdex1bcfun,x,t);
% A surface plot is often a good way
to study a solution.
surf(x,t,sol)
title('Numerical solution computed
with 10 mesh points.')
xlabel('Distance z')
ylabel('Time t')
% Defining the PDE-------------------
function [c,f,s] = pdefun(x,t,u,DuDx)
c = 1;
f = DuDx;
s = 0;
% Defining the IC-------------------
function u0 = pdex1ic(x)
```

```
%u0 = 2*(x>=1/2).*(x-1/2);
u0 = 2*(x>=1/2).*(1-x)+2*(x< 1/2).*x;
% Defining the BC------------------------
function [pl,ql,pr,qr] = pdex1bcfun(xl,ul,xr,ur,t)
pl = ul;
ql = 0;
pr = ur;
qr = 0;
% The analytical solution------------------
function ansolution = solutionpde(x, t)
sumseries = 0
for n = 1:20
 term = sin(n*pi/2)*sin(n*pi*x)*exp(-t*(n*pi)^2)/(n*n);
 sumseries = sumseries+term;
end
ansolution = sumseries*8/pi^2.
```

Appendix D: Optimal Control

D.1 Setting up the Optimal Stopping Problem

We begin by replacing the performance measure (3.4) with the equivalent $\sum_{s=t}^{\tau} F(s, X^{t,x}(s))$ and then redefine $\mathbb{X}$ by adding one singleton $\{k\}$, with $k \notin \mathbb{X}$, which plays the role of a "flag" for the stopping time τ. The set of controls $\mathcal{U}_{t,T}$ is given by the canonical basis in $\mathbb{R}^{T-t}$, i.e., $u = (u_t, \ldots, u_{T-1}) \in \mathcal{U}_{t,T}$ and a single entry u_s is equal to 1 ("stopping at time s") and all the others equal 0. Then, we set dynamics as

$$p(dx; s, y, u, \omega) = \begin{cases} p(dx; s, X(s, \omega), \omega) & \text{if } u = 0, \\ \delta_k(dx) \text{ (Dirac delta mass on } \{k\}) & \text{if } u = 1 \text{ or } y = k, \end{cases}$$

or, alternatively, as

$$X^{\mathbf{u}}(s+1, \omega) = \begin{cases} f\big(s, X^{\mathbf{u}}(s, \omega), W(s, \omega)\big) & \text{if } u = 0, \\ k & \text{if } u = 1, \end{cases}$$

$$f(s, k, w) = k \quad \text{for all } s \text{ and } k.^{1}$$

This means that the system is trapped into a steady state k once the control has been activated ($u = 1$). Finally, the cost/reward function vanishes on the steady state $\{k\}$:

$$F(s, y, u) = \begin{cases} F(s, y) & \text{if } y \neq k, \\ 0 & \text{if } y = k. \end{cases}$$

The equivalence of this formulation to an optimal stopping problem can now be easily proven as an exercise. Notice that the artificial state k signals the "death" of the system. For practical purposes, there is no need to cast an optimal stopping problem in this framework.

---

[1] We remark the difference between symbols "ω" and "w". The former denotes a generic elementary event in a sample space Ω. You may think of it as a sample path itself, as is the case of the so-called canonical process $X(t, \omega) = \omega(t)$, defined on $[0, T] \times \mathbb{R}^{[0,T]}$. The latter is a real value assumed by the random noise $W(t)$.

D.2 Proof of the Bellman Principle of Optimality

The two control policies $\mathbf{u}$ and $\mathbf{u}'$ give rise to the same exact dynamics on $\{t, \ldots, s\}$. Therefore their performance coincides on $\{t, \ldots, s-1\}$. Note that they need not be equal at s because the performance index in general depends on the control and the two control policies $\mathbf{u}$ and $\mathbf{u}'$ may not match at time s. This observation reduces the problem to the one of showing that the performance stemming from the *residual* dynamics $(X^{t,x,\mathbf{u}}(s), \ldots, X^{t,x,\mathbf{u}}(T))$ resulting from applying $\mathbf{u}$ is no greater than the reward deriving from the residual dynamics obtained by using $\mathbf{u}'$ instead. Control $\mathbf{u} = (u(t, \cdot), \ldots, u(T, \cdot))$ applied to the system dynamics for the remaining period gives rise to a reward assessment:

$$J^* := \sum_{i=s}^{T-1} F\left(i, X^{t,x,\mathbf{u}}(i), u\left(i, X^{t,x,\mathbf{u}}(i)\right)\right) + \Psi\left(X^{t,x,\mathbf{u}}(T)\right).$$

The uniqueness of the solution of a dynamic system implies the flow property:

$$X^{t,x,\mathbf{u}}(i) = X^{s, X^{t,x,\mathbf{u}}(s), \mathbf{u}|_s}(i),$$

where $t \leq s \leq i \leq T$, and $\mathbf{u}|_s$ is the truncated control $(u(s), \ldots, u(T))$ resulting from restricting $\mathbf{u}$ on the remaining period $\{s, \ldots, T\}$. By applying, in order, the flow property, the optimality of $\hat{\mathbf{u}}^s$ on $\{s, \ldots, T\}$, and the flow property again, we have:

$$J^* = \sum_{i=s}^{T-1} F\left(i, X^{s, X^{t,x,\mathbf{u}}(s), \mathbf{u}|_s}(i), u\left(i, X^{s, X^{t,x,\mathbf{u}}(s), \mathbf{u}|_s}(i)\right)\right)$$

$$+ \Psi\left(X^{s, X^{t,x}(s), \mathbf{u}|_s}(T)\right)$$

$$\leq \sum_{i=s}^{T-1} F\left(i, X^{s, X^{t,x,\mathbf{u}}(s), \hat{\mathbf{u}}^s}(i), \hat{u}\left(i, X^{s, X^{t,x,\mathbf{u}}(s), \hat{\mathbf{u}}^s}(i)\right)\right)$$

$$+ \Psi\left(X^{s, X^{t,x,\mathbf{u}}(s), \hat{\mathbf{u}}^s}(T)\right)$$

$$= \sum_{i=s}^{T-1} F\left(i, X^{t,x,\mathbf{u}'}(i), \hat{u}\left(i, X^{t,x,\mathbf{u}'}(i)\right)\right) + \Psi\left(X^{t,x,\mathbf{u}'}(T)\right).$$

This last term represents the performance generated by residual dynamics $(X^{t,x,\mathbf{u}'}(s), \ldots, X^{t,x,\mathbf{u}'}(T))$ corresponding to the control policy $\mathbf{u}'$.

D.3 Proof of the Dynamic Programming Algorithm

We need to prove that $\mathbf{u}^B$ dominates any other control policy in $\mathcal{U}_{t,T}$. We apply the Bellman principle of optimality at each step in the preceding backward induction. At time T, if the reached state is y, the generated performance is uncontrollable and matches $\Psi(y)$. This holds for any control policy and thus for the optimal one. At time $T-1$, whatever is the control policy $\mathbf{u} = (u(t), \ldots, u(T-2))$ adopted on the elapsed period $\{t, \ldots, T-2\}$, the policy

$$\left(u(t, \cdot), \ldots, u(T-2, \cdot), \hat{u}^{T-1}(T-1, \cdot)\right),$$

with $\hat{u}^{T-1}(T-1, \cdot) = \arg\max_{\mathcal{U}_{T-1,T}} J(T-1, \cdot, \mathbf{u})$, dominates any other control policy sharing the same first $T-2$ components. This is because each entry of the control policy contributes to the overall performance J *additively*. Therefore, the optimal control policy must have $\hat{u}(T-1, \cdot)$ as its $(T-1)$th component. At time $T-2$, the Bellman principle of optimality states that the control policy:

$$\left(u(t, \cdot), \ldots, \hat{u}^{T-2}(T-2, \cdot), \hat{u}^{T-2}(T-1, \cdot)\right),$$

with $(\hat{u}^{T-2}(T-2, \cdot), \hat{u}^{T-2}(T-1, \cdot)) = \arg\max_{\mathcal{U}_{T-2,T}} J(T-2, \cdot, \mathbf{u})$, dominates any other control policy sharing the same first $T-3$ components. But the previous step says that the optimal control policy must have $\hat{u}^{T-1}(T-1, \cdot)$ as its $(T-1)$th component, i.e.,

$$\left(u(t, \cdot), \ldots, \hat{u}^{T-2}(T-2, \cdot), \hat{u}^{T-1}(T-1, \cdot)\right)$$
$$\geq \left(u(t, \cdot), \ldots, u(T-3, \cdot), \hat{u}^{T-2}(T-2, \cdot), \hat{u}^{T-2}(T-1, \cdot)\right)$$
$$\geq \left(u(t, \cdot), \ldots, u(T-3, \cdot), u(T-2, \cdot), u(T-1, \cdot)\right).$$

Proceeding this way, we come up to the control policy:

$$\mathbf{u}^B = \left(\hat{u}^t(t, \cdot), \ldots, \hat{u}^{T-2}(T-2, \cdot), \hat{u}^{T-1}(T-1, \cdot)\right)$$
$$\geq \left(\hat{u}^t(t, \cdot), \ldots, \hat{u}^t(T-2, \cdot), \hat{u}^t(T-1, \cdot)\right) = \hat{\mathbf{u}}.$$

But $\hat{\mathbf{u}} = \arg\max_{\mathcal{U}_{t,T}} J(t, \cdot, u)$ dominates all control policy, in particular $\mathbf{u}^B$, i.e., $\hat{\mathbf{u}} \geq \mathbf{u}^B$. Therefore, $\hat{\mathbf{u}} = \mathbf{u}^B$.

Bibliography

Abate, J., Choudhury, G.L., Whitt, W. (1996). On the Laguerre Method for Numerically Inverting Laplace Transforms. *Informs. Comput.* 8, 413–427.

Abate, J., Choudhury, G.L., Whitt, W. (1998). Numerical Inversion of Multidimensional Laplace Transforms by the Laguerre Method. *Performance Evaluation*, 31, 229–243.

Abate, J., Whitt, W. (1992a). Numerical Inversion of Probability Generating Functions. *Operations Research Letters* 12, 245–251.

Abate, J., Whitt, W. (1992b). The Fourier-Series Method for Inverting Transforms of Probability Distributions. *Queueing Systems Theory Appl.* 10, 5–88.

Abken, P.A. (2000). An Empirical Evaluation of Value at Risk by Scenario Simulation. *Journal of Derivatives* 7, 12–30.

Abramowitz, M., Stegun, I.A. (1965). *Handbook of Mathematical Functions.* National Bureau of Standards, Washington, DC. (Reprinted by Dover, New York.)

Abramowitz, M., Stegun, I.A., Kampen, J. (1993). *Handbook of Mathematical Functions* (3rd ed.). John Wiley.

Acworth, P., Broadie, M., Glasserman, P. (1998). A Comparison of Some Monte Carlo and Quasi-Monte Carlo Methods for Option Pricing. In: Hellekaled, P., Larcher, G., Niederreiter, H., Zinterhof, P. (Eds.), *Monte Carlo and Quasi-Monte Carlo Methods.* Springer-Verlag.

Aitsahlia, F., Lai, T. (1997). Valuation of Discrete Barrier and Hindsight Options. *The Journal of Financial Engineering* 6(2), 169–177.

Aït-Sahalia, Y., Lo, A.W. (1998). Non-Parametric Estimation of State-Price Densities Implicit in Financial Asset Prices. *Journal of Finance* 53(2), 499–548.

Akahori, J. (1995). Some Formulae for a New Type of Path-Dependent Options. *Annals of Applied Probability* 5, 383–388.

Akesson, F., Lehoczky, L. (2000). Path Generation for Quasi-Monte Carlo Simulation of Mortgage-Backed Securities. *Management Science* 46, 1171–1187.

Albrecher, H., Mayer, P., Schoutens, W., Tistaert, J. (2007). The Little Heston Trap. *Wilmott Magazine*, January, 83–92.

Alexander, C. (2001). *Market Models: A Guide to Financial Data Analysis*, Wiley Series in Financial Engineering. John Wiley & Sons.

Altman, E., Resti, A., Sironi, A. (2004). Deafault and Recovery Rates in Credit Risk Modeling: A. Review of the Literature and Empirical Evidence. *Journal of Finance Literature* 1.

Altman, E., Resti, A., Sironi, A. (2005). The Link Between Default and Recovery Rates: Theory, Empirical Evidence and Implications. *The Journal of Business* 78(6), 2203–2228.

Amin, K.I. (1993). Jump Diffusion Option Valuation in Discrete Time. *Journal of Finance* 48(5), 1883–1863.

Andersen, L.B.G. (2007). Efficient Simulation of the Heston Stochastic Volatility Model. (Available at: http://ssrn.com/abstract=946405).

Andersen, T.G. (1995). Simulation and Calibration of the HJM Model. *Working Paper*, General Re Financial Products, New York.

Andersen, T.G., Broadie, M. (2001). A Primal–Dual Simulation Algorithm for Pricing Multi-Dimensional American Options. *Working Paper*, Columbia Business School, New York.

Andersen, L., Brotherton-Ratcliffe, R. (1996). Exact Exotics. *Risk* 9, 85–89.

Andricopoulos, A.D., Widdicks, M., Duck, P.W., Newton, D.P. (2003). Universal Option Valuation Using Quadrature Methods. *Journal of Financial Economics* 67, 447–471.

Antia, H.M. (2002). *Numerical Methods for Scientists and Engineers* (2nd ed.). H.M. Birkhäuser.

Antonov, I.A., Saleev, V.M. (1980). An Economic Method of Computing lp_τ-Sequences. *USSR Comput. Maths. Math. Phys.* 19, 252–256.

Aparicio, S., Hodges, S. (1998). Implied Risk-Neutral Distribution: A Comparison of Estimation Methods. *FORC preprint* PP98-95, Warwick University.

Appel, G., Hitschler, F. (1980). *Stock Market Trading Systems*. Traders Press.

Applebaum, D. (2004). *Lévy Processes and Stochastic Calculus*. Cambridge University Press.

Asmussen, S., Rosinski, J. (2001). Approximations of Small Jumps of Lévy Processes with a View Towards Simulation. *Journal of Applied Probability* 38, 482–493.

Atkinson, K.E. (1989). *An Introduction to Numerical Analysis*. John Wiley and Sons.

Atkinson, C., Fusai, G. (2004). Discrete Extrema of Brownian Motion and Pricing of Lookback Options. *Working Paper*, Dipartimento SEMEQ, Università del Piemonte Orientale.

Atlan, M., Geman, H., Yor, M. (2005). Options on Hedge Funds under the High Water Mark Rule. *Working Paper.*

Avellaneda, M., Buff, R., Friedman, C., Grandchamp, N., Kruk, L., Newman, J. (2001). Weighted Monte Carlo: A New Technique for Calibrating Asset-Pricing Models. *International Journal of Theoretical and Applied Finance* 4(1), 1–29.

Avellaneda, M., Gamba, R. (2000). Conquering the Greeks in Monte Carlo: Efficient Calculation of the Market Sensitivities and Hedge-Ratios of Financial Assets by Direct Numerical Simulation. In: Avellaneda, M. (Ed.), *Quantitative Analysis in Financial Markets, Vol. III.* World Scientific, Singapore, 336–356.

Avellaneda, M., Paras, A. (1994). *Dynamic Hedging with Transaction Costs: From Lattice Models to Nonlinear Volatility and Free-Boundary Problems.* Unpublished manuscript.

Avellaneda, M., Scherer, K. (2002). All for One and One for All: A Principal Components Analysis of the Latin American Brady Bond Market from 1994 to 2000. *International Journal of Theoretical and Applied Finance* 5(1), 79–107.

Avellaneda, M., Wu, L. (1999). Pricing Parisian-Style Options with a Lattice Method. *International Journal of Theoretical and Applied Finance* 2(1), 1–17.

Baccara, M., Battauz, A., Ortu, F. (2005). Effective Securities in Arbitrage-Free Markets with Bid-Ask Spreads at Liquidation: A Linear Programming Characterization. *Journal of Economic Dynamics and Control* 30(1), 55–79.

Bahra, B. (1997). Implied Risk-Neutral Probability Density Functions from Option Prices: Theory and Application. *Working Paper* 66, Bank of England, 1368–5562.

Bakshi, G.S., Cao, C., Chen, Z.W. (1997). Empirical Performance of Alternative Option Pricing Models. *Journal of Finance* 52, 2003–2049.

Baldi, P., Caramellino, L., Iovino, G. (1999). Pricing General Barrier Options: a Numerical Approach Using Sharp Large Deviations. *Mathematical Finance* 9(4), 293–321.

Baldick, R., Kolos, S., Tompaidis, S. (2003). Valuation and Optimal Interrruption for Interruptible Electricity Contracts. *Working Paper*, University of Texas.

Ballotta, L. (2001). Lévy Processes, Option Valuation and Pricing of the Alpha-Quantile Option. *Ph.D. Thesis*, Università Cattolica Sacro Cuore, Milan.

Ballotta, L., Kyprianou, A. (2001). A Note on the Alpha-Quantile Option. *Applied Mathematical Finance* 8, 137–144.

Bandi, F.M., Nguyen, T.H. (1999). Fully Nonparametric Estimators for Diffusions: A Small Sample Analysis. *Working Paper*, University of Chicago.

Bandi, F.M., Nguyen, T.H. (2003). On the Functional Estimation of Jump-Diffusion Processes. *Journal of Econometrics* 116, 293–328.

Barbieri, A., Garman, M.B. (1996). Putting a Price on Swings. *Energy and Power Risk Management* 1(6).

Barbieri, A., Garman, M.B. (1997). Ups and Downs of Swings. *Energy and Power Risk Management* 2(1).

Barlow, M.T. (2002). A Diffusion Model for Electricity Prices. *Mathematical Finance* 12, 287–298.

Barone-Adesi, G. (2005). The Saga of the American Put. *Journal of Banking and Finance* 29(11), 2909–2918.

Barone-Adesi, G., Whaley, R. (1987). Efficient Analytic Approximation of American Option Values. *Journal of Finance* 42(2), 301–320.

Barrett, R., Berry, M., Chan, T.F., Demmel, J., Donato, J., Dongarra, J., Eijkhout, V., Pozo, R., Romine, C., Van der Vorst, H. (1994). *Templates for the Solution of Linear Systems: Building Blocks for Iterative Methods*. SIAM, Philadelphia (http://www.netlib.org/templates/Templates.html).

Barton, D.E., Dennis, K.E.R. (1952). The Conditions Under Which Gram–Charlier and Edgeworth Curves are Positive Definite and Unimodal. *Biometrika* 39, 425–427.

Barucci, E. (2003). *Financial Markets Theory: Equilibrium, Efficiency and Information*. Springer Finance. Springer-Verlag, Berlin–Heidelberg–New York.

Bates, D.S. (1991). The Crash of '87: Was It Expected? The Evidence from Options Markets. *Journal of Finance* 46(3), 1009–1044.

Bates, D. (1996). Jumps and Stochastic Volatility: Exchange Rate Processes in Deutschemark Options. *Review of Financial Studies* 9, 69–108.

Battauz, A. (2002). Change of Numéraire and American Options. *Stochastic Analysis and Applications* 20(4), 709–730.

Baxter, M., Rennie, A. (1996). *Financial Calculus: An Introduction to Derivative Pricing*. Cambridge University Press.

Baz, J., Das, S.R. (1996). Analytical Approximations of the Term Structure for Jump-Diffusion Processes: A Numerical Analysis. *Journal of Fixed Income* 78–86.

Bazaraa, M.S., Sherali, H.D., Shetty, C.M. (1993). *NonLinear Programming, Theory and Algorithms* (2nd ed.). Wiley.

Beaglehole, D., Dybvig, P., Zhou, G. (1997). Going to Extremes: Correcting Simulation Bias in Exotic Option Valuation. *Financial Analysts Journal* 53, 62–68.

Beamon (1998). Competitive Electricity Prices. *Working Paper*, Energy Information Administration.

Bedendo, M., Anagnou, I., Hodges, S.D., Tompkins, R. (2005). Forecasting Accuracy of Implied and GARCH-Based Probability Density Functions. *Review of Futures Markets* 14(1), Summer.

Bellman, R. (1957). *Dynamic Programming*. Dover Publications.

Bellman, R., Kalaba, R.E., Lockett, J. (1966). *Numerical Inversion of the Laplace Transform. Application to Biology, Economics, Engineering and Physics*. American Elsevier, New York.

Benth, F.E., Dahl, L.O., Karlsen, K.H. (2003). Quasi Monte Carlo Evaluation of Sensitivities of Options in Commodity and Energy Markets. *International Journal of Theoretical and Applied Finance* 6(8), 865–884.

Benth, F.E., Kallsen, J., Meyer-Brandis, T. (2007). A Non-Gaussian Ornstein–Uhlenbeck Process for Electricity Spot Price Modeling and Derivatives Pricing. *Applied Mathematical Finance* 14(2), 153–169.

Benth, F.E., Saltyte-Benth, J. (2006). Analytical Approximation for the Price Dynamics of Spark Spread Options. *Studies of Nonlinear Dynamics & Econometrics* 10(3).

Bertsekas, D.P. (2005). *Dynamic Programming and Optimal Control*, Vol. 1 and 2 (2nd ed.). Athena Scientific.

Bessembinder, H., Chan, K. (1995). The Profitability of Technical Trading Rules in the Asian Stock Markets. *Pacific-Basin Finance Journal* 3, 257–284.

Bessembinder, H., Chan, K. (1998). Market Efficiency and the Returns to Technical Analysis. *Financial Management* 27(2), 5–17.

Best, M.J., Grauer, R. (1991). On the Sensitivity of Mean-Variance Efficient Portfolios to Changes in Asset Means: Some Analytical and Computational Results. *Review of Financial Studies* 4, 315–342.

Bhanot, K. (2000). Behavior of Power Prices. *Journal of Risk* 2, 43–62.

Biffis, E., Millossovich, P. (2006). The Fair Value of Guaranteed Annuity Options. *Scandinavian Actuarial Journal* 1, 23–41.

Björk, T. (2004). *Arbitrage Theory in Continuous Time* (2nd ed.). Oxford University Press.

Block, S.B. (1999). A Study of Financial Analysts: Practice and Theory. *Financial Analysts Journal* 55(4), 86–95.

Blume, M.E. (1975). Betas and Their Regression Tendencies. *Journal of Finance* 30(3), 785–789.

Borodin, A.N., Salminen, P. (2002). *Handbook of Brownian Motion: Facts and Formulae* (2nd ed.). Birkhäuser.

Botterud, A., Bhattacharyya, B., Ilic, M. (2002). Futures and Spot Prices, an Analysis of the Scandinavian Electricity Market. *Working Paper*, MIT.

Bouchard, B., Ekeland, I., Touzi, N. (2004). On the Malliavin Approach to Monte Carlo Approximation of Conditional Expectations. *Finance and Stochastics* 8(1), 45–71.

Bouchaud, J.P., Potters, M., Sestovic, D. (2000). Hedged Monte Carlo: Low Variance Derivative Pricing with Objective Probabilities. *Working Paper*, Science & Finance Capital Fund Management, France.

Boudoukh, J., Whitelaw, R.F., Richardson, M., Stanton, R. (1997). Pricing Mortgage-Backed Securities in a Multifactor Interest Rate Environment: A Multivariate Density Estimation Approach. *Review of Financial Studies* 10, 405–446.

Box, G.E.P., Muller, M.E. (1958). A Note on the Generation of Random Normal Deviates. *Annals of Mathematical Statistics* 29, 610–611.

Boyle, P. (1977). Options: A Monte Carlo Approach. *Journal of Financial Economics* 4(3), 323–338.

Boyle, P., Broadie, M., Glasserman, P. (1997). Monte Carlo Methods for Security Pricing. *Journal of Economic Dynamics and Control* 21, 1267–1321.

Boyle, P., Evnine, J., Gibbs, S. (1989). Numerical Evaluation of Multivariate Contingent Claims. *Review of Financial Studies* 2(2), 241–250.

Boyle, P., Kolkiewicz, A. (2002). Pricing American Derivatives Using Simulation: A Biased Low Approach. In: Fang, K.T., Hickernell, F.J., Niederreiter, H. (Eds.), *Monte Carlo and Quasi-Monte Carlo Methods.* Springer-Verlag, Berlin–Heidelberg–New York.

Boyle, P.P., Lau, S.H. (1994). Bumping Up Against the Barrier with the Binomial Method. *Journal of Derivatives* 1, 6–14.

Boyle, P., Tan, K.S. (1997). Quasi-Monte Carlo Methods. *Working Paper*, University of Waterloo.

Boyle, P., Tian, Y.S. (1998). An Explicit Finite Difference Approach to the Pricing of Barrier Options. *Applied Mathematical Finance* 5, 17–43.

Brandt, M.W. (2006). Portfolio Choice Problems. In: Ait-Sahalia, Y., Hansen L.P. (Eds.), *Handbook of Financial Econometrics.* Elsevier, forthcoming.

Brandt, M.W., Santa Clara, P. (2002). Simulated Likelihood Estimation of Diffusions with an Application to Exchange Rate Dynamics in Incomplete Markets. *Journal of Financial Economics* 63, 161–212.

Breeden, D., Litzenberger, R.H. (1978). Prices of State-Contingent Claims Implicit in Option Prices. *Journal of Business* 51(4), 621–651.

Brennan, M., Schwartz, E. (1977). The Valuation of American Put Options. *Journal of Finance* 32(2), 449–463.

Brennan, M., Schwartz, E. (1978). Finite Difference Methods and Jump Processes and the Pricing of Contingent Claims: A Synthesis. *Journal of Financial and Quantitative Analysis* 13, 461–474.

Brennan, M., Schwartz, E. (1985). Determinants of GNMA Mortgage Prices. *Journal of the American Real Estate and Urban Economics Association* 13, 209–228.

Brigo, D., Mercurio, F. (2006). *Interest Rate Models – Theory and Practice: With Smile, Inflation and Credit* (2nd ed.). Springer Finance. Springer-Verlag, Berlin–Heidelberg–New York.

Brigo, D., Mercurio, F., Rapisarda, F. (2004). Smile at the Uncertainty. *Risk* 97–100.

Britten-Jones, M. (1999). The Sampling Error in Estimates of Mean-Variance Efficient Portfolio Weights. *Journal of Finance* 54(2), 655–671.

Briys, E., Bellalah, M., Mai, H.M., De Varenne, F. (1998). *Options, Futures and Exotic Derivatives.* Wiley, England.

Broadie, M., Detemple, J. (1996). American Option Valuation: New Bounds, Approximations, and a Comparison of Existing Methods. *Review of Financial Studies* 9, 1211–1250.

Broadie, M., Detemple, J. (1997). Valuation of American Options on Multiple Assets. *Review of Financial Studies* 7, 241–286.

Broadie, M., Glasserman, P. (1996). Estimating Security Price Derivatives Using Simulation. *Management Science* 42, 269–285.

Broadie, M., Glasserman, P. (1997). Pricing American-Style Securities Using Simulation. *Journal of Economic, Dynamics and Control* 21, 1323–1352.

Broadie, M., Glasserman, P. (1997). A Stochastic Mesh Method for Pricing High-Dimensional American Options. *Working Paper*, Columbia University.

Broadie, M., Glasserman, P., Jain, G. (1997). Enhanced Monte Carlo Estimates of American Options Prices. *Journal of Derivatives* 4, 25–44.

Broadie, M., Glasserman, P., Kou, S. (1997). A Continuity Correction for Discrete Barrier Options. *Mathematical Finance* 7(4), 325–349.

Broadie, M., Glasserman, P., Kou, S. (1999). Connecting Discrete and Continuous Path-Dependent Options. *Finance and Stochastics* 3, 55–82.

Broadie, M., Kaya, Ö. (2006). Exact Simulation of Stochastic Volatility and Other Affine Jump Diffusion Processes. *Operations Research* 54, 217–231.

Brock, W., Lakonishok, J., LeBaron, B. (1992). Simple Technical Trading Rules and the Stochastic Properties of Stock Returns. *Journal of Finance* 47(5), 1731–1764.

Brooks, R.D., Faff, R.W., McKenzie, M.D. (1998). Time-Varying Beta Risk of Australian Industry Portfolios: A Comparison of Modelling Techniques. *Australian Journal of Management* 23, 1–22.

Bruner, R.F., Eades, K.M., Harris, R.S., Higgins, R.C. (1998). Best Practices in Estimating the Cost of Capital: Survey and Synthesis. *Financial Practice and Education* 27, 13–28.

Bruti-Liberati, N., Platen, E. (2006). Strong Approximations of Stochastic Differential Equations with Jumps. *Journal of Computational and Applied Mathematics* 205(2), 982–1001.

Bruti-Liberati, N., Platen, E. (2007). Approximation of Jump Diffusions in Finance and Economics. *Computational Economics* 29(3/4), 283–312.

Bruti-Liberati, N., Nikitopoulos-Sklibosios, C., Platen, E. (2006). First Order Strong Approximations of Jump Diffusions. *Monte Carlo Methods and Applications* 12(3), 191–209.

Burlisch, R., Stoer, J. (1992). *Introduction to Numerical Analysis* (2nd ed). Springer-Verlag, Berlin–Heidelberg–New York.

Buser, S.A. (1986). Laplace Transforms as Present Value Rules: A Note. *Journal of Finance* XLI(1), 243–247.

Cai, N., Kou, S.G. (2007). Pricing Asian Options via a Double-Laplace Transform. Working paper, Columbia University.

Cairns, A.J. (2004). *Interest Rate Models: An Introduction*. Princeton University Press.

Campa, J.M., Chang, K., Reider, R. (1997). ERM Bandwidths for EMU and After: Evidence from Foreign Exchange Options. *Economic Policy* 24, 55–89.

Campa, J.M., Chang, K., Reider, R. (1998). Implied Exchange Rate Distributions: Evidence from OTC Option Markets. *Journal of International Money and Finance* 17(1), 117–160.

Campbell, J.Y., Lo, A.W., MacKinlay, A.C. (1997). *The Econometrics of Financial Markets*. Princeton University Press, Princeton, NJ.

Carmona, R., Dayanik, S. (2003). Optimal Multiple Stopping of Linear Diffusions and Swing Options. *Working Paper*, Princeton University.

Carr, P. (1998). Randomization and the American Put. *Review of Financial Studies* 11(3), 597–626.

Carr, P. (2000). Deriving Derivatives of Derivative Securities. *Journal of Computational Finance* 4(2), 5–29.

Carr, P., Geman, H., Madan, D.H., Yor, M. (2003). Stochastic Volatility for Lévy processes, *Mathematical Finance* 13(3), 345–382.

Carr, P., Geman, H., Madan, D.H., Wu, L., Yor, M. (2005). Option Pricing Using Integral Transforms. *Working Paper.*

Carr, P., Hirsa, A. (2003). Why Be Backward? Forward Equations for American Options. *Risk* 16(1), 103–107.

Carr, P., Jarrow, R., Myneni, R. (1992). Alternative Characterizations of American Put Options. *Mathematical Finance* 2(4).

Carr, P., Madan, D.H. (1999). Option Evaluation Using the Fast-Fourier Transform. *Journal of Computational Finance* 2(4), 61–73.

Carr, P., Schröder, M. (2004). Bessel Processes, the Integral of Geometric Brownian Motion and Asian Options. *Theory of Probability and its Applications* 71(1), 113–141.

Carr, P., Yang, G. (1997). Simulating Bermudan Interest Rate Derivatives. *Working Paper*, Morgan Stanley, New York.

Carr, P., Yang, G. (1998). Simulating American Bond Options in an HJM Framework. *Working Paper*, Morgan Stanley, New York.

Carrière, J. (1996). Valuation of the Early-Exercise Price for Options Using Simulations and Nonparametric Regression. *Insurance: Mathematics and Economics* 19(1), 19–30.

Cartea, Á., Figueroa, M.G. (2005). Pricing in Electricity Markets: A Mean Reverting Jump Diffusion Model with Seasonality. *Applied Mathematical Finance* 12(4), 313–335.

Cartea, Á., Williams, T. (2007). UK Gas Markets: the Market Price of Risk and Applications to Multiple Interruptible Supply Contracts, forthcoming in *Energy Economics*.

Carverhill, A., Pang, K. (1998). Efficient and Flexible Bond Option Valuation in the Heath, Jarrow and Morton Framework. In: Dupire, B. (Ed.), *Monte Carlo: Methodologies and Applications for Pricing and Risk Management*. Risk Publications, London.

Cathcart, L. (1998). The Pricing of Floating Rate Instruments. *Journal of Computational Finance* 1(4), 31–51.

Cerny, A. (2003). *Mathematical Techniques in Finance: Tools for Incomplete Markets*. Princeton University Press.

Chan, K.C., Karoly, G.A., Longstaff, F.A., Sanders, A.B. (1992). The Volatility of Short Term Interest Rates: An Empirical Comparison of Alternative Models of the Term Structure of Interest Rates. *Journal of Finance* 47, 1209–1227.

Chapman, D., Pearson, N. (2000). Is the Short Rate Drift Actually Non Linear? *Journal of Finance* 55(1), 355–388.

Chen, R.R., Scott, L. (1993). Maximum Likelihood Estimation for a Multifactor Equilibrium Model of the Term Structure of Interest Rates. *Journal of Fixed Income* 3, 14–31.

Chen, R., Scott, L. (1995). Interest Rate Options in Multifactor Cox–Ingersoll–Ross Models of the Term Structure. *Journal of Derivatives* 3, 53–72.

Cherubini, U., Luciano, E. (2001). Value-at-Risk Trade-Off and Capital Allocation with Copulas. *Economic Notes* 30(2), 235–256.

Cherubini, U., Luciano, E., Vecchiato, W. (2004). *Copula Methods in Finance*. Wiley Finance Series. John Wiley & Sons.

Chesney, M., Cornwall, J., Jeanblanc-Picqué, M., Kentwell, G., Yor, M. (1997). Parisian Pricing. *Risk* 10(1), 77–80.

Chesney, M., Jeanblanc-Picqué, M., Yor, M. (1995). Brownian Excursion and Parisian Barrier Options. *Advances in Applied Probability* 29, 165–184.

Cheuk, T., Vorst, T. (1996). Complex Barrier Options. *Journal of Derivatives*, Fall, 8–22.

Chopra, V., Ziemba, T. (1993). The Effect of Errors in Means, Variances and Covariances on Optimal Portfolio Choice. *Journal of Portfolio Management* 19(3).

Choudhury, G.L., Lucantoni, D.M., Whitt, W. (1994). Multidimensional Transform Inversion with Applications to the Transient M/G/1 Queue. *Annals of Applied Probability* 4(3), 719–740.

Christoffersen, P.F. (1998). Evaluating Interval Forecast. *International Economic Review* 39, 841–862.

Christoffersen, P., Jacobs, K. (2004). The Importance of Loss Function in Option Evaluation. *Journal of Financial Economics* 72, 291–318.

Churchill, R.V., Brown, J.W. (1989). *Complex Variables and Applications* (5th ed.). McGraw-Hill Companies.

Clewlow, L., Carverhill, A. (1994). On the Simulation of Contingent Claims. *Journal of Derivatives* 2, 66–74.

Clewlow, L., Strickland, C. (1997). Monte Carlo Valuation of Interest Rate Derivatives Under Stochastic Volatility. *Journal of Fixed Income* 7(3), 35–45.

Clewlow, L., Strickland, C. (1998). *Implementing Derivative Models*. Wiley & Sons, London.

Clewlow, L., Strickland, C. (2000). *Energy Derivatives: Pricing and Risk Management*. Lacima Group Publications.

Clewlow, L., Strickland, C., Kaminski, V. (2001). Risk Analysis of Swing Contracts. *Energy and Power Risk Management*.

Cochrane, J.H. (2001). *Asset Pricing*. Princeton University Press, Princeton, NJ.

Connor, G., Herbert, N. (1999). Estimation of the European Equity Model. *Horizon: The Barra Newsletter* 169.

Cont, R., Tankov, P. (2004). *Financial Modelling with Jump Processes*. Chapman & Hall/CRC Press.

Conze, A., Viswanathan, R. (1991). Path-Dependent Options: The Case of Lookback Options. *Journal of Finance* 46(5), 1893–1907.

Cook, R.D., Weisberg, S. (1982). *Residuals and Influence in Regression*. Chapman and Hall, New York.

Corielli, F. (2006). Hedging with Energy. *Mathematical Finance* 16(3), 495–517.

Courtadon, G. (1982). A More Accurate Finite Difference Approximation for the Valuation of Options. *Journal of Financial and Quantitative Analysis* 17(5), 697–703.

Cox, J.C. (1996). The Constant Elasticity of Variance Option Pricing Model. *Journal of Portfolio Management*, Special Issue, 15–17.

Cox, J.C., Ingersoll, J.E., Ross, S.A. (1985). A Theory of the Term Structure of Interest Rates. *Econometrica* 53(2), 385–407.

Cox, J.C., Ross, S.A. (1976). The Valuation of Options for Alternative Stochastic Processes, *Journal of Financial Economics* 3(1-2), 145–166.

Cox, J.C., Ross, S.A., Rubinstein, M. (1979). Option Pricing: A Simplified Approach. *Journal of Financial Economics* 7, 229–264.

Craddock, M., Heath, D., Platen, E. (2000). Numerical Inversion of Laplace Transforms: A Survey of Techniques With Applications to Derivatives Pricing. *Journal of Computational Finance* 4(1).

Craig, I., Thompson, A.M. (1994). Why Laplace Transforms are Difficult to Invert Numerically? *Computers in Physics* 8(6), 648–654.

Crump, K. (1976). Numerical Inversion of Laplace Transform using Fourier Series Approximation. *J. Assoc. Comp. Mach.* 23, 89–96.

Cryer, C.W. (1971). The Solution of a Quadratic Programming Problem Using Systematic Overrelaxation. *SIAM J. Control* 9, 385–395.

Dahl, L.O., Benth, F.E. (2002). Fast Evaluation of Asian Basket Option by Singular Value Decomposition. In: *Monte Carlo and Quasi-Monte Carlo Methods*. Spinger, 201–213.

Das, S.R. (1997a). A Direct Discrete-Time Approach to Poisson–Gaussian Bond Option Pricing in the Heath–Jarrow–Morton Model. *Working Paper*, Harvard Business School.

Das, S.R. (1997b). Poisson–Gaussian Processes and the Bond Markets. *Working Paper*, Harvard University.

Das, S.R., Sundaram, R.K. (1999). Of Smiles and Smirks: A Term Structure Perspective. *Journal of Financial and Quantitative Analysis* 34(2), 211–239.

Dassios, A. (1995). The Distribution of the Quantile of a Brownian Motion with Drift and the Pricing of Related Path-Dependent Options. *Annals of Applied Probability* 4, 719–740.

Davies, B., Martin, B.L. (1970). Numerical Inversion of Laplace Transforms: A Critical Evaluation and Review of Methods. *Journal of Computational Physics* 33, 1–32.

Davis, P., Rabinowitz, P. (1975). *Methods of Numerical Integration*. Academic Press, New York.

Davydov, D., Linetsky, V. (2001a). Pricing and Hedging Path-Dependent Options under the CEV Process. *Management Science* 47, 949–965.

Davydov, D., Linetsky, V. (2001b). Structuring, Pricing and Hedging Double Barrier Step Options. *Journal of Computational Finance* 5(2), 55–87.

D'Ecclesia, R.L., Zenios, S.A. (1994). Risk Factor Analysis and Portfolio Immunization in the Italian Bond Market. *Journal of Fixed Income*, 51–60.

De Jong, C., Huisman, R. (2002). Option Formulas with Mean Reverting Power Prices with Spikes. *Working Paper*, Erasmus University, Rotterdam.

Demange, G., Rochet, J.-C. (1997). *Methodes Mathématiques de la Finance*. Economica.

Dempster, M.A., Hutton, J.P. (1999). Pricing American Options by Linear Programming. *Mathematical Finance* 9(3).

Deng, S. (1999). Financial Methods in Competitive Electricity Markets. *Ph.D. Thesis*, University of California at Berkeley.

Deng, S., Johnson, B., Sogomonian, A. (1999). Spark Spread Options and the Valuation of Electricity Generation Assets. *Proceedings of the 32nd Hawaii International Conference on System Sciences*.

Den Iseger, P. (2006). Numerical Inversion of Laplace Transforms Using a Gaussian Quadrature for the Poisson Summation Formula. *Probability in the Engineering and Informational Sciences* 20(1), 1–44.

Derman, E., Kani, I. (1994). Riding on a Smile. *Risk* 7(2), 32–39.

Detry, P.J., Grégoire, P. (2001). Other Evidences of the Predictive Power of Technical Analysis: The Moving Average Rules on European Indexes. *Working Paper*, EFMA 2001 Lugano Meeting.

Devroye, L. (1986). *Non-Uniform Random Variate Generation*. Springer-Verlag, Berlin–Heidelberg–New York.

D'Halluin, Y., Forsyth, P.A., Vetzal, K.R. (2003). Robust Numerical Methods for Contingent Claims under Jump Diffusion Processes. *IMA Journal on Numerical Analysis* 25, 87–112.

D'Halluin, Y., Forsyth, P.A., Labahn, G. (2005). A Semi-Lagrangian Approach for American Asian Options Under Jump Diffusion. *SIAM Journal on Scientific Computing*, 27, 315–345.

Di Graziano, G., Rogers, L.C.G. (2005). A New Approach to the Modelling and Pricing of Correlation Credit Derivatives. *Working Paper*, Cambridge University.

Doetsch, G. (1970). *Introduction to the Theory and Application of the Laplace Transformation*. Springer-Verlag, Berlin–Heidelberg–New York.

Dongarra, J., Bunch, J., Moler, C., Stewart, G. (1979). *LINPACK User's Guide*. SIAM Pub., Philadelphia.

Douady, R. (1998). Model Calibration in the Monte Carlo Framework. In: Dupire, B. (Ed.), *Monte Carlo: Methodologies and Applications for Pricing and Risk Management*. Risk Publications, London.

Douglas, M., Simin, T. (2003). Outlier-Resistant Estimates of Beta. *Financial Analysts Journal* 59(5), 56–69.

Duan, J. (1995). The Garch Option Pricing Model. *Mathematical Finance* 5, 13–32.

Duan, J. (1996). Cracking the Smile. *Risk* 9, 55–59.

Duan, J.-C., Dudley, E., Gauthier, G., Simonato, J.-G. (2003). Pricing Discretely Monitored Barrier Options by a Markov Chain. *Journal of Derivatives* 10(4), Summer, 9–32.

Duan, J.-C., Gauthier, G., Simonato, J.-G. (2001). Asymptotic Distribution of the EMS Option Price Estimator. *Management Science* 47(8), 1122–1132.

Duan, J., Simonato, J.G. (1998). Empirical Martingale Simulation for Asset Prices. *Management Science* 44(9), 1218–1233.

Dubner, H., Abate, J. (1968). Numerical Inversion of Laplace Transforms by Relating them to the Finite Fourier Cosine Transform. *J. ACM* 15(1), 115–123.

Duffy, D.A. (1993). On the Numerical Inversion of Laplace Transforms: Comparison of Three new Methods on Characteristic Problems from Applications. *ACM Trans. on Math. Soft.* 19(3), 333–359.

Duffie, D. (2001). *Dynamic Asset Pricing Theory* (3rd ed.). Princeton University Press.

Duffie, D., Glynn, P. (1995). Efficient Monte Carlo Simulation of Security Prices. *Annals of Applied Probability* 5, 897–905.

Duffie, D., Pan, J., Singleton, K.J. (1998). Transform Analysis and Asset Pricing for Affine Jump-Diffusions. *Econometrica* 68(6), 1343–1376.

Duffie, D., Singleton, K.J. (1993). Simulated Moments Estimation of Markov Models of Asset Prices. *Econometrica* 61, 929–952.

Dumas, B., Luciano, E. (1991). An Exact Solution to a Dynamic Portfolio Choice Problem with Transaction Costs. *Journal of Finance* 46(2), 577–595.

Dunn, K.B., McConnell, J.J. (1981a). A Comparison of Alternative Models for Pricing GNMA Mortgage-Backed Securities. *Journal of Finance* 36(2), 471–490.

Dunn, K.B., McConnell, J.J. (1981b). Valuation of GNMA Mortgage-Backed Securities. *Journal of Finance* 36(3), 599–617.

Dumas, B., Fleming, J., Whaley, R. (1998). Implied Volatility Functions: Empirical Test. *Journal of Finance* 53, 2059–2106.

Dupire, B. (1994). Pricing with a Smile. *Risk* 7(1), 18–20.

Dupire, B. (Ed.) (1998). *Monte Carlo: Methodologies and Applications for Pricing and Risk Management*. Risk Publications, London.

Dupire, B., Savine, A. (1998). Dimension Reduction and Other Ways of Speeding Monte Carlo Simulation. In: *Risk Handbook*. Risk Publications, 51–63.

Dyke, P.P. (1999). *An Introduction to Laplace Transforms and Fourier Series*. Springer-Verlag, Berlin–Heidelberg–New York.

Efron, B. (1979). Bootstrap Methods: Another Look at the Jackknife. *Annals of Statistics* 7, 1–26.

Efron, B., Tibshirani, R. (1986). Bootstrap Methods for Standard Errors, Confidence Intervals, and Other Measures of Statistical Accuracy. *Statistical Science* 1, 54–77.

Emanuel, D., MacBeth, J. (1982). Further Results on the Constant Elasticity of Variance Call Option Pricing Model. *Journal of Financial and Quantitative Analysis* 17, 533–554.

Embrechts, P., Klüppelberg, C., Mikosch, T. (1997). *Modelling Extremal Events for Insurance and Finance*. Springer-Verlag, Berlin–Heidelberg–New York.

Embrechts, P., Lindskog, F., McNeil, A. (2001). Modeling Dependence with Copulas and Applications to Risk Management. *Working Paper*, ETH, Zurich.

Engle, R.F., Ng, V. (1993). Measuring and Testing the Impact of News on Volatility. *Journal of Finance* 48, 1749–1779.

Escribano, Á., Peña, J.I., Villaplana, P. (2002). Modeling Electricity Prices: International Evidence. *Working Paper* 02-27, Economic Series 08, Departamento de Economia, Universidad Carlos III de Madrid.

Evans, M., Swartz, T. (2000). *Approwimating Integrals via Monte Carlo and Deterministic Methods*. Oxford Statistical Science Series 20. Oxford University Press.

Eydeland, A., Wolyniec, K. (2002). *Energy and Power Risk Management: New Developments in Modeling, Pricing and Hedging*. Wiley, Chicago.

Falloon, W., Turner, D. (1999). The Evolution of a Market. In: *Managing Energy Price Risk*, RiskBooks, London.

Fama, E. (1969). Efficient Capital Markets: A Review of Theory and Empirical Work. *Journal of Finance* 25(2) 383–417.

Fama, E., Blume, M. (1966). Filters Rules and Stock-Market Trading. *Journal of Business* 39(1), 226–241.

Fama, E., French, K. (1992). The Cross-Section of Expected Stock Returns. *Journal of Finance* 47, 427–465.

Feller, W. (1951). Two Singular Diffusion Problems. *Annals of Mathematics* 54(1), 173–182.

Fermanian, J.D., Scaillet, O. (2004). Some Statistical Pitfalls in Copula Modeling for Financial Applications. *Research Paper* 108, FAME, Université de Genève.

Fermanian, J.D., Wegkamp, M. (2004). Time-Dependent Copulas. *Working Paper*, CREST, Paris.

Figlewski, S., Gao, B. (1999). The Adaptive Mesh Model: A New Approach to Efficient Option Pricing. *Journal of Financial Economics* 53, 313–351.

Fiorenzani, S. (2005). Load-Based Models for Electricity Prices. *Working Paper*, EDISON Trading.

Fiorenzani, S. (2006a). Financial Optimization and Risk Management in Refining Activities. *International Journal of Global Energy*. Special Issue on Energy Finance 26(1), 62–82.

Fiorenzani, S. (2006b). Pricing Illiquidity in Energy Markets. *Energy Risk* (May), 65–75.

Fiorenzani, S. (2006c). *Quantitative Methods for Electricity Trading and Risk Management: Advanced Mathematical and Statistical Methods for Energy Finance*. Palgrave Macmillan Trading.

Fishman, G.S. (1996). *Monte Carlo: Concepts, Algorithms, and Applications*. Springer-Verlag, Berlin–Heidelberg–New York.

Fleming, W.H., Rishel, R.W. (1975). *Deterministic and Stochastic Optimal Control*. Springer-Verlag, Berlin–Heidelberg–New York.

Fournié, E., Lasry, J.-M., Touzi, N. (1997). Monte Carlo Methods for Stochastic Volatility Models. In: Rogers, L.C.G., Talay, D. (Eds.), *Numerical Methods in Finance*. Cambridge University Press.

Fournié, E., Lasry, J.M., Lebuchoux, J., Lions, P.L., Touzi, N. (1999). Applications of Malliavin Calculus to Monte Carlo Methods in Finance. *Finance and Stochastics* 3, 391–412.

Fréchet, M. (1951). Sur les Tableaux de Corrélation dont les Marges sont Données. *Annales Universitaires Lyon Sc.* 4, 53–84.

Freedman, D.A., Peters, S.C. (1984). Bootstrapping a Regression Equation: Some Empirical Results. *Journal of the American Statistical Association* 79, 97–106.

Fu, M.C., Madan, D., Wang, T. (1998). Pricing Continuous Asian Options: A Comparison of Monte Carlo and Laplace Transform Inversion Methods. *Journal of Computational Finance* 2(1), 49–74.

Fusai, G. (2000). Corridor Options and Arc-Sine Law. *Annals of Applied Probability* 10(2), 634–663.

Fusai, G. (2001). Applications of Laplace Transform for Evaluating Occupation Time Options and Other Derivatives. PhD. Thesis, University of Warwick.

Fusai, G. (2004). Pricing Asian Options via Fourier and Laplace Transforms. *Journal of Computational Finance* 7(3).

Fusai, G., Abrahams, I.D., Sgarra, C. (2006). An Exact Analytical Solution for Discrete Barrier Options. *Finance and Stochastics* 10, 1–26.

Fusai, G., Recchioni, M.C. (2001). Analysis of Quadrature Methods for Pricing Discrete Barrier Options. *Working Paper*, Financial Options Research Center Preprint, 2001/119, Warwick Business School, to appear in *Journal of Economics Dynamics and Control*.

Fusai, G., Sanfelici, S., Tagliani, A. (2002). Practical Problems in the Numerical Solution of PDE's in Finance. *Rendiconti per gli Studi Economici Quantitativi*, Università Ca' Foscari Venezia, 105–132.

Fusai, G., Tagliani, A. (2001). Pricing of Occupation Time Derivatives: Continuous and Discrete Monitoring. *Journal of Computational Finance* 5(1), 1–37.

Gabbi, G., Sironi, A. (2005). Which Factors Affect Corporate Bonds Pricing: Empirical Evidence from Eurobonds Primary Market Spreads. *The European Journal of Finance* 11(1), 59–74.

Galiani, S. (2003). Copula Functions and Their Application in Pricing and Risk Managing Multiname Credit Derivative Products. *MSc Dissertation*, King's College, University of London.

Galluccio, S., Le Cam, Y. (2006a). Implied Calibration of Stochastic Volatility Jump Diffusion Models. *Working Paper* (downloadable at ssrn.com).

Galluccio, S., Le Cam, Y. (2006b). Modelling Hybrids with Jumps and Stochastic Volatility. *Working Paper* (downloadable at ssrn.com).

Galluccio, S., Roncoroni, A. (2006). A New Measure of Cross-Sectional Risk and Its Empirical Implications for Portfolio Risk Management. *Journal of Banking and Finance*, forthcoming. (Preprint, available on www.ssrn.com).

Gander, W., Gautschi, W. (2000). Adaptive Quadrature-Revisited. *BIT* 40(1), 84–101.

Garbow, B.S., Giunta, G., Lyness, J.N., Murli, A. (1988a). Software for an Implementation of Weeks' Method for the Inverse Laplace Transform Problem. *ACM Trans. Math. Software* 14, 163–170.

Garbow, B.S., Giunta, G., Lyness, J.N., Murli, A. (1988b). Algorithm 662: A FORTRAN Software Package for Numerical Inversion of the Laplace Transform Based on Weeks'Mmethod. *ACM Trans. Math. Software*, 14, 171–176.

Garcia, D. (2003). Convergence and Biases of Monte Carlo Estimates of American Option Prices Using a Parametric Exercise Rule. *Journal of Economic Dynamics and Control* 27, 1855–1879.

Gardner, D., Zhuang, Y. (2000). Valuation of Power Generation Assets: A Real Options Approach. *Algo Research Quarterly* 3, 2–20.

Gatheral, J. (2006). *The Volatility Surface: A Practitioner's Guide*. Wiley Finance.

Gatti, S., Rigamotti, A., Saita, F., Senati., M. (2006). Measuring Value at Risk in Project Finance Transactions. *European Financial Management* (forthcoming).

Gaver, D.P. Jr. (1966). Observing Stochastic Processes and Approximate Transform Inversion. *Operations Research* 14(3), 444–459.

Geman, H., El Karoui, N., Rochet, J.C. (1995). Changes of Numéraire, Changes of Probability Measure and Option Pricing. *Journal of Applied Probability* 32, 443–458.

Geman, H., Eydeland, A. (1995). Domino Effect. *Risk* 8(4), 65–67.

Geman, H., Roncoroni, A. (2006). Understanding the Fine Structure of Electricity Prices. *Journal of Business* 79(3), forthcoming (Preprint available on www.ssrn.com).

Geman, H., Yor, M. (1993). Bessel Processes, Asian Options and Perpetuities. *Mathematical Finance* 3(4), 349–375.

Geman, H., Yor, M. (1996). Pricing and Hedging Double Barrier Options: A Probabilistic Approach. *Mathematical Finance* 6(4), 365–378.

Gentle, J.E. (1998). *Random Number Generation and Monte Carlo Methods*. Springer-Verlag, Berlin–Heidelberg–New York.

Gerber, H.U., Shiu, E.S. (1994). Option Pricing by Esscher Transforms. *Transactions of the Society of Actuaries* XLVI, 99–140.

Gibson, M.S., Pristsker, M. (2000). Improving Grid-Based Methods for Estimating Value at Risk of Fixed-Income Portfolios. *Working Paper*, Federal Reserve Board, Washington.

Gihman, I.I., Skorohod, A.V. (1979). *Controlled Stochastic Processes*. Springer-Verlag, Berlin–Heidelberg–New York.

Gitman, L.J., Mercurio, V.A. (1982). Cost of Capital Techniques Used by Major U.S. Firms: Survey and Analysis of Fortune's 1000. *Financial Management* 14(4), 21–29.

Glasserman, P. (2004). *Monte Carlo Methods in Finance.* Springer-Verlag, Berlin–Heidelberg–New York.

Glasserman, P., Heidelberger, P., Shahabuddin, P. (1999a). Importance Sampling in the Heath-Jarrow-Morton Framework. *Journal of Derivatives* 6, 32–50.

Glasserman, P., Heidelberger, P., Shahabuddin, P. (1999b). Stratification Issues in Estimating Value-At-Risk. In: *Proceedings of the Winter Simulation Conference.* IEEE Press, New York.

Glasserman, P., Heidelberger, P., Shahabuddin, P. (2000). Variance Reduction Techniques for Estimating Value-at-Risk. *Management Science* 46, 1349–1364.

Glasserman, P., Zhao, X. (1999). Fast Greeks by Simulation in Forward LIBOR Models. *Journal of Computational Finance* 3, 5–39.

Glynn, P.W., Iglehart, D.L. (1989). Importance Sampling for Stochastic Simulations. *Management Science* 35, 1367–1392.

Glynn, P.W., Whitt, W. (1992). The Efficiency of Simulation Estimators. *Operations Research* 40, 505–520.

Gobet, E., Munos, R. (2002). Sensitivity Analysis Using Itô–Malliavin Calculus and Martingales: Application to Stochastic Optimal Control. *Report* 498, Centre de Mathématiques Appliquées, Ecole Polytechnique, Palaiseau, France.

Gocharov, Y., Pliska, S.R. (2003). Optimal Mortgage Refinancing with Endogenous Mortgage Rates. *Working Paper*, University of Illinois at Chicago.

Goldenberg, D. (1991). A Unified Method for Pricing Options on Diffusion Processes. *Journal of Financial Economics* 29(1), 3–34.

Goldman, M.B., Sosin, H.B., Gatto, M.A. (1979). Path-Dependent Options: Buy at the Low, Sell at the High. *Journal of Finance* 34, 1111–1127.

Golub, G., Van Loan, C. (1996). *Matrix Computations.* John Hopkins Studies in Mathematical Sciences, Baltimore.

van den Goorbergh, R.W.J., Genest, C., Werker, B. (2003). Multivariate Option Pricing Using Dynamic Copula Models. *Working Paper* 2003-122, Center, Tilburg University.

Gourieroux, C., Monfort, A. (1996). *Simulation Based Econometric Methods.* Oxford University Press.

Greene, W.H. (2002). *Econometric Analysis* (5th ed.). Prentice Hall, New Jersey.

Grigoriu, M. (2003). *Stochastic Calculus.* Birkhäuser.

Grüne, L., Semmler, W. (2004). Solving Asset Pricing Models with Stochastic Dynamic Programming. *Working Paper* 54, CEM, Bielefeld University.

Guiotto, P., Roncoroni, A. (2001). Theory and Calibration of HJM with Shape Factors. In: Geman et al. (Eds.), *Mathematical Finance – Bachelier Congress 2000.* Springer-Verlag, Berlin–Heidelberg–New York, 407–426.

Hageman, L.A., Young, D.M. (1981). *Applied Iterative Methods.* Academic Press, New York.

Hampel, F.R. (1986). *Robust Statistics.* Wiley, New York.

Harvey, A.C. (1994). *Forecasting, Structural Time Series Models and the Kalman Filter.* Cambridge University Press.

Harvey, D.I., Leybourne, S.J., Newbold, P. (1999). Forecast Evaluation Tests in the Presence of ARCH. *Journal of Forecasting* 18, 343–445.

Herold, U., Maurer, R. (2002). Portfolio Choice and Estimation Risk: A Comparison of Bayesian Approaches to Resampled Efficiency. *Working Paper* 94, Johann Wolfgang Goethe Universitat, Frankfurt.

Hestenes, M.R., Stiefel, E. (1952). Methods of Conjugate Gradients for Solving Linear Systems. *Journal Research National Bureau of Standard* 49, 409–436.

Heston, S. (1993). A Closed-Form Solution for Options with Stochastic Volatility with Application to Bond and Currency Options. *Review of Financial Studies* 6, 327–343.

Heynen, R.C., Kat, H.M. (1995). Lookback Options with Discrete and Partial Monitoring of the Underlying Price. *Applied Mathematical Finance* 2, 273–284.

Hinz, Y. (2003). Modelling Day-Ahead Electricity Prices. *Applied Mathematical Finance* 10, 149–161.

Hirsa, A., Madan, D. (2003). Pricing American Options under Variance Gamma. *Journal of Computational Finance* 7(2), 63–80.

Hoeffding, W. (1940). Massstabinvariante Korrelationstheorie. *Schriften der Mathematischen Seminars und Instituts für Angewandte Mathematik der Universität Berlin* 5, 181–233.

Hörfelt, P. (2003). Extension of the Corrected Barrier Approximation by Broadie, Glasserman, and Kou. *Finance and Stochastics* 7, 231–243.

Hong, H.S., Hickernell, F.J. (2000). Implementing Scrambled Digital Nets. *Unpublished Technical Report*, Hong Kong Baptist University.

Hu, T., Müller, A., Scarsini, M. (2003). Some Counterexamples in Positive Dependence. Applied Mathematics *Working Paper*, Series 28/2003, ICER, Torino.

Huber, P. (1981). *Robust Statistics*. Wiley, New York.

Hudson, R., Dempsey, M., Keasey, K. (1996). A Note on the Weak Form Efficiency of Capital Markets: The Application of Simple Technical Trading Rules to UK Stock Prices – 1935 to 1994. *Journal of Banking and Finance* 20, 1121–1132.

Hugonnier, J.N. (1999). The Feynman–Kac Formula and Pricing of Occupation Time Derivatives. *International Journal of Theoretical and Applied Finance* 2(2), 153–178.

Hui, C.H., Lo, C.F., Yuen, P.H. (2000). Comment on Pricing Double Barrier Options Using Laplace Transforms by Antoon Pelsser. *Finance and Stochastic* 4, 105–107.

Huisman, R., Mahieu, R. (2003). Regime Jumps in Electricity Prices. *Energy Economics* 25, 425–434.

Hull, J.C. (2005). *Options, Futures and Other Derivatives* (6th ed.). Prentice-Hall.

Hull, J.C., White, A. (1987). The Pricing of Options on Assets with Stochastic Volatilities. *Journal of Finance* 42(2), 281–300.

Hull, J.C., White, A. (1990). Valuing Derivative Securities Using the Explicit Finite Difference Method. *Journal of Financial and Quantitative Analysis* 25(1), 87–100.

Hull, J.C., White, A. (2003). Valuation of a CDO and an N-th to Default CDS without Monte Carlo Simulation. *Working Paper*, University of Toronto.

Hsu, M. (1998). Spark Spread Options Are Hot! *Journal of Electricity* 11, 28–39.

Imai, J., Tan, K.S. (2002). Enhanced Quasi-Monte Carlo Method with Dimension Reduction. *Proceedings of the 2002 Winter Simulation Conference*, 1502–1510.

Ince, E.L. (1964). *Ordinary Differential Equations*. Dover Publications, Inc., New York.

Ingersoll, J.E. (1986). *Theory of Financial Decisions Making*. Rowman & Littlefield Publishers, Inc.

Isakov, D., Hollistein, M. (1999). Application of Simple Technical Rules to Swiss Stock Prices: Is it profitable? *Finanzmarket and Portfolio Management* 13(1), 9–26.

Jackwerth, J. (1999). Option Implied Risk-Neutral Distributions and Implied Binomial Trees: A Literature Review. *Journal of Derivatives* 7(2), 66–82.

Jackwerth, J., Rubinstein, M. (1996). Recovering Probability Distributions from Option Prices. *Journal of Finance* 51(5), 1611–1631.

Jacod, J., Protter, P. (1998). Asymptotic Error Distributions for the Euler Method for Stochastic Differential Equations. *Annals of Probability* 26, 267–307.

Jacod, J., Shiryaev, A. (1988). *Limit Theorems for Stochastic Processes*. Springer-Verlag, Berlin–Heidelberg–New York.

Jaillet, P., Ronn, E., Tompaidis, S. (2003). Valuation of Commodity Based Swing Options. *Management Science* 50(7), 909–921.

James, W., Stein, C. (1961). Estimation with Quadratic Loss. *Proceedings of the Fourth Berkeley Symposium on Mathematical Statistics and Probability*. University of California Press, Berkeley, 361–379.

James, J., Webber, N. (2000). *Interest Rate Modelling*. Wiley Series in Financial Engineering, John Wiley & Sons.

Jamshidian, F. (1987). Pricing of Contingent Claims in the One-Factor Term Structure Model. *Working Paper*, Merryl Linch, New York. In: *Vasicek and Beyond* (1996), Risk Publications.

Jamshidian, F. (1989). An Exact Bond Option Formula. *Journal of Finance* 44, 205–209.

Jamshidian, F. (1990). The Preference-Free Determination of Bond and Option Prices from the Spot Interest Rate. *Advances in Futures and Options Research* 4, 51–67.

Jamshidian, F. (1991a). Bond and Options Evaluation in the Gaussian Interest Rate Model. *Research in Finance* 9, 131–170. Appeared also in: *Vasicek and Beyond* (1996), Risk Publications.

Jamshidian, F. (1991b). Forward Induction and Construction of Yield Curve Diffusion Models. *Journal of Fixed Income* 1, 62–74.

Jamshidian, F. (1991c). Commodity Option Evaluation in the Gaussian Futures Term Structure Model. *Review of Futures Markets* 10(2), 324–346.

Jamshidian, F. (1992). An Analysis of American Options. *Review of Futures Markets* 11(1), 73–80.

Jamshidian, F. (1993). Options and Futures Evaluation with Deterministic Volatility. *Mathematical Finance* 3(2), 149–159.

Jamshidian, F. (1995). A Simple Class of Square-Root Models. *Applied Mathematical Finance* 2, 61–72.

Jamshidian, F. (1996). Bond, Futures and Option Evaluation in the Quadratic Interest Rate Model. *Applied Mathematical Finance* 3, 93–115.

Jamshidian, F. (1997). Libor and Swap Market Model and Measures. *Finance and Stochastics* 1, 293–330.

Jamshidian, F. (1997). A Note on Analytical Valuation of Double Barrier Options. *Working Paper*, Sakura Global Capital.

Jamshidian, F. (1999). Libor Market Model with Semimartingales. In: *Option Pricing, Interest Rates and Risk Management* (2001), Cambridge University.

Jamshidian, F., Zhu, Y. (1997). Scenario Simulation: Theory and Methodology. *Finance and Stochastics* 1, 43–67.

Jarrow, R.A. (1986). The Pricing of Commodity Options with Stochastic Interest Rates. *Advances in Futures and Options Research* 2, 19–45.

Jarrow, R.A., Rudd, A. (1982). Approximate Option Valuation for Arbitrary Stochastic Processes. *Journal of Financial Economics* 10, 347–369.

Jensen, M., Benington, G. (1970). Random Walks and Technical Theories: Some Additional Evidence. *Journal of Finance* 25, 469–482.

Jobson, J.D., Korkie, B. (1980). Estimation of Markowitz Efficient Portfolios. *Journal of the American Statistical Association* 75, 544–554.

Joe, H. (1997). *Multivariate Models and Dependence Concepts*. Monographs on Statistics and Applied Probability. Chapman and Hall, London.

Joe, H., Xu, J.J. (1996). The Estimation Method of Inference Functions for Margins for Multivariate Models. Unpublished *Working Paper*, University of British Columbia.

Johannes, M. (1999). Jumps in Interest Rates: A Nonparametric Approach. *Working Paper*, University of Chicago.

Johannes, M. (2004). The Statistic and Economic Role of Jumps in Interest Rates. *Journal of Finance* 59, 227–260.

Johnson, N.L., Kotz, S. (1995). *Continuous Univariate Distributions*, Vol. 1 and 2 (2nd ed.). Wiley Series in Probability and Statistics.

Jolliffe, L. (1986). *Principal Components Analysis. Series in Statistics.* Springer-Verlag, Berlin–Heidelberg–New York.

Joskow, P., Kahn, J. (2001). A Quantitative Analysis of Pricing Behavior in California Wholesale Electricity Market During Summer 2000. *Working Paper*, MIT.

Jouanin, J.F., Riboulet, G., Roncalli, T. (2003). Financial Applications of Copula Functions. *Working Paper*, Crédit Lyonnais.

Ju, N. (2002). Pricing Asian and Basket Options via Taylor Expansion. *Journal of Computational Finance* 5(3), 79–103.

Judge, G.G., Hill, R.C., Griffiths, W.E., Lütkepol, H., Lee, T.C. (1988). *Introduction to the Theory and Practice of Econometrics* (2nd ed.). Wiley, New York.

Kahl, C., Jäckel, P. (2005). Not-so-Complex Logarithms in the Heston Model. *Wilmott Magazine*, Sept., 94–103.

Kahl, C., Jäckel, P. (2006). Fast Strong Approximation Monte Carlo Schemes for Stochastic Volatility Models. *Quantitative Finance* 6(6), 513–536.

Kallsen, J., Tankov, P. (2004). Characterization of Dependence of Multidimensional Lévy Processes Using Lévy Copulas. *Working Paper*, Ecole Polytechnique, France.

Karatzas, I., Lehoczky, J.P., Sethi, S.P., Shreve, S.E. (1986). Explicit Solution of a General Consumption/Investment Problem. *Math. Operations Research* 111, 261–294.

Karatzas, I., Lehoczky, J.P., Shreve, S.E. (1987). Optimal Portfolio and Consumption Decisions for a "Small Investor" on a Finite Horizon. SIAM *Journal of Control and Optimization* 25, 1557–1586.

Karatzas, I., Shreve, S.E. (1997). *Brownian Motion and Stochastic Calculus* (2nd ed.). GTM Collection, Springer-Verlag, Berlin–Heidelberg–New York.

Kat, H.M. (2001). *Structured Equity Derivatives: The Definitive Guide to Exotic Options and Structured Notes*. Wiley Finance, London.

Këllezi, E., Webber, N. (2004). Valuing Bermudian Options when Asset Returns Are Lévy Processes. *Quantitative Finance* 4, 87–100.

Kendall, M. (1994). *Advanced Theroy of Statistics* (6th ed.). Edward Arnold, London, Halsted Press, New York.

Keppo, J. (2004). Pricing Electricity Swing Options. *Journal of Derivatives* 11, 26–43.

Kimberling, C.H. (1974). A Probabilistic Interpretation of Complete Monotonicity. *Adequationes Math.* 10, 152–164.

Kloeden, P.E., Platen, E. (2000). Numerical Solution of Stochastic Differential Equations. *Applications of Mathematics Collection*. Springer-Verlag, Berlin–Heidelberg–New York.

Knez, P.J., Ready, M.J. (1997). On the Robustness of Size and Book-to-Market in Cross-Sectional Regressions. *Journal of Finance* 52(4), 1355–1382.

Knittel, C.R., Roberts, M.R. (2001). An Empirical Examination of Deregulated Electricity Prices. *Working Paper*, Boston University.

Koehler, J.R., Owen, A. (1996). Computer Experiment. In: *Handbook of Statistics*, Design and Analysis of Experiments.

Kou, S. (2002). A Jump-Diffusion Model for Option Pricing. *Management Science* 48, 1086–1101.

Krylov, N.V. (1980). *Controlled Diffusion Processes*. Springer-Verlag, Berlin–Heidelberg–New York.

Kunitomo, N., Ikeda, M. (1992). Pricing Options with Curved Boundaries. *Mathematical Finance* 2, 275–298.

Kupiec, P. (1995). Techniques for Verigying the Accuracy of Risk Measurement Models. *Journal of Derivatives*, 2, 173–184.

Kushner, H.J. (1967). *Stochastic Stability and Control*. Academic Press, New York.

Kushner, H.J., Dupuis, P. (1992). *Numerical Methods for Stochastic Control Problems in Continuous Time*. Springer-Verlag, Berlin–Heidelberg–New York.

Kwok, Y.-K. (1998). *Mathematical Models of Financial Derivatives*. Springer-Verlag, Berlin–Heidelberg–New York.

Kwok, Y.-K., Barthez, D. (1989). An Algorithm for the Numerical Inversion of Laplace Transforms. *Inverse Problems* 5, 1089–1095.

Lacoste, V., El Karoui, N., Jeanblanc, M. (2005). Optimal Portfolio Management with American Capital Guarantee. *Journal of Economic Dynamics and Control* 29, 449–468.

Lambert, J.D. (1991). *Numerical Methods for Ordinary Differential Systems. The Initial Value Problem*. John Wiley & Sons.

Lamberton, D., Lapeyre, B. (1996). *Introduction to Stochastic Calculus Applied to Finance*. Chapman & Hall, London.

Lari Lavassani, A., Simchi, M., Ware, A. (2000). A Discrete Valuation of Swing Options. *Canadian Applied Mathematics* 9, 35–74.

Lavely, J., Wakefield, G., Barrett, B. (1980). Toward Enhancing Beta Estimates. *Journal of Portfolio Management* 6(4), 43–46.

Lax, P.D., Richtmyer, R.D. (1956). Survey of the Stability of Linear Finite Difference Equations. *Comm. Pure Appl. Math.* 9, 267–293.

L'Ecuyer, P. (1988). Efficient and Portable Combined Random Number Generators. *Communications of the ACM* 31.

L'Ecuyer, P. (1994). Uniforml Random Number Generation. *Annals of Operations Research* 53, 77–120.

L'Ecuyer, P., Simard, R., Wegenkittl, S. (2002). Sparse Serial Tests of Uniformity for Random Number Generators. *SIAM Journal of Scientific Computing* 24, 652–668.

Leblanc, B., Scaillet, O. (1998). Path Dependent Options on Yields in the Affine Term Structure Model. *Finance and Stochastics* 2, 349–367.

Lehmann, E. (1966). Some Concepts of Dependence. *Annals of Mathematical Statistics* 37, 1137–1153.

Levy, E. (1992). Pricing European Average Rate Currency Options. *Journal of International Money and Finance* 11, 474–491.

Lewis, A. (2000). *Option Valuation Under Stochastic Volatility*. Finance Press, Newport Beach.

Lewis, A. (2002). Asian Connections. *Wilmott Magazine* 57–63.

Lewis, P.A.W., Shedler, G.S. (1979). Simulation of Nonhomogeneous Poisson Processes by Thinning. *Naval Logistics Quarterly* 26, 403–413.

Lì, X.D. (2000). On Default Correlation: A Copula Approach. *Journal of Fixed Income* 9, 43–54.

Li, A., Ritchken, P., Sankarasubramanian, L. (1995). Lattice Methods for Pricing American Interest Rate Claims. *Journal of Finance* 50, 719–737.

Linetski, V. (1999). Step Options. *Mathematical Finance* 9(1), 55–96.

Linetsky, V. (2004). Spectral Expansions for Asian (Average Price) Options. *Operations Research* 52, 856–867.

Lipton, A. (1999). Similarities via Self-Similarities. *Risk* 12(9), 101–105.

Lipton, A. (2001). *Mathematical Methods for Foreign Exchange*. World Scientific.

Litterman, R., Scheinkman, J. (1991). Common Factors Affecting Bond Returns. *Journal of Fixed Income* 1, 54–61.

Lo, A., MacKinlay, A.C., Zhang, J. (1997). Econometric Models of Limit Order Executions. *Working Paper* 6257, NBER.

Longin, F. (1996). The Asymptotic Distribution of Extreme Stock Market Returns. *Journal of Business* 69, 383–408.

Longin, F. (2000). From VaR to Stress Testing: The Extreme Value Approach. *Journal of Banking and Finance* 24, 1097–1130.

Longin, F., Bouyé, E., Legras, J., Soupé, F. (2001). Correlation and Dependence in Financial Markets. HSBC CCF. *Quants* 41.

Longin, F., Solnik, B. (2001). Correlation Structure of International Equity Markets During Extremely Volatile Periods. *Journal of Finance* 46, 649–676.

Longstaff, F.A. (1993). The Valuations of Options on Coupon Bonds. *Journal of Banking and Finance* 17(1), 27–42.

Longstaff, F.A. (2002). Optimal Recursive Refinancing and the Valuation of Mortgage-Backed Securities. *Working Paper*, UCLA.

Longstaff, F.A., Schwartz, E.S. (2001). Valuing American Options by Simulation: A Simple Least Squares Approach. *Review of Financial Studies* 14, 113–147.

Lord, R., Koekkoek, R., Van Dijk, D. (2006). Comparison of Biased Simulation Schemes for Stochastic Volatility Models. Discussion Paper No. 06-046/4, Tinbergen Institute.

Lucia, F., Schwartz, E. (2002). Electricity Prices and Power Derivatives. *Review of Derivative Research* 5, 5–50.

Luciano, E., Marena, M. (2002). Copulae as a New Tool in Financial Modelling. *Operational Research: An International Journal* 2, 139–155.

Luenberger, D.G. (1989). *Linear and Nonlinear Programming* (2nd ed.). Addison-Wesley.

Lund, A., Ollmar, F. (2003). Analyzing Flexible Load Contracts. *Working Paper*.

MacMillan, L.W. (1986). An Analytical Approximation for the American Put Prices. *Advances in Futures and Options Research* 1, 119–139.

Maddala, G.S., Li, H. (1996). Bootstrap Based Tests in Financial Models. In: Maddala, G.S., Rao, C.R. (Eds.), *Handbook of Statistics. Statistical Methods in Finance* 14. Elsevier.

Manoliu, M., Tompaidis, S. (2002). Energy Futures Prices: Term Structure Models with Kalman Filter Estimation. *Applied Mathematical Finance* 9, 21–43.

Martin, R.D., Simin, T. (1999). Robust Estimation of Beta. *Technical Report* 350, Department of Statistics, University of Washington.

Marsaglia, G. (1972). The Structure of Linear Congruential Generators. In: Zaremba, S.K. (Ed.), *Applications of Number Theory to Numerical Analysis*, 249–286. Academic Press, New York.

Marsaglia, G., Bray, T.A. (1964). A Convenient Method for Generating Normal Variables. *SIAM Review* 6, 260–264.

Maspero, D., Saita, F. (2005). Risk Measurement for Asset Managers: A Test of Relative VaR. *Journal of Asset Management* 5(5), 338–350.

McCauley, R., Melick, W. (1996a). Risk Reversal. *Risk* 9(11), 54–57.

McCauley, R., Melick, W. (1996b). Propensity and Density. *Risk* 9(12), 52–54.

McKean, H.P. (1967). Appendix: A Free Boundary Problem for the Heath Equation Arising From a Problem in Mathematical Economics. *Industrial Management Review* 6, 32–39.

Melick, W., Thomas, C.P. (1997). Recovering an Asset's Implied PDF from Option Prices: An Application to Crude Oil During the Gulf Crisis. *Journal of Financial and Quantitative Analysis* 32(1), 91–115.

Melino, A., Turnbull, S. (1990). Pricing Foreign Currency Options with Stochastic Volatility. *Journal of Econometrics* 45, 239–265.

Meneguzzo, D., Vecchiato, W. (2004). Copula Sensitivity in Collateralized Debt Obligations and Basket Default Swaps. *Journal of Futures Markets* 24(1), 37–70.

Merton, R. (1971). Optimum Consumption and Portfolio Rules in a Continuous-Time Model. *Journal of Economics Theory* 3, 373–413. Erratum: *ibidem* 6 (1973), 213–214.

Merton, R. (1974). On the Pricing of Corporate Debt: The Risk Structure of Interest Rates. *Journal of Finance* 29, 449–470.

Merton, R. (1976). Option Pricing when Underlying Stock Returns Are Discontinuous. *Journal of Financial Economics* 3(1/2), 125–144.

Meucci, A. (2005). *Risk and Asset Allocation*. Springer-Verlag, Berlin–Heidelberg–New York.

Michaud, R.O. (1998). *Efficient Asset Management*. Harvard Business School Press, Boston.

Mikusinski, P., Sherwood, H., Taylor, M.D. (1992). Shuffles of Min. *Stochastica* 13, 61–74.

Milevsky, M.A., Posner, S.E. (1998). Asian Options, The Sum of Lognormals and the Reciprocal Gamma Distribution. *Journal of Financial and Quantitative Analysis* 33(3), 409–422.

Miltersen, K. (1999). Pricing Interest Rate Contingent Claims: Implementing a Simulation Approach. *Working Paper*, Odense University.

Mitchell, A.R., Griffiths, D.F. (1980). *The Finite Difference Method in Partial Differential Equations*. John Wiley (Corrected reprinted edition, 1994).

Moorthy, M. (1995a). Numerical Inversion of Two-Dimensional Laplace Transforms Fourier Series Representation. *Applied Numerical Mathematics* 17, 119–127.

Moorthy, M.V. (1995b). Inversion of the Multi-Dimensional Laplace Transform – Expansion by Laguerre Series. *Z. Angew. Math. Phys.* 46, 793–806.

Morton, K.W., Mayers, D.F. (1994). *Numerical Solution of Partial Differential Equations*. Cambridge University Press.

Murphy, J. (1999). Technical Analysis of the Financial Markets. *Report*, New York Institute of Finance.

Musiela, M., Rutkowski, M. (1997). *Martingale Methods in Financial Modelling*. Applications of Mathematics 36. Springer-Verlag, Berlin–Heidelberg–New York.

Nahum, E. (1998). On the Distribution of the Sumpremum of the Sum of a Brownian Motion with Drift and a Marked Point Process, and the Pricing of Lookback Options. *Technical Report* N. 516, Dept. of Statistics, Berkeley.

Nelsen, R.B. (1999). *An Introduction to Copulas*. Lectures Notes in Statistics. Springer-Verlag, Berlin–Heidelberg–New York.

Niederreiter, H. (1992). *Random Number Generation and Quasi-Monte Carlo Methods*. CBMS-NSF 63, SIAM.

Oksendal, B. (2003). *Stochastic Differential Equations: An Introduction with Applications*. Springer-Verlag, Berlin–Heidelberg–New York.

Oksendal, B., Sulem, A. (2004). *Applied Stochastic Control of Jump Diffusions*. Springer-Verlag, Berlin–Heidelberg–New York.

Owen, A. (1998). Latin Supercube Sampling for Very High-Dimensional Simulations. *ACM Transaction on Modelling and Computer Simulation* 8, 71–102.

Owen, A. (2002). Variance and Discrepancy with Alternative Scramblings. *ACM Transactions on Computational Logic*, Vol. V.

Pacelli, G., Recchioni, M.C., Zirilli, F. (1999). A Hybrid Method for Pricing European Options Based on Multiple Assets. *Applied Mathematical Finance* 6, 61–85.

Pelsser, A. (2000). Pricing Double Barrier Options Using Laplace Transforms. *Finance and Stochastics* 4, 95–104.

Pedersen, A. (1995). A New Approach to Maximum Likelihood Estimation for Stochastic Differential Equations Based on Discrete Observations. *Scandinavian Journal of Statistics* 22, 55–71.

Piazzesi, M. (2001). An Econometric Model of the Yield Curve with Macroeconomic Jumps Effects. *Working Paper*, University of California, Los Angeles.

Picoult, E. (1999). Calculating Value-at-Risk with Monte Carlo Simulation. In: Dupire, B. (Ed.). *Monte Carlo: Methodologies and Applications for Pricing and Risk Management* 209–229. Risk Publications, London.

Pilipovich, D., Wengler, J. (1998). Getting into the Swing. *Energy and Power Risk Management* 2(10).

Pitsianis, N., Van Loan, C. (1993). Approximation with Kronecker Products. In: *Linear Algebra for Large Scale and Real Time Application*. Kluwer Academic Publishers, 293–314.

Platzman, L.K., Ammons, J.C., Bartholdi, J.J. (1988). A Simple and Efficient Algorithm to Compute Tail Probabilities from Transforms. *Oper. Res.* 26, 137–144.

Poncet, P., Gesser, V. (1997). Volatility Patterns: Theory and Some Evidence from the Dollar-Mark Option Market. *Journal of Derivatives* 5(2).

Portait, R., Bajeux-Besnainou, I., Jordan, J. (2001). An Asset Allocation Puzzle: Comment. *American Economic Review* 91(4), 1170–1180.

Portait, R., Bajeux-Besnainou, I., Jordan, J. (2003). Dynamic Asset Allocation for Stocks, Bonds and Cash. *Journal of Business* 76(2), 263–287.

Portait, R., Nguyen, P. (2002). Dynamic Mean Variance Efficiency and Asset Allocation with a Solvency Constraint. *Journal of Economics Dynamics and Control*.

Prakasa-Rao, B.L.S. (1999). *Semimartingales and Their Statistical Inference*. Chapman & Hall/CRC.

Press, W.H., Teukolsky, S.A., Vetterling, W.T., Flannery, B.P. (1992). *Numerical Recipes in C: The Art of Scientific Computing*. Cambridge University Press.

Protter, P. (2005). *Stochastic Integration an Differential Equations* (2nd ed.). Springer-Verlag, Berlin–Heidelberg–New York.

Quarteroni, A., Sacco, R., Saleri, F. (2000). *Numerical Mathematics*. Springer-Verlag, Berlin–Heidelberg–New York.

Rebonato, R. (1998). *Interest Rate Option Models* (2nd ed.). Wiley & Sons.

Rebonato, R. (1999). *Volatility and Correlation in the Pricing of Equity, FX and Interest-Rate Options*. Wiley Series in Financial Engineering, John Wiley & Sons.

Ribeiro, C., Webber, N. (2003). Valuing Path Dependent Options in the Variance-Gamma Model by Monte Carlo with a Gamma Bridge. *Journal of Computational Finance* 7.

Ribeiro, C., Webber, N. (2005). Correcting for Simulation Bias in Monte Carlo Methods to Value Exotic Options in Models Driven by Lévy Processes. *Applied Mathematical Finance*.

Rich, D.R. (1994). The Mathematical Foundations of Barrier Option Pricing Theory. *Advances in Futures and Options Research* 7, 267–371.

Ritchken, P. (1995). On Pricing Barrier Options. *Journal of Derivatives* 3, 19–28.

Richtmyer, R.D., Morton, K.W. (1967). *Difference Methods for Initial Value Problems* (2nd ed.). Wiley-Interscience, New York.

Rogers, C. (2000). Evaluating First-Passage Probabilities for Spectrally One-Sided Lévy Processes. *Journal of Applied Probability* 37(4), 1173–1180.

Rogers, L.C.G. (2002). Monte Carlo Valuation of American Options. *Mathematical Finance* 12, 271–286.

Rogers, L.C.G., Shi, Z. (1992). The Value of an Asian Option. *Journal of Applied Probability* 32, 1077–1088.

Rogers, L.C.G., Talay, B. (Eds.) (1997). *Numerical Methods in Finance.* Cambridge University Press.

Rogers, L.C.G., Williams, D. (1987). *Diffusions, Markov Processes and Martingales*, Vol. 2, Ito Calculus. Wiley.

Roncoroni, A. (1995). A Trade-off Optimal Choice Problem arising in the Financial Economic Policy of Developing Countries: Private Sector Credit Demand Incentives under Constrained Debt Recovery Policy. *"Laurea" Degree Dissertation*, Bocconi University, Milan.

Roncoroni, A. (1997). Principal Component Analysis for Finite and Infinite Dimensional Dynamical Models. *Working Paper*, Courant Institute of Mathematical Sciences, New York.

Roncoroni, A. (1999). Infinite Dimensional HJM Dynamics for the Term Structure of Interest Rates. *Working Paper* 9903, ESSEC.

Roncoroni, A. (2000). The S Option – An Alternative to the Surrender Option in Mortgage Backed Securities. *Working Paper*, CEREG, Université Paris Dauphine.

Roncoroni, A. (2002). Essays in Quantitative Finance: Modelling and Calibration in Interest Rate and Electricity Markets. *Ph.D. Dissertation*, Université Paris IX Dauphine, France.

Roncoroni, A. (2004). Models for Risk Management in the Energy Markets and the Italian "Nuovo Mercato Elettrico". *Technical Report*, The Italian Stock Exchange, Milan.

Roncoroni, A., Galluccio, S., Guiotto, P. (2003). Shape Factors and Cross-Sectional Risk. *Working Paper*, ESSEC Business School, France.

Roncoroni, A., Moro, A. (2006). Flexible-Rate Mortgages. *International Journal of Business* 11(2).

Roncoroni, A., Zuccolo, V. (2004). The Optimal Exercise Policy of Volumetric Swing Options with Penalty Constraints. *Working Paper*, ESSEC Business School.

Ross, S. (1997). *Simulation.* Academic Press, San Diego.

Rousseeuw, P.J., Leroy, A.M. (1987). *Robust Regression and Outlier Detection.* Wiley, New York.

Rubinstein, R. (1981). *Simulation and the Monte Carlo Method.* John Wiley & Sons, New York.

Rubinstein, M. (1994). Implied Binomial Trees. *Journal of Finance* 49(3), 771–818.

Rubinstein, M., Reiner, E. (1991). Breaking Down the Barriers. *Risk* 8, 28–35.

Sankaran, M. (1963). Approximations to the Non-Central Chi-Square Distribution. *Biometrika* 50, 199–204.

Salminen, P., Wallin, O. (2005). Perpetual Integral Functionals of Diffusions and Their Numerical Computations. *Working Paper.*

Salopek, D.M. (1997). *American Put Options.* Chapman & Hall, CRC.

Samuelson, P.A. (1967). Rational Theory of Warrant Pricing. *Industrial Management Review* 6, 13–31.

Sankaran, M. (1963). Approximations to the Non Central Chi-Square Distribution. *Biometrika* 50, 199–204.

Sato, K.I. (2000). *Lévy Processes and Infinitely Divisible Distributions.* Cambridge University Press.

Sbuelz, A. (1999). A General Treatment of Barrier Options and Semi-Static Hedges of Double Barrier Options. *Working Paper*, London Business School.

Sbuelz, A. (2005). Hedging Double Barriers with Singles. *International Journal of Theoretical and Applied Finance* 8, 393–407.

Scaillet, O. (2000). Nonparametric Estimation of Copulas for Time Series. *Journal of Risk* 5(4), 25–54.

Scherer, B. (2002). Portfolio Resampling: Review and Critique. *Financial Analysts Journal* 58(6), 98–109.

Schonbucher, P.J. (2003). *Credit Derivatives Pricing Models*. Wiley Finance, London.

Schoutens, W. (2003). *Levy Processes in Finance*. Wiley.

Schroder, M. (1989). Computing the CEV Option Pricing Formula. *Journal of Finance* 44, 211–219.

Schwager, J.D. (1996). *Schwager on Futures: Technical Analysis*. Wiley & Sons, New York.

Schwartz, E.S. (1997). The Stochastic Behavior of Commodity Prices: Implications for Valuation and Hedging. *Journal of Finance* 52, 923–973.

Schwartz, E.S., Torous, W.N. (1989). Prepayment and the Valuation of Mortgage-Backed Securities. *Journal of Finance* 44, 375–392.

Schwartz, E.S., Torous, W.N. (1992). Prepayment, Default, and the Valuation of Mortgage Pass-Through Securities. *Journal of Business* 65, 221–239.

Schweizer, B., Wolff, E. (1981). On Non Parametric Measures of Dependence for Random Variables. *Annals of Statistics* 9, 879–885.

Selby, M.J.P. (1983). The Application of Option Theory to the Evaluation of Risky Debt. *Ph.D. Thesis*, London Business School.

Seydel, R.U. (2006). *Tools for Computational Finance* (3rd ed.). Springer Universitext.

Shampine, L.F., Reichelt, M.W. (1997). The MATLAB ODE Suite. SIAM *Journal on Scientific Computing*, 18, 1–22.

Sharpe, W.F., Alexander, G.J., Bailey, J.V. (1999). *Investments* (6th ed.). Prentice-Hall International.

Shimko, D. (1991). *Finance in Continuous Time: A Primer*. Kolb Publishing Company.

Shimko, D. (1993). Bounds of Probability. *Risk* 6(4), 33–37.

Singhal, K., Vlach, J. (1975). Computation of Time Domain Response by Numerical Inversion of the Laplace Transform. *Journal of the Franklin Institute* 2, 110–127.

Shiryaev, A.N. (1978). *Optimal Stopping Rules*. Springer-Verlag, Berlin–Heidelberg–New York.

Shreve, S. (2004). *Stochastic Calculus for Finance II: Continuous-Time Models*. Springer-Verlag, Berlin–Heidelberg–New York.

Siegmund, D. (1976). Importance Sampling in the Monte Carlo Study of Sequential Tests. *Annals of Statistics* 4, 673–684.

Silverman, B.W. (1986). *Density Estimation for Statistics and Data Analysis*. Chapman & Hall.

Singhal, K., Vlach, J. (1975). Computation of Time Domain Response by Numerical Inversion of the Laplace transform. *Journal of the Franklin Institute* 2, 110–127.

Singhal, K., Vlach, J., Vlach, M. (1975). Numerical Inversion of Multidimensional Laplace Transforms. *Proc. IEEE* 63, 1627–1628.

Skeel, R., Berzins, M. (1990). A Method for the Spatial Discretization of Parabolic Equations in One Space Variable. *SIAM Journal on Scientific and Statistical Computing* 11, 1–32.

Sklar, A. (1959). Fonctions de Repartition à n Dimensions et leurs Marges. *Publications de l'Institut de Statistique de l'Université de Paris* 8, 229–231.

Smith, G.D. (1985). *Numerical Solution of Partial Differential Equations: Finite Difference Methods*. Oxford University Press.

Söderlind, P., Svensson, L. (1997). New Techniques to Extract Market Expectations from Financial Instruments. *Journal of Monetary Economics* 2(40), 383–429.

Stanton, R. (1995). Rational Prepayment and the Valuation of Mortgage-Backed Securities. *Review of Financial Studies* 8, 677–708.

Stanton, R. (1997). A Nonparametric Model of Term Structure Dynamics and the Market Price of Interest Rate Risk. *Journal of Finance* 7(5), 1973–2002.

Steeley, J.M. (1990). Modelling the Dynamics of the Term Structure of Interest Rates. *Economic and Social Review* 21, 337–361.

Stehfest, H. (1970). Algorithm 368: Numerical inversion of Laplace Transform. *Communication of the ACM* 13(1), 47–49.

Stevenson, T. (2001). Filtering and Forcasting Spot Electricity Prices. *Working Paper*, UTS, Sydney.

Stewart, G. (1973). *Introduction to Matrix Computations*. Academic Press, New York.

Stoer, J., Bulirsch, R. (1980). *Introduction to Numerical Analysis*. Springer-Verlag, Berlin–Heidelberg–New York.

Strauss, W.A. (1992). *Partial Differential Equations: An Introduction*. Wiley & Sons, Chichester, England.

Sullivan, M.A. (2000). Pricing Discretely Monitored Barrier Options. *Journal of Computational Finance* 3(4) 35–52.

Sullivan, R., Timmermann, A., White, H. (1999). Data-Snooping, Technical Trading Rule Performance, and the Bootstrap. *Journal of Finance* 54(5), 1647–1691.

Sweeney, R.J. (1988). Some New Filter Rule Tests: Methods and Results. *Journal of Financial and Quantitative Analysis* 23(3), 285–300.

Talay, D. (1982). How to Discretize Stochastic Differential Equations. In: *Lecture Notes in Mathematics* 972. Springer-Verlag, Berlin–Heidelberg–New York, 276–292.

Talay, D. (1984). Efficient Numerical Schemes for the Approximation of Expectations of Functionnals of the Solutions of a S.D.E., and Applications. In: *Lecture Notes in Control and Information Sciences* 61. Springer-Verlag, Berlin–Heidelberg–New York, 294–313.

Talay, D. (1995). Simulation and Numerical Analysis of Stochastic Differential Systems: A Review. In: Kree, P., Wedig, W. (Eds.), *Probabilistic Methods in Applied Physics*, Lecture Notes in Physics 451, Springer-Verlag, Berlin–Heidelberg–New York, 63–106.

Talbot, A. (1979). The Accurate Numerical Inversion of Laplace Transforms. *J. Inst. Math. Appl.* 23(1), 97–120.

Tankov, P. (2005). Simulation and Option Pricing in Lévy Copula Model. In: Avellaneda, M., Cont, R. (Eds.), *Mathematical Modelling of Financial Derivatives*, IMA volumes in Mathematics and Applications, Springer-Verlag, Berlin–Heidelberg–New York.

Tavella, D., Randall, C. (2000). *Pricing Financial Instruments: The Finite Difference Method*. Financial Engineering, Wiley.

Thompson, A.C. (1995). Valuation of Path-Dependent Contingent Claims with Multiple Exercise Decisions Over Time: The Case of Take-or-Pay. *Journal of Financial and Quantitative Analysis* 30(2), 271–293.

Thompson, G.W.P. (1998). Fast Narrow Bounds on the Value of Asian Options. *Working Paper*, University of Cambridge.

Thorp, W.A. (2000). The MACD: A Combo of Indicators for the Best of Both Worlds. *AAII Journal* 30–34.

Tian, Y. (1999). Pricing Complex Barrier Options Under General Diffusion Processes. *Journal of Derivatives*, Winter, 11–30.

Titman, S., Tompaidis, S., Tsyplakov, S. (2004). Market Imperfections, Investment Flexibility and Default Spreads. *Journal of Finance* 59(1), 165–205.

Tompkins, R., D'Ecclesia, R.L. (2006). Unconditional Return Disturbances: A Non Parametric Simulation Approach. *Journal of Banking and Finance* 30(1), 287–314.

Topper, J. (2005). *Financial Engineering with Finite Elements*. The Wiley Finance Series.

Trigeorgis, L. (1991). A Log-Transformated Binomial Numerical Analysis Method for Valuing Complex Multi-Option Investments. *Journal of Financial and Quantitative Analysis* 26(3), 309–326.

Turnbull, S., Wakeman, L. (1991). A Quick Algorithm for Pricing European Average Options. *Journal of Financial and Quantitative Analysis* 26, 377–389.

Varga, R. (1962). *Matrix Analysis*. Prentice-Hall, Englewood Ckiffs, NJ.

Vasicek, O.A. (1973). A Note on Using Cross-Sectional Information in Bayesian Estimation of Security Betas. *Journal of Finance* 28(5), 1233–1239.

Vasicek, O. (1977). An Equilibrium Characterization of the Term Structure. *Journal of Financial Economics* 5, 177–188.

Vecer, J. (2001). A New PDE Approach for Pricing Arithmetic Average Asian Options. *Journal of Computational Finance* 4(4), 105–113.

Vetzal, K.R. (1998). An Improved Finite Difference Approach to Fitting the Initial Term Structure. *Journal of Fixed Income* 7 (March), 62–81.

Villeneuve, S., Zanette, A. (2002). Parabolic ADI Methods for Pricing American Options on Two Stocks. *Math. Oper. Res.* 27(1), 121–149.

Vlach, J., Singhal, K. (1993). *Computer Methods for Circuit Analysis and Design* (2nd ed.). Van Nostrand Reinhold Company, New York.

Wang, S.S. (1999). Aggregation of Correlated Risk Portfolios: Models and Algorithms. Preprint, CAS Committee on Theory of Risk.

Webber, N., Kuan, G. (2003). Valuing Barrier Options in One-factor Interest Rate Models. *Journal of Derivatives* 10, 33–50.

Weeks, W. (1966). Numerical Inversion of Laplace Transforms Using Laguerre Functions. *Journal ACM* 13(3), 419–429.

Weideman, J.A.C. (1999). Algorithms for Parameter Selection in the Weeks Method for Inverting the Laplace Transform. *SIAM J. Sci. Comput.* 21(1), 111–128.

Wilmott, P., Dewynne, J.N., Howison, S. (1993). *Option Pricing: Mathematical Models and Computation*. Oxford Financial Press.

Yohai, V.J., Stahel, W.A., Zamar, R.H. (1991). A Procedure for Robust Estimation and Inference in Linear Regression. In: Stahel, W., Weisberg, S. (Eds.), *Directions in Robust Statistics and Diagnostics*. Springer-Verlag, Berlin–Heidelberg–New York, 365–374.

Yor, M. (1991). *On Exponential Functionals of Brownian Motion and Related Processes*. Springer-Verlag, Berlin–Heidelberg–New York.

Yor, M. (2001). *Exponential Functionals of Brownian Motion and Related Processes*. Springer-Verlag, New York.

Young, D.M. (1971). *Iterative Solution of Large Sparse Systems*. Academic Press.

Zauderer, E. (2006). *Partial Differential Equations of Applied Mathematics*. Pure and Applied Mathematics: A Wiley-Interscience Series of Texts, Monographs and Tracts, 3rd ed.

Zhang, J.E. (2001). A Semi-Analytical Method for Pricing and Hedging Continuously Sampled Arithmetic Average Rate Options. *Journal of Computational Finance* 5(1), 59–79.

Zhang, X.L. (1997). Numerical Analysis of American Option Pricing in a Jump-Diffusion Model. *Mathematics of Operations Research* 22, 668–690.

Zhu, Y.I., Wu, X., Chern, I.L. (2005). *Derivative Securities and Difference Methods*. Springer-Verlag, Berlin–Heidelberg–New York.

Zvan, R., Forsyth, P.A., Vetzal, K.R. (1998a). Penalty Methods for American Options with Stochastic Volatility. *J. Comput. Appl. Math.* 91, 199–218.

Zvan, R., Forsyth, P.A., Vetzal, K.R. (1998b). Robust Numerical Methods for PDE Models of Asian Options. *Journal of Computational Finance* 1, 39–78.

Zvan, R., Vetzal, K.R., Forsyth, P.A. (2000). PDE Methods for Pricing Barrier Options. *Journal of Economic Dynamics and Control* 24, 1563–1590.

Index

Printing: Krips bv, Meppel, The Netherlands
Binding: Stürtz, Würzburg, Germany